Lifting the Scientific Veil
Science Appreciation for the Nonscientist

PAUL SUKYS

An Ardsley House Book

ROWMAN & LITTLEFIELD PUBLISHERS, INC.
Lanham • Boulder • New York • Oxford

ROWMAN & LITTLEFIELD PUBLISHERS, INC.

Published in the United States of America
by Rowman & Littlefield Publishers, Inc.
4720 Boston Way, Lanham, Maryland 20706

12 Hid's Copse Road
Cumnor Hill, Oxford OX2 9JJ, England

Copyright © 1999 by Paul Sukys

All rights reserved. No part of this publication may be reproduced, stored in a retrieval system, or transmitted in any form or by any means, electronic, mechanical, photocopying, recording, or otherwise, without the prior permission of the publisher.

British Library Cataloguing in Publication Information Available

Library of Congress Cataloging-in-Publication Data

Sukys, Paul.
    Lifting the scientific veil : Science appreciation for the nonscientist / Paul Sukys.
      p. cm.
    Includes bibliographical references and index.
    ISBN 0-8476-9600-6 (pa. : alk. paper)
    1. Science—Popular works. I. Title.
Q162.S84 1999
500—dc21   99-12159
    CIP

Printed in the United States of America

♾™The paper used in this publication meets the minimum requirements of American National Standard for Information Sciences—Permanence of Paper for Printed Library Materials, ANSI/NISO Z39.48–1992.

To my mother, Catherine Louise,
    who nurtured my past.
        To my wife, Susan,
            who enriches my present.
                To my daughters, Jennifer, Ashley, and Megan,
                  who are the future.

Contents

Preface / xv
Acknowledgments / xxi

Unit I Introduction to Science — 1

CHAPTER 1
The Need for Science Appreciation — 1
Commentary: The Canals of Mars and the Theory of Relativity / 1
Chapter Outcomes / 2

1-1 **Why Study Science?** / 2
 The Paradox of Science Education / 3
 The Cause of Science Illiteracy / 5
 The Consequences of Science Illiteracy / 5
 The Goal of Science Appreciation / 6

1-2 **How Science Works: An Overview** / 7
 The Nature of Scientific Truth / 7
 The Traditional Scientific Method / 9

1-3 **Creativity and Cultural Influences in Science** / 14
 Creativity in Science / 14
 Social and Cultural Influences in Science / 18
 The Role of the Mathematical Model in Science / 22

Conclusion / 24
 Review / 25
 Understanding Key Terms / 26
 Review Questions / 26
 Discussion Questions / 27

Analyzing *Star Trek* / 27
 Background / 27
 Special Instructions / 29
 Viewing Assignment / 29
 Thoughts, Notions, and Speculations / 30

Notes / 31

Unit II The Origin of the Universe — 33

CHAPTER 2
The Mystery Called the Big Bang — 33
Commentary: When Past Becomes Future / 33
Chapter Outcomes / 34

2-1 **Why Study the Big Bang?** / 34

2-2 **Misconceptions Concerning the Big Bang** / 36
 The Big Bang Was Not an Explosion / 37
 The Big Bang Theory Is Not Antireligious / 38

2-3 **The Nature of the Big Bang** / 41
 Implications of the Red Shift Discovery / 43
 The Subatomic Universe / 48

2-4 **A Description of the Big Bang** / 52
 The Problem of Causation / 53
 A Big Bang Chronology / 57

2-5 **Evidence Supporting the Big Bang** / 58
 The Expanding Nature of the Universe / 59
 The Expansive Blanket of Background Radiation / 59

The Wealth of Hydrogen and Helium / 60

Conclusion / 60
Review / 60
Understanding Key Terms / 62
Review Questions / 62
Discussion Questions / 63

Analyzing *Star Trek* / 64
Background / 64
Viewing Assignment / 64
Thoughts, Notions, and Speculations / 64

Notes / 65

CHAPTER 3
Requiem for the Big Bang 68
Commentary: A Trip into the Luminiferous Ether / 68
Chapter Outcomes / 69

3-1 **Is the Big Bang Really a New Paradigm? / 70**
The Expanding Universe Revisited / 70
Kuhn and Scientific Revolutions / 71

3-2 **Problems with the Big Bang Model / 74**
The Age of the Universe / 74
The Mystery of Galactic Movement / 77
The COBE Discovery / 81

3-3 **Alternative Theories / 82**
The Steady State Theory / 82
Quasi–Steady State Cosmology / 85
Plasma Cosmology / 86

3-4 **The Nonscientist and the Origin of the Universe / 89**
Unlocking the Theory of Everything / 89
Particle Accelerators as the Hope for the Future / 91
The Theory of Everything: Myth or Reality? / 91
The Rebirth of the Cosmological Constant / 94

Conclusion / 95
Review / 95
Understanding Key Terms / 96
Review Questions / 96
Discussion Questions / 97

Analyzing *Star Trek* / 98
Background / 98
Viewing Assignment / 98
Thoughts, Notions, and Speculations / 98

Notes / 99

Unit III The Mystery of Quantum Physics 103

CHAPTER 4
The Emerging Quantum Universe 103
Commentary: The Quest for the Holy Grail / 103
Chapter Outcomes / 104

4-1 **Why Study Quantum Theory? / 104**

4-2 **The Historical Development of Quantum Theory / 106**
The Ancient View of Reality / 106
The Birth of Atomic Theory / 109
The Theory of Transubstantiation / 109

4-3 **The Development of Atomic Theory / 110**
Newtonian Atomic Theory / 110
The Contributions of Chemistry / 112

4-4 **The Debate on Atomic Structure / 113**
Maxwell's Theory of Electromagnetism / 113
Thomson's Plum Pudding Model / 114
Rutherford's Solar System Model / 115
Constructivism vs. Positivism Revisited / 116

4-5 **The Advent of Quantum Theory / 118**
Max Planck and the Birth of the Quanta / 119
Bohr's Solution to the Unstable Atom / 120
The Structure of de Broglie's Atom / 121
Schrödinger's Mathematical Refinement / 123

Conclusion / 124
Review / 124
Understanding Key Terms / 127
Review Questions / 127
Discussion Questions / 127

Analyzing *Star Trek* / 128
Background / 128
Viewing Assignment / 129
Thoughts, Notions, and Speculations / 129

Notes / 130

CHAPTER 5
Quantum Weirdness Verified — 133

Commentary: Quantum Physics in Wonderland / 133
Chapter Outcomes / 134

5-1 **The Puzzle of Quantum Weirdness** / 134

5-2 **Probabilities, Uncertainty, and Complementarity** / 135
Heisenberg's and Schrödinger's Equations / 135
Born's Probability Waves / 136
Heisenberg's Uncertainty Principle / 137
Complementarity and Quantum Weirdness / 138

5-3 **The Copenhagen Interpretation** / 140
The Essentials of the Copenhagen Interpretation / 141
The Mystery of Schrödinger's Cat / 144
Einstein's Interpretation of Quantum Weirdness / 145
The Essential Focus of the Debate / 147

5-4 **The Einstein-Podolsky-Rosen Thought Experiment** / 148
The Nature of the Einstein-Podolsky-Rosen Thought Experiment / 149
The Continuing Debate / 152

5-5 **Experimental Verification of Quantum Weirdness** / 152
Bell's Interconnectedness Theorem / 153
The Aspect Verification Experiment / 154

5-6 **The Implications of Nonlocality** / 155
The Quantum School of Nonlocality and Unity / 155
The Classical School of Locality and Reality / 157

5-7 **The Children of the Copenhagen Interpretation** / 159
The Princeton Interpretation / 160
The Subquantal Reality Theory / 160
The Many-Worlds Theory / 162

Conclusion / 163
Review / 163
Understanding Key Terms / 166
Review Questions / 167
Discussion Questions / 167

Analyzing Star Trek / 168
Background / 168
Viewing Assignment / 168
Thoughts, Notions, and Speculations / 169

Notes / 170

Unit IV The Theory of Relativity — 173

CHAPTER 6
The Special Theory of Relativity — 173

Commentary: The Assassination of Adolf Hitler / 173
Chapter Outcomes / 174

6-1 **The Relative Nature of Relativity** / 174
Psychological Relativity / 175
The Ancient Greek Notion of Stability / 175
Relativity According to Galileo and Newton / 177

6-2 **Light and the Mystery of the Ether** / 180
Electricity, Magnetism, and Light / 181
The Mystery of the Ether / 182
The Conflict Caused by Light / 184

6-3 **The Special Theory of Relativity** / 185
Restructuring the Universe / 187
The Resolution of the Conflict / 188

6-4 **The Alteration of Some Basic Assumptions** / 189
The Lorentz Transformation / 189
The Time Dilation Effect / 190
The Relativity of Simultaneity / 195
Bergson's Objections / 198

Conclusion / 199
Review / 200
Understanding Key Terms / 201
Review Questions / 201
Discussion Questions / 201

Analyzing *Star Trek* / 202
Background / 202
Viewing Assignment / 202
Thoughts, Notions, and Speculations / 203

Notes / 204

CHAPTER 7
Past and Future Time Travel 206

Commentary: The Premature Death of John F. Kennedy / 206
Chapter Outcomes / 207

7-1 Travel into the Future / 207
Shortcomings of Time-Dilation Time Travel / 208
The General Theory of Relativity / 210
Black Holes and Time Travel into the Past / 213
Quasars, Active Galaxies, and Primordial Black Holes / 216

7-2 Travel into the Past / 220
Rotating Black Hole Time Machines / 220
Gott's Temporal Distortion Twist Theory / 225
Morris and Thorne's Traversable Wormholes / 227

7-3 Time-Travel Paradoxes / 230
Contradictory-Event Paradoxes / 231
Information Paradoxes / 236

7-4 Alternate Time Lines / 237
Parallel Universes and Alternate Time Lines / 237
Science Fiction and Alternate Universes / 238

Conclusion / 240
Review / 240
Understanding Key Terms / 241
Review Questions / 242
Discussion Questions / 242

Analyzing *Star Trek* / 242
Background / 242
Viewing Assignment / 243
Thoughts, Notions, and Speculations / 243

Notes / 244

Unit V The Social and Cultural Impact of the Theory of Relativity 247

CHAPTER 8
The Philosophic Side of Relativity 247

Commentary: Relativity and the Nature of Albert Einstein / 247
Chapter Outcomes / 248

8-1 Einstein's Influence on Philosophy / 248
The Nature of Einstein's Influence / 250
The Union of Physics and Philosophy / 252

8-2 Positivism and Relativity / 254
Kant's Ideas on Space and Time / 254
The Einstein-Kant Debate / 255

8-3 Idealism and the Simultaneous Existence of Time / 257
Idealism and Relativity / 258
The Eternal Space-Time Cube / 260
Lifelines, Light Cones, and Elsewhen / 265

8-4 Philosophy and the Space-Time Cube / 270
The Principle of the Fixity of Time / 270
Volition and the Space-Time Cube / 271
The Theory of Relativity vs. Quantum Theory / 272

Conclusion / 274
Review / 276
Understanding Key Terms / 278
Review Questions / 278
Discussion Questions / 278

Analyzing *Star Trek* / 279
Background / 279
Viewing Assignment / 279
Thoughts, Notions, and Speculations / 280

Notes / 281

Unit VI Life and Evolution 283

CHAPTER 9
The Origin of Life 283

Commentary: Extraterrestrial Life in the Fast Lane / 283
Chapter Outcomes / 284

9-1 Life vs. Nonlife / 285
    The Characteristics of Life / 286
    Life at the Subatomic Level / 287

9-2 Alternative Theories on the Nature of Life / 288
    Vitalism and Reductionism / 288
    Emergentism and Complexification / 291

9-3 Prelife and Life / 294
    Conditions and Forces during the Prelife Era / 294
    From Prelife to Life / 294
    Experimental Verification / 296
    Alternate Theories / 298

9-4 Making Life from Nonlife / 299
    Synthetic Protolife / 299
    The Handmade Cell / 299
    The Extraterrestrial Origin of Life / 300

9-5 Life: An Accident or an Inevitability? / 301
    The Birth of a Star / 302
    The Origin of Life / 302

    Conclusion / 303
    Review / 304
    Understanding Key Terms / 305
    Review Questions / 305
    Discussion Questions / 306

    Analyzing *Star Trek* / 306
    Background / 306
    Viewing Assignment / 307
    Thoughts, Notions, and Speculations / 307

    Notes / 308

CHAPTER 10
Evolution and the Emergence of Humanity   310

Commentary: The Age of Rocks or the Rock of Ages? / 310
Chapter Outcomes / 311

10-1 What Is Human Life? / 312
    The Concept of Acquired Humanness / 312
    Vitalism and Reductionism / 313
    Emergentism and Acquired Humanness / 314

10-2 The Theory of Evolution / 315
    The Concept of Evolution / 315
    Genetic Variants, Reproduction, and Mutations / 318
    The Appearance of New Species / 320
    The Appearance of Humanity / 321

10-3 Decimation, Contingency, and Mass Extinctions / 328
    The Decimation/Contingency View of Evolution / 328
    The Mass Extinction View of Evolution / 331

    Conclusion / 335
    Review / 335
    Understanding Key Terms / 337
    Review Questions / 337
    Discussion Questions / 338

    Analyzing *Star Trek* / 339
    Background / 339
    Viewing Assignment / 339
    Thoughts, Notions, and Speculations / 340

    Notes / 341

Unit VII Procreation, Biotechnology, and the Law   343

CHAPTER 11
Juriscience and Human Life   343

Commentary / 343
Chapter Outcomes / 344

11-1 Science, Ethics, and Law / 344

11-2 What Is the Law? / 345
    The Ethical Roots of the Law / 346
    Contemporary Sources of American Law / 346

11-3 The Intersection of Science and the Law / 348
    Similarities between Science and Law / 350

The Intersection between Science
   and Law / 352

11-4 Juriscience and Procreative Freedom / 354
   The Law, Science, and Procreative
     Freedom / 355
   The Acquisition-of-Humanness Standard / 357
   Other Standards of Humanness / 359

11-5 Juriscience and Procreative Rights / 365
   Procreative Freedom and the Right
     to Privacy / 365
   Procreative Freedom and Responsibilities / 366
   Juriscience and Contragestive Agents / 367

   Conclusion / 371
   Review / 371
   Understanding Key Terms / 373
   Review Questions / 373
   Discussion Questions / 374

   Analyzing *Star Trek* / 375
   Background / 375
   Viewing Assignment / 375
   Thoughts, Notions, and Speculations / 376

   Notes / 377

CHAPTER 12
The Regulatory Environment and Genetic Engineering   380
Commentary: The Law and the Meaning
   of Manufactured Life / 380
Chapter Outcomes / 381

12-1 How the Law Regulates Science / 381
   The Basic Plan of the Constitution / 382
   The Regulation of Basic Research Funding / 384
   The Science Court Proposal / 388

12-2 How the Law Regulates Technology / 390
   How Science and Technology
     Are Different / 391
   The Regulation of Technology / 392

**12-3 Genetics and the Frontiers
   of Juriscience / 394**
   An Introduction to Genetics / 395
   A Primer on Genetic Engineering / 399
   The Regulation of Genetic Engineering / 404

Conclusion / 408
Review / 409
Understanding Key Terms / 410
Review Questions / 411
Discussion Questions / 411

Analyzing *Star Trek* / 412
Background / 412
Viewing Assignment / 413
Thoughts, Notions, and Speculations / 413

Notes / 414

Unit VIII Science: The Final Frontier   417

CHAPTER 13
The Implications of Science for the Future   417
Commentary: Historical Study of New York
   City, 2022 / 417
Chapter Outcomes / 418

**13-1 The Future According
   to Heilbroner / 419**

13-2 The Challenge of Population / 420
   The Growing Population / 420
   Causes of the Population Explosion / 422
   The Consequences of an Unchecked
     Population / 423

13-3 The Challenge of War / 426
   The New World Order / 426
   The Global Realignment
     of Nations / 430
   Available Strategic Options / 434

13-4 The Challenge of the Environment / 436
   Evidence of Global Warming / 436
   Consequences of Global Warming / 438

Conclusion / 439
Review / 440
Understanding Key Terms / 441
Review Questions / 441
Discussion Questions / 442

Analyzing *Star Trek* / 442
Background / 442

Viewing Assignment / 443
Thoughts, Notions, and Speculations / 443

Notes / 444

CHAPTER 14
The History of the Future 448
Commentary: "Both: I Will Have Them Both!" / 448
Chapter Outcomes / 449

14-1 The Role of Science and Technology / 449

14-2 Science, Technology, and the Population Problem / 450
New Reproductive Technologies / 451
Decentralization through Communication and Transportation / 453

14-3 Science, Technology, and the Threat of War / 457
Alternative Energy Sources / 457
Biotechnology and the Green Revolution / 458
Nanotechnology and the Redistribution of Wealth / 460
The Dangers of a World without War / 464

14-4 Science, Technology, and the Environment / 465
Scientific and Technological Practicality / 465
The Earth Summit and Other Conferences / 466

14-5 Toward a Global Science-and-Technology Policy / 468
Principles for the Creation of a Science-and-Technology Policy / 468

Practical Steps toward a Global Science-and-Technology Policy / 475

Conclusion / 481
Review / 481
Understanding Key Terms / 483
Review Questions / 483
Discussion Questions / 484

Analyzing *Star Trek* / 484
Background / 484
Viewing Assignment / 485
Thoughts, Notions, and Speculations / 485

Notes / 486

APPENDIX 1
Time Line of Scientific Thought 489

APPENDIX 2
Brief Biographies of Scientists Discussed 497

Glossary 504

Bibliography 540

Index 558

About the Author 581

Preface

Have you ever noticed that it is perfectly acceptable, even expected, for scientists and engineers to have an interest in art, music, philosophy, and poetry, but completely atypical for an artist, musician, philosopher, or poet to have an interest in science and technology? I have several friends in the Engineering and the Health Sciences divisions at my institution who play musical instruments, paint, sculpt, and even write poetry. Most of these individuals are amazed to discover that a humanities instructor like myself has an interest in science and scientific pursuits. It astonishes them to learn that I am a member of the American Association for the Advancement of Science and the New York Academy of the Sciences, that I subscribe to *Scientific American*, *Science*, *Science News*, *Physics Today*, *Astronomy*, and *The American Scientist*, and that my bookshelves are lined with texts by Stephen Hawking, Albert Einstein, Niels Bohr, and others of equal repute.

My interest in science has followed a strange and unlikely path. It originally stems from my youth, when I was fascinated by Captain Video, Commando Cody, Mr. Wizard, and Tom Swift, Jr. Unfortunately, that interest waned in my teens when my high school chemistry teacher told me I was the worst student he ever had. (Actually, "dumbest" was the word that he used.) My interest was renewed quite by accident almost three decades later, when I received a copy of Stephen Hawking's work *A Brief History of Time* as a gift. What I discovered as I read Hawking's book was that the world of science had rushed forward at light speed, leaving me thirty years in the past, harboring some amazingly primitive ideas about the nature of reality. Recent discoveries in fields such as quantum physics, cosmology, and chaos theory had left me as ignorant as a seventeenth-century peasant who still believed that Earth was the center of the universe.

I have since found out that my experience is not unique. Many people feel the same alienation from the world of science. I'm not certain exactly why this is so, but I suspect that the educational system must share a good portion of the blame. As far as I can ascertain, nobody actually comes right out and says that the nonscientist cannot comprehend or appreciate the intricacies of science. Nonetheless, this attitude is apparent. Once a student declares a major in business, literature, or philosophy, that student's exposure to science during his or her academic career is severely curtailed. Nonscience students are directed toward introductory courses that are often taught in a dull and uninteresting fashion. As a result, science becomes a difficult chore with an apparent irrelevance to their lives.

This situation is disheartening but it could be overlooked as an unfortunate but inescapable byproduct of the educational system, if it were not for two very important facts. The first is that nonscientists are just as capable of appreciating science as scientists are capable of appreciating nonscientific subjects like music, art, or literature. No one expects a physics major to write the great American novel, but that does not prevent her from cultivating an interest in prose, poetry, and drama by taking a course or two in literary appreciation. Why shouldn't we give the same type of opportunity to the English major? Just because he may not unravel the theory of everything doesn't mean that he can't appreciate the difficulties and the benefits involved in the study of quantum physics. If we do not give the nonscience student the same meaningful opportunity that we give the science major to explore areas beyond his or her chosen field of study, then we are cheating young minds of their right to delve into all areas of knowledge, not just the ones that they seem best suited for at this particular moment in their young lives.

The second item that elevates this situation beyond the level of an unfortunate but unavoidable spin-off of the educational system is the fact that nonscientists run the rest of the world. They make the laws. They run the corporations. They finance the research programs. They sit on the zoning boards. They buy the products. They donate the money. They hold the political offices. And they vote. If the citizens of a nation look at science through an impenetrable fog that separates their everyday lives from the pursuit of science, then they are likely to view science as the work of a few pampered scientists who deserve financial support only after everything else has been funded. Science will not flourish if the nonscientists, through no fault of their own, fail to appreciate the value of science and to understand the significant impact that it has on their everyday lives.

The reality of these two inescapable facts led to the writing of *Lifting the Scientific Veil: Science Appreciation for the Nonscientist*. The idea for the book arose from my experiences over a nine-year period teaching a science-appreciation course. It is designed to introduce nonscientists to four of the most significant scientific theories of the twentieth century in an informative and stimulating manner. The hope is that they will see science as something that is not only a part of their lives, and therefore deserving of their support, but also as something that is fascinating in its own right.

The four major theories that are covered in the the text are:

The big bang theory of the origin of the universe
The theory of quantum physics
The theory of relativity
The theory of the evolution of life on Earth

Each unit is devoted to one of these theories. The units are organized in a way that leads the reader first to an understanding of the theory and then to an exploration of how that theory impacts upon their individual lives. In places the book explores the legal and political impact of a new theory. At other times it examines the philosophical and theological implications of a particular scientific breakthrough. And at times, it simply explains how the everyday life of a typical citizen might be changed by the introduction of advanced technology.

The writing style is deliberately informal in an attempt to put the reader at ease. The features at the beginning and the end of the chapters are meant to supplement the text and to provide options for further study and thought. These features include, at the beginning of each chapter, an opening commentary and a set of chapter outcomes. The opening commentary presents a short vignette designed to whet the reader's appetite by setting the overall tone of the chapter and by introducing its central theme. The chapter outcomes give the reader an overview of the major points covered in each chapter and in doing so they establish a road map through the material.

The features at the end of each chapter include a summary, a list of key terms, a set of review questions, and a series of discussion questions. The summary at the end of each chapter is organized around the subdivisions that mark the organizational scheme of that chapter. The review questions parallel the chapter outcomes and thus provide an efficient way to reexamine the main ideas discussed in each chapter. Finally, the discussion questions give the reader a chance to explore the themes that spin off from the material within each chapter.

Another key feature of the book is its direct tie-in with the television series *Star Trek: The Next Generation*. One reason for using *Star Trek* is the belief that the popularity of this series is ample evidence of the optimistic attitude that many people share as they look to the future. It is interesting to note, therefore, that the optimistic future depicted in *Star Trek: The Next Generation* is made possible by advances in science and technology, the same science and technology that many nonscientists either fear or reject. Curiously, many people who accept the concepts depicted in *Star Trek* also question the wisdom of such technological marvels as the now-defunct superconducting supercollider in Texas or the Human Genome Project.

Of course, such an attitude is not difficult to understand. We simply need to remember that *Star Trek* episodes are written by perfectly normal twentieth-century writers, not twenty-fourth century Star Fleet officers. Consequently, the subconscious (and sometimes conscious) sociological and cultural notions that enter many of the episodes represent twentieth-century attitudes toward science and technology. Ironically, some of these attitudes are decidedly antiscientific; others are unrepentantly in favor of science and technology.

In this book, we shall explore both ends of this spectrum by examining an episode of *Star Trek: The Next Generation* at the close of each chapter. Each of these episodes is followed by a feature entitled "Thoughts, Notions, and Speculations." The questions in this feature generally reflect an effort to relate the events depicted in the episode with the science and technology discussed in the chapter. The episode may be a feasible extension of the science discussed in the chapter or it may depict a scientific or technological inconsistency. The idea is not to point out minor errors in the episodes but to explore the implications of the science discussed within the chapter. All of the *Star Trek* episodes that are featured in the text are available on videocassette. If possible, it is helpful to view these episodes before any discussion begins.

Several additional features appear at the end of the book. To help the reader trace the development of modern science, the first appendix presents a time line of scientific discoveries. The scientific biographies found in the second appendix can help the reader develop an appreciation for the context in which many scientific discoveries took place. In addition, the glossary provides a series of definitions that will facilitate the reader's understanding of the concepts that form the core of the book.

Moreover, because *Lifting the Scientific Veil* is also intended for use as a science appreciation textbook it is accompanied by an *Instructor's Manual*, which includes several features designed to enhance the academic usefulness of the text. For each chapter in the text a corresponding chapter appears in the manual, which includes several suggestions for the presentation of the material in the (text) chapter, a chapter outline, a list of performance expectations, a suggested time period for the presentation of the material, central points to be emphasized within the chapter, and answer keys for the review questions, the questions for discussion, and the questions following each *Star Trek* episode. In addition, the manual includes a sample test bank and a sample final examination.

At this point one final issue must be mentioned. My educational background falls primarily within law, literature, philosophy, and theology rather than science. Although these areas make up a good portion of the text, the main subject, of course, is science. Consequently, I have sought the advice and assistance of a variety of professional scientists, who have helped me explain the scientific and the technological concepts that lie at the heart of this book. I would like to extend my thanks to these individuals, each of whom took the time to read the manuscript, often several times, and whose kind assistance made this book the effective collaboration that it became. Where *Lifting the Scientific Veil* succeeds in making science accessible to the nonscientist, much of the credit goes to these individuals. Where the text fails in this regard, the fault is mine.

Thanks go to Steve Abedon, professor of microbiology at The Ohio State University; William Despain, professor of biology at North Central State College; Ralph Hunt, amateur astronomer and professor of English at The Ohio State University; Saverio Pascazio, professor of physics at the University of Bari, Italy; James L. Pazun, professor of chemistry and physics at Pfeiffer University; William Protheroe, professor emeritus of astronomy at The Ohio State University; and Janet Tarino, professor of chemistry at The Ohio State University. A grateful thank you also goes to Dr. Nigel A. Sharp of the National Optical Astronomy Observatories for his kind help with several images provided by NOAO, The Association of Universities for Research Astronomy, and the National Science Foundation. I must also extend special thanks to Tim Berra, professor of zoology at The Ohio State University, for the exceptional effort that he made in the development of the chapters on the origin of life and the evolutionary process.

In addition, I would like to express my appreciation to Gordon Brown, professor of law at North Shore Community College, who reviewed the legal concepts that appear at various points within the text, and who, with unerring accuracy, revealed where I had gone too far in some of my conclusions. Philosophical aspects of the treatment were reviewed by Theodore Gracyk of Moorhead State University, Ellen R. Klein of Whitman College, Eric Palmer of Allegheny College, and Doren Recker of Oklahoma State University. I must also thank Professor Wayne Ramsey of New York University and Nassau Community College for his information and eye-opening lectures during my stay at NYU. Thanks must also go to Karyn Bianco, of Ardsley House, whose critical eye and keen perception helped mold a text that is much more accurate, much more understandable, and much less biased than it was in its earlier incarnations. Thanks also to Martin Zuckerman, editor at Ardsley House, who maintained confidence in me and in the project, and who consistently offered tough but honest criticism when and where it was needed the most.

I would also like to thank my late father, Vitus John, who accepted his young son's bookish temperament even though I am certain he would have preferred to raise a shortstop rather than a scholar, and my mother, Catherine Louise, who, with kindness and love, encouraged me in the best of times and helped me through the worst of times. I must also thank my daughters: Jennifer, who understood when I didn't always remember to call, and Ashley and Megan, who were patient during the long hours I was buried in work.

Most importantly, I must extend the most heartfelt thanks to my wife, Susan, whose expert proofreading created the most flawless manuscript I have ever produced, whose sharp insight pointed out many subtle inconsistencies and contradictions in the original manuscript, and whose love, patience, understanding, and confidence pulled me through some of the most difficult and darkest stages in the writing of this book. If this book is a success, it is her doing as much as mine.

Let me add one final note for your consideration. As a nonscientist, I am acutely aware of those areas of science which cause the most difficulty for my fellow nonscientists. An understanding of these aspects of science is often taken for granted by scientists, who seem at times unaware of the layperson's bewilderment in the face of these issues. For example, many nonscientists are puzzled by the relationship between mathematical proofs and scientific theory. They view any page of equations as some sort of latter-day magical incantation that must be taken "on faith" but which really does not "prove" anything. This is certainly not the case. Because scientific scholars take such knowledge for granted, I have found that it is often more appropriate for a nonscientist to explain these concepts to a layman.

In this way I approach much of the material in this book, not so much as a teacher who instructs but as a fellow traveler who, as the book develops, explores new ideas along with the reader. Like the old soldier who can talk to new recruits with first-hand experience about the difficulties they will have in battle, I believe that I can relate to my fellow nonscientists the problems and the frustrations they will encounter as they attempt to understand and apply these theories. I also believe that I can pass on to them the wonder, awe, and delight that nonscientists experience when they finally open their eyes and say:

Ah ha! So that's what those guys meant all along!

Acknowledgments

Excerpts from STAR TREK: THE NEXT GENERATION © 1996 by Paramount Pictures Corporation. All Rights Reserved. Used with permission.

CHAPTER 1

Excerpt from "Enemies of Promise," by J. Michael Bishop. Reprinted from *The Wilson Quarterly*, Summer 1995. Copyright © 1995 by J. Michael Bishop.

Excerpt from "After the Big Crunch," by David Goodstein. Reprinted from *The Wilson Quarterly*, Summer 1995. Copyright © 1995 by David Goodstein.

Excerpt from *Structure of Scientific Revolutions* by Thomas Kuhn. Reprinted with permission of The University of Chicago Press, Copyright 1962.

Excerpt from *Making Science: Between Nature and Society* by Stephen Cole. Copyright © 1992 by the President and Fellows of Harvard College. Reprinted by permission of Harvard University Press.

CHAPTER 2

Excerpt from *God and the Astronomers* by Robert Jastrow. Reprinted with permission of Robert Jastrow. Copyright © 1978 by Reader's Library, Inc.

Excerpt from *Science and Christ* by Pierre Teilhard de Chardin. Copyright © 1965 by Editions du Seuil. Reprinted by permission of Georges Borchardt.

Excerpt from Evry Schatzman, *Our Expanding Universe*. Copyright © 1989. Reprinted with permission of The McGraw-Hill Companies.

Excerpt from *The Shadows of Creation: Dark Matter and the Structure of the Universe* by Riordan and Schramm. © 1991 by Michael Riordan and David Schramm. © 1991 W. H. Freeman and Company.

Excerpt from "The Self-Reproducing Inflationary Universe," by Andrei Linde, *Scientific American*, November 1994.

CHAPTER 3

Excerpts from *The Genesis Machine* by James P. Hogan. Copyright © 1978 by James Patrick Hogan. Reprinted by permission of Ballantine Books, a Division of Random House, Inc.

Excerpts from *Dreams of a Final Theory* by Steven Weinberg. Copyright © 1992 by Steven Weinberg. Reprinted by permission of Pantheon, a Division of Random House, Inc.

Excerpt from *The End of Physics: The Myth of a Final Theory* by David Lindley. Copyright © 1993 by David Lindley, reprinted by permission of HarperCollins Publishers, Inc.

Excerpt from *Hyperspace: A Scientific Odyssey Through Parallel Universes, Time Warps, and the 10th Dimension* by Michio Kahu. Copyright © 1994 Oxford University Press. Reprinted by permission of Oxford University Press.

CHAPTER 4

Excerpt from *A Brief History of Time: From the Big Bang to Black Holes* by Stephen Hawking. Copyright © 1988 by Stephen Hawking. Reprinted by permission of BANTAM, a division of Bantam Doubleday Dell Publishing Group, Inc.

Excerpt from *An Introduction to the Historiography of Science* by Helge S. Kragh. Copyright © 1991 Cambridge University Press.

Excerpts from *The History of Greek Philosophy* by W. K. C. Guthrie. Copyright © 1962 Cambridge University Press.

Excerpts from *Opticks or a Treatise of the Reflections, Inflections and Colours of Light* by Sir Isaac Newton. Copyright © 1952. Reprinted by permission of Dover Publications Inc.

Excerpt from *A Treatise on Electricity and Magnetism* by James Clerk Maxwell. 2 vols. 1954. Reprinted by Permission of Dover Publications.

Excerpt from *The World within the World* by John Barrow. Copyright © 1988 by John D. Barrow. Oxford University Press. Reprinted by permission of Oxford University Press.

Excerpt from *The Particle Garden* by Gordon Kane (page 206). Copyright © 1995 Gordon Kane. Reprinted by permission of Addison-Wesley Longman Inc.

CHAPTER 5

Excerpt from *The Physical Principles of Quantum Theory* by Werner Heisenberg. 1949. Reprinted by permission of Dover Publications, Inc.

Excerpt from *The World within the World* by John Barrow. Copyright © 1988 by John D. Barrow. Oxford University Press. Reprinted by permission of Oxford University Press.

Excerpts from *Atomic Physics and Human Knowledge* by Niels Bohr. Copyright, Aage Bohr, Courtesy, Niels Bohr Archive, Copenhagen.

Excerpt from "My View of the World," by Erwin Schrödinger. Reprinted by permission of Ox Bow Press, Woodbridge, Connecticut.

Excerpt from *The Physics: The History of a Scientific Community in Modern America*, Copyright © 1995 by Daniel J. Kevles. Harvard University Press. Reprinted by permission of Alfred A. Knopf, Inc. a Division of Random House, Inc.

Excerpt from *Speakable and Unspeakable in Quantum Mechanics* by John S. Bell. Copyright © 1987 Cambridge University Press.

Excerpt reprinted with permission from *At Home in the Universe* by John A. Wheeler (Woodbury, NY: American Institute of Physics, 1996). Copyright 1996 American Institute of Physics.

CHAPTER 6

Excerpt from *The Meaning of Relativity: Including Relativistic Theory of the Non-Symmetric Field* by Albert Einstein. Copyright © 1922, 1945, 1950, 1953 by Princeton University Press, Copyright © 1956 by the Estate of Albert Einstein, Copyright © 1984 by the Hebrew University of Jerusalem. Reprinted by permission of Princeton University Press.

Excerpt from *Aristotle's Poetics* translated by Richard Hope. University of Nebraska Press, 1961.

Excerpt from *Mathematical Principles of Natural Philosophy and His System of the World* by Isaac Newton, ed. Florian Cajori. Edited/Translated by Andrew Motte. Copyright © 1934 and 1962 Regents of the University of California.

Excerpts from *A Treatise on Electricity and Magnetism* by James Clerk Maxwell. 2 vols, 1954. Reprinted by permission of Dover Publications, Inc.

Excerpt from *Out of My Later Years* by Albert Einstein. Copyright © 1956, 1984 by Estate of Albert Einstein. Carol Publishing Group.

Excerpt from *Relativity: The Special and General Theory* by Albert Einstein. Copyright © 1961 by the Estate of Albert Einstein. Permission granted by the Albert Einstein Archives, the Hebrew University of Jerusalem, Israel.

CHAPTER 7

Excerpts from Albert Einstein, *Ideas and Opinions*. Copyright © 1954, 1982 by Crown Publishers, Inc.

Excerpt from A. J. Friedman and Carol Donley *Einstein as Myth and Muse*. Copyright © 1985 Cambridge University Press.

Excerpt reprinted with permission of Simon and Schuster *About Time: Einstein's Unfinished Revolution* by Paul Davies. Copyright © 1995 by Orion Productions.

Excerpt from *Cosmic Time Travel: A Scientific Odyssey* by Barry Parker. Copyright © 1991 Barry Parker. Plenum Press a division Plenum Publishing Corporation.

Excerpt from "The Arrow of Time." © 1990 by Peter Coveney and Roger Highfield. Published by Virgin Publishing Ltd.

Excerpt from "The Quantum Physics of Time Travel," By David Deutsch and Michael Lockwood, *Scientific American*, March 1994.

CHAPTER 8

Excerpts from Abraham Pais, Subtle is the Lord: *The Science and the Life of Albert Einstein*. Copyright © 1982 by permission of Oxford University Press.

Excerpt from "The Philosophical Significance of the Theory of Relativity," by Hans Reichenbach, in *Albert Einstein: Philosopher-Scientist*, ed. Paul Arthur Schlipp. Copyright © 1949, 1951, 1969, 1970 by the Library of Living Philosophers, Inc. Reprinted by permission of Open Court Publishing Company, a division of Carus Publishing.

Excerpts from *The Fourth Dimension: A Guided Tour of the Higher Universes* by Rudy Rucker. Copyright © 1985 by Rudy Rucker. Reprinted by permission of Houghton Mifflin Co. All rights reserved.

Excerpt from "A Remark about the Relationship Between Relativity Theory and Idealistic Philosophy," by Kurt Gödel, in *Albert Einstein: Philosopher-Scientist*, ed. Paul Arthur Schlipp. Copyright © 1949, 1951, 1969, 1970 by the Library of Living Philosophers, Inc. Reprinted by permission of Open Court Publishing Company, a division of Carus Publishing.

Excerpts from "The Fate of Philosophy in the Twentieth Century," by Henry Aiken. First published in *The Kenyon Review* (Spring, 1962), OS Vol. XXIV, No. 2. Copyright The Kenyon Review.

CHAPTER 9

Excerpts from *God and the New Physics* reprinted with the Permission of Simon and Schuster from *God and the New Physics* by Paul Davies. Copyright © 1983 by Paul Davies.

Excerpts from *Seven Clues to the Origin of Life: A Scientific Detective Story* by A.G. Cairns-Smith. Copyright © 1985 Cambridge University Press.

Excerpt from Raymond Daudel, *The Realm of Molecules*. Copyright © 1993 McGraw-Hill Companies. Reproduced with permission of The McGraw-Hill Companies.

Excerpts from *From Atoms to Quarks* by James Trefil. Copyright © 1994 by James Trefil. Reprinted by permission of Doubleday, a division of Bantam Doubleday Dell Publishing Group, Inc.

Excerpt from "Introduction" to *The Chemistry of Life* by Martin Olumucki. Copyright © 1993 McGraw-Hill Companies. Reproduced with permission of the McGraw-Hill Companies.

Excerpt from *Philosophy of Biology Today* by Michael Ruse. Copyright © State University of New York.

Excerpt from *At Home in the Universe: The Search for Laws of Self-Organization and Complexity* by Stuart Kauffman. Copyright © by Stuart Kauffman. Reprinted by permission of Oxford University Press.

Excerpt from *The History of the Earth: An Illustrated Chronicle of an Evolving Planet* by William K. Hartmann and Ron Miller. Copyright 1991 by William K. Hartmann. Reprinted with permission of Workman Publishing.

Excerpt from *Signs of Life*. Copyright © 1994 by Robert Pollack. Reprinted by permission of Houghton Mifflin Co. All rights reserved.

Excerpts from *Interpreting Evolution* by H. James Birx (Amherst, NY: Prometheus Books). Copyright 1991. Reprinted by permission of the publisher.

Excerpt from "Molding the Metabolism," by David Freedman, *Discover*, August 1992. Reprinted by permission of Discover Syndication.

CHAPTER 10

Excerpts from *The Facts of Life: Science and the Abortion Controversy* by Harold J. Morowitz and James Trefil. Copyright © 1992 by Harold J. Morowitz and James Trefil. Reprinted by permission of Oxford University Press.

Excerpts from *The Phenomenon of Man* by Pierre Teilhard de Chardin. Copyright © 1955 by Editions de Seuil. Copyright 1959 in the English translation by William Collins Sons & Co. Ltd., London and Harper and Row, Publishers, Inc., New York.

Excerpt from "Forward: Taking Freedom Seriously," by Robin L. West, *Harvard Law Review*. Copyright © 1990 by The Harvard Law Review Association. Reprinted by permission of *Harvard Law Review*. Reprinted by permission of Robin L. West, Professor of Law, Georgetown University Law Center.

Excerpt reprinted from *The Battle of the Beginnings* by Del Ratzsch. © 1996 by Del Ratzsch. Used by permission of Inter Varsity Press, P.O. Box 1400, Downers Grove, IL 60515.

Excerpt from *Evolution and the Myth of Creationism* by Tim Berra. Copyright © 1990 by the Board of Trustees of the Leland Stanford Junior University. Reprinted by permission of the Stanford University Press.

Excerpts from Robert Foley, *Humans before Humanity*. Copyright © 1995 by Robert Foley. Reprinted by permission of Blackwell Publishers, Ltd.

Excerpt from *The Origin of Humankind* by Richard Leakey. Copyright © 1994 by Basic Books, HarperCollins Publishers, Inc.

Excerpts from *Wonderful Life: The Burgess Shale and the Nature of History* by Stephen Jay Gould. Copyright © by Stephen Jay Gold Reprinted by permission of W. W. Norton & Company, Inc.

CHAPTER 11

Excerpt from *The Facts of Life: Science and the Abortion Controversy* by Harold J. Morowitz and James Trefil. Copyright © 1992 by Harold J. Morowitz and James Trefil. Reprinted by permission of Oxford University Press.

Excerpt from "Forward: Taking Freedom Seriously," by Robin L. West, *Harvard Law Review*. Copyright © 1990 by The Harvard Law Review Association. Reprinted by permission of *Harvard Law Review*. Reprinted by permission of Robin L. West, Professor of Law, Georgetown University Law Center.

CHAPTER 12

Excerpt from *Culture Clash: Law and Science in America* by Steven Goldberg, New York University Press, Copyright © 1994 by New York University.

Information on the unforeseen implications of refrigeration, the automobile, and television reported in "The Art of Forecasting." Copyright © 1991 The World Future Society, 4916 Saint Elmo Avenue, Bethesda, Maryland 20814 U.S.A. (301) 656-8274.

Excerpt from "Asilomar and Recombinant DNA: The End of the Beginning," by Donald S. Fredrickson, *Biomedical Politics*, National Academy Press, Copyright © 1991 by the National Academy of Sciences.

Excerpt from Commission on Hereditary Disorders 43 Md. Code Ann. Section 817 in *Law, Science and Medicine* by Judith Areen. Copyright © 1984 by The Foundation Press.

CHAPTER 14

Excerpt from "The True Blue American" by Delmore Schwartz, from *Selected Poems: Summer Knowledge*. Copyright © 1959 by Delmore Schwartz. Reprinted by permission of New Directions Publishing Corp.

Unit 1 An Introduction To Science

Chapter 1
The Need for Science Appreciation

COMMENTARY: THE CANALS OF MARS AND THE THEORY OF RELATIVITY

In 1910, Percival Lowell published *Mars the Abode of Life*. In this book and its predecessor, *Mars and Its Canals*, Lowell, in vivid and often captivating prose, demonstrates not only the existence of canals on the surface of Mars but also that those canals were constructed by intelligent beings. The book is filled with maps, graphs, spectrograms, equations, and charts, all of which purport to prove Professor Lowell's claim. For perfectly good reasons, many people believed Lowell at the time. After all, his observations seemed to be validated by his work and by the work of those who shared his vision; moreover, he was a respected member of the academic community. He was the director of the observatory at Flagstaff, Arizona, a nonresident professor of astronomy at the Massachusetts Institute of Technology (MIT), and the winner of many awards, not the least of which was the Janssen Medal of the Societe Astronomique de France. Consequently, Lowell was eminently qualified and, therefore, easy to believe. The same could not be said for another, slightly more obscure scientific author. Five years earlier, in 1905, Albert Einstein, a clerk in the Swiss Patent

Office, had published three papers in the *Annalen der Physik*. Einstein had neither the position nor the prestige of Lowell. In fact, if a clerk holding the same type of job that Einstein held in 1905 were to attempt to have a single paper published in one of today's scientific journals, that paper would undoubtedly be rejected. Yet, almost a century later, Einstein's theory of relativity stands as one of the two cornerstones of modern physics, while Lowell's belief in Martian canals has been discredited. The question that we address here is not why Lowell was wrong and Einstein right. That type of a specialized explanation is beyond the scope of this text. Instead, we undertake a far more difficult task. We will help the nonscientist appreciate the accomplishments of science without surrendering the right to evaluate those accomplishments as objectively as possible.

CHAPTER OUTCOMES

After reading this chapter, the reader should be able to accomplish the following:

1. Outline the causes of scientific illiteracy.
2. Predict some of the consequences of scientific illiteracy.
3. Explain the goal of science appreciation.
4. Describe the nature of scientific truth.
5. Describe the traditional scientific method.
6. Explain Kuhn's model of scientific inquiry.
7. Identify the situations that allow for creativity in science.
8. Define positivism and constructivism.
9. Discuss the social and cultural influences on science.
10. Describe the role of mathematics within the scientific process.

1-1   WHY STUDY SCIENCE?

In 1957, scientists in what was then known as the Soviet Union launched Sputnik, the world's first artificial satellite. This event sent shock waves throughout the entire Western world. However, the event was particularly disturbing to the United States, which, up to that point, had considered itself far superior to the Soviet Union in science and science education. Clearly, the educators in the United States had a lot of work to do. The job of catching up and producing top-grade scientists in the United States was attacked with enormous energy, and the results were highly admirable. The federal government responded to the challenge by creating the office of special assistant to the president for science and technology and by giving that position to James Killian, then president of MIT. Federal funds were poured into American high schools and colleges in an effort to improve science education. In addition, federal money was also dedicated to scientific research and development programs to such an extent that the next ten years saw the yearly budget for research and development in the United States reach 11 percent of the entire federal budget. This meant that by 1967, the United States government was spending more than 16 billion dollars on scientific research each year.[1]

The Paradox of Science Education

The success of American scientific education and scientific research following this influx of federal dollars is legendary. The National Aeronautics and Space Administration was established in 1958 and, not only began to outstrip the Soviet program, but eventually landed a man on the Moon in 1969, one year ahead of President Kennedy's original target date.[2] American institutions of higher education became one of the leading sources of young engineers throughout the Western world. By 1989, American universities were producing about 65,000 engineering graduates each year.[3] Success in research and development also saw growth in the production of new products and the development of advanced technologies. A list of accomplishments during this period of history would include not only exotic achievements like the development of jet airliners, nuclear reactors, computers, intercontinental ballistic missiles, nuclear submarines, communication satellites, and the space shuttle, but also inventions like hand-held calculators, microwave ovens, and video recorders, which have changed our lives.[4]

Clearly, the United States has benefited from a successful research and development program and a healthy portion of the world's leading scientists and engineers. Nonetheless, the United States also leads the world in scientific illiteracy. Contemplate the following statistics: "In recent international testing, U.S. high school students finished 9th in physics, among the top 12 nations, 11th in chemistry, and dead last in biology."[5] Or consider a report issued by the Committee on Education and Human Resources of the Federal Coordinating Council for Science, Engineering, and Technology, which disclosed that 50 percent of the adults examined in a recent survey had no idea that the Earth orbited the Sun in a year's time. The report commented that this situation is extremely discouraging because of the many educational and informational opportunities that Americans have today.[6]

A recent study by the National Science Foundation (NSF), aimed at determining the level of science literacy in the United States, produced results that are clearly in line with the foregoing statistics. The report began by noting that a basic level of science literacy would require an individual to have a fundamental understanding of scientific language and ideas, a basic grasp of the scientific method, and an appreciation of the influence of science on culture. Given this criteria the report concluded that less than seven percent of the population of

Figure 1.1 Astronaut Harrison "Jack" Schmitt Collecting Geological Samples during the Apollo 17 Lunar Landing Mission in December of 1972.

The ability of the American system to focus its educational, industrial, governmental, and financial resources successfully on a scientific goal was demonstrated by NASA's Apollo program. In 1961, at a time when the American space program had succeeded in placing only a single manned spacecraft in a suborbital flight, President John F. Kennedy challenged the nation to place an American on the Moon by 1970. In fact, the Apollo program did this.

Photo Credit: National Aeronautics and Space Administration

the United States is literate in science. These results are in line with similar studies conducted by the NSF since 1979, indicating that the situation is certainly not improving with the passage of time.[7]

Interestingly enough, neither this ignorance nor bewilderment is reserved for the nonscientist. In their book, *Science Matters: Achieving Scientific Literacy*, Robert Hazen and James Trefil of George Mason University reported that only three of twenty-four physicists and geologists that they surveyed could explain the difference between DNA and RNA. Later in the same book, Trefil and Hazen reported that, in a private questionnaire administered to graduating seniors at their own institution, they discovered that half of those surveyed could not explain the difference between an atom and a molecule.[8] When I tried a similar survey at my institution among the graduating seniors, I found that only 43 of the 138 polled (31 percent) could identify the nine planets of the solar system in correct order. The same survey revealed that less than 45 percent of those graduates knew the distance of the Earth from the Sun, and fewer than half had any idea how many moons orbit the planet Mars.[9]

The Cause of Science Illiteracy

Americans are not incapable of understanding **science**. On the contrary, no group of people that has accomplished what the people of the United States have accomplished can be anything but highly competent in science. Therefore, if science illiteracy shows up, the explanation must lie elsewhere. In this case, the explanation lies within the educational system itself. The sad fact of the matter is that the American educational system does very little to promote an understanding of or an appreciation for science in nonscience students. In an article entitled "After the Big Crunch," David L. Goodstein, vice provost and professor of physics and applied physics at the California Institute of Technology (Caltech), has described the process of science education in this country as a "mining-and-sorting operation designed to discover and rescue diamonds in the rough, ones capable of being cleaned and cut and polished into glittering gems."[10] Later Goodstein goes on to say that "all the other human rocks and stones are indifferently tossed aside in the course of the operation. Thus, science education at all levels is largely a dreary business, a burden to student and teacher alike—until the happy moment arrives when a teacher-miner finds a potential peer, a real, if not yet gleaming, gem. At this point, science education becomes, for the few involved, exhilarating and successful."[11]

The process that Goodstein describes begins in elementary school. The scientific preparation of many elementary school instructors is woefully inadequate. Often elementary education programs in our colleges and universities do not include any science at all in the curriculum. In fact, some students decide on elementary education as a career because they know that by doing so they can escape science altogether. Is it any wonder then that students are not introduced to an appreciation of science at the elementary school level?[12] Figures compiled by the National Science Foundation indicate that half of the children in American elementary schools have lost all interest in science before they enter seventh grade.[13] This neglect, according to Goodstein, continues into high school. As an example, Goodstein cites the fact that most of the nation's 22,000 high schools do not employ a qualified physics teacher but instead delegate physics courses to someone with a background in one of the other sciences, or in some instances, to someone with no formal science education at all.[14]

The Consequences of Science Illiteracy

This neglect is regrettable, but it would be nothing more than a piece of very bad news, if it were not for two very important facts. The first is that the nonscientist can appreciate science just as much as the scientist can appreciate the humanities. No one expects physicists to write Shakespearean drama, but that does not stop them from attending dramatic presentations at the local theater. Why shouldn't nonscientists seek similar opportunities to appreciate science? Just because nonscientists will not develop a cold fusion reactor or unravel the Theory of Everything, doesn't mean that they cannot appreciate the difficulties and the benefits involved in the study of physics. If nonscientists are not encouraged to pursue scientific interests the same way that scientists are encouraged to study the humanities, then we are denying people the right to explore all areas of knowledge, not just the ones they have chosen as the focus of their careers.

The second item that elevates this situation beyond the level of an unfortunate bit of bad news is the fact that nonscientists run the rest of the world. Nonscientists make the laws. They run the corporations. They finance research programs. They sit on zoning boards. They buy the products. They donate money. They hold political office. And they vote. If the citizens of a nation look at science through an impenetrable fog, if they see their lives as separate from the pursuit of science, then they are likely to view science as the work of a few scientists who deserve financial support only after everything else has been funded. This is precisely the point that John Ziman makes in his study, *Teaching and Learning about Science and Society*, when he notes that the picture that our society has of science is formed in the classroom. Regrettably, according to Ziman, this picture of science is largely neglected, with the result that most nonscientists are taught almost nothing about science, and what they are taught rarely has any practical use. Unfortunately, the opinions and attitudes of nonscientists are extremely important in shaping policy decisions regarding science and scientific projects. Consequently, according to Ziman, it is absolutely essential that nonscientists learn more about science and scientific matters.[15] Ziman's point seems to be that science will not flourish if the nonscientists of a nation fail, through no fault of their own, to appreciate the value of science and to understand the significant impact that it has on their everyday lives.

The Goal of Science Appreciation

The goal of science appreciation is to help nonscientists understand some of the basic concepts of science and to lead them to an appreciation of the impact that science has on all aspects of their individual lives. This goal is one of the recommendations set forth by the Commission on Excellence in Education in its study, *A Nation at Risk*.[16] It is also an objective that Ziman focuses on in *Teaching and Learning about Science and Society* when he says that general education must reach beyond the fundamentals of reading and writing and give students a basic understanding of the world in which they live. Ziman's point is that an understanding of the modern world cannot be separated from the scientific breakthroughs that shaped that world. Consequently, in a general education curriculum, the study of science is just as essential as an examination of politics, sociology, law, and economics.[17]

Accordingly, this text will introduce students to four of the most important theories of the twentieth century and attempt to show how those theories impact society and culture. These four theories include: (1) the Big Bang theory of the origin of the universe, (2) quantum physics, (3) the theory of relativity, and (4) the evolution of life on the planet Earth. Each unit in the text is devoted to one of these theories. The units are organized in a way that leads the readers first to an understanding of each theory, and then to an exploration of how that theory impacts upon their individual lives. Sometimes the book explores the economic or political impact of a new theory. At other times, it examines the philosophical, literary, ethical or legal implications of a particular scientific breakthrough. And at others, it simply explains how the everyday life of the typical citizen might be changed by the introduction of advanced technology. The hope is that at the end of the text, readers will see science as something that is not only a part of their lives, and therefore deserving of their support, but also as something that is fascinating in its own right.

1-2 HOW SCIENCE WORKS: AN OVERVIEW

Before we can delve into these theories, however, we must pause and lay some essential ground work. Consequently, the rest of this introductory chapter is devoted to four objectives. First, we investigate how science works. Second, we examine creativity within science. Third, we explore how society and culture affect science. Finally, we take some time at the close of the chapter to scrutinize the role of mathematics in scientific research.

The Nature of Scientific Truth

Curiosity about the universe is the driving energy behind science. Questions about how the world works, what it is made of, how the individual parts relate to the whole, and how we can use that knowledge to shape our world are all factors in science. Yet, scientists do not simply compile data like stamp collectors fill an album. Instead, they also interpret that data so that the world becomes an orderly and predictable place.[18] Science, then, helps us to make sense of the universe. Of course, the same could be said of many other disciplines. Consequently, it might be helpful at this point to examine how science differs from other fields of study. To accomplish this, we focus first on two disciplines with which the nonscience major may be more familiar. Those two disciplines are **law** and literature.

The Goals and Process of Law. The law has a social function. It is designed to maintain order, stability, and justice within society. Because the goal of the law is so crucial to a proper functioning of the society, it is often viewed as more important than the process by which the goal is obtained. In the United States, for example, the process of creating and enforcing the law is complex and diverse. There are different types of law made by different parts of the governmental system. For example, the legislature creates statutory law that prohibits people from engaging in some pursuits, such as those involving criminal activity, while demanding that they participate in others, such as paying taxes. The courts, on the other hand, provide a forum whenever there is a dispute that involves a violation of those statutes. For the dispute to be settled, the participants in the court system must follow a particular procedure.[19]

In the United States that procedure is known as the adversarial system. According to the rules of the adversarial system, each side in a dispute has its champion, the attorney of record, who goes into battle against the other side's champion. The battle, either a civil lawsuit or a criminal prosecution, ensues, with victory going to the champion who convinces the judge or the jury of the veracity of his or her case. Not all countries adhere to this process, however. Some very viable legal systems, many of which are in Western Europe, are based on a fact-finding approach rather than an adversarial system. In the fact-finding system, the major objective of a trial is to uncover the truth. Many people find this approach preferable to the adversarial system, which has victory as its primary objective.[20] However, both the adversarial and the fact-finding systems have the same goal—to establish order, predictability, and justice within their respective societies. Can one system truly be said to be better than the other? Which of the two systems will consistently arrive at the "right" judgments? There is no satisfactory way to answer this question. The choice of legal systems depends on a series of factors so diverse and so unpredictable that the question cannot be answered objectively without an in-depth sociological and historical study.

The Goals and Process of Literary Criticism. The reliance on subjectivity can also be seen if we examine literary criticism. A work of literature can be defined as a written composition, whether of verse or prose, that has as its objective the communication of the author's feelings, ideas, or experiences. Moreover, the author of a literary work attempts to communicate with the reader on both an intellectual and an emotional level. If we accept this definition, the next question is how a reader can determine whether a particular piece of literature is worthwhile. To answer this question, some literary critics would choose to concentrate on the theme that the writer attempts to convey. They might ask whether the writer has attempted to say something significant about the human condition. Or they might focus on the writer's creation of a tragic dilemma for the protagonist. Other critics might choose instead to focus on the writer's style. Has the writer demonstrated a talented manipulation of concrete and specific language? Has he or she made appropriate use of connotation and denotation? Has there been a healthy application of figurative language within the work? Still other critics might look for a social message reflecting the age in which the author lived, whereas another group might see the work as a psychological profile on the author's upbringing. The point here is that the standards used and the processes by which those standards are chosen and applied are subjective.

The Goals and Process of Science. As noted, the goal of science is to uncover the truth about the universe. However, that does not mean that scientists compile data and conclusions like a librarian lines the stacks of a library with new books. Instead, scientists seek to interpret that data and those conclusions in an attempt to uncover an understanding of how the universe operates.[21] Yet, as we have seen, law and literature also seek an understanding of at least a portion of the universe. What then is the difference? The difference lies in the process by which that truth is sought. The choice of systems by which the law of a particular country operates and the methods adopted by literary critics are subjective and arbitrary. In applying the law, the United States may elect to use an adversarial system, and France may choose a fact-finding system, and no one can say, with much confidence, that either system is better than the other. Similarly, a literary critic at an American university can apply one critical approach to the interpretation of a poem, while a French critic can apply another, and no one can argue, with much credibility, which approach is the better one. In contrast, a scientist in the United States and another in France must be able to convince one another that the methods that they have used to validate or disprove a theory follow acceptable **objective standards**. Objective standards are criteria that are designed to reach conclusions that are not distorted by bias or prejudice. **Science** then is the study of nature using a generally accepted set of objective standards. These standards must be in line with what is known as the universal scientific method.

The Traditional Scientific Method

The historical development of the scientific method is a legendary tale that has been detailed in numerous books and articles. Rather than belabor that point, we simply focus on explaining the way that science is practiced today. Prior to the Scientific Revolution in the seventeenth and eighteenth centuries in western Europe, scholars relied on a combination of pure thought and the dogmatic pronouncements of past authority in their attempts to prove what rules govern the operation of the universe. As can well be imagined, this technique did little to advance scientific knowledge in the West. It was not until men like Galileo, Kepler, and Newton moved their efforts away from a reliance on past authorities and pure thought and turned instead to observation, experimentation, and verification that the modern scientific method was born.[22]

At the risk of oversimplifying the technique, we can say that the **scientific method** is an objective process by which scientists use observation and experimentation to verify theoretical explanations about the operation of nature. Understandably, many scientists do not like the term *scientific method* because it suggests a single, rigid way to do science. Moreover, the label scientific method implies that only a stereotypical *scientific mind* can use the scientific method. Such is not the case. In science, as in any profession, there are many different individuals all of whom may add their own peculiar twist to the scientific method.

Nevertheless, certain generalities can be made about the scientific method. For example, it involves several carefully executed steps. First, an observation or a series of observations is made. Some of these observations can be made directly, as when we note the regularity of the sunrise, the colors in a rainbow, the rate at which trees grow, or the direction in which a river flows. Other observations can be made by instruments that enhance our natural abilities. This is what Galileo did when he turned his telescope toward the heavens and discovered the mountains and craters on the Moon. Second, in an attempt to understand the phenomenon that has been observed, someone will propose an as yet unproven **hypothesis** to explain the phenomenon. This hypothesis is generally stated in the form of an empirical principle. The goal of this hypothesis is to impose order on the observed phenomena, so that they no longer appear as random events but instead become part of a system. The hypothesis also makes testable predictions about the observed phenomena. Third, to verify (or disprove) the hypothesis, controlled experiments are made to test the predictions. Sometimes the experiments duplicate the efforts of other researchers. At other times, however, the experiments add new dimensions to the research designed to further the researcher's understanding of the empirical principle proposed by the hypothesis. Of course, observations continue to play a role here as the researchers calibrate instruments, compile computer printouts, and analyze other research data.[23]

Fourth, if the experiments verify the empirical principle proposed by the hypothesis, that hypothesis becomes a **law of nature** or a **natural law**. Exactly when such a threshold has been reached is, at best, problematic, and it is safe to say that most hypotheses do not become laws of nature. Nevertheless, some hypotheses do reach that status. Finally, at some point someone begins to see the relationships among a variety of these laws of nature. These relationships work together to predict intricate patterns within a large portion of the natural world. This interdependent network of laws is labeled a **theory**.[24] The theory will also make predictions that can be tested by observation and experiment. Of course, this is not the end of the story because science is an energetic process that is constantly changing as new observations are made, new hypotheses are proposed, and new experiments are conducted to test those hypotheses.[25]

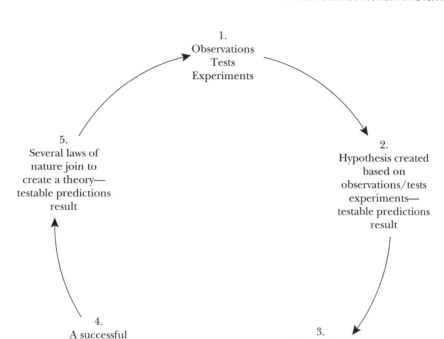

Figure 1.2 The Cycle of the Scientific Method.
The scientific method is an objective process by which scientists use observation and experimentation to verify theoretical explanations about the operation of the universe. This diagram emphasizes the cyclical nature of the scientific method. Observations, tests, and experiments lead to a hypothesis that must be verified by further observations, tests, and experiments. A successful hypothesis becomes a law of nature, which may eventually become a part of a theory. Theories are also subject to verification or falsification by objective observations, tests, and experiments, as the cycle begins again.

A Baseball Analogy. In one of his now famous lectures on basic physics, the legendary physicist Richard Feynman of Caltech, used an extensive analogy in which he sought to explain the scientific method by comparing it to an attempt to discover the rules of chess by observing the process of an actual game. The universe, Feynman said, is like a game of chess conducted by the unseen gods, and scientists are observers who are simply trying to uncover the rules by watching the game.[26]

This idea can be simplified somewhat, and applied to baseball. Anyone who has attempted to explain the game of baseball to the uninitiated will soon learn that the task is a formidable one. We cannot possibly observe the progress of an entire baseball game and deduce all of the rules. To make our job manageable, we may choose to focus on only one small part of the game. Let us say that we decide to focus on the moves of the pitcher. After countless observations, we propose the following hypothetical proposition: "The pitcher always throws the ball to the catcher." Now, we test the hypothesis by observing the pitcher over many innings. If possible, we set up a controlled experiment or two to test the empirical principle proposed by the hypothesis. After these experiments have verified our understanding of the role of the pitcher, we are convinced that we understand his job in the game. The preceding hypothetical proposition is thus our first empirical principle of baseball.

We are comfortable, perhaps even complacent in the certainty of our newly found knowledge about baseball. In fact, we expect that our understanding of the role of the pitcher will soon be accepted as a law of nature. And then the unexpected happens. The batter suddenly hits the ball on the ground. The pitcher is able to field the ball. However, much to our astonishment, he does not throw it to the catcher. Instead, he throws it to the first baseman. Suddenly, a new observation has caused us to doubt the first empirical law of baseball. We now have several choices. We can abandon the first empirical law. We can doubt our senses. We can pretend we did not see the errant throw. Or we can admit to the strange behavior, continue to observe to see if it happens again and, if it does, readjust the first empirical law of baseball. If we chose the last route, we are much more likely to take a step closer to understanding the complete game of baseball.

Verification and Falsification. The scientific method works in much the same manner. However, the process is never ending. We can never know all of the truth. Every time a new theory is suggested, tested, and verified, it joins that body of knowledge we call science. However, its membership in that body of knowledge is guaranteed only as long as it has not been falsified by a new, equally objective test. Paradoxically, it is within the falsification process that we find the true credibility of the scientific method. A single experimental result that contradicts a hypothesis, a law of nature, or an entire theory for that matter places that hypothesis, law, or theory at risk. Yet, that is the beauty of the system. Hypotheses, laws, and theories must be tested and verified, and although the need for testing and verification does indeed place those hypotheses, laws, and theories at risk, it also makes them more and more credible with each successful test.[27]

Often, we value science not so much because of the results, as because of the process used to obtain those results. This is why Bertrand Russell insisted that the work of a scientist is recognized not so much by its truth as by the reason that a scientist believes in that truth.[28] A crucial part of this process is that any theory that purports to describe accurately a portion of nature must meet three criteria: continuity, clarity, and predictability. **Continuity** or **correspondence**, as it is sometimes called, requires that a theory evolve logically from that portion of scientific knowledge that directly affects the new theory. For example, as Jeremy Bernstein points out in *Cranks, Quarks, and the Cosmos*, a theory that proposes to solve the problem of perpetual motion must begin by pointing out flaws or shortcomings in the Second Law of Thermodynamics.[29] Ignoring the Second Law of Thermodynamics in the development of perpetual motion is a little like offering to explain the governmental system of the United States without bothering to read the Constitution.

Second, a successful theory must have **clarity** and be open to public scrutiny. To be of value, a theory must be concise enough to be tested according to objective criteria. Theories that require some sort of hidden knowledge, special talent, or mysterious magic to duplicate are immediately suspect. Theories must also be public. The public revelation of a theory demonstrates a willingness to submit that theory to impartial testing. There must be no fear that a public revelation of the theory will somehow threaten or damage the theory.[30]

Finally, an effective theory will be characterized by **predictability**. **Predictability** means that the theory will make verifiable forecasts about phenomena associated with that theory. If a theory accurately represents the natural universe, then the portion of the natural universe that falls under the direct influence of that theory must adhere to that theory's properly established limits. For instance, if we understand the theory of gravity, then we can predict what will happen when we spill a drink, throw a baseball, drop a bomb, or launch a rocket. Theories that cannot make verifiable predictions will not pass into that body of knowledge called science.[31]

Positivism. The belief that the process of scientific research is objective and unbiased is called **positivism**. According to positivism, science can be distinguished from law, literature, and most other areas of knowledge because scientific truth is derived from an objective evaluation of carefully controlled observation and experimentation. The positivist believes that objective truth exists and is discoverable by properly executed scientific techniques. Moreover, scientific theories are considered accurate reflections of the real world precisely because they have been tested according to objective techniques that can be duplicated and verified.[32] As James Trefil notes in *Reading the Mind of God*, right answers can be found in science.[33] This is often one of the most difficult concepts for the nonscientist to accept. Many nonscientists are comfortable with the subjectivity of law, literature, philosophy, and religion. Consequently, they are suspicious of a discipline that claims to have the "right" answers. This is as it should be. The nonscientist must never surrender his or her right to demand proof from any scientist who claims to have the answer to a question. Nevertheless, the nonscientist should also be open-minded enough to appreciate the fact that most scientists have as their goal the discovery of objective truth.

This open-minded attitude means that nonscientists cannot reject a verified theory simply because they do not like it. Unless they have verifiable evidence to the contrary or at least enough verifiable evidence to cast doubt on that theory, they cannot reject it outright. Instead, they must ask, "Did this scientist, in his or her attempt to verify the theory in question, use an objective, universally accepted technique?" If so, the next step might be to ask whether the researcher had any personal bias that might have tainted the results of his or her research. Once nonscientists are convinced of a researcher's honesty and objectivity, they can ask whether other reputable scientists have had the chance to validate those results. If the theory has been validated, then the nonscientist must consider the truthfulness of the proven proposition, even if it is unappealing or unsettling. Conversely, a nonscientist should not choose to support a pet theory that is aesthetically, ethically, or religiously appealing if the experimental evidence has proven it incorrect. I do not mean to imply that theories are not at times incorrect, that experimental methods are not at times flawed, or that scientists are not at times sloppy in their work. Nor do I mean to imply that changes do not occur within the body of accepted scientific knowledge. This would be an unreasonable position. Moreover, it would be a position that flatly contradicts the ideas examined in the next section. However, the scientific method does demand that scientists and nonscientists alike give any theory that has been verified a level of respect and credibility that goes far beyond simply labeling it an "opinion." The theory need not be carved in stone and declared unchangeable, but it should be recognized as an approximation of nature that stands a much better chance of accuracy than an opinion.

1-3   CREATIVITY AND CULTURAL INFLUENCES IN SCIENCE

Despite this optimistic outlook, many nonscientists still shy away from science altogether. To nonscientists science can look sterile and rigid. The psychologist and historian may look at science as somehow isolated from cultural and social influences, and political scientists, economists, and business people may wonder if scientists ever worry about the realities of the sociopolitical system. Fortunately, the reality of science has enough creativity to satisfy the most fanciful artist, enough cultural influence to delight the most skeptical psychologist, and enough intrigue to interest the most cynical political scientist.

Creativity in Science

The emphasis on objectivity and verification within the scientific method does not, in any way, eliminate creativity from the scientific process. On the contrary, creativity is actually one of the most critical of all ingredients. To understand the creative nature of science, however, we turn to the work of Thomas Kuhn of MIT, who in 1962 wrote a landmark study of the scientific process, entitled *The Structure of Scientific Revolutions*. The process of scientific research that we have discussed thus far generally falls within what Kuhn would label normal science. **Normal science** involves "research firmly based upon one or more past scientific achievements, achievements that some particular scientific community acknowledges for a time as supplying a foundation for its further practice."[34] The achievements of which Kuhn speaks must share two characteristics. First, they must be so innovative that they cause a large number of researchers to abandon any previous, rival theories and adhere firmly to the new one. Second, they must be incomplete enough to give those researchers the chance to work on a variety of problems aimed at developing a more complete understanding of the new achievement. Any achievement that reaches this level of sophistication Kuhn labels a **paradigm**.[35] A good example of this occurred when Copernicus proposed the idea that the Sun rather than the Earth might be at the center of the planetary system. This is the type of achievement that meets Kuhn's two requirements. First, the idea that the Sun might be at the center of the planetary system was so innovative that it caused many researchers to abandon the previous idea that the Earth is at the center. Second, the idea of the Copernican system was open-ended enough to present a number of unsolved problems to which researchers could devote their energies.

Kuhn goes on to explain that scientists who are immersed within a certain paradigm share identical assumptions and practice the scientific method according to the same rules. This solid dedication to a particular pattern of scientific activity is the basis for normal science.[36] Within an established paradigm, normal science takes place in a relatively calm and predictable context. Experiments and observations are carried out in an attempt to verify the underlying theoretical basis for the paradigm, to uncover hitherto unknown aspects of the paradigm, and to attempt to fill in any of the missing pieces of that theoretical puzzle. A case in point is the Copernican system. When Copernicus first offered the idea that the Sun might be at the center of the known universe, the proposition disturbed the accepted paradigm of an Earth-centered universe. However, once established, the Copernican paradigm ushered in a relatively calm period during which researchers further explored the theory, by extending certain principles and by adding to the body of facts that support the theory.

Nevertheless, as we have noted repeatedly, the world of science is dynamic. It changes constantly, sometimes in unpredictably startling ways. Such an unpredictable event is referred to as an anomaly. An **anomaly** occurs when a natural event of some sort appears that challenges the accepted paradigm. In other words, the paradigm predicts that one thing should happen, and an experiment or an observation produces an unexpected, contradictory result.[37] For instance, an anomaly appeared within the Copernican theory when the German astronomer Johannes Kepler observed that the motion of Mars did not match Copernican predictions. If the Copernican paradigm was to survive, the motion of Mars as observed by Kepler, had to be explained.[38] Often an anomaly can be clarified within the context of the accepted paradigm. Thus, a technician may have made an error in calibrating the instruments, the equipment itself may be faulty, or the observation may be compromised by the bias of the experimenter. In such situations, the observed anomaly was not really an anomaly at all. Rather, it was either an unavoidable mistake caused by faulty equipment, a case of poor methodology on the part of the researcher, or some other unforeseen technical problem.[39] None of these explanations accounted for Kepler's measurement of the Martian orbit, however. Kepler's work on the Mars project lasted over eight years, during which he tested over seventy circular orbits of Mars, matching those calculations with the measurements of his mentor the Danish astronomer Tycho Brahe. No mistake was found in any of those calculations. Therefore, the answer had to lie elsewhere.[40]

When an anomaly like the orbit of Mars cannot be explained by a technical error, then it must be interpreted in some other manner. Sometimes the anomaly may be explained by readjusting the paradigm in some way. Such adjustments preserve the paradigm, while, at the same time, accounting for the anomaly. Still, such adjustments are not to be seen as insignificant events because they rarely involve simply adding a concept to the existing paradigm. Instead, they compel the investigator to create a new and different understanding of the universe. This is what occurred when Kepler adjusted the Copernican model by assuming that the orbit of Mars and, therefore the orbits of the other planets, were ellipses rather than perfect circles. Such a solution resulted only when Kepler realized that he had to view the entire paradigm from a different perspective.[41] The explanation for the anomaly, in this case the elliptical orbits of the planets, then becomes part of the predictive pattern of the paradigm. These types of anomalies result in emerging discoveries that readjust the shape of the paradigm. Some parts of the paradigm may be destroyed, but other aspects are constructed in its place. The overall paradigm benefits because the emerging discovery has shaped it in such a way that it is now a much more accurate representation of nature.[42]

According to Kuhn, paradigms are not easily disturbed by anomalies. Many researchers within a particular paradigm expect and even welcome anomalies. In fact, Kuhn noted that it is virtually impossible for a paradigm to solve all of its own problems.[43] Those who adhere to a paradigm stubbornly resist surrendering that paradigm. This is the nature of its strength. If this were not the case, then every anomaly would distract experimenters so often and to such an extent that the main work of normal science would crack and crumble under the strain.[44] However, at times an anomaly is so persistent, so inexplicable, and so troublesome that it leads to a crisis within the accepted paradigm. At this point the anomaly becomes more than a mere discrepancy in experimental results. It is more than a simple mystery. It becomes so serious that it may threaten to undermine the essential groundwork upon which the entire existing paradigm is based. This type of an anomaly leads to a crisis situation within the paradigm. The anomaly is no longer a side issue that can be addressed by slightly altering the paradigm, like the orbit of Mars could be addressed by altering the Copernican assumption of circular orbits. Instead, the anomaly becomes the central issue of the paradigm, and most scientists in the field are compelled to focus on it as the primary objective of their research.[45]

The expanding crisis prompted by the anomaly triggers a distressing period of uncertainty within the field. This period of uncertainty, or "crisis" as Kuhn called it, is fertile ground for the creative impulses within the scientific community. This is a creative period of **revolutionary science**, during which the conventional procedures and the customary rules of normal science are relaxed. New and different approaches to the problem are encouraged. Radical solutions to the problems are entertained as the most creative minds of science tackle the difficulties in an attempt either to alter the old paradigm radically or to introduce a new one. For a while during this period of revolutionary science, researchers live like artists and poets in a world slightly out of focus and slightly out of phase with their normal experience.[46]

Crisis and the Emergence of Quantum Physics. It was this type of world that physicists lived in during the early part of the twentieth century, as a series of troubling anomalies challenged the existing paradigm of classical physics. These anomalies involved the study of the laws of nature as they affect matter and energy at the atomic and subatomic levels, a study that became known as quantum physics. By the early 1920s, these anomalies had become so prevalent that they became the primary focus of the profession. It was impossible to work in physics at that time without becoming embroiled in the crisis that was brewing within that paradigm. In 1925, John Van Vleck, an American physicist and one of the primary contributors to the development of the contemporary theory of electromagnetic systems, described the crisis by comparing the attempts of various physicists to reconcile classical and quantum effects with the twists and turns of a contortionist.[47]

What Van Vleck was referring to was one of the most famous anomalies that ever threatened an existing paradigm, the anomaly described as wave–particle duality. Up to that time, observation and experimentation demonstrated that radiation sometimes acts as if it were a wave and at other times as if it were a particle. This dual nature of radiation challenged the essential foundation of classical physics that clearly viewed nature as having an independent, verifiable existence, unaffected by the observation of the experimenter. Even more incredible was the evidence that a researcher could construct the experiment in such a way that the experimenter seemed to be more than just a passive observer. In fact, it appeared as if the reality of the particle or the wave was determined by the action of the experimenter. In his book *The Physicists*, Daniel Kevles of Caltech portrayed the situation in physics of the 1920s in terms that clearly reflect the same type of crisis situation that Kuhn described. As Kevles explained, the situation was so serious that, at the time, it appeared to threaten, indeed practically to destroy, the entire domain of classical Newtonian physics.[48] Kuhn also pinpointed this same moment as a crisis situation for physics, when he noted that at this time in history Einstein wrote, "It was as if the ground had been pulled out from under one, with no firm foundation to be seen anywhere, upon which one could have built."[49] Later Kuhn attributed similar sentiments to the Austrian physicist Wolfgang Pauli who in 1925 noted, "At the moment physics is again terribly confused. In any case, it is too difficult for me, and I wish I had been a movie comedian or something of the sort and had never heard of physics."[50]

Crisis situations in science cannot last forever, however. Eventually, something must be resolved about the errant anomaly, no matter how long it takes. As Kuhn notes, a crisis can lead to three possible outcomes. The most obvious outcome occurs if the original paradigm is somehow reconciled with the anomaly, despite the difficulty, integrating the two diverse positions. Another resolution occurs when researchers in the field conclude that, given the present state of knowledge and the current state of technology, the solution to the anomaly will not be found now but will have to wait for developments in the far future. The final, and perhaps most radical resolution of the anomaly, comes with the birth of a new paradigm that then competes with the old one for approval.[51] Of course, the possibility for creativity does not end with the resolution of a crisis. On the contrary, creativity is certainly possible during periods of normal science. However, in crisis periods, creativity becomes the norm, rather than the exception.

Social and Cultural Influences in Science

Just as science cannot avoid the need for creativity, it also cannot escape the strong influences of society and culture. Certainly the question of how science shapes society and culture is of interest to nonscience majors in history, law, sociology, and political science. Accordingly, many of the important issues involving the influence of science on society and culture are integral to the mission of this text. However, the opposite question is also of interest to sociologists, legal scholars, historians, and political scientists alike. That question asks how society and culture shape the processes and the results of science. The influence of culture and society on science has been the source of much debate in recent years. In fact, as noted earlier, there are enough social and cultural influences within science to delight the most skeptical sociologist.

Positivism vs. Constructivism. In a very real sense, everything that we have touched on thus far has in some way involved the sociology of science. The first question is "Does science shape society and culture?" This question can be answered and is explored later in this book. Before concluding this chapter, however, we address the other question, "Is science shaped by society and culture?" To answer this question, we examine two divergent points of view on how science, society, and culture interact. These are the positivist approach and the constructivist approach.

The positivist approach is based on the belief that natural law exists as an objective set of rules, which can be discovered by objective scientific processes. This approach, discussed earlier, clearly adheres to the scientific method. According to positivism, science can be distinguished from most other areas of knowledge because scientific truth is derived from an objective evaluation of carefully controlled observation and experimentation. Positivism, then, is the way that most practicing researchers would describe their work, and it is the position that will guide much of the discussion in this text.[52] However, positivism is not without its opponents. The chief opponent of positivism is constructivism.

Constructivism declares that all approximations of the truth are created within a particular social and cultural setting. Therefore, the influence of that social and cultural setting is inescapable. As a result, all truth, even scientific truth, is subject to that social and cultural context. Certainly, there is much accuracy in this statement. It is a fact that many decisions and choices affecting the direction and the results of scientific research are caused by the social and cultural climate.[53] For example, there can be no doubt that decisions on which areas of science receive financial and political support depend upon social and cultural forces that are clearly beyond the control of any individual or any group of researchers. Can there be any doubt, for instance, that the federal money poured into the development of the atomic bomb during World War II would not have been available in such abundance had it not been for the threat posed by the supposed German advances in nuclear research? Similarly, the speed at which a particular area of science moves forward depends on social and cultural forces. Would the United States have succeeded in landing a man on the Moon so quickly had it not been for President Kennedy's pledge to do so in the early 1960s? And would Kennedy have even thought to make such a pledge had it not been for the perceived threat of Russian superiority in space science?

These types of cultural and social influences on science are not open for debate, except by those who would deny such influences in any field of study. However, the constructivist claim goes far beyond these examples of cultural and social influence. According to the constructivist, even scientific truth is determined by social and cultural influences. Scientific truth, then, loses much of its objectivity. As Stephen Cole, professor of sociology at the State University of New York at Stony Brook, writes in his book, *Making Science*:

The position of the constructivists is that the content of solutions to scientific problems is developed in a social context and through a series of social processes. In this sense the content of science is socially constructed. Each new discovery is a result not of the application of a set of rational rules for evaluating empirical evidence but of chance occurrences, the availability of particular equipment, the availability by chance of particular substances, and social negotiations among people inside the laboratory and sometimes between those inside the laboratory and those outside the laboratory.[54]

The difference between the positivist view and the constructivist view is clearly not one of degree. Rather, the two views represent extreme positions in regard to scientific truth. Positivism supports the inevitability of correct conclusions in scientific research, and constructivism denies that possibility because of the overriding influence of society and culture. Moreover, these are not the only two views. In *Making Science*, Cole points out that Thomas Kuhn's theories in relation to normal science and revolutionary science represent a step away from positivism and toward constructivism, without fully embracing the complete relativism of the constructivist view. In addition, Cole offers his own view on the nature of scientific knowledge. This view, which is based on the differences between core knowledge and frontier research in science, represents an attempt to create a balance between positivism and constructivism.[55] The applicability of both positivism and constructivism are explored in a series of questions at the end of this chapter and at various points throughout the rest of the text. For now, however, we look at an example and see how the two positions interpret the situation.

In the summer of 1994, the New York Academy of Sciences sponsored a conference on quantum physics at the University of Maryland Baltimore County (UMBC). (At this point, it might be helpful to read the material beginning on page 48 in Chapter 2 concerning the nature of the subatomic universe.) Essentially two schools of thought dominated the conference. One school supported nonlocality, whereas the other clearly opposed it. Nonlocality is a theoretical property of subatomic particles that holds that even widely separated subatomic systems may still act as if they are entangled and may, in fact, be somehow connected with one another. Thus, an observer may interact with a local subatomic system and affect a distant one that is still entangled with the nearby system. The distant system is affected by a nonlocal event. The nonlocal nature of this event explains why the phenomenon is called nonlocality.[56]

The fact that the physicists disagree about nonlocality is not at all surprising. What is interesting, however, is noting how the scientists aligned themselves in this dispute. Alaine Aspect of the Institut d'Optique Theorique et Appliquee, Orsay, France, is the chief supporter of nonlocality. It was Aspect who in 1982, in a series of experiments performed at the University of Paris, gave the most convincing proof, up to that point, of the validity of nonlocality.[57] At the UMBC conference, the position demonstrated by Aspect was supported theoretically and experimentally by many of the physicists in attendance. One of those supporters was Marlan Scully of the physics department at Texas A & M University. Although Scully's paper discussed matters other than the principle of nonlocality, his comments and questions throughout the conference indicated his support of nonlocality in general and of Aspect's results in particular.[58]

Looking at the other side of the ledger, we find several theoretical and experimental physicists who were opposed to the notion of nonlocality. Chief among the antinonlocality school of thought was David Klyshko of Moscow State University, whose paper demonstrated disdain not only for the principle of nonlocality but also for all interpretations of quantum physics that do not attempt to see quantum physical phenomena in terms of classical physics.[59] Klyshko seemed to be saying that most, if not all, quantum phenomena can be explained in terms of classical physics or at the very least by using classical language. In fact, Klyshko goes a step further by saying that the current fad of using metaphysical and supernatural language to explain quantum physics is pointless and absurd. According to Klyshko, the metaphysical language and all metaphysical aspects of quantum theory have no usefulness whatsoever.[60]

The point here is not to understand who offered the best experimental proof for these theories or whose equations were the most thorough or the most accurate. Rather, the point is to note that in these cases the French and the American scientists tended to support nonlocality, whereas the Russian scientist did not. The positivist would interpret this debate by noting that one of these two positions is more accurate than the other and that the proper application of the scientific method will eventually compel the scientific community to adopt one or the other view. Naturally, this will take an enormous amount of time, countless experimental observations, and hundreds of published papers, but in the end, as with any scientific position, the conclusion that most accurately represents nature will triumph.

The constructivist would see this as a typical example of how social and cultural influences shape consensus within science. The constructivist would see no luck or coincidence in the fact that the French and the American scientists tended to support nonlocality, whereas the Russian did not. The constructivist would argue that it should be no surprise to learn that the Russian physicist, who came from the strictly regulated, highly autocratic, no-nonsense culture of the former Soviet Union, is the one who sees no need for the use of metaphysical language, let alone the actual application of any metaphysical or supernatural explanations in quantum physics. Nor is it strange to see that the French and the American scientists, who are used to the notions of freedom and revolution, would gravitate toward a theory that promotes not only a free and flexible interpretation of quantum phenomena but also one that is highly revolutionary. The social and cultural milieu of each group of scientists determines how they will respond to and interpret the data. The truth, the constructivist would say, is, therefore, relative.

The Role of the Mathematical Model in Science

One of the most formidable concepts for the nonscientist to master is the role that mathematics plays in the scientific method. Part of the difficulty here is rooted in the fact that, unless a mathematical problem involves something concrete, like dollars and cents, or something unambiguous, like weights and measures, many nonscientists frequently see mathematics as either hopelessly abstract or incredibly complicated. Frequently, nonscientists are so bewildered by the relationship between mathematical models and experimental truth that they simply discount the mathematical model as having little or nothing to do with their understanding of science. Often nonscientists ask in all sincerity whether a particular theory is "really true" or "just mathematics."

To help overcome this prejudice against mathematical models, let us look at the role that mathematics plays in the scientific method. A starting point might be an understanding of scientific models in general. A **model** is a representation that helps promote the understanding of a concept because it simplifies the complex aspects of that concept and in this way makes it easier for us to comprehend the intricacies of that concept.[61] For example, an architect's model of a house helps the potential homeowner to visualize the arrangement of the rooms, the relative sizes of the house and the lot on which it is built, and the amount of space available within each room of the structure. This enables the new homeowner to make plans before the day of actual occupancy. Many parts of the actual home, such as the plumbing schematics and wiring diagrams, are, by necessity, left out of the model. These omissions are necessary because it is impossible to include in the model everything within the actual building.

The same shortcomings apply to a narrative model, that is, one constructed entirely of words. A good example of a narrative model is the baseball analogy that was used earlier. The objective of this baseball model was to draw an approximation of the scientific method and, by analogy, lead us to a better understanding of that real world of science. Unfortunately, in the interests of simplicity, the baseball model focuses on only a part of the game and thus leaves out many of the aspects of the game itself. For instance, it does not include the facts that the pitcher sometimes throws strikes and sometimes balls, that the batter may actually swing the bat and hit the ball, or that the catcher may throw the ball to the shortstop in an attempt to stop a runner who is trying to steal second base. Including all of these extraneous facts would simply confuse the issue.

Even more important is the fact that the baseball model also ignores some of the characteristics of the real world that it strives to represent.[62] The baseball model does not explain how controlled experiments are conducted, nor does it account for the number of experiments needed for verification. The baseball model also ignores the problem of how to deal with experimental anomalies like the attempted steals described previously. Still the baseball model does accomplish its objective. It simplifies the complex process of the scientific method and makes that method easier to comprehend.

As enlightening as a narrative model like the baseball analogy may be, it is not nearly as concise as a mathematical model. The objective behind the use of a **mathematical model** is to create a simulation of nature with abstract mathematical symbols. Mathematical models are created by inductive logic. Those models are then applied using deductive logic to predict future occurrences.[63] **Inductive logic** involves observing specific instances and individual pieces of evidence to arrive at a general conclusion. On the other hand, **deductive logic** involves applying generalized principles to arrive at specific conclusions about a particular situation.[64] For example, suppose as a child, I observe that my father, upon returning home from work, enters the house shaking water from his raincoat. I then hear him say, "It's raining outside. If I hadn't worn my raincoat today I would be soaking wet by now." From this experience I can conclude, first, that it is raining and, second, that wearing a raincoat protected my father on that particular day. This would be the inductive part of my reasoning process. Suppose further that, after a series of similar experiences, I conclude that, as a general principle, that I should dress appropriately whenever adverse weather threatens. Today I learn that the Weather Channel has predicted heavy rain for my area of the country. Applying my general principle to this particular situation, I decide that today would be a good day to wear my raincoat. This is the deductive part of my reasoning process.

In the construction of a mathematical model, the theoretician also uses both inductive and deductive logic examples of real-world phenomena. From those observations, the theoretician constructs a mathematical model that reflects those observations. This is the inductive portion of the process. The mathematical model, generally embodied in a set of general principles or **theorems**, can then be applied as if it were a master key for unlocking a code. From a careful application of this key, the theoretician can predict what will happen in the natural world. In a sense, then, the theoretician has doubled back and returned to the starting point, the observable, real world. This is the deductive portion of the process.[65]

Now at this point many nonscientists legitimately ask why scientists even bother with mathematical models. The answer lies in the reliability of the mathematical model. According to John Casti of the Technical University of Vienna, scientists must use mathematical models, despite the indirect nature of the technique, because such models represent the most organized and dependable way to make testable predictions about natural phenomenon. Therefore, Casti concludes that until a new and more direct technique is discovered, scientists will continue to use mathematical models.[66] Another way to say this is to note that it is difficult to argue with an equation that follows precise rules and always comes to the same conclusion.

This is not the end of the story, however. There are problems with the use of mathematical models. First, mathematical models make most aspects of modern science incomprehensible to the nonscientist;[67] this drives a solid wedge between the scientist and the nonscientist. The solution here might be to remind the nonscientist that it is rarely necessary to comprehend every nuance of the mathematical foundation of a theory to understand the abstract concepts, practical applications, and philosophical ramifications of that theory. We do not need to understand the mathematical foundation of quantum theory to comprehend the revolutionary effects that theory has had on the world of science.

The second difficulty is that the language of mathematics is not always as universal as it at first appears. Sometimes theoreticians coin their own symbols and formulate their own relationships to validate a theory. One theoretician, who asked to remain anonymous, explained the situation to me rather well on the last day of the UMBC quantum physics conference. I had remarked to him in an off-handed way that I had a difficult time following most of the presentations whenever the speakers began to project their equations on the screen. He reassured me by noting that, "Some of these guys make up their own symbols and then apply those symbols in complex equations that no one understands. But the rest of us nod our heads and wait our turn at the overhead projector." I am still not quite certain whether I found comfort or bewilderment in this comment. Whatever the case, it clearly demonstrates the difficulty that some nonscientists have with the need for mathematical models in science. Their position is simple. If even the experts disagree on the nature and efficacy of mathematical models, then why should they be expected to accept a particular mathematical model as an approximation of the natural world, let alone as absolute truth? A moderate degree of skepticism like this is healthy. It helps the nonscientist avoid blind, unquestioned faith in science. Still, like most things, even skepticism can go too far. In this case skepticism should be balanced by an open mind that is willing to explore the mysteries of science without being overwhelmed by them.

CONCLUSION

The objective of this chapter is to introduce the nonscientist to the crisis in science education and to investigate the need for science appreciation among the nonscientific population. In addition, the chapter attempts to introduce the nonscientist to the scientific method, as well as to some of the contemporary interpretations of the operation of science. As the chapter suggests, science need not be seen as incomprehensible or bewildering. Nor should it be seen as sterile, rigid, and unimaginative. Instead, it should be viewed as a dynamic, creative force in society that both shapes and is shaped by social and cultural forces. In the next chapter, we will explore one of the most influential and controversial theories of the twentieth century, the Big Bang Theory of the origin of the universe.

Review

I-1. The United States is blessed with a successful research and development program and with a healthy supply of the world's leading scientists and engineers. Despite this success, the American educational system does almost nothing at all to promote an understanding of or an appreciation for science by nonscientists. This neglect should be avoided because nonscientists are just as capable of appreciating science as the scientist is capable of appreciating the humanities. The science education of nonscientists should also not be neglected because nonscientists play a significant role in running the economic, social, and political life of the planet. The goal of science appreciation is to help nonscientists understand some of the most significant scientific theories of the modern age and to lead them to an appreciation of the impact that science has on all aspects of their individual lives.

I-2. Scientists do not simply compile data. Instead, scientists attempt to interpret that data according to the scientific method so that we can see the world as an orderly and predictable place. The scientific method is an objective process by which scientists use observation and experimentation to verify theoretical explanations regarding the operation of nature. The process involves several carefully executed steps. First, an observation or a series of observations is made. Second, in an attempt to understand the phenomenon that has been observed, someone will propose a hypothesis to explain the phenomenon. Third, to verify (or disprove) the hypothesis, controlled experiments are made. Fourth, if the experiments verify the hypothesis, it becomes a law of nature. Finally, the relationships among these laws of nature become apparent. These relationships work together to predict intricate patterns within a large portion of nature. This interdependent network is labeled a theory. Every time a new theory is suggested, tested, and verified, it joins that body of knowledge known as science. However, its membership in that body of knowledge is guaranteed only as long as it has not been falsified by a new, equally objective test.

I-3. Creativity is one of the most critical ingredients in science. Most of the process of scientific research falls within normal science, which includes research that is based upon the accepted paradigm. An anomaly occurs when a natural event of some sort appears that challenges the accepted paradigm. At times an anomaly is so persistent that it leads to a crisis within the accepted paradigm. The crisis triggers a distressing period of revolutionary science within the field. This period of revolutionary science is fertile ground for the creative impulses within the scientific community. Just as science cannot avoid the need for creativity, it also cannot escape the strong influences of society and culture. A central question involving social influences on science is whether scientific theories are shaped by society and culture. Two answers to this question are found within the positivist and the constructivist approaches to science. The positivist approach is based on the belief that laws of nature exist as objective rules that can be discovered by objective scientific processes. Constructivism declares that all approximations of the truth are created within a particular social and cultural setting and that the influence of that social and cultural setting, therefore, is inescapable. Mathematical models are used in scientific research to predict future occurrences. In the construction of a mathematical model, the theoretician observes examples of real-world phenomena and from those observations constructs a mathematical model that reflects those observations. The mathematical model, generally embodied in a set of theorems, can then be applied as if it were a key for unlocking a master code. From a careful application of this key, the theoretician can predict what will happen in the natural world.

Understanding Key Terms

anomaly	law of nature	positivism
clarity	mathematical model	predictability
constructivism	model	revolutionary science
constructivist	natural law	science
continuity (correspondence)	normal science	scientific method
deductive logic	objective standards	theorem
hypothesis	paradigm	theory
inductive logic		

Review Questions

1. What are the causes of scientific illiteracy?
2. What are some possible consequences of scientific illiteracy?
3. What is the goal of a course in science appreciation?
4. How do law and literature differ from science?
5. How does the scientific method operate?
6. What is Kuhn's model of scientific inquiry?
7. What situations allow for creativity in science?
8. What is positivism? What is constructivism?
9. How do social and cultural influences affect science?
10. What is the role of mathematics within the scientific process?

Discussion Questions

1. Do you agree or disagree with the value judgments made in this chapter concerning the state of science education in this country at this time? What has your personal experience been in this regard? Do you see any value in a science-education course? Explain each of your answers.
2. Do you tend to support the positivist view or the constructivist view of science? Explain why you hold this position.
3. Can you pinpoint any examples of the shaping of science by social or cultural forces? What do you think of the feminist charge that much of scientific research is dominated and, therefore, shaped by a male-dominated society? Can you think of any directions that science would not have taken had science not been male-dominated? What about the reverse situation? Can you think of any directions that science might have taken, had it been more female-dominated in the past? Explain your response to each question.
4. What do you think of the charge that much of scientific research is dominated and, therefore, shaped by the military? Or the position that says that most research today is dominated by the multinational corporations? Explain your response to each question.
5. The discussion in the chapter points out that there are problems with the use of mathematical models in science. First, mathematical models make most modern science incomprehensible to the nonscientist. Second, the language of mathematics is not always as universal as it at first appears. Sometimes theoreticians coin their own symbols and formulate their own relationships to validate a pet theory. What do you think of each of these criticisms? Do you find that you are bewildered and that you, therefore, distrust scientific theories based upon mathematical models? Explain. Do you feel skeptical about the veracity of such mathematical models? Explain.

ANALYZING *STAR TREK*

Background

Before examining the following episode of *Star Trek: The Next Generation*, it would be helpful to get an overview of the series. *Star Trek: The Next Generation* is set in the twenty-fourth century at a time when a united planet Earth has enjoyed three hundred years of peace. In the twenty-first century a man named Zephram Cochrane invented a method of space travel known as warp drive, which allows a starship to travel at many times the speed of light, making interstellar travel a reality. The Earth is one of the charter members the United Federation of Planets, a group of star systems united for the peaceful exploration of the galaxy. The space-faring arm of the United Federation of Planets is known as Star

Fleet Command. The focus of *Star Trek: The Next Generation* is the Starship Enterprise, a galaxy-class starship that carries over 2000 crew members and passengers, including not only working Starfleet personnel but also their families. The commanding officer of the USS Enterprise is Captain Jean Luc Picard. The mission of the Enterprise is "to explore strange new worlds, to seek out new life and new civilizations, to boldly go where no one has gone before."

Other crew members of the Enterprise include first officer, Commander Will Riker; chief medical officer, Dr. Beverly Crusher; chief engineer, Lt. Commander Geordi LaForge; counselor, Commander Deanna Troi; chief of security, Lt. Commander Tasha Yar; tactical officer, Lt. Worf; science officer, Lt. Commander Data; acting ensign Wesley Crusher, Dr. Crusher's son; transporter chief, Chief Miles O'Brien; chief medical officer during the second season, Dr. Katherine Pulaski; and Guinan, the operator of a bar and recreational area on the Enterprise, known as Ten Forward.

Chief allies of the Federation include the Vulcans, a race of superior humanoids who have based their civilization on logic; and the Klingons, a warlike people who, until recently, were the Federation's greatest enemies. The chief antagonists to the Federation are the Romulans, another warlike humanoid race; the Cardassians, a third warlike humanoid race; and the Borg, a race of robotic creatures tied together by a hive mentality. Periodically, the Enterprise is also plagued by a super powerful being known only as "Q."

Other key Star Trek concepts include the transporter, away teams, the replicator, the holodeck, impulse drive, phasers, and the Prime Directive. The transporter is a device that can dematerialize matter and move it to a distant position instantly. The process is usually referred to as "beaming." It is generally used to send an away team to the surface of a planet. An away team is a group of crew members assigned to visit a planet to accomplish the objectives of a mission. The replicator is a device that uses a form of advanced **nanotechnology**. Nanotechnology allows for the creation of objects by rearranging their molecular structure. **Replicators** are used to assemble food items, clothing, and other material supplies needed on a starship. The holodeck creates three-dimensional scenes by the combined application of holography and replicator technology. The holodeck is generally used for recreational activities.

Impulse drive is a supplementary propulsion system used by starships. Impulse drive is powered by fusion reactors and generally operates only at sublight velocities. Phasers are weapons of pure energy beams. Some phasers are handheld portable weapons; others are used in defense of the ship. **Photon** torpedoes are matter–**antimatter** projectiles used as weapons. As the name suggests, the Prime Directive is the principle rule of the Federation. The Prime Directive forbids Starfleet personnel from interfering with the internal development of pre-warp-drive cultures. The idea is based on historical evidence which indicates that whenever an advanced culture encounters a less developed culture, the advanced culture inevitably exploits the other.

Special Instructions

The following episode from *Star Trek: The Next Generation* reflects some of the issues that are presented in this chapter. The episode has been carefully chosen to represent several of the most interesting aspects of the chapter. When answering the questions at the end of the episode, you should express your opinions as clearly and openly as possible. You may also want to discuss your answers with others and compare and contrast those answers. Above all, you should be less concerned with the "right" answer and more with explaining your position as thoroughly as possible.

Viewing Assignment—*Star Trek: The Next Generation*, "Contagion"

The USS Enterprise 1701-D has been contacted by its sister ship, the USS Yamato. Captain Varley of the Yamato has reported to Captain Picard of the Enterprise that the Yamato has experienced a number of inexplicable systems failures. Unfortunately, before the Enterprise can arrive to help, the Yamato self-destructs. Captain Picard decides that the crew of the Enterprise will take over the mission of the Yamato. As it turns out, that mission was to find the home planet of the Iconians, a long-dead race which, according to legend, had developed an advanced state of technology. By following the clues left by Varley on his log, the Enterprise does indeed locate the Iconian home planet. However, the Enterprise also undergoes the same type of systems failures that destroyed the Yamato. Chief engineer LaForge discovers that the systems failures on the Enterprise are being caused by an Iconian program that entered the central computer on board the Enterprise when the log of the Yamato was downloaded into the computer. The Iconian program had entered the Yamato's system when that ship was scanned by an Iconian probe. The Iconian program is now struggling with the central computer for control of the Enterprise. To complicate matters further, the Iconian home planet is located in the Neutral Zone that separates Federation space from the Romulan Empire. Consequently, when the Enterprise arrives at Iconia, a Romulan starship is waiting. Because of his background in archeology, Picard heads the away team that beams down to Iconia. The captain is afraid that the advanced technology of the Iconians will come under the control of the Romulans, and so he elects to destroy the central control station on Iconia. Later, LaForge saves the Enterprise by shutting down the central computer and restarting it. This action purges the computer of the dangerous Iconian program.

**Thoughts,
Notions,
and Speculations**

1. Note the dual nature of the conflict in "Contagion." Both sources of conflict are caused by technological problems. The first problem is the systems failures caused by the Iconian program. The problem reveals just how dependent the crew of the Enterprise is on the smooth functioning of their technology. Naturally, a space vehicle of any sort will be dependent on technology. However, as we learn in this episode, ninety percent of the systems on board the ship are automated. The captain cannot even have a cup of tea without activating the computer-controlled replicator system. Are the authors of this episode suggesting that we in the twentieth century are too dependent upon technology, or are they simply warning us that this is the direction in which we are headed? Does the initial conflict presented in this episode relate to the issue of science education as presented in this chapter? Why or why not? What lesson in regard to science education can be learned from the initial conflict presented in this episode? Explain your response to each question.

2. The second conflict in the episode is caused by the captain's fear that the advanced technology of the Iconians will fall into the hands of the Romulans. The subtle implication and the hidden message here is that technology can be dangerous. Of course, there is not much that is new in this message. What is interesting here is that Captain Picard elects to solve the problem by destroying technology. Is Picard practicing the philosophy of the Luddites, who in the nineteenth century destroyed the automated textile factories that they believed were robbing them of employment? How does this conflict relate to the issue of science education as presented in this chapter? Explain your response. Could Picard's destruction of the Iconian technology be considered the actions of a positivist or a constructivist? Explain.

3. Another curious aspect of the episode "Contagion" is found in LaForge's solution to the systems failures. Recall that LaForge shut down and restarted the central computer on the Enterprise to purge the system of the Iconian Program. This action is heralded as a brilliant inspiration. Yet, such a solution would be the first course of action that any computer operator would take today. Even a novice computer owner would know to take this simple step. Yet, it does not occur to LaForge until he witnesses the android Data undergo a similar shut-down, start-up process. Are the writers suggesting here that an overdependence on technology is frequently accompanied by an ignorance about how that technology actually works? How does this position relate to the issue of science education? Explain.

NOTES

1. Daniel J. Kevles, "The Changed Partnership," 45–46.
2. Ibid., 46–47.
3. D. Allan Bromley, *By the Year 2,000*, 7.
4. Kevles, 47.
5. J. Michael Bishop, "Enemies of Promise," 64–65.
6. Bromley, 8.
7. Jon D. Miller, *The Public Understanding of Science and Technology in the United States, 1990*, 13–15.
8. Robert M. Hazen and James Trefil, *Science Matters*, xiii–xiv.
9. This survey was compiled with the assistance and cooperation of the administration of North Central State College in Mansfield, Ohio. The poll was conducted among the 1996 graduating class, all of whom were awarded associate degrees that year.
10. David L. Goodstein, "After the Big Crunch," 56.
11. Ibid.
12. Ibid., 56–57.
13. Bromley, 6.
14. Goodstein, 56–57.
15. John Ziman, *Teaching and Learning about Science and Society*, 13.
16. National Commission on Excellence in Education, *A Nation at Risk*, 25.
17. Ziman, 13.
18. Leslie Stevenson and Henry Byerly, *The Many Faces of Science*, 2–3.
19. Gordon W. Brown and Paul A. Sukys, *Business Law with UCC Applications*, 9th ed., 677.
20. Ibid.
21. Stevenson and Byerly, 2–3.
22. Roger Newton, *What Makes Nature Tick?* 11; James Trefil, *Reading the Mind of God*, 32–34.
23. John L. Casti, *Searching for Certainty*, 25–27.
24. Ibid.
25. Stevenson and Byerly, 2–3.
26. Richard P. Feynman, *Six Easy Pieces*, 24–27.
27. Steven Goldberg, *Culture Clash*, 8.
28. Stevenson and Byerly, 3.
29. Jeremy Bernstein, *Cranks, Quarks, and the Cosmos*, 19–20.
30. Casti, 29.
31. Bernstein, 17, 20–24.
32. Stephen Cole, *Making Science*, 3–7.
33. Trefil, 35.
34. Thomas S. Kuhn, *The Structure of Scientific Revolutions*, 10.
35. Ibid.
36. Ibid., 11.
37. Ibid., 52–53.
38. Timothy Ferris, *Coming of Age in the Milky Way*, 78–79.
39. Cole, 8; Kuhn, 79–80. Kuhn explains that this is the case because the objective of most normal science is to confirm the existing paradigm. Thus, experiments assume the validity of the paradigm and results that contradict the paradigm are generally not viewed as a threat to it. The initial reaction to such contradictions is to blame the researcher, not the paradigm. A researcher who achieves contradictory results and then condemns the paradigm is generally discredited as "a carpenter who blames his tools."
40. Ferris, 78–79.
41. Ferris, 79; Kuhn 52–53.
42. Kuhn, 52–53, 66.
43. Ibid., 79.
44. Ibid., 65, 82.
45. Ibid., 5–6, 66–68, 80–83.
46. Ibid., 5–6, 79.
47. Daniel J. Kevles, *The Physicists*, 159.
48. Ibid.
49. Kuhn, 83.
50. Ibid.
51. Ibid.
52. Cole, 3–7.
53. Ibid., 10–14.
54. Ibid., 12.
55. Ibid., 7–17.
56. John D. Barrow, *The World within the World*, 146–47.
57. Ibid., 147.
58. Marlan O. Scully, Ulrich W. Rathe, and Susanne F. Yelin, "Second-Order Photon-Photon Correlations and Atomic Spectroscopy," 28–39.
59. D. N. Klyshko, "Quantum Optics: Quantum, Classical, and Metaphysical Aspects," 13–27.
60. Ibid.
61. Casti, 31.
62. Ibid.
63. Ibid., 31–33.
64. H. Ramsey Fowler and Jane E. Aaron, *The Little, Brown Handbook*, 137.
65. Casti, 31–33.
66. Ibid., 32–33.
67. Newton, 17.

Unit II The Origin of the Universe

Chapter 2
The Mystery Called the Big Bang

COMMENTARY: WHEN PAST BECOMES FUTURE

On April 23, 1992, George Smoot, the head of the research team responsible for the **Cosmic Background Explorer (COBE)**, announced that the explorer had located temperature variations in the background radiation of the early universe that were sufficient to account for the evolution of the universe according to the **big bang theory**, the orthodox scientific explanation for the creation of the universe. At a press conference held to announce the news, Smoot said the discovery was the scientific equivalent of "looking at the face of God." Smoot's inadvertent, almost off-hand statement was grabbed eagerly by the news media and flashed around the globe. In fact, Smoot's comment actually seemed to cause more of a stir than the discovery itself. Some members of the scientific community, charged him with stepping beyond the limits imposed by science, and some members of the religious community accused him of blasphemy. Why did the focus of this revolutionary announcement shift from the discovery itself to Smoot's inadvertent statement? Many reasons could be offered as an explanation for this shift in focus. However, one reason may have been that the discovery disturbed many people. Yet, Smoot had done nothing more than what all scientists have done since Copernicus and Galileo. Like Copernicus and Galileo, Smoot was helping to sharpen our understanding and appreciation of the universe. Understandably, this sharpened vision makes many people uncomfortable. However, the fact that such a vision makes many people uncomfortable will not make that vision go away. The sooner we learn to accept and appreciate the sharpened image of the universe given to us by science, the sooner we will be in harmony with that universe. This and other similar ideas are addressed in this chapter.

CHAPTER OUTCOMES

After reading this chapter the reader will be able to accomplish the following:

1. Identify reasons that a nonscientist might have for exploring the origin of the universe.
2. Explain why the big bang should not be thought of as an explosion.
3. Explain why the big bang theory is not antireligious.
4. Describe the implications of Edwin Hubble's discovery regarding galactic movement.
5. Outline the contributions to the theory of the expanding universe by Slipher, Lemaitre, Friedmann, and Eddington.
6. Outline the number and types of fundamental particles that make up the subatomic universe.
7. Identify the four fundamental forces of the universe.
8. Contrast the reproducing universe theory with the evolving universe theory.
9. Identify the stages in the chronological story of the big bang.
10. Explain the primary evidence that supports the big bang theory.

2-1 WHY STUDY THE BIG BANG?

How did the universe begin? Has the universe always existed in its present state, or did it have a starting point billions of years in the past? Is the universe that we inhabit the one and only universe? Or is it but a single universe in a never-ending stream of periodic, self-recreating universes? Are we living in a planned universe that is evolving toward a meaningful future, or are we drifting in an accidental universe that is doomed to inevitable destruction? Each culture in history, from the mighty empires of the ancient world to the technological civilization of the twentieth century, has a story that tells how the universe began. The Hindu scriptures, for example, have no less than three creation myths that attempt to explain the origin of the universe. The Jainists, on the other hand, hold that the universe was not created at all, but instead has existed forever, as an eternal series of endless, repetitive cycles.

Today, many of our ideas about the origin of the universe are a curious blend of religion, science, and personally held beliefs. Some people strongly believe in the story of creation as outlined in *Genesis*, the first book of the Bible. Others prefer a more scientific explanation. Still others prefer to unite religion and science. Whatever your belief system might be, it will be beneficial for you to have an understanding of the standard scientific model for the creation of the universe. If you are satisfied with *Genesis* or some other religious interpretation of creation, then learning about the scientific model may help you appreciate your own beliefs even more than you do at the present time. If you are convinced that science can unravel the secrets of the origin of the universe, then you almost certainly see the value of exploring the scientific model of creation. However, even if you are not particularly concerned about the origin of the universe, learning about the scientific model will, at the very least, contribute to your storehouse of knowledge about the universe. Such knowledge will help you to separate fact from fiction as you sift through the morass of information and misinformation thrown at you by the various news media whenever they discuss the big bang.

Unfortunately, the media, which are the primary sources of scientific information for many nonscientists, frequently misinterpret scientific truth. Of course, we must be careful not to lay too much blame on the journalists for this shortcoming. After all, it is not their job to spread scientific truth. Rather, their job is to report the news as accurately, efficiently, and dramatically as possible. Unfortunately, accuracy is sometimes sacrificed for efficiency and drama. Often in search of a flashy headline or a catchy sound byte, journalists focus on the more sensational or more controversial aspects of a story, rather than on the dull, but fairly substantial facts that make up the bulk of that story. This approach can lead to inaccuracies, discrepancies, and, at times, a lot of confusion about science.

Of particular interest to us here is how the news media have treated the theory of the big bang. As noted in the commentary at the beginning of the chapter, a case in point involves the remarks made by George Smoot, the head of the COBE research team. Most of the news stories about the COBE discovery did little to explain how that discovery related to the big bang theory. Instead, those stories zeroed in on the controversial words of Smoot. Unfortunately, this preference for the sensational aspects of the story misrepresented the real nature of the big bang and led many people to a distorted view of the standard scientific model of the origin of the universe. Accordingly, these inaccurate accounts left a lot of people ignorant about some very fundamental truths concerning the origin of the universe. These inaccuracies can also drive a wedge between the scientist and the nonscientist, something that should be avoided whenever possible. Consequently, the rest of this chapter is devoted to setting the record straight about the standard scientific model for the origin of the universe.

As noted above, the standard model for the origin of the universe is called the big bang theory. Unfortunately, the name itself can be somewhat misleading. Nevertheless, all of the scientific literature concerned with the topic uses the name, "big bang," so we will use it as well. This chapter has three objectives: (1) to set the record straight by explaining what the big bang is not; (2) to explore what the big bang really is as well as provide a detailed chronological outline of the critical events of the big bang; and (3) to look at some of the most convincing evidence supporting the big bang.

In the next chapter, we take a cue from the work of Thomas Kuhn (see Chapter 1) and examine some of the anomalies that threaten to crack the solid edifice of the big bang theory. After that, we examine some alternate theories that are presently competing with the big bang. Included in this will be some questions about the social and cultural influences on these theories, as well as on the big bang theory itself. Finally, we explore the question of why a non-scientist should care about any of this. In doing so, we look not only at the theological and philosophical reasons for studying the origin of the universe, but also at some of the more practical aspects of that study.

2-2   MISCONCEPTIONS CONCERNING THE BIG BANG

As noted, the first step in our examination of the big bang origin theory is to dispel some of the common misconceptions that have plagued the theory since it was first suggested over half a century ago. In other words, before we can examine what the big bang is, we must first examine what it is not. To state it quite directly, there are three common misconceptions about the big bang that must be addressed. The first misconception is the idea that the big bang was an explosion. The second is the belief that the big bang was proposed by scientists as a replacement for religious explanations for the creation of the universe. The final misconception is the belief that the big bang theory is an unsupported, unsubstantiated "guess" rather than a solid theory that has passed quite a few verification tests over the last half century. (*Note:* This is not to say that the big bang theory is unopposed. On the contrary, as is true of any established theory, there are anomalies (which we will examine in the next chapter) that threaten its continued acceptance as the dominant theory. The first two misconceptions are addressed immediately. The final misconception is covered at the end of this chapter after we have established what the big bang really is.

The Big Bang Was Not an Explosion

The first problem in relation to the big bang theory is the fact that many people have the mistaken notion that the big bang was an explosion. Of course, it is easy to see why people get this idea. The name, big bang, suggests an explosion of some sort. Moreover, many people, even some who know better, perpetuate this inaccurate image by repeating it in books, articles, interviews, and lectures. For instance, in an otherwise excellent book entitled *God and the New Physics*, Paul Davies, professor of natural philosophy at the University of Adelaide, in South Australia, uses the word "explosion" to describe the big bang.[1] The idea of an exploding universe implied by the name "big bang" is more than just an unfortunate label. The designation "big bang" confuses and misleads nonscientists to such an extent that they often reject the theory as nonsense. After all, they argue, how can a universe that is clearly built of intricate and complex structures be the result of a destructive event like an explosion? The image of an explosion "jump-starting" the universe leaves nonscientists understandably bewildered. It is no wonder that they walk away shaking their heads in disbelief at the sad, if somewhat comic, illogic of the scientific community.

Unfortunately, this bit of confusion will remain unless the nonscientist is shown a more accurate picture of the big bang. What the big bang theory actually states is that the universe began some 15 to 20 billion years ago when an infinitely dense subatomic entity, known as the **singularity**, appeared, inflated suddenly, and, as the inflation slowed and the universe cooled, coalesced into the cosmos as we know it today.[2] This image is quite different from the common view that the universe began with an explosion. The image of an explosion suggests that the singularity burst into fragments that scattered throughout existing space. Rather, the theory implies that the singularity actually was the universe in its simplest, purest, and most symmetric form. Then suddenly, for reasons not yet fully explained, this singularity began to inflate and cool. This implies that space itself, was and is still expanding. Galaxies, stars, and planets are not expanding into space. Rather, *space itself is expanding*, and the galaxies, stars, and planets are going along for the ride.[3] Later in this chapter, we will see a more complete picture of this process. For now it should be enough to dispel the idea that the big bang is an explosion.

If the big bang is not an explosion, where did the unfortunate name "big bang" come from, and why has it persisted for so long? The name "big bang" was actually crafted by the British astronomer Fred Hoyle, who invented the term during a radio broadcast in which he was trying to explain the difference between the two most widely accepted theories concerning the origin of the universe: the steady state theory and the expanding universe theory. In his attempt to distinguish between the two theories he called the expanding universe theory the "big bang." Much to Hoyle's dismay, the colorful name, "big bang," became extremely popular, and the theory itself became more widely known than ever before.[4]

The Big Bang Theory Is Not Antireligious

Many nonscientists also have the mistaken notion that the big bang is somehow antireligious. It is difficult to trace the source of this hostility, especially because of the key roles that several churchmen have played in support of the theory. If it were not for the Belgian astronomer, cosmologist, and priest, Georges Lemaitre, for example, the theory of the expanding universe might not have gotten the initial support that pushed it to the forefront of modern cosmology. Moreover, this hostility may be symptomatic of the distrust that many people have toward science in general. Or it may have to do with the fact that the big bang theory does reach back to the origin of the universe, a topic that many people feel might be best left to religious teachers rather than scientists.

The Neutrality of the Big Bang Theory. The first point that must be made in defense of the big bang theory is that it has no religious pretensions whatsoever. It is, like any theory, simply a way to explain the evidence that exists in the universe today. As we shall see later in this chapter, many of the scientists who are responsible for the big bang theory were not actively searching for an origin theory when they made their contributions. Instead, they were observing the universe, making measurements, and coming to conclusions that seemed to flow logically from that evidence.

Edwin Hubble, for example, the astronomer who is generally credited with being one of the first to present evidence supporting the notion of an expanding universe, was observing the light that reached the Earth from distant galaxies. He was not attempting to explain the creation of the universe. Similarly, in 1964 when Robert Wilson and Arno Penzias of Bell Laboratories discovered the existence of background radiation, which supplies further evidence for the big bang, they were not searching for the origin of the universe. They were trying to eliminate what they thought was a strange interference in their instrumentation. Like Galileo before them, Hubble, Wilson, and Penzias were curious about how the universe works. The fact that their discoveries helped support the big bang theory was simply because of the logical conclusions that flowed from the data that they had compiled.

Religious Overtones of the Big Bang Theory. The second point to make in defense of the big bang theory is that even though the theory itself has no religious pretensions, some experts have attempted to exploit the religious overtones of the theory. This tendency to focus on God and religion confuses the issue; on the one hand, many scientists deny that any theory should be seen as an attempt to explain religious issues, while, on the other hand, some scientists do precisely that. Surprisingly, many of those who have detected religious implications in the big bang theory have actually used the theory in support of

Genesis and other creation stories, rather than to oppose those creation stories. For example, Robert Jastrow, founder of NASA's Goddard Institute for Space Studies, states rather unequivocally in his book *God and the Astronomers* that, "the astronomical evidence leads to a biblical view of the origin of the world. The details differ, but the essential elements in the astronomical and biblical accounts of Genesis are the same: the chain of events leading to man commenced suddenly and sharply at a definite moment in time, in a flash of light and energy."[5] A second individual who supports the religious overtones of the big bang is Stanley Jaki, a priest and historian of science, who in his work, *Cosmos and Creator* explains that the existence of the finite universe leads inevitably to the need for a Creator.[6] According to Jaki, the coherent nature of the universe demands the conclusion that it was given to us by a Creator.[7] In fact, Jaki goes so far as to say that the acceptance of a Creator is a prerequisite to a belief in the existence of the universe itself. Otherwise, humanity's underlying sense of purpose and direction is without point.[8]

Another priest and scientist, Pierre Teilhard de Chardin, takes the implications of the expanding universe even further. Teilhard does not simply state that the expanding, evolutionary universe can be reconciled with religion. Instead, he argues that religion in general and Christianity in particular can reach its true potential *only* within an expanding, evolutionary universe. As Teilhard explains in his controversial work *Science and Christ*, "Experience shows that traditional Christology can accept an evolutionary world-structure; but, what is even more, and what contradicts all predictions, it is within this new organic and unitary ambience, and by reason of this particular curve of linked Space-Time, that it can develop most freely and fully. It is there [i.e., within the expanding universe] that Christology takes on its true form."[9]

At the opposite end of the spectrum from Jastrow, Jaki, and Teilhard are those scientists who oppose the big bang theory not for scientific reasons but on philosophical or theological grounds. Interestingly enough, within the scientific world, many opponents of the big bang theory base their opposition on the fact that the big bang model invites a kind of metaphysical speculation about the origin of the universe. In fact, the Austrian-born astronomers Hermann Bondi and Thomas Gold, along with the British cosmologist Fred Hoyle, especially dislike the idea that the universe had a beginning because, according to their point of view, such a theory reaches outside the physical world and invites metaphysical speculation about the origin of the universe.[10] As a result, Bondi, Gold, and Hoyle propose an alternative theory that they call **steady state cosmology**. To account for the fact that the universe in a steady state would look the same at all times for all observers, Bondi, Gold, and Hoyle propose that matter is continuously created spontaneously throughout all regions of space. According to this version of the steady state model, the amount of new matter necessary to account for this would amount to about one atom per year for every one hundred cubic meters of space.[11] Since he first suggested the steady state model in the 1940s, Hoyle has modified the theory to account for several discoveries that have occurred since that time. Both versions of steady state theory will be explored in the next chapter.

Creationism and the Big Bang. To review then, the idea that the big bang theory is antireligious must be seriously doubted, first, because the theory itself makes no religious claims, and second because when religious interpretations of the big bang are offered, those interpretations generally call upon the big bang to support, rather than discount, any spiritual dimension to the creation event. However, before leaving this topic, we must point out that there is at least one sense in which the big bang theory could be seen as incompatible with religion. This approach, generally referred to as **creationism** or **creation science**, insists upon a literal reading of the Bible. A literal reading of *Genesis* would indicate that the universe was created in six days, that the state of the universe has been unchanged since those original six days, and that the universe itself is a little over 6000 years old. Since the big bang contradicts each of these Biblical facts, creationists see the big bang theory as antireligious.[12]

When creationists are faced with evidence of a universe far older than 6000 years, their reply is that, for some reason, perhaps to test our faith, God has made the universe appear billions of years old when in fact it is much younger. The same argument can be offered for evidence of an evolving universe and for evidence of a much longer creation process than the six days laid out in Genesis.[13] However, if this were true, it is difficult to see how the creationists can blame scientists who have been successfully fooled by this divinely manipulated evidence. This would seem to be especially true since, as we have seen, some of these experts, like Jastrow, Teilhard, and Jaki, have come to the conclusion that the big bang theory demonstrates supernatural intervention in the creation of the universe.

Still, it is unlikely that the creationists and those who believe in the big bang theory will agree on the origin of the universe. This impasse should not, however, stop the creationists from learning about the big bang theory or from expressing their disagreement with that theory, any more than it should stop the big bang adherents from learning about creationism or from expressing their point of view in that regard. After all, in the Star Trek episode, "Where No One Has Gone Before," at the end of this chapter, even though Captain Picard is skeptical about Kozinski's work, that does not prevent him from learning about the project or from expressing how he feels about it. It also, by the way, led to a new understanding of the nature of reality, thereby proving that knowledge and discourse are always better avenues than ignorance and self-censorship.

2-3   THE NATURE OF THE BIG BANG

Now that we have spent some time examining several common misconceptions about the big bang, we move on to examine its true nature. To do this with any degree of accuracy, we must visit the observatory of Edwin Hubble, a former law student turned astronomer, whose work in the 1920s suggested the notion of the expanding universe. In 1923, at the Mount Wilson Observatory in California Hubble discovered that the Andromeda nebula, once thought to be an enormous cloud of dust and gas at the edge of the Milky Way, was actually a separate grouping of billions of stars, located far outside of the Milky Way. These enormous groups of stars are now known as **galaxies**.[14] Galaxies come in three main shapes: spiral, elliptical, and irregular. Spiral galaxies are generally shaped like flat disks. Andromeda and the Milky Way are typical spiral galaxies, each containing between 200 and 300 billion stars. Elliptical galaxies, on the other hand, are generally shaped like spheres, and irregular galaxies are those that have no set shape.[15] While Hubble was studying several distant galaxies, he discovered that most of those galaxies were emitting light that is shifted toward the red end of the **spectrum**.[16] Hubble used this data to conclude that most of these galaxies are moving away from us. Hubble came to this conclusion by interpreting the red shift data as a manifestation of the **Doppler effect**, which is named after J. Christian Doppler, the Austrian mathematician who discovered the phenomenon in 1842.[17]

Doppler actually used the effect to explain the nature of sound waves, like those emitted from the siren of a police car (or, in Doppler's case, a musical band on a moving train) as it approaches, passes, and then moves away from a pedestrian. As the police car moves toward a group of pedestrians by the side of the road, the frequency with which the sound waves reach them increases. This causes the pitch of the siren to rise. As the police car passes and moves away from the pedestrians, the frequency with which the sound waves reach the pedestrians falls. Consequently, the pedestrians experience a corresponding fall in the pitch of the siren. A similar effect is noted with light waves. If a light source, such as a galaxy, is moving toward an observer, the frequency with which the light waves reach the observer will increase. This will result in light shifted toward the blue end of the spectrum. If the light source is moving away from the observer, then the frequency with which the waves reach the observer decreases. The result is light shifted toward the red end of the spectrum. This **red shift** convinced Hubble that the galaxies he had observed were moving away from our position in the universe.[18] Evry Schatzman, an astrophysicist and a member of the French Academy of Sciences, explained Hubble's discovery in this way: "(o)bserving the spectrum of the galaxies, Hubble noted that these galaxies were 'shifting' toward the red: they appeared to be redder than they were. We know that red light has a longer wavelength than the average

wavelength of visible light. Thus we must conclude that the galaxies are moving away from us."[19]

Implications of the Red Shift Discovery

Actually Hubble was not the first astronomer to observe the red shift phenomenon. In 1909, an astronomer at the Lowell Observatory in Arizona named Vesto Slipher began a project aimed at measuring the rotation speed of spiral nebulae. At the beginning of the twentieth century, the accepted view was that nebulae were located close to the Sun. After many observations and measurements, Slipher, having worked with data on thirteen nebulae, discovered that all but two demonstrated spectra shifted toward the red end of the spectrum. Even more astonishing was the fact that the higher red shifts were found to be emanating from the fainter objects. Slipher even found one nebula that had a red shift indicating a speed of 1000 km/sec. Because such high velocities were well beyond that of any stars, Slipher concluded that these nebulae were far outside our galaxy.[20]

As noted above, Hubble's later observations clearly confirmed Slipher's conclusions. However, Hubble's work also went beyond the observations of Slipher. In addition to confirming the red shift observations, Hubble also used the data that he had compiled to determine the distances to those galaxies and, as a result, was also able to estimate the age of the universe. Hubble concluded that the rate of movement of a galaxy away from the Milky Way Galaxy increases in proportion to its distance from our galaxy. This observation came to be known as Hubble's law, and the proportion—that is, the ratio of velocity to distance—as the Hubble constant.[21]

Hubble's initial assessment of the age of the universe was inaccurate. This inaccuracy was caused by errors in his analysis of the distances to the galaxies. Since the time that Hubble made his original measurements, advanced

Figure 2.1 A Spiral Galaxy.

M31, NGC 224, the Andromeda Galaxy, is a Type Sb spiral galaxy in the constellation Andromeda. It is the most easily visible and nearest spiral galaxy, about 2.2 million light years distant from Earth. The lines across the left side of the image are the result of a satellite trail. *Note:* The prefix "M" before a numbered astronomical object refers to a cataloging system initiated in the eighteenth century by the astronomer Charles Messier. Similarly, the prefix NGC before a numbered astronomical object stands for that object's designation in the New General Catalog. The term "Sb" refers to the shape of the galaxy. The letter "S" indicates that the galaxy has a spiral shape. The lower case letter following the "S" refers to how tightly the arms of the spiral are wound around its center. An "a" designation refers to a galaxy with very tightly drawn arms, whereas a "c" designation indicates loosely wound arms. In this case the designation "Sb" means that M31, NGC 224 is a spiral galaxy with arms that are drawn around the galaxy to a moderate degree. A spherical galaxy with an apparent bar of stars intersecting its core is referred to as a barred spiral. The letters "SB" indicate that a galaxy is a barred spiral. Again, the lower case "a," "b," and "c" are used to indicate how tightly the arms of the barred spiral are wound around the central core of the galaxy.

Photo Credit: Copyright, Association of Universities for Research in Astronomy Inc. (AURA), all rights reserved. AURA operates The National Optical Astronomy Observatories for The National Science Foundation.

• 44 • Chapter 2 The Mystery Called the Big Bang

Figure 2.2 An Elliptical Galaxy.
M87, NGC 4486, Virgo A, 3C 274, Type E0 is a giant elliptical galaxy in the constellation Virgo. This galaxy has a jet that is also a radio source. The whole object is the fifth brightest in the sky (in radio frequencies). *Note:* The term "E0" refers to the shape of the galaxy. The letter "E" indicates that the galaxy has an elliptical shape. The number "0" following the "E" refers to how elliptical the galaxy actually is. A "0" designation indicates that the galaxy is essentially a sphere. At the opposite end of the spectrum a "7" would indicate that the galaxy in question is flattened. In this case the designation "E0" means that M87, NGC 4486 is an elliptical galaxy with an essentially spherical shape.
Photo Credit: Copyright, Association of Universities for Research in Astronomy Inc. (AURA), all rights reserved. AURA operates The National Optical Astronomy Observatories for The National Science Foundation.

astronomical techniques have allowed cosmologists to reach more accurate estimates of the age of the universe.[22] However, despite Hubble's initial errors, those who supported his vision of the universe maintained faith in the implications of his observations. One reason for this fact was that Hubble's data tied directly into the theoretical predictions of several other cosmologists, notably the Belgian astronomer, and priest Georges Lemaitre, and the Russian cosmologist and geophysicist, Alexander Friedmann. Both Lemaitre and Friedmann saw

Figure 2.3 An Irregular Galaxy.
NGC 3077 is a Type II irregular galaxy in the constellation Ursa Major. This galaxy has dust lanes that do not follow the usual spiral pattern. *Note:* Galaxies that do not fall under the elliptical, spiral, or barred spiral designations are said to be "irregular galaxies."
Photo Credit: Copyright, Association of Universities for Research in Astronomy Inc. (AURA), all rights reserved. AURA operates The National Optical Astronomy Observatories for The National Science Foundation.

unnecessary difficulties in Einstein's image of the universe.[23] Einstein saw the universe as static and unchanging. This predisposition is seen in the two assumptions upon which he based his **theory of relativity**: "1. There exists an average **density** of matter in the whole of space which is everywhere the same and different from zero,"[24] and "2. The magnitude ('radius') of space is independent of time."[25] To make certain that these two assumptions maintained a consistent relationship within the theory, Einstein added a **cosmological constant** to the equations. The idea of the cosmological constant was to include a repulsion factor or "negative **pressure**" that would balance the attractive effects of **gravity**, thus creating the static or steady state universe that Einstein preferred.[26] He later characterized this addition of the cosmological constant as "a complication of the theory, which seriously reduces its logical simplicity."[27]

Neither Lemaitre nor Friedmann found this approach palatable. Lemaitre was convinced that Einstein's cosmological constant did not create the steady state that Einstein sought. In fact, Lemaitre demonstrated that only a slight deviation in the equations would move Einstein's static universe into either a contracting or an expanding universe. From this notion, Lemaitre conceived of a cosmological model based on the idea that the universe began as a "primeval atom," balanced between repulsive and attractive forces. When the repulsive force exerted slightly more power than the attractive force, it created a sudden imbalance that resulted in an expanding universe.[28]

Friedmann also found fault with Einstein's cosmological constant. Friedmann demonstrated that it was feasible to preserve Einstein's first assumption, that matter was spread evenly throughout space, provided that one was willing to drop the second assumption, that the radius of space was not related to time. With this interpretation in mind, Friedmann developed a cosmological model

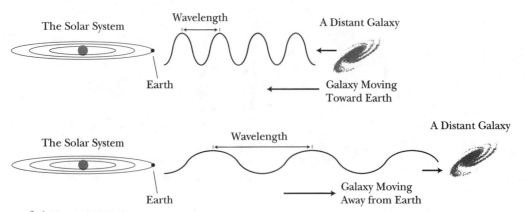

Figure 2.4 The Red Shift Phenomenon.
The wavelength of light radiated from a galaxy moving toward Earth is shorter than that radiated from a galaxy moving away from Earth. On Earth we see the spectrum of the latter type as shifted to the red end of the spectrum.

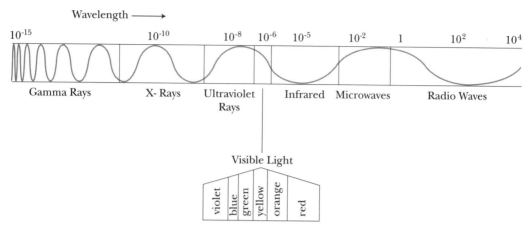

Figure 2.5 The Full Spectrum.
Pictured is the full spectrum of electromagnetic radiation. The waves on the right are extremely long, whereas those on the left are very short. The central area represents visible light.

of the universe that was nonstatic in nature. His model was based on the idea that the universe expanded from an initial state of very high density.[29]

In the 1930s, the British astronomer, Sir Arthur Eddington, reluctantly saw a relationship between the theoretical predictions of Lemaitre and Friedmann and the observations of Hubble. By moving backward in time, the universe would be more and more organized until a point was reached at which maximum organization would rule it. This is the end point of the journey backward in time, but it is also the beginning of the universe from our perspective in the present. From this notion, and from the ideas and observations of Slipher, Hubble, Einstein, Lemaitre, and Friedmann, came the idea of the expanding universe.[30]

As we rewind the videocassette of the universe, the **galactic clusters** get closer together. The universe also becomes hotter because all energy and all matter are being squeezed into a tinier and tinier amount of space. As this backward trend continues, matter also becomes much denser. The heat begins to break the matter into its constituent parts. The universe, at this point, becomes a soup of subatomic particles. When we reach the earliest split second in time, the entire universe is merged into a symmetrical entity usually referred to as the **singularity**. And yet this states it poorly. The singularity is not really a "thing" or an "entity." Instead, the singularity should be viewed as that point of existence at which the entire universe was crystallized into a uniform, symmetrical dividing line between existence and nonexistence. It is at this boundary that cosmology and physics merge.[31] In their book, *The Shadows of Creation*, Michael Riordan, staff scientist at the Universities Research Association, and David Schramm, professor of physics at the University of Chicago, explain the process this way: "(t)he history of the Universe from this earliest instant has been a saga of ever-growing asymmetry and increasing complexity. As space expanded and temperatures cooled, particles began collecting and structures started forming from this ultrahot, featureless plasma. Eventually clusters and galaxies, stars, planets, and even life itself emerged."[32]

As Riordan and Schramm point out, the history of the universe immediately after the appearance of the initial singularity is a history of "particles . . . collecting" and "structures . . . forming."[33] To understand this process, we must have an accurate picture of the basic building blocks of the universe. This requires that we pause in our study of the origin of the universe to examine the structure of the subatomic universe.

The Subatomic Universe

To simplify our approach to this subject, we must focus on the structure of matter and the nature of the four fundamental forces. As a first step, we must visualize the universe on the smallest scale possible. This can be accomplished by imagining an observer using a powerful microscope to examine the subatomic structure of a single grain of sand. After passing the molecular level, the observer's point of view would enter the realm of the atom. Most people are familiar with the **solar system model** of the atom that adorns the covers of many high school science textbooks. Although, as we shall see later in this text, this stylized model is somewhat inaccurate, nevertheless it is helpful in constructing a workable image of the subatomic universe. In place of the Sun at the center of this miniature solar system is the **nucleus**. The nucleus, however, is constructed of smaller units called protons and neutrons. **Protons** are positively charged subatomic entities; **neutrons** are neutral in charge.[34]

The protons and neutrons, long thought to be indivisible, are also made up of even smaller building blocks, namely **quarks**.[35] Quarks were named by Murray Gell Mann of Caltech who pulled the term from "Three quarks for Muster Mark," a line in *Finnegans Wake* by James Joyce.[36] Each proton is made of three quarks: two up quarks and one down quark. Each neutron also consists of three quarks. This time, however, the combination is two down quarks and one up quark. No one knows for certain, but some physicists believe that quarks may be made of even smaller subatomic entities.[37] Surrounding the nucleus are **electrons**. Electrons are freely moving, negatively charged matter particles. If we maintain the solar system model of the atom, we might say that the electrons **orbit** the nucleus. This description is, of course, inaccurate because the electrons do not orbit the nucleus like the planets orbit the Sun. Planets are not limited to orbits located at predetermined places in space. For example, the orbit of Mercury is located approximately 35,991,000 miles from the Sun. However, conceivably, the orbit could have been located at 30,000,000 miles or at 40,000,000 miles from the Sun. There is no law of nature that says that planets can orbit a star at only certain predetermined points in space.[38]

The orbitals (not orbits) of electrons are quite different. According to Heisenberg's **uncertainty principle** (see Chapters 4 and 5), it is impossible to determine the exact position of an electron. Focusing on the energy of the electron rather than on the position means that we cannot know the exact location of the electron. We can know only the probability of finding an electron within a region of space within the atom. This is known as the probability distribution of the electron. The region of space where the probability is high of finding the electron is called the **orbital**.[39]

Electrons can exist only at certain predetermined energy levels within the space of the atom. Moreover, the maximum number of electrons that can exist at each level is severely limited. For instance, the limit at the lowest level is two, whereas at the second level the limit is eight.[40] To move from one energy level to another, the electron must emit or absorb energy, in the form of a photon. A **photon** is the force-carrying particle responsible for the transmission of

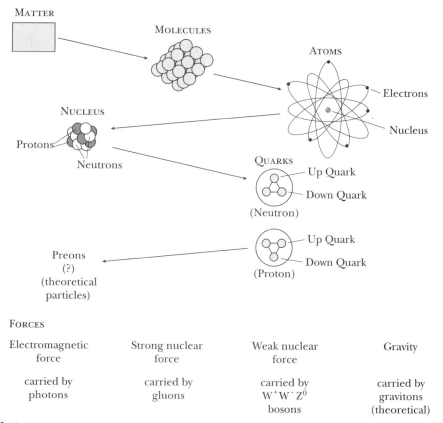

Figure 2.6 The Structure of the Subatomic Universe.
The entire universe can be thought of as consisting of matter particles together with the four fundamental forces. The forces determine how matter particles interact with one another.

electromagnetic radiation or light.[41] Of course, since energy levels are limited, the electron can move only from one predetermined level to another. In fact, the word *move* is somewhat inaccurate. It might be better to think of this movement as a nonspatial transition from one energy level to another.[42]

Although difficult to visualize, the idea of a nonspatial transition from one energy level to another is not as mysterious as it may seem. Because electrons exist only at certain energy levels (which sometimes are referred to as quantized states), a change in energy level corresponds to the emission or the absorption of a photon. When the electron moves from a higher energy level to a lower one, for example, it emits a photon that has an energy level equal to the difference between the higher level at which the electron began and the lower level at which it ends up.[43]

As a guide, it might be helpful to imagine the energy level of an electron flashing as a numeral on a digital clock. When the energy level of our imaginary electron changes, the number would jump from one value to the next without having to display any intervening numbers. Or, to use another analogy, imagine that the temperature of an object, as measured on a digital thermometer, jumps from 70 to 73°F without passing from 71 to 72°F first. If we were to apply this to a hydrogen atom, it would work like this. In its normal state, the single electron of a hydrogen atom occupies the lowest energy state possible. This is referred to as the **ground state** of the atom. In the hydrogen atom, the electron would be situated at the lowest energy level (E_1) which is equivalent to -13.6 electron volts (eV). In an excited and, therefore, unstable state, the electron would occupy only certain energy levels. The next highest energy level (E_2) would be -3.4 eV. The next highest (E_3) would be -1.5 eV. If an electron situated at E_3 were to emit a photon, it would "fall" from E_3 to E_2 and register -3.4 eV at E_2. The photon would have an energy level equivalent to 1.9 eV, the difference between the energy levels at E_3 and E_2.[44] Despite the fact that this image of the atom explains many things, it still needs further refinement. It is sufficient for now to understand that the electron is an integral part of the structure of the atom.

The term **quantum physics** comes from this tendency of the subatomic universe to come in distinct **quanta** or packets of energy. The term *quantum* was invented by the German physicist, Max Planck in 1900 in his attempt to explain a puzzling phenomenon involving the radiation of heat from a hot object. The quantized state of an electron corresponds to such distinct quanta.[45] Recent advances in quantum physics will be explored at length in Chapters 4 and 5. For now, however, it is sufficient to remember that terms like *quantum effects* and *quantum phenomena* refer to events at the subatomic level.

Electrons are part of a larger family of subatomic entities known as leptons. **Leptons** are distinguished from other matter particles, such as quarks, because the leptons usually are not associated with the nucleus of an atom.[46] Another prominent member of the lepton family is the neutrino. **Neutrinos** are nearly massless particles that have no electric charge but swarm about the universe, rarely interacting with any other particles. A neutrino can fly though a million miles of solid iron and never touch another particle. Billions of neutrinos are gliding through this book at this very second.[47] Because the neutrino is so light, until 1998 many physicists believed it might have no mass at all. However, in that year a group of physicists working together at a research facility in Japan discovered that the neutrino does indeed possess mass. Although that mass is quite small, it must be taken into account in any cosmological model that professes to explain the origin and destiny of the universe.[48]

Before moving on to the four fundamental forces of nature, it is helpful to try to visualize the size of an atom, as well as the relative sizes of its parts. According to Robert Hazen and James Trefil in their book, *Science Matters: Achieving Scientific Literacy*, atoms are so small that one billion billion are needed to fill the head of a pin.[49] Visualizing the relative sizes of the parts of an atom is just as interesting. In their book, *The Shadows of Creation*, Michael Riordan and David Schramm suggest expanding an atom to the size of a football stadium. At that magnitude, the atom would be one trillion times its normal size. Moreover, the nucleus would be flying around like a housefly at the 50-yard line, while up in the bleachers, the electrons would be hopping around like fleas.[50]

The matter particles make up only part of the picture of the subatomic universe. The remainder of the picture is supplied by the four fundamental forces of nature. These four forces include the electromagnetic force, gravity, the strong nuclear force, and the weak nuclear force.[51] Each of the four forces is carried by messenger or force-carrying particles. The **electromagnetic force** is found in the sunlight that wakes us in the morning and the electricity that runs the television by which we fall to sleep at night. The force-carrying particle that is responsible for transmitting the electromagnetic force is the photon.[52]

The other force that people deal with on a regular basis is gravity. Gravity differs from the other three forces because it always attracts, and because, although it manages to operate over great distances, it is relatively weak. It may seem strange to describe gravity as a relatively weak force, because, from our perspective at least, gravity appears to be quite strong. At the subatomic level, however, gravity is very weak when compared to the power exerted by the electromagnetic force. The force-carrying particle responsible for gravity is called the **graviton**. However, since the gravitational attraction exerted by the graviton is extremely weak, it is difficult to detect. For this reason, up until the present at least, scientists have been unable to detect gravitons. Nevertheless, so much is known about how gravity operates that it is possible for scientists to predict what properties gravitons would possess. According to these predictions gravitons should have neither **mass** nor charge and they should move at **light speed**.[53]

The **strong nuclear force** is the force found in the heart of the atom. In fact, it is the quantum "glue" that holds the universe together. It is conveyed by the **gluon**, an appropriately named force-carrying particle that binds the quarks together making protons and neutrons. Without the strong nuclear force nothing, not even the pages of this book or the neurons of your brain, would cling together.[54] The weak nuclear force is the least familiar of the four forces. The **weak nuclear force** is responsible for transforming protons into neutrons during the process of fusion within stars. This process of changing a proton into a neutron is called **beta decay**. The force-carrying particle responsible for this process comes in three forms. Two are charged particles known as **W^+** and **W^- particles**, and one is neutral, the **Z^0 particle**. These three particles along with photons, gravitons, and gluons, belong to a category of particles called bosons. This is why the W^+, W^-, and Z^0 particles are often referred to as the **W^+, W^-, and Z^0 bosons**.[55] Now that we have a general understanding of the structure of the subatomic universe as it exists today, we can examine the process by which that universe came into existence, according to the big bang theory.

2-4 A DESCRIPTION OF THE BIG BANG

As noted earlier, the big bang is the generally accepted scientific explanation for the origin of the universe. Today anomalies in observational data are creating a crisis of the type predicted by Thomas Kuhn for the big bang theory. New, recently revealed measurements of far-distant galaxies have given us confusing data that must either be explained within the context of the big bang or lead to a new, more satisfactory theory. Still, because the big bang remains the orthodox theory, it is essential that the nonscientist have an understanding of the various stages in the big bang scenario. With these stages in mind, it will be easier to discuss both the anomalous evidence contradicting the theory, as well as the practical and philosophical implications of the theory and its challengers. In the next chapter, the anomalous evidence challenging the big bang is explored in depth, and the practical and philosophical implications of the theory are also examined. For now, it is crucial to understand a chronological account of the big bang. Therefore, for the remainder of the chapter, our exploration of the big bang will be divided into three parts. First, we look at the question of what caused the big bang. Second, we examine a step-by-step chronological account of the big bang. Finally, we review the most critical evidence in support of the big bang.

The Problem of Causation

Conventional wisdom places the origin of the universe at sometime between 15 and 20 billion years ago. According to the big bang theory, some 15 to 20 billion years ago the singularity abruptly snapped into existence from the nothingness of the void. Just exactly how the singularity snapped from nonexistence into actuality is yet to be explained. The idea that the universe began from nothing at all can be a major stumbling block for the nonscientist in his or her attempt to understand the big bang theory. How is it possible for anything, let alone the entire universe, to come from nothing? Some nonscientists would argue convincingly that such a question falls outside the realm of science and should, thus, be left to the philosophers and theologians. There are some cosmologists, however, who have begun offering some theoretical explanations for the mysterious snap. Each of these theories accounts for the creation of the singularity without resorting to a spiritual or supernatural agency. In other words, they provide an explanation for the generation of the singularity within the physical universe. The two primary theories in this regard are the reproducing universe and the evolving universe.

The Reproducing Universe. The theory of the **reproducing universe** suggests that the present universe is one of a long line of self-replicating universes. This theory has been proposed by several prominent physicists and cosmologists, including Andrei Linde of Stanford University. To explain the origin of the singularity, Linde returns to the world of the quantum universe. He notes, as his premise, that space is not empty. On the contrary, space is filled with scalar fields, which are an assortment of fields that crisscross the universe and interact with subatomic particles. The Higgs Field, for example, is a scalar field that interacts with W and Z particles, giving them mass.[56] Linde believes that a proper understanding of scalar fields will allow us to comprehend the inflationary nature of our universe and may provide us with an understanding of how the universe began.[57]

According to Linde, all scalar fields experience quantum fluctuations which act very much like waves. Initially, these waves might take many different paths; but eventually, they would rest on top of one another, thus magnifying the intensity of that particular scalar field in some areas while lessening its intensity in others. In certain extraordinary instances a wave in the scalar field might inflate so rapidly that it would outdistance all other parts of the field. A wave of sufficient strength might create fresh inflationary regions, any one of which could blossom into a self-sustaining offspring or baby universe.[58] What would this offspring universe be like? On this point Linde is unclear. At times it seems that he sees these offspring universes separated from the parent universe within their own space-time continuum. According to this view, the offspring universe would exist completely outside the realm of the parent universe.[59] On the other hand, at times Linde envisions the offspring universe as an "inflationary bubble" that sprouts from a creative region within the single, unified universe.[60]

It may be that Linde intends the term "universe" to mean all existence, including the offspring universes. If that is the case, then his references to our part of the universe can probably be taken to mean the observable universe that we usually think of when considering the big bang.[61] Linde also points out that each of these offspring universes may be self-reproducing. According to this twist in the theory, an unending succession of self-reproducing offspring universes gives rise to an infinite line of successive universes. Thousands, even millions of such offspring universes could inhabit their own unique space-time fragments.[62]

Linde puts it this way: "one inflationary universe sprouts other inflationary bubbles, which in turn produce other inflationary bubbles . . . In this scenario the universe as a whole is immortal. Each particular part of the universe may stem from a singularity somewhere in the past, and it may end up in a singularity somewhere in the future. There is, however, no end for the evolution of the entire universe."[63] Would these inflationary bubbles ever collide with one another? According to Linde the answer is problematic. It may be that the offspring universes must remain eternally isolated from one another, each surrounded by vast regions of nothingness. On the other hand, such universes might converge with one another. However, such a collision might be catastrophic if the two universes are governed by incompatible principles.[64]

What exactly does this theory have to do with the generation of the singularity that serves as the origin point of the present universe, the so-called parent universe? If the origin of our universe could be traced to a previously existing parent universe (or to a creative region of the single universe) then without resorting to spiritual or supernatural causes, we have an explanation of how our universe could have a beginning and a finite age. Thus, we no longer need speak of *the big bang*. Instead, there have been uncounted succeeding big bangs giving birth to an endless line of offspring universes.[65] The inescapable conclusion of Linde's reproducing universe theory is that the present universe may have had a physical cause. The mere fact that this cause cannot be traced to an event within our own space-time continuum does not make it any less real or any less physical. It just places the scalar field wave phenomenon that created this universe beyond observation and measurement. Admittedly, this takes a leap of faith—but no more of a leap than that required to accept the notion that the Earth is in a constant state of motion. If the motion of the Earth can be accepted by most people without direct experience or observation, why not a quantum fluctuation origin for our universe?[66]

The Evolving Universe. Still, the notion that a random, subatomic event, such as a quantum fluctuation could serve as the origin of our vast universe is very hard for many nonscientists to accept, if only because the belief in quantum phenomenon is, itself, difficult to grasp. For these skeptics, Lee Smolin, a Syracuse University physicist, offers the theory of the evolving universe. The **evolving universe theory** explains the origin of the universe in the life cycle of the stars. All stars are essentially enormous fusion reactors that create light and heat energy by fusing helium from hydrogen. Each star, including the Sun, lives its long life in a state of stability. This **stability** is created when the fusion energy pushes outward while gravity pushes inward, creating an ideal balance between the two rival forces and forming a perfect state of harmony for the star.[67]

Even this ideal condition of stability cannot last forever. At some point, the hydrogen fuel is depleted, and the fusion energy can no longer resist the inward push of gravity. As a result, the star begins to collapse in upon itself. If the star has a mass equal to that of the Sun, its collapse will stop once it reaches the electron barrier. This means that, as the electrons are pushed together by the force of gravity, they are compelled to move at a more rapid pace. Eventually, this causes an outward pressure that halts the collapse of the star. The result of this collapse is a **white dwarf** star about the size of the planet Earth. If the mass of the deteriorating star is somewhere between 1.4 and 5 times greater than that of the Sun, then the star will collapse beyond the electron barrier. The immense crush of gravity destroys the repulsive force of the electrons, and the nuclei begin to crash into each other. Neutrons and protons within the nuclei also separate from one another. The electrons collide with the protons, creating more and more neutrons, until the star is dominated by neutrons. The result is a **neutron star**, about the size of an average American city.[68]

A few stars will have a mass that is far greater than five times that of the Sun. Such a star will collapse well beyond the electron and gluon barriers, crushing even neutrons as it forms a black hole. A **black hole** is so massive that it creates a gravitational field powerful enough so that nothing, not even light, can escape from its event horizon.[69] Some scientists, notably Smolin, believe that the intense gravitational field might create a bulge in the space-time continuum creating a baby universe. The baby universe would remain connected to our universe until quantum activity causes the evaporation of the black hole at which time the offspring would separate from the mother universe.[70]

While this theory sounds a lot like Linde's reproducing universe, there is a key difference. Linde's universe depends upon an unobservable quantum fluctuation of the scalar field whereas Smolin's depends upon the collapse of a massive star. Still, the two events are inexorably intertwined. According to Smolin's theory, the quantum fluctuations of the scalar field predicted by Linde may be sparked by the energy created as a collapsing star reaches the infinitely dense state of a singularity.[71]

• 56 • Chapter 2 The Mystery Called the Big Bang

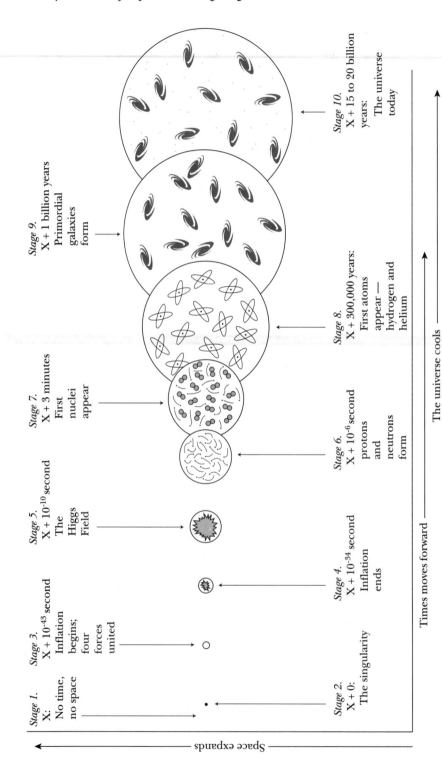

Figure 2.7 A Big Bang Chronology.
According to the theory of the big bang, the universe began some 15 to 20 billion years ago, when the singularity appeared out of nowhere. Thus, the absence of time and space would represent the first stage of a big bang chronology, and the appearance of the singularity would be the second stage. Succeeding stages are depicted as moving from left to right.

Similarly, both Smolin and Linde predict that the offspring universe (or inflationary bubble) might contain mutations that would somehow subtly alter the laws of nature as they are experienced in the offspring (or bubble). Thus, an offspring might be slightly different from its parent. The immediate offspring might look and act quite similar to its parent. However, as mutation follows mutation down through the successive generations of offspring universes, a new line of universes would evolve from the original parent, so that, generations later, the descendant universe would be virtually unrecognizable by its great ancestor universe.[72]

These differences would, of course, make no difference to the ancestor universe because it would probably never intersect with the offspring universe. Smolin, like Linde, can thus explain how this universe can have a beginning and a finite age without capitulating to spiritual or supernatural causes. Again it is no longer necessary to refer to *the big bang*. Instead, it would be much more suitable to discuss *a big bang*, because there has probably been an endless series of big bangs caused by contracting stars, thus giving birth to a progression of evolving universes.[73]

A Big Bang Chronology

As noted previously, the theory of the big bang states that some 15 to 20 billion years ago the universe began when a singularity appeared, regardless of the cause. The steps of the big bang chronology would appear as follows:

Stage 1   (X) In the beginning there existed no time and no space as we know them within this universe.

Stage 2   (X+0) At the instant of creation, the singularity appears. It is subatomic in size. It is infinitely dense. Quantum effects must be taken into account. In fact, since the entire universe is subatomic, quantum effects rule all aspects of that early universe. The present laws of nature do not yet exist.

Stage 3   ($X + 10^{-43}$ second) The universe is still quite tiny, perhaps the size of a grain of salt. The four fundamental forces of the universe are all one force. The unified forces infuse this primitive state of the universe with a tremendous energy concentration which flashes outward as a repulsive antigravity wave creating more space and, as a result, even more intense antigravity flashes, which cause an exponential expansion of the early universe. Under the effects of this inflation, the universe grows to twice its previous size every 10^{-38} second. The *inflationary epoch*, as it has been called, despite its relatively short duration, magnifies the quantum oscillations, permitting the particles to *cluster*. Because gravity is associated with mass, this clustering allows gravity to assert itself as a separate force. The appearance of gravity counteracts the expansion, slowing it and stopping the rapid inflationary phase of the universe. However, gravity did not completely halt the expansion caused by the inflationary epoch. Instead, it simply slowed the expansion rate to the steady outward movement we witness today. The effects of gravity account for the present rate of expansion and also lead to the extraordinarily large and astonishingly complex multigalactic structures present in the contemporary universe.[74]

Stage 4   ($X + 10^{-34}$ second) The inflationary epoch ends. At this point, the strong force becomes recognizable as an independent force. The electroweak force is left in its unified state. The simplest matter particles, leptons and quarks, are in abundance.[75]

Stage 5   ($X + 10^{-10}$ second) The Higgs boson or Higgs field appears, somehow giving mass to W and Z bosons but not to photons, thus breaking the symmetry of the electroweak force. The electromagnetic and the weak nuclear forces are no longer unified.[76]

Stage 6   ($X + 10^{-6}$ second) The entire primeval universe has reached the dimensions of the solar system. It swarms with a sea of basic subatomic particles which only now begin to coalesce into protons and neutrons.[77]

Stage 7   ($X + 3$ minutes) Protons and neutrons begin to coalesce into rudimentary atomic nuclei. These include hydrogen and helium nuclei. However, the temperature is still too hot for the formation of atoms.[78]

Stage 8   ($X + 300,000$ years) Now the universe witnesses the appearance of the first atoms. These are the simple atoms of hydrogen and helium.[79]

Stage 9   ($X + 1$ billion years) Primordial galaxies and their component stars are in the formation process at this point.[80]

Stage 10   ($X + 15$ to 20 billion years [that is, to the present moment]) At this time the universe is still expanding from its initial primordial inflationary epoch to the supergalactic structures evident today and toward its mysterious yet inevitable future.[81]

2-5   EVIDENCE SUPPORTING THE BIG BANG

The big bang theory of the origin of the universe has remained in the forefront of modern cosmology for a long time because the evidence verifying its validity has been overwhelmingly in its favor. There are many diverse pieces of evidence supporting the big bang theory. However, the three most credible parcels of evidence are the expanding nature of the universe, the existence of the cosmic background radiation, and the abundance of the primordial elements of hydrogen and helium throughout the universe.

2-5 Evidence Supporting the Big Bang

The Expanding Nature of the Universe

The first and most easily understood piece of evidence in support of the big bang model is Hubble's expanding universe. As noted earlier, Hubble discovered that the galaxies do not fall together and do not float about at random because they are rushing apart. This means that the universe itself is expanding. If the universe is expanding, much like a balloon being blown up, then at some time in the distant past it must have been tinier and more compact. At its origin point in the past, the universe must have been as small and compact as a singularity.[82]

The Expansive Blanket of Background Radiation

A second impressive affirmation of the big bang is the existence of an expansive blanket of cosmic background radiation enveloping the entire universe. The big bang theory is based on the assumption that the universe began in an extremely hot state. However, as soon as the early universe formed, it began to cool. Soon the temperature of the universe had dropped sufficiently to allow for the formation of protons and neutrons. Even at that point, however, it was still too hot for the creation of atoms. Although photons were free to travel, they could move only a slight distance before colliding with a free electron. As a result, the universe at this time was opaque to radiation (see Stages 6 and 7).[83]

Eventually, the universe cooled to a temperature of approximately 3000 degrees above absolute zero (3000 degrees Kelvin or 3000 K). This occurred at about 300,000 years after the instant of creation. At this time things had cooled enough to permit the formation of hydrogen and helium nuclei atoms (see Stage 8). The universe was now transparent to radiation. Consequently, since the electrons were now bound up in the hydrogen and helium atoms, the vast majority of photons could move freely without colliding with those electrons. The photons could, in fact, continue to travel forever. Many of these photons are detected today as the **cosmic background radiation**. Because the expansion of the universe has red-shifted the photons to lower energies, that is to longer wavelengths, the night sky is not brilliantly bright, as might ordinarily be expected.[84]

The possibility of this radiation was predicted as early as 1920. In 1948, the American physicists Ralph Alpher and Robert Herman predicted that the radiation would appear as if it were coming from a body of only 5 K. This estimate was based on the red-shifted nature of the radiation. Unfortunately for Alpher and Herman, they could not confirm their prediction, because at that time no devices could detect radiation of such low energy.[85] Consequently, the existence of this cosmic background radiation was not confirmed until 1964, by Arno Penzias and Robert Wilson, two physicists who were at Bell Laboratories having difficulty with a very precise microwave detector.[86] Penzias and Wilson discovered radiation at a temperature of about 3 K, which was in essential agreement with predictions made by Alpher and Herman.[87] The necessary variations in the temperature of the cosmic microwave background radiation have since been confirmed by the now famous COBE discovery recounted earlier in this chapter.[88] The accepted figure today for the background radiation is 2.73 K.[89]

The Wealth of Hydrogen and Helium

The final and perhaps most definitive bit of evidence endorsing the big bang theory is the wealth of hydrogen and helium remaining in the universe at the present time. Working backward from the structure of the universe as it exists today, scientists have concluded that the early universe was so hot that the major activity at that time was the fusion of hydrogen nuclei to form helium. If these calculations are accurate, the present structure of the universe should be approximately 99 percent hydrogen and helium. Moreover, the total mass of the universe should be measured at approximately 25 percent helium and 75 percent hydrogen. This is, in fact, the case.[90]

CONCLUSION

In our attempt to extend science appreciation to the nonscientist, we have begun at the origin of the universe. Our initial study has explored the standard model of the origin. We have indicated several times in this chapter, however, that the present theory of the big bang is not without anomalies that challenge its orthodoxy. It is to those anomalies that we now turn our attention. In addition, in the next chapter we explore some of the more practical applications of research into the origin of the universe.

Review

2-1. The standard model for the origin of the universe is called the big bang theory. This chapter had three objectives. First, it explained what the big bang was not. Then it approached the problem from the opposite perspective and attempts to explore what the big bang really was. This explanation included not only the basic ingredients of the big bang but also a chronological outline of the critical events of the big bang in each of ten stages. The final part of this chapter was devoted to a look at some of the most convincing evidence supporting the big bang.

2-2. The first problem with the big bang theory is the fact that many people have the mistaken notion that the big bang was an explosion. Once the theory is properly understood the nonscientist can see that the big bang was not really an explosion after all. What the big bang theory actually states is that the universe began some 15 to 20 billion years ago when an infinitely dense subatomic entity, known as the singularity, appeared, inflated suddenly, and, as the inflation slowed and the universe cooled, coalesced into the cosmos as we know it today. Many nonscientists also have the mistaken notion that the big bang is somehow antireligious. The first point that must be made in defense of the big bang theory is that it has no religious pretensions whatsoever. It is, like any theory, simply a way to explain the evidence that exists. The second point to make in defense of the big bang theory is that, even though the theory itself has no religious pretensions, some people have seen religious overtones within the theory. Many of those who have detected religious implications in the big bang theory have actually used the theory in support of *Genesis* and other creation stories, rather than to oppose those creation stories. There is at least one sense in which the big bang theory could be seen as incompatible with religion. This approach, generally referred to as creationism or creation science, insists upon a literal reading of the Bible.

2-3. While studying several distant galaxies, Hubble discovered that most of those galaxies are emitting light that were shifted toward the red end of the spectrum. Hubble used this data to conclude that most of these galaxies are moving away from us. Hubble came to this conclusion by interpreting the redshift data as a manifestation of the Doppler effect. Hubble was not the first astronomer to observe the red shift phenomenon. Hubble's later observations, however, clearly confirmed the earlier observations. Hubble's work also went beyond the earlier observations. In addition to confirming the red shift observations, Hubble also used the data that he had compiled to determine the distances to those galaxies and, as a result, was also able to estimate the age of the universe. Hubble's data was later tied into the theoretical predictions of several other cosmologists, notably Georges Lemaitre and Alexander Friedmann. Lemaitre and Friedmann both predicted an expanding universe. To understand how the universe expands, it is necessary to have an accurate picture of the basic building blocks of the subatomic universe.

2-4. According to the big bang theory, the singularity abruptly snapped into existence from the nothingness of the void. Just exactly how the singularity snapped from nonexistence into actuality is yet to be explained. Some cosmologists, however, have begun offering some theoretical explanations for the mysterious snap from oblivion to reality. Each of these theories accounts for the creation of the singularity without resorting to a spiritual or supernatural agency. The theory of the reproducing universe suggests that the present universe is one of a long line of self-replicating universes. This theory has been proposed by several prominent physicists, including Andrei Linde of Stanford University. Lee Smolin, a Syracuse University physicist, offers the theory of the evolving universe, which sees the origin of the universe in the life cycle of the stars.

2-5. The big bang theory has remained in the forefront of modern cosmology for a long time because, at least until lately, the evidence verifying its validity has been overwhelmingly in its favor. There are many diverse pieces of evidence supporting the big bang theory, the most credible parcels of which are the expanding nature of the universe, the existence of the cosmic background radiation, and the abundance of the primordial elements of hydrogen and helium throughout the universe.

Understanding Key Terms

beta decay
big bang
Cosmic Background Explorer (COBE)
cosmic background radiation
cosmological constant
creationism
density
Doppler effect
electromagnetic force
electrons
electroweak force
evolving universe theory
galactic cluster
galaxy
gluons
gravitons
gravity
ground state
Higgs field
Hubble constant
Hubble's law
leptons
light speed
mass
neutrinos
neutrons
neutron star
nucleus
orbit
orbital
photons
protons
quantum
quarks
red shift
relativity theory
scalar field
singularity
solar system model
space-time continuum
spectrum
steady state cosmology
strong nuclear force
uncertainty principle
W^+, W^-, and Z^0 particles
weak nuclear force
white dwarf

Review Questions

1. Why should nonscientists explore the origin of the universe?
2. Was the big bang an explosion? Explain.
3. Is the big bang theory antireligious? Explain.
4. What are the implications of Hubble's discovery regarding galactic movement?
5. How did Slipher, Lemaitre, Friedmann, and Eddington contribute to the theory of the expanding universe?
6. What are the number and types of fundamental particles that make up the subatomic universe?
7. What are the four fundamental forces of the universe?
8. What are the differences between the reproducing universe and the evolving universe theories?

9. What are the stages in the chronological story of the big bang?
10. What primary evidence supports the big bang theory?

Discussion Questions

1. In Chapter 1, we explored two theories concerning the cultural and social influences on the development of scientific theories. These two theories, positivism and constructivism, present conflicting ideas about the sociological and cultural aspects of science. Of all the scientists of the twentieth century none is more respected than Albert Einstein. Yet, as we learned in this chapter, Einstein added the cosmological constant to his theory of relativity to preserve the idea of a static universe. (To be fair, we must also remember that Einstein later withdrew that idea from the theory.) How does Einstein's decision to add the cosmological constant support the constructivist view of science? Explain.

2. It could be argued that Einstein's experiences in Germany before, during, and immediately after the First World War may have shaped his need to believe in a static and unchanging universe. Such a universe might have given him the stability and predictability that a Germany at war and in economic and political collapse had failed to provide. Do you tend to support this interpretation? Why or why not? Speculate on why a Belgian priest like Lemaitre and a Russian geophysicist like Friedmann would come up with the idea of an expanding and, therefore, impermanent universe. Defend your ideas.

3. The British astronomer Fred Hoyle has battled the big bang theory for most of his professional life. Speculate on what social and/or cultural factors might have contributed to compel Hoyle to oppose the big bang with such determination. Think about the religious aspects of the theory. Might this have something to do with Hoyle's opposition? In light of this, consider why Hoyle, who opposes the idea of a God who created the universe, and the creationists, who support the need for God as the creator of the universe, actually find themselves on the same side in their opposition to the big bang theory.

4. Review the details of the reproducing universe of Linde and the evolving universe of Smolin. What do these two theories hold in common? How are they opposed to each other? Speculate on any cultural or social influences that might have caused each of these cosmologists to construct a theory allowing for both a starting point to the universe and an infinite line of recreating universes. What might Hoyle think about these two theories? How would a creationist react to the reproducing universe of Hawking and Linde? What about the evolving universe of Smolin?

5. In the next chapter, we will be introduced to two women who have made their professional reputations in cosmology. Speculate on why, up until the present time, only a few women have made cosmology a career.

ANALYZING *STAR TREK*

Background

The following episode from *Star Trek: The Next Generation* reflects some of the issues that are presented in this chapter. The episode has been carefully chosen to represent several of the most interesting aspects of the chapter. When answering the questions at the end of the episode, you should express your opinions as clearly and openly as possible. You may also want to discuss your answers with others and compare and contrast those answers. Above all, you should be less concerned with the "right" answer and more with explaining your position as thoroughly as possible.

Viewing Assignment—*Star Trek: The Next Generation*, "Where No One Has Gone Before"

In this episode, the Enterprise plays host to a warp-drive-propulsion expert named Kozinski and his mysterious assistant, the Traveler. Kozinski plans to improve the warp-drive capabilities of the Enterprise. Unfortunately, his tampering with the warp-drive engines sends the ship millions of light years into deep space. In this region of space, strange things begin to happen aboard the Enterprise as the lines between reality and unreality begin to break down.

Thoughts, Notions, and Speculations

1. The origin of the universe is linked to an understanding of the subatomic universe. In this episode we see that the Traveler can tap into the fabric of space-time. Why does this allow him to propel the ship such great distances? Explain. Identify some of the more practical applications of the Traveler's abilities.
2. In the first two chapters, we have spent some time discussing the interaction of social and cultural influences on a person's individual attitudes. How do the background and the abilities of the Traveler help shape his reaction to Wesley? Explain. If the crew of the Enterprise would have known the true nature of the Traveler from the beginning, would they have welcomed him or feared him? Explain. The abilities of the Traveler make him seem almost magical. How is the crew of the Enterprise like a group of medieval scholars meeting a person from the twentieth century with a telephone, television, fax machine, hand-held calculator, microwave oven, videocassette recorder, and desktop computer? How are they unlike those medieval scholars? Explain.
3. Kozinski's work with the warp-drive theory and the revelation of the Traveler's abilities reveal a new understanding about the nature of space and time. Kozinski did not consciously plan to create a new understanding of reality. He was not trying to make a philosophical or a theological statement when he developed his new warp-drive technique. Yet his work leads the crew of the Enterprise into an area of deep space where their experiences compel a redefinition of the

nature of reality. As such, these discoveries are bound to have an effect on theology and philosophy. In this way Kozinski's discoveries are roughly analogous to the discovery of the temperature variation in the background radiation of the early universe discovered by COBE. Should scientists like Kozinski and Smoot be permitted to make theological and philosophical statements about the implications of their work or should they be silent on such matters? Explain.

4. As discussed in this chapter, the big bang model invites a kind of metaphysical speculation about the origin of the universe. Some scientists, notably, Hermann Bondi, Thomas Gold, and Fred Hoyle, especially dislike metaphysical aspects of this theory. As a result, they propose an alternative theory that they call steady state cosmology. Kozinski's work with the warp-drive theory and the revelation of the Traveler's abilities reveal a new understanding of space and time which also invites metaphysical speculation. Suppose other scientists in the Federation oppose Kozinski's work on metaphysical grounds. Would this justify a search for a replacement theory or should the opponents of Kozinski's work be compelled to accept his evidence? Explain.

5. Kozinski's tampering with the warp-drive engines sends the Enterprise into a region of space where strange things begin to happen as the lines between reality and unreality begin to break down. Speculate on the relationship between this strange area of the space-time continuum and the baby universes discussed by Linde and Smolin? Could it be that the Enterprise is in one of the parallel universes mentioned by Linde and Smolin? Explain.

NOTES

1. Paul Davies, *God and the New Physics*, 10.
2. Precise figures on the age of the universe vary. Some experts will admit to a much larger range—from 10 to 20 billion years. Other prefer a range between 12 and 18 billion years. Still others will only say that the universe began at least 15 billion years ago. Terence Dickinson, *From the Big Bang to Planet X*, 16–17, 22–23; John Gribbin, *The Omega Point*, 1–2; Evry Schatzman, *Our Expanding Universe*, 73.
3. Dickinson, 16–17, 22–23.
4. John Boslough, *Masters of Time*, 39.
5. Robert Jastrow, *God and the Astronomers*, 14.
6. Stanley Jaki, *Cosmos and Creator*, 45.
7. Ibid. Also see Stanley Jaki, *Is There a Universe?* In this work, Jaki gives a comprehensive overview of his own philosophical beliefs along with a historical account of the development of modern cosmology.
8. Jaki, *Cosmos*, xii.
9. Pierre Teilhard de Chardin, *Science and Christ*, 189.
10. Davies, 22–23; Willem B. Drees, *Beyond the Big Bang: Quantum Cosmologies and God*, 22–23; Jaki, *Cosmos*, 20; Eric J. Lerner, *The Big Bang Never Happened*, 144–45.
11. Jaki, *Cosmos*, 20; William J. Kaufmann III, *Relativity and Cosmology*, 116–17; Lerner, 144–45.
12. Drees, 21–22; Timothy Ferris, *The Whole Shebang: A State of the Universe(s) Report*, 172.
13. Drees, 22. See also, Tim M. Berra, *Evolution and the Myth of Creationism*.

14. William K. Hartmann, *The Cosmic Voyage Through Time and Space*, 378; David Millar, ed. et al., *The Cambridge Dictionary of Scientists*, 163; Sybil P. Parker, ed. *McGraw-Hill Concise Encyclopedia of Science and Technology*, (1994), 826.
15. Hartmann, 375, 384.
16. A spectrum presents the light from an object according to wavelength or photon energy. Light with shorter wavelengths (or higher energy photons) are found at the blue end of the spectrum. Light with longer wavelengths (or lower energy photons) are found at the red end of the spectrum. Beyond the visible spectrum, shorter wavelengths (or higher energy photons) are found in gamma radiation, x-rays, and ultraviolet light, all of which are invisible to human vision. Similarly, beyond the visible spectrum, longer wavelengths (or lower energy photons) are found in infrared radiation, microwave radiation, and radio waves. Hartmann, 65–66.
17. Michael Rowan-Robinson, *Ripples in the Cosmos*, 39; Schatzman, 27–30.
18. Rowan-Robinson, 39; Schatzman, 27–30.
19. Schatzman, 29–30.
20. Paul LaViolette, *Beyond the Big Bang*, 256–57.
21. John Barrow and Joseph Silk, *The Left Hand of Creation*, 8–9; Alan Lightman, *Time for The Stars: Astronomy in the 1990's*, 80; Joseph Silk, *The Big Bang*, 56–57.
22. Barrow and Silk, 10–11.
23. Schatzman, 32–33; Lightman, 78–79.
24. Albert Einstein, *Relativity: The Special and General Theory*, 152.
25. Ibid.
26. Albert Einstein, *Relativity*, 152; Albert Einstein, *The Meaning of Relativity: Including Relativistic Theory of the Nonsymmetric Field* (1956, 1984), 109–12.
27. Einstein, *Meaning*, 111.
28. LaViolette 258–59; Schatzman, 32–33.
29. Einstein, *Relativity*, 152–53; LaViolette, 257–58; Lightman, 79; Schatzman, 33.
30. Barrow and Silk 8–9; Schatzman, 30, 33–34. It is interesting to note that Eddington knew of Lemaitre's theory as early as 1927 but had rejected it as esthetically unappealing. Apparently, to Eddington the idea that the universe began abruptly at some finite point in the past added an element of discontinuity that he found unsettling. Consequently, he simply dismissed Lemaitre's 1927 paper outlining the "primeval atom" theory. In 1932, Eddington, now convinced of the theory's validity, resurrected Lemaitre's paper and had it printed in the Royal Astronomy Society's proceedings. In essence, the belated publication of Lemaitre's paper, marked the official approval of the expanding universe theory. George Smoot and Keay Davidson, *Wrinkles in Time*, 54–57.
31. Barrow and Silk, 36–38; Drees, 223–24; Lightman, 81–82; Michael Riordan and David Schramm, *The Shadows of Creation*, 18.
32. Riordan and Schramm, 18.
33. Ibid.
34. Stephen Hawking, *A Brief History of Time* (1988), 64–65; Riordan and Schramm, 29–35.
35. Hawking, 65; Robert M. Hazen and James Trefil, *Science Matters*, 126; Riordan and Schramm, 34; Trinh Xuan Thuan, *The Birth of the Universe*, 74.
36. Hawking, 65; Hazen and Trefil, 126; Riordan and Schramm, 34; Thuan, 74.
37. Hawking, 65; Hazen and Trefil, 126–27; Riordan and Schramm, 35.
38. Amit Goswami, *The Self-Aware Universe*, 29–30; Hazen and Trefil, 58–60; Hans Christian von Baeyer, *Taming the Atom*, 33–34. Although there is no law of nature that says that planets can orbit a star at only certain predetermined points in space, the planets of our system do orbit the Sun at regularly spaced intervals that seem to follow a pattern. This rule, known as Bode's rule, after Johann Bode, who popularized a concept developed by the German astronomer Johann Titius, describes the regular positions of the planets in mathematical terms. According to Bode's rule, we must first write down a series of 4s, each representing a planet. To this row of 4s we add the numerical sequence 0, 3, 6, 12, 24, 48, and so on. The sum of each number is then divided by 10. This will give us the number of astronomical units (AU's) between the Sun and each individual planet. (*Note:* An astronomical unit is equal to the distance between the Earth and the Sun: 93 million miles.) Bode's rule simply seems to be a way to describe the relationships of the planets to the Sun. Moreover, there is one exception to the rule—Neptune. According to Bode's rule, Neptune should be 38.8 AU's from the Sun. In reality, it is 30 AU's from the Sun. See Hartmann, 46–47.
39. Alan Isaacs, ed., *A Dictionary of Physics*, 295–97; Roger S. Jones, *Physics for the Rest of Us*, 156–62; Janet Tarino, The Ohio State University—Mansfield, June 28, 1996.
40. Karl F. Kuhn, *Basic Physics*, 57–59.
41. Hawking, 70–71; Hazen and Trefil, 128.
42. Goswami, 29–30; Hazen and Trefil, 58–60; Jones, 149.

43. Goswami 29–30; Hazen and Trefil 58–60; Tony Hey and Patrick Walters, *The Quantum Universe*, 46–48; Jones, 148–50.
44. Hey and Walters, 46–48; Kuhn, 218–26; Leon M. Lederman and David N. Schramm, *From Quarks to the Cosmos: Tools of Discovery*, 39–40.
45. Isaac Asimov, *Atom: Journey across the Subatomic Universe*, 111–13; Goswami, 29–30; Hawking, 54; Hazen and Trefil, 58–60; Fred Alan Wolf, *Taking the Quantum Leap: The New Physics for Non-Scientists*, 78–82.
46. Hazen and Trefil, 126.
47. Harris, *The Creation of the Universe*, video cassette; Riordan and Schramm, 37–39; James Trefil, *From Atoms to Quarks*, 188.
48. Malcolm W. Browne, "Mass Found in Elusive Particle; Universe May Never Be the Same; Discovery on Neutrino Rattles Basic Theory About All Matter," *New York Times*, Friday, 5 June 1998, A1.
49. Hazen and Trefil, 55.
50. Riordan and Schramm, 30.
51. Hazen and Trefil, 127–29.
52. Hawking, 70–71; Hazen and Trefil, 128; Jones, 288–91.
53. Hawking, 70; Hazen and Trefil, 128–29; Jones, 290.
54. Hawking, 72–73; Hazen and Trefil, 128.
55. Jones, 290–91, 304.
56. Andrei Linde, "The Self-Reproducing Inflationary Universe," 50–53.
57. Linde, 49–53; Ferris, 239–41, 258–59.
58. Linde, 54–55; Ferris, 258–59.
59. Drees, 50–51.
60. Linde, 54–55; Ferris, 260–61,
61. Drees, 50–51.
62. Linde, 54–55; Ferris, 260–61.
63. Linde, 54.
64. Ferris, 261–62.
65. Drees, 50; Ferris, 262–64.
66. Drees, 50; Ferris, 262–64.
67. Hartmann, 286–88, 290–91.
68. Ibid, 290–91, 296–98, 302–4.
69. Ibid., 304–7. Until recently, the existence of black holes had remained unconfirmed. However, several astronomers now believe they have uncovered a way to distinguish between neutron stars and genuine black holes. Accordingly, some astronomical objects, which were previously labeled black hole candidates have been redesignated as confirmed black holes. For details, see Jean-Pierre Lasota, "Unmasking Black Holes," 41–47 and Jeffrey E. McClintock, "A Black Hole Caught in the Act," 45.
70. Paul Davies, *The Mind of God: The Scientific Basis for a Rational World*, 70–72, 221.
71. Ibid., 70–72, 221–22.
72. Davies, *Mind*, 70–72, 221; Drees, 50–51; Ferris, 261–62.
73. Davies, *Mind*, 221–22; Drees, 50; Ferris, 262–64; Linde, 54–55.
74. Smoot and Davidson, 181, 188–89. The negative exponent in these figures represents the number of decimal places. Thus, 10^{-3} = 0.001 of a second. Hartmann, 463–64.
75. James Trefil, *The Dark Side of the Universe*, 47–48.
76. Ferris, 216; Riordan and Schramm, 165; Trefil, *The Dark Side of the Universe*, 48.
77. Thuan, 73–77.
78. Smoot and Davidson, 181; Trefil, *The Dark Side of the Universe*, 48.
79. Trefil, *The Dark Side of the Universe*, 49; Thuan, 74–77.
80. Thuan, 77.
81. Smoot and Davidson, 181; Thuan, 77.
82. Davies, *God and the New Physics*, 12–13; Jeff Kanipe, "Beyond the Big Bang," 130–31.
83. William Protheroe, Professor Emeritus, The Ohio State University, July 25, 1996; Thuan, 69–75; Trefil, *Dark Side*, 47–50.
84. William Protheroe, The Ohio State University, July 25, 1996; Smoot and Davidson, 285; Thuan, 69–75; Trefil, *Dark Side* 47–50. (The figure 3000 degree Kelvin or 3000K represents temperature measured in degrees centigrade above 0K, which is absolute zero. Absolute zero is reached when all the heat from a body has been removed. Such a state is, of course, impossible to reach. For scientific work, however, absolute zero still stands as the lowest temperature possible. See Smoot and Davidson, 75; Parker, 4.)
85. William Protheroe; Smoot and Davidson, 75.
86. Davies, *God and the New Physics*, 21; Hawking, 41; Smoot and Davidson, 81–84.
87. William Protheroe; Smoot and Davidson, 81–84; Trefil, *Dark Side*, 51.
88. Ferris, 33–35.
89. Richard Talcott, "COBE's Big Bang," 144.
90. Davies, *God and the New Physics*, 21–22; Smoot and Davidson, 58.

Chapter 3
Requiem for the Big Bang

COMMENTARY: A TRIP INTO THE LUMINIFEROUS ETHER

In 1929, a science fiction writer named Roy Rockwood published an entertaining novel entitled *By Air Express to Venus*. The novel recounts the adventures of three travelers who are taken aboard a strange, interplanetary airship, the Blue Streak, and flown to a rendezvous on the planet Venus. Initially, the Venusian captain of the airship shares as little as possible with the Earthmen about the propulsion system that drives his powerful craft. However, he does explain many of the other wonders of interplanetary space as the ship makes its remarkable voyage to Venus. The captain becomes especially vocal whenever the airship runs into danger during the flight. One unexpected danger comes in the form of a series of powerful "ether blasts" that rock the Blue Streak and threaten either to destroy the craft or drive it off course.[1] The existence of these ether blasts takes the modern reader by surprise. This is because the entire concept of ether as a medium for lightwaves in space has passed from our collective *memory*. Yet, in 1929, Rockwood was quite certain that his readers would be as familiar with the interstellar ether as they were with radio. This was true despite the fact that the need to postulate the existence of ether had been shown to be superfluous by the Michelson–Morley experiment at the Case School of Applied Science in Cleveland and later by Einstein in his now-famous article on relativity. This gap, however, should not be surprising. Many nonscientists have a mistaken vision of science as a monolithic body of knowledge that all scientists share without question. The nonscientist is often unaware of the dynamic nature of scientific research. Except for those who regularly read *Scientific American* or some other journal designed for the amateur scientist, most nonscientists are surprised to learn that there is widespread disagreement within the scientific community about many theories, some of which the nonscientist simply assumes enjoy universal acceptance. An excellent case in point is the big bang theory. As we saw in the last chapter, there is much evidence in support of this theory. However, we also hinted at the fact that there is a growing body of evidence that the theory may not be the last word on the origin of the universe. Is this body of evidence simply an anomaly that will be explained within the big bang paradigm, or will the mounting evidence usher in an era of crisis followed by the type of revolutionary science predicted by Thomas Kuhn? These questions and others like them form the focal point of this chapter.

CHAPTER OUTCOMES

After reading this chapter the reader will be able to accomplish the following:

1. Discuss whether the expanding universe theory has ushered in a new paradigm.
2. Explain how to measure the age of the universe.
3. Identify an alternative explanation for the red shift phenomenon.
4. Explain the problems caused by the discovery of "the streaming motion."
5. Identify the significance of the discoveries of the Great Attractor and the Great Wall.
6. Explain the implications of recent measurements made by the Cosmic Background Explorer.
7. Discuss the theories of the steady state universe and quasi-steady state cosmology.
8. Outline the elements of the plasma cosmology explanation for the origin of the universe.
9. Discuss some of the practical applications of origin research.
10. Explain reasons for and against continuing to search for the "theory of everything."

3-1 IS THE BIG BANG REALLY A NEW PARADIGM?

Can the development of the big bang theory, as discussed in the last chapter, be seen as the creation of a new paradigm within science? At first blush, the answer to this question may seem to be an unqualified "yes." The introduction of the expanding universe theory and the big bang standard model appears to have all of the indicators of a new paradigm. After all, the idea of an expanding universe clearly changed our conception of the origin and ultimate fate of the universe. Some people have even gone so far as to compare the discovery of the expanding universe to the introduction of the Copernican system during the Scientific Revolution. However, as we are about to see, this conclusion may be a bit premature. To examine the question of whether the expanding universe and the big bang theory can really be considered an established paradigm, we first review the events leading to that theory and then put those events within the context of Thomas Kuhn's ideas on revolutionary science.

The Expanding Universe Revisited

The logical starting point in the examination of a new paradigm is a statement of the old one, which in this case, saw the universe as permanent and eternal. Space and time were stable conditions that did not change. This prevailing paradigm was based upon two fundamental assumptions about the inherent nature of the universe. The first was that matter was evenly distributed throughout the universe. The second was that the size of the universe bore no relationship to time. These assumptions were, in fact, the two hypotheses upon which Einstein based the theory of relativity. Einstein framed these assumptions in the following way:

1. There exists an average density of matter in the whole of space which is everywhere the same and different from zero.
2. The magnitude ("radius") of space is independent of time.[2]

These assumptions are consistent with the prevailing view of the universe at the turn of the century as static and unchanging.[3] As we saw in the last chapter, reinterpretations of the theory of relativity and observational data concerning the red shift led to a new picture of the universe, that challenged the static, unchanging model. The observations that proved so crucial to the readjustment of the paradigm included those of Vesto Slipher, who, after working with data on thirteen nebulae, discovered that all but two emitted spectra shifted toward the red end of the spectrum.[4] Later Edwin Hubble observed the same red shift phenomena and used that data to determine the distances to those galaxies and, as a result, the age of the universe. Hubble concluded that the rate of movement of a galaxy away from the Milky Way increases in proportion to its distance from our galaxy. This observation came to be known as Hubble's law, and the proportion, that is the ratio of velocity to distance, came to be known as the Hubble constant.[5]

The reinterpretations of relativity that contributed to the new view of the universe were introduced first by Georges Lemaitre and later by Alexander Friedmann. As we explored in the last chapter, both Lemaitre and Friedmann were responding to the addition of a cosmological constant to Einstein's theory of relativity. Einstein had added the cosmological constant to maintain a static, unchanging universe. The cosmological constant added a repulsion factor or "negative pressure" that would balance the attractive effects of gravity, thus creating the static or steady state universe that Einstein preferred.[6]

Both Lemaitre and Friedmann found this approach unnecessary. Lemaitre believed that Einstein's cosmological constant failed to maintain the steady state that Einstein wanted. Lemaitre then went on to show that a small deviation in the equations would move Einstein's static universe into either a contracting or an expanding universe. From this idea, Lemaitre created a new cosmological model based on the idea that the universe began as a "primeval atom," balanced between repulsive and attractive forces. When the repulsive force exerted slightly more power than the attractive force, it created a sudden imbalance that resulted in an expanding universe.[7]

Friedmann also found fault with Einstein's cosmological constant. Friedmann showed that it was possible to preserve Einstein's first assumption, that matter was spread evenly throughout space, provided he was willing to drop the second assumption, that the radius of space was not related to time. From his concept, Friedmann developed a cosmological model of the universe that was nonstatic in nature. His model was based on the idea that the universe expanded from an initial state of very high density.[8] It remained for Sir Arthur Eddington to show that the observational data compiled by Slipher and Hubble lent support to the expanding universe concept promoted by Lemaitre and Friedmann. As John Barrow and Joseph Silk declare in *The Left Hand of Creation*, it was this insight that led to the new paradigm, the paradigm of the big bang.[9] Gone was the static, steady, unchanging universe, and in its place was the concept of an expanding universe.

Kuhn and Scientific Revolutions

All of this sounds quite convincing, but are Barrow and Silk correct in their conclusion? Has the old paradigm of the static universe really been replaced by the expanding universe, or are we, instead, within the throes of a crisis situation, the type of crisis that precipitates a scientific revolution? To answer this question, let us review the events leading to the expanding universe one more time within the context of Thomas Kuhn's theory on the development of scientific revolutions. As Kuhn explains in his study, *The Structure of Scientific Revolutions*, normal science takes place in a relatively calm and predictable context within an established paradigm. If we apply Kuhn's hypothesis to the development of the big bang model, the original, established paradigm would have been the belief in the static, unchanging universe.[10] The observations carried out by Slipher and Hubble would represent attempts within normal science to explore the extent of this paradigm further. Had their observations verified the underlying assumption of the static universe, there would have been no crisis and no revolution. Normal science would have moved forward, and the paradigm would have been verified. Moreover, the work of Slipher and Hubble would have been a mere footnote in the history of cosmology. That is not what happened, however. Instead, the observations of Slipher and Hubble produced an anomaly.

Recall that an anomaly occurs when an observation of some sort appears that challenges the accepted paradigm. In other words, the paradigm predicts that one thing should happen and an observation produces a contradictory result.[11] Sometimes an anomaly can be explained within the context of the accepted paradigm. However, at times an anomaly is so troublesome that it can lead to a crisis within the accepted paradigm. If a crisis is precipitated, then the anomaly becomes more than a simple oddity. It becomes, instead, the primary focus of research in that area until one of three results occurs:

1. despite the initial problem, the anomaly is resolved within the context of the original paradigm;
2. the solution to the anomaly is postponed for a later generation of scientists; or
3. a new paradigm is introduced to replace the old, discredited paradigm.[12]

The persistence and the consistency with which the expanding universe and the big bang theory are put forth by the media might lead the nonscientist to the conclusion that Barrows and Silk are correct when they proclaim in *The Left Hand of Creation* that the expanding universe has become a paradigm.[13] This conclusion is likely fueled by the desire for stability within our own personal-belief system, a stability that is provided by a consistent theory like the big bang. It is also likely to be perpetuated by the news media and by the writers of popular science texts because it is a theory that has, quite frankly, won the hearts of many of the experts in the field. Pope Pius XII's announcement in 1951 in an address before the Pontifical Academy of Sciences that the big bang theory supported the Biblical account of creation also helped the new theory to be acceptable to many people who might otherwise have rejected it as antireligious.[14]

Nevertheless, it may still be somewhat premature to characterize the big bang and the expanding universe as parts of a new paradigm. It would probably be a lot more accurate to characterize the present situation as one of Kuhn's crisis situations in science. Recall that a crisis situation is a creative period in science, during which the conventional procedures and the customary rules of normal science are relaxed. New and different approaches to the problem are encouraged. Radical solutions to the problems are entertained as the most creative minds of science tackle the difficulties in an attempt either to alter the old paradigm radically or to introduce a new one. During this period of revolutionary science, researchers live in a strange and uncertain world that changes rapidly as observational anomalies are added to the mass of evidence that something may be wrong with the existing paradigm.

If we are within a period of revolutionary science, we should expect conflict. We might even see conflict as natural and commonplace. We would also expect to see scientists making creative, perhaps even radical, attempts to solve the existing problem. The observational evidence might bounce back and forth, at one time supporting the old paradigm, at another time suggesting a change in it, and at still other times promoting the elimination of that paradigm in favor of a new one that is more in line with experimental data. As we shall see, cosmology does seem to be in a revolutionary period. Because of the discoveries made by Slipher and Hubble, the equations proposed by Lemaitre and Friedmann, and the conclusions suggested by Eddington, cosmology has been in the midst of a crisis during which the old paradigm of the static, unchanging universe has been and continues to be challenged by the expanding universe model.

For the remainder of the chapter, we enter this strange and uncertain world of revolutionary science. Far from being dull and unexciting, revolutionary science can be as captivating as an unsolved murder mystery. The mystery that is the focus of this period of revolutionary science is the origin of the universe. Like all murder mysteries, this puzzle has its share of victims and potential victims. In this case the potential victims are the static universe and the big bang model. The mystery also includes several suspects, numerous red herrings that lead us in the wrong direction, various clues that point us toward the correct explanation, and an unnamed scientific detective who at some point in the future will be clever enough to uncover the ultimate solution.

One difference between an ordinary murder mystery and the mystery of the cosmos is, of course, that astronomers and cosmologists are presently facing the most baffling "whodunit" in history, the question of how the universe began. In an effort to understand this crisis in science and in an attempt to solve the mystery, we explore some of the observational data (some recently discovered *clues*) that have surfaced in relation to the origin question. We also interpret this data in light of two alternative theories (two new *suspects*) that have been proposed as solutions to the puzzle of creation. Finally, we address the issue of purpose and practicality and attempt to answer the following question: Why should a nonscientist be concerned with research into the origin of the universe? Or, to put it another way, why should a nonscientist be interested in "whodunit"?

3-2 PROBLEMS WITH THE BIG BANG MODEL

Our starting point will be to examine two perplexing problems that have persisted in relation to the expanding universe and the big bang model. The first of these problems has been present since Hubble first made his original observations of the red shift phenomenon back in the 1920s. This problem concerns the age of the universe. The second problem is a relatively new one. It concerns the mounting evidence that some of the galaxies within the universe are not moving as they should in order to be in harmony with the expanding nature of the universe, which we call the mystery of galactic movement.

The Age of the Universe

The big bang theory places the age of the universe at between 15 and 20 billion years. Recent observations and measurements made by the Hubble Space Telescope have, however, begun to call into question the accuracy of that estimate. In 1995, a team of astronomers led by Wendy Freedman of the Carnegie Observatories in California began to use the Hubble Space Telescope to measure the age of the universe.[15] The advent of the Hubble Space Telescope has allowed these astronomers to make more accurate measurements than ever before because the telescope is beyond the Earth's atmosphere, which tends to blur and distort the images of astronomical bodies sent toward the Earth.[16]

The hope was that by measuring the distance from the Earth to a distant galaxy, Freedman and her coworkers would be able to determine how much the universe has expanded since the beginning of time and, therefore, how old the universe is today. The process is a little like measuring the duration of a car trip. If you know the average **velocity** of the car and the distance it has traveled, you can figure out how long the trip has taken. If the car is traveling at an average of 50 miles per hour and it has traveled 600 miles, by a simple act of division, you can calculate that the trip has taken 12 hours. Another way of saying this is that the trip would be 12 hours old.[17]

Such measurements, however, are pointless when cosmologists and astronomers choose to focus upon nearby galaxies such as Andromeda. This is because Andromeda and the Milky Way are so close to one another that their gravitational fields are inextricably intertwined, causing them not only to follow the expansive movement of the universe, but also to stream toward one another.[18] So instead of using a relatively nearby galaxy, Freedman chose to focus on a far distant cluster of galaxies called the **Coma cluster**. A galactic cluster, or cluster of galaxies as it is sometimes called, is a collection of galaxies that are comparatively near to one another and that generally relate to one another gravitationally.[19] Measuring the distance to the Coma cluster, however, could not be done directly. To calculate the distance to the Coma cluster, Freedman first had to measure the distance to the Virgo cluster, a group of galaxies that was somewhat closer to the Milky Way but not as close as Andromeda. Her team could then use the distance to the Virgo cluster as the base measurement for their determination of the distance to the Coma cluster. Instead of using all of the galaxies within the Virgo cluster, however, Freedman chose to focus on one target galaxy, M 100, as a starting point. The choice proved to be fortuitous because M 100 was found to be rich with Cepheid variable stars.[20] **Cepheid variable stars** are the best choice for such calculations because they vary in brightness at regular intervals making comparison measurements easier.[21]

To measure the distance to a Cepheid, Freedman's team first had to determine the period of variation in the star's brightness. With the period of variation in hand, they could determine the average luminosity of the Cepheid. The **absolute magnitude** or **luminosity** of a star is the amount of energy that the star is radiating. The determination of a Cepheid's absolute magnitude can be quite precise because the interval of oscillation in the Cepheid's brightness is directly related to its absolute magnitude. The absolute magnitude of the Cepheid is then compared to its apparent magnitude. The **apparent magnitude** is the brightness the star seems to have from our vantage point. A comparison of the star's absolute brightness with its apparent brightness will yield its distance from the solar system.[22] (The technique is similar to determining the relative distances to two light bulbs. Let us say the two light bulbs appear to us to have equal brightness, yet one is a 25 watt bulb and the second one is a 150 watt bulb. Clearly, if they appear to have the same brightness from our point of view then the second bulb, with the much higher degree of real brightness, must be farther away from us than the first bulb.)

Freedman and her team searched throughout M 100 for Cepheid variables until, after considering some 40,000 candidates, they located a total of 20 Cepheids. After making the appropriate measurements, they concluded that M 100 is about 56 million light years from the solar system.[23] The rest of Freedman's measurements are based on two assumptions. The first is that the distance to M 100 is the average distance to the other spiral galaxies within the **Virgo cluster**. However, since the team could not be absolutely certain of this assumption, their measurements may be off by a factor of 20 percent. The second assumption is that the spiral galaxies within the Coma cluster, which was their ultimate target, possessed the same intrinsic brightness as the spirals within the Virgo cluster. Using these two assumptions Freedman determined that the Coma cluster is 5.5 times farther away from the Milky Way than is the Virgo cluster. The actual figure that the Freedman team settled on was in the neighborhood of 300 million light years.[24]

The next step in Freedman's study was to recalculate the Hubble constant, based upon these new figures. Recall that Edwin Hubble had concluded that the rate of movement of a galaxy away from the Milky Way increases in proportion to its distance from our galaxy. Freedman arrived at a new estimate of the Hubble constant by calculating the ratio of the velocity of the Coma cluster to its distance from the Milky Way. The new figure that Freedman's team arrived at was 80 (±17) kilometers per second. The previous figure set the Hubble constant at 50 kilometers per second. The faster rate of expansion means that the universe is much younger than it was once believed to be. If Freedman's figures are correct, then the universe is only about 8 to 12 billion years old.[25]

This conclusion contradicts the prediction made by the big bang theory which, as noted earlier, places the origin of the universe at sometime between 15 and 20 billion years ago. The mystery deepens, however, when we consider that the oldest stars in the Milky Way Galaxy are estimated to be about 15 billion years old. Since it is impossible for the stars to be older than the universe, something is clearly wrong.[26] Moreover, Freedman's measurement of the Hubble constant seems to be in line with other recent measurements supporting the higher figure, thus confirming the conclusion that the universe may be much younger than was first thought. For example, Michael Pierce of Indiana University, operating a ground-based telescope outfitted with a computer-operated mirror designed to correct distortions caused by the atmosphere, used several Cepheids within NGC 5471 and another galaxy within the Virgo cluster to arrive at a Hubble constant of 87.[27] Another astronomer, John Tonry of MIT, found a Hubble constant of 80, and Robert Kirshner of Harvard University concluded that the Hubble constant is 73. Each of these figures is within the Freedman's range of 80 (±17) for the Hubble constant.[28]

Recall, however, that the observations that led to the big bang explanation for the expanding universe theory also led to the present period of revolutionary science. At least, that is our present proposition, and so it would be quite surprising if Freedman's conclusions about the Hubble constant and the age of the universe went unchallenged for any length of time. It is possible, perhaps even likely, that Freedman's measurements will turn out to be one of the red herrings that lead us temporarily astray as we attempt to solve the mystery of the origin of the universe. As is true in any crisis situation in science, however, the puzzling nature of the mystery often leads to some highly original suggestions as to how to solve the problem. In this case, several astronomers have suggested using a star cluster located south of the Milky Way plane rather than Virgo which lies north of it. An ideal galactic cluster lying south of the galactic plane is the **Fornax cluster**.[29]

The Fornax cluster has several advantages over the Virgo cluster. For example, it is about 85 percent smaller than the Virgo cluster. This smaller size means that the galaxies within Fornax are much closer together. This higher density will allow for a more accurate calculation of the Hubble constant than any measurement that uses the Virgo cluster as a base. Moreover, using a galactic cluster located south of the galactic plane will allow for a comparison with Virgo which lies north of the plane. This offers some intriguing possibilities since examining Fornax will result in figures for the expansion of the universe in the direction opposite that used in measuring Virgo.[30] Unfortunately, the Fornax cluster is not without its disadvantages. For instance, it contains fewer spiral galaxies than Virgo, and therefore, fewer Cepheid variables. This is because only spiral galaxies contain Cepheids. Moreover, not all these spirals are adequate candidates. If a spiral galaxy faces the earth edgewise, it is difficult for astronomers to identify the Cepheids within the dense galactic center. Nevertheless, the Fornax team members assigned to the Hubble telescope have settled on two galaxies and are currently at work attempting to determine the value of the Hubble constant. If the value uncovered by the Fornax team matches Freedman's figure, the new figures will lend support to the idea of a younger universe.[31] If this happens, it may become necessary to rethink our conclusions about the age of the stars or perhaps even the idea of the expanding universe itself. The fact that the jury is still out on this issue should not surprise us, however. It is all a part of the dynamic atmosphere to be expected during a period of revolutionary science.

The Mystery of Galactic Movement

The age-of-the-universe mystery is not, of course, the only problem currently plaguing the theories related to the origin of the universe. A second enigma involves gravity and the movement of the galaxies. The problem can be stated simply. If everything in the universe is governed by the attractive force of gravity, why isn't everything in the universe falling together. In fact, why didn't all the matter in the universe fall together eons ago? Vesto Slipher and Edwin Hubble's work with the red shift phenomenon appeared to have answered these questions over sixty years ago. According to the work of Slipher and Hubble, the galactic red shift demonstrated that the universe had not yet collapsed because the galaxies are rushing away from one another.[32]

We have seen how this discovery and the work of Lemaitre and Friedmann led to the concept of the expanding universe. Moreover, we have seen that the idea of the expanding universe is supported by a substantial amount of evidence. Still, much of this evidence is based upon two essential assumptions. The first is that Hubble's red shift does indeed mean that the galaxies are rushing away from one another. The second is that most of the galaxies cooperate with this expansion. Recent observations, have thrown doubt on both of these assumptions.[33]

The Red Shift Phenomenon. The first assumption, that the red shift phenomenon means that the galaxies are rushing away from one another, has been placed in doubt by a series of observations made by William Tifft, an astronomer at the University of Arizona. Tifft observed that the red shifts coming from galactic structures are not dispersed in an irregular manner but, instead, appear to be distributed at distinct quantum levels.[34] If we maintain the idea that the red shift phenomenon is related to galactic velocity, then the data compiled by Tifft indicates that the galactic red shifts are not steady. Instead, they tend to make quantum leaps of 45 miles per second.[35] Subsequent measurements forced Tifft to recalculate his findings so that the intervals between red shifts related to some galactic types dropped to a third or a half of the original figure of 45 miles per second.[36]

However, the evidence of galactic quantum velocity shifts still sounds strangely like the quantum energy levels that occur within the subatomic universe (see Chapters 2, 4, and 5). Recall that in the quantum model of the atom, the electrons exist at predetermined energy levels within the structure of the atom. Moreover, these electrons can exist at only those energy levels. Thus, when photons are emitted or absorbed by the atom, the electrons move from one level to the next. Tifft correlates this phenomenon to galactic structures, theorizing that those structures also exist at predetermined energy levels, each of which may correspond to a particular quantized red shift.[37] If a galaxy's red shift is related to the energy state it occupies, then the big bang theory may be in trouble. The big bang is predicated upon the assumption that the universe is expanding. This expansion is demonstrated by the red shift phenomenon. If the red shift is found to result from galactic energy states rather than the expansion rate, then an important piece of evidence supporting the big bang evaporates.[38]

Moreover, according to Tifft's findings it is also possible that the red shift associated with particular galaxies may actually evolve with age. Thus, the difference between the red shift of nearby galaxies as opposed to the red shift associated with more distant galaxies may be due to the age of those galaxies rather than to their movement.[39] If all of this is true, it means that we must, at the very least, rethink the relationship between the red shift phenomenon and the expansion of the universe. It may turn out that the first and foremost evidence for the expansion of the universe, the red shift phenomenon, is actually related to the types and the ages of the galaxies rather than to the expansion of the universe. Only time and future observations will answer these questions.

The Streaming Motion. Tifft's red shift finding is not the only recently discovered piece of evidence contributing to the present period of revolutionary science. In the late seventies, a group of scientists using balloons rather than ground-based detectors set about measuring the tiny deviations in the cosmic

Figure 3.1 A Galactic Cluster.
This is the Perseus cluster of galaxies, a cluster similar to the Local Group, of which the Milky Way is a member. The Local Group consists of two large spiral galaxies, the Milky Way, our home galaxy, and the Andromeda galaxy, along with 24 other galaxies.
Photo Credit: Copyright, Association of Universities for Research in Astronomy Inc. (AURA), all rights reserved. AURA operates The National Optical Astronomy Observatories for The National Science Foundation.

background radiation. The scientists involved in this project uncovered a new mystery. Instead of measuring a uniform red shift, they uncovered regions that were blue-shifted.[40]

Even more puzzling was the later discovery that the Local Group of galaxies had a similar heading.[41] The **Local Group** consists of two large spiral galaxies, the Milky Way and the Andromeda, along with twenty-four other galactic structures. Most of these other galaxies are grouped around either the Milky Way or Andromeda.[42] This motion, later dubbed the **streaming motion**, is roughly analogous to the type of movement that might be experienced by a group of white water rafters who are attempting to row to a nearby shore.

Because the current is so strong, it continues to carry the rafters downstream despite their effort to row to the shoreline.[43]

Later studies confirmed that the Local Group was not alone. Observations demonstrated that the Local Group had been joined by the Virgo cluster, another group of galaxies that was a part of the streaming motion. Eventually, evidence was accumulated to demonstrate that the Local Group and the Virgo cluster were being pulled toward an even more immense group of galaxies, the **Hydra-Centaurus supercluster**. A **supercluster** is a collection of galactic clusters. As if that were not enough, all three galactic clusters, the Local Group, Virgo, and Hydra-Centaurus, were moving at 600 km/sec toward an even more gigantic gravitational source located more than 150 million light years away from the Milky Way.[44]

In effect, this **Great Attractor**, as it was eventually christened, was tugging at the Local Group, Virgo, and Hydra-Centaurus and pulling every one of those vast galactic clusters off course.[45] The Great Attractor was acting like a strong current in a river, pulling a raft downstream, while the rafters were trying desperately to row toward the shore. The race to find the Great Attractor then began in earnest in the 1980s. Eventually, it was discovered by Alan Dressler of the Carnegie Institute in Washington, DC and Sandra Faber of the University of California at Santa Cruz.[46] What Faber and Dressler found, however, was even more astounding than anyone had anticipated. The Great Attractor was not one, but two enormous superclusters of galaxies. These two superclusters had a combined mass that exceeded that of the Milky Way Galaxy by a factor of 20,000. The Great Attractor occupied a portion of space beyond Hydra-Centaurus that covered 300 million light years.[47]

The discovery of the Great Attractor was followed quickly by the discovery of another even more colossal extragalactic structure, the **Great Wall**. This new structure, which was discovered by John Huchra and Margaret Geller, both of the Smithsonian Center for Astrophysics, consists of thousands of galactic structures and stretches some 500 million light years in length and 15 million light years in thickness.[48] The discovery of the Great Wall led Huchra and Geller to a new understanding of the structure of the intergalactic universe. Instead of existing in a random, but relatively even distribution throughout the entire universe, as was previously imagined, the galaxies now appear to be collected into unimaginably huge extragalactic sheets or bubbles consolidating thousands of galaxies together. In between these monumental extragalactic sheets, such as the Great Attractor and the Great Wall, lie vast deserts of barren space. The two metaphors usually connected with this image are that of a honeycomb and that of a bathtub filled with soap bubbles. In the latter image, the filmy soap bubbles would represent the sheets of extragalactic structures, and the space in between the bubbles would represent the desolate between-extragalactic wastelands. These enormous zones of emptiness are over 400 million light years across and devoid of any coherent structures.[49]

The discovery of these extragalactic sheets has presented another challenge to the big bang. According to the time frame established under the big bang model, there simply was not enough time for these extragalactic structures to have formed. Yet, the structures are there for all to see. Clearly, something has to give. Either the theory must be modified (or abandoned) or some new evidence gathered to explain the presence of these enormous extragalactic structures.[50]

The COBE Discovery

As noted previously, revolutionary science sets the stage for radical innovations. The atmosphere created by revolutionary science fuels the imagination and stimulates individuals to try things that they might otherwise not attempt during periods of normal science. For instance, they may be stimulated to use a new technique, to observe previously ignored phenomena, or to lobby for governmental support of programs that might otherwise not be pursued. One such innovation is the Cosmic Background Explorer (COBE) satellite that was sent into a polar orbit by NASA on Saturday, November 18, 1989.[51] The COBE mission was to study the background radiation that provides scientists with evidence in support of the big bang. As explained on page 59, this radiation represents the leftover glow of the enormous heat that permeated the universe in the early moments of the big bang. Just as the embers of an extinguished campfire let us know that the fire was once glowing hot, the background radiation reveals traces of the enormous heat generated by the big bang in the early stages of the universe.

One of the problems with past measurements of the background radiation from ground-based instrumentation was that the radiation was too smooth. For enormous extragalactic networks, such as the Great Wall and the Great Attractor, to have formed, some regions of the early universe had to be denser than other regions. The denser regions would tend to pull together under the influence of gravity. Gradually, over time, these dense regions would become the extragalactic structures that exist today.[52] If the dense regions did in fact exist in the early universe, then today we should be able to measure small temperature fluctuations in the background radiation. These temperature variations would correspond to different levels of density in the early universe. This is because radiation loses energy to resist the stronger gravitational pull of the dense region. The loss of energy can be measured as a drop in temperature. Dense regions would, therefore, be cooler than areas of lower density.[53] Only a small temperature fluctuation in this early radiation was necessary. Once this small fluctuation occurred, gravity gradually increased in strength so that the structures and superstructures now present in the universe could begin to form.[54]

No ground-based measurements had turned up the temperature variations needed to confirm the inherent roughness of the early universe. In fact, COBE's first measurements also failed to reveal the necessary fluctuations. Then on April 23, 1992 the picture changed. Dr. George Smoot of Lawrence Berkeley Laboratory and a chief COBE researcher announced at a Washington meeting of the American Physical Society that the inherent roughness that was needed to support the big bang theory had been mapped by COBE. Many researchers hailed the discovery as proof positive of the big bang. Their hope and the hope of all big bang supporters was, of course, that the COBE discovery of the temperature variations would finally end the debate.[55]

That hope, unfortunately, was premature. There are still some problems with the COBE discovery, not the least of which is the minuteness of the temperature variations uncovered by the satellite. In fact, they measure only about one part in 100,000. Still, this is enough of a fluctuation to account for the present-day structures that we have observed throughout the universe.[56]

3-3   ALTERNATIVE THEORIES

Naturally, the COBE discovery has not been conclusive enough to end the crisis and confirm that the big bang is, in fact, a new paradigm. In fact, just the opposite has occurred. Several astronomers and cosmologists have suggested alternative theories that pose a direct challenge to the big bang model. Two of those theories represent the paradigm of the eternal universe. These theories, respectively referred to as the steady state theory and quasi-steady state cosmology, attempt to reconcile the evidence that has been used to support the expanding universe and the big bang model with the traditional paradigm of the eternal universe. The third theory, plasma cosmology, is so exotic that it may actually represent the introduction of yet a third paradigm concerning the origin of the universe.

The Steady State Theory

The steady state theory in its original form was first proposed by Hermann Bondi and Thomas Gold of Cambridge University as an alternative explanation to the big bang theory. Bondi and Gold were joined by Fred Hoyle, who, at about the same time, suggested a slightly different version of the same theory. According to Bondi the steady state theory has two distinct advantages over the big bang theory. First, the steady state theory eliminates any problems concerning the age of the universe, a concern which, as we have seen, can give cosmologists recurring nightmares. Second, the steady state theory avoids the question of where the universe came from in the first place, because according to the steady state theory, the universe is eternal.[57] Later commentators have pointed out that the steady state theory is also much more appealing philosophically because it predicts a much more optimistic future for the universe than the ultimate end foretold by the big bang. The big bang theory, after all, gives us nothing to look forward to other than the heat death predicted by the inevitable movement of the universe toward a state of total entropy.[58]

At the heart of the steady state theory of the universe is the so-called **perfect cosmological principle**, which states that, from a large scale point of view, the universe should look the same regardless of the observer's position in space and time.[59] This principle also supports the assumption that the laws of physics operate in the same way at all positions in space and at all points in time throughout the universe. The idea that the laws of physics are immutable stands in marked contrast to the big bang theory, which suggests that the laws of physics operated quite differently in the first moments after the birth of the universe. Bondi explains quite convincingly that this adherence to the immutable laws of nature is one of the most attractive features of the theory.[60]

According to Bondi, the steady state theory is also attractive because it is compatible with the expanding universe. At first look, the idea that the steady state theory and the expanding universe are compatible may seem contradictory. After all, how can an eternal universe be reconciled with the idea that the universe is expanding? The apparent contradiction is resolved once we understand that a steady state universe does not imply a static one. In a static universe absolutely no change takes place. In contrast, in a steady state universe, change takes place but in a predictable and regular fashion. The difference between steady conditions and a static environment can be seen throughout nature. An air current, for example, might move in a steady fashion as experienced by a stationary observer. The current, while steady, would not, however, be static. In fact, individual particles within the air might accelerate and decelerate at different speeds. These small individual variations would not, however, detract from the overall, steady movement of the air current.[61]

The same would be true of a steady state universe. According to this approach, the universe may be expanding and yet it may still be eternal. The problem with this position is that it implies that the distribution of matter within the universe should be thinning out as the universe expands. This, however, violates the perfect cosmological principle, which says that the universe should look the same regardless of the observer's position in space and time. If the universe exists in a steady state, yet is also expanding, it will look different as the matter within the universe spreads out. According to Bondi, this is true only if the amount of matter remains constant. If, however, matter continuously replenishes itself at all times, then as the universe expands there is more and more matter to fill the space created by that expansion. Thus, the universe will continue to look the same at all points in space and time. The idea that matter is continuously created throughout space is perhaps the most controversial aspect of the steady state theory.[62]

Two fundamental weaknesses can be ascribed to the idea that matter is undergoing a continuous creation throughout the universe. First, continuous creation violates the principle of the conservation of matter or mass. Bondi dismisses this objection by noting that the continual creation hypothesis has yet to be contradicted by observational evidence.[63] Moreover, as it turns out, the rate of creation required by the perfect cosmological principle is incredibly small. According to Bondi, the perfect cosmological principle would be satisfied if every 10^9 years one atom is generated per every liter of space.[64] With these figures in mind, Bondi concludes that, since such a rate of creation is unobservable, there is no evidence that continual creation contradicts the principle of conservation of matter. The two principles of continual creation and the conservation of energy principle, therefore, are not contradictory.[65]

The second problem with Bondi's continuous creation of matter theory is that neither he nor Gold can offer an explanation for the mechanism behind this creative process. This is especially crucial since Bondi is adamant that the creation of matter within the steady state universe is the creation of matter from absolutely nothing.[66] Bondi does, however, calculate the initial velocity and the initial temperature of each newly created particle. He also demonstrates that the newly created matter particles are most likely hydrogen atoms, although he does not eliminate the possibility that matter might be created as single electrons, protons, and neutrons.[67] Moreover, he also suggests that the creation of matter is balanced by the disappearance of energy within the total universe, thus creating a perfect state of equilibrium.[68] All of this is impressive, but it does not substitute for the fact that Bondi cannot explain the actual creative process. Consequently, even opponents of the big bang who might want to believe Bondi are left with the vaguely unsettling feeling that he has done exactly what he accuses the big bang proponents of doing. He has fallen back on a metaphysical explanation for the continual creation of matter.

Given these weaknesses, it should not be surprising that most cosmologists today have rejected the steady state theory. Interestingly enough, however, these weaknesses by themselves did not lead to the end of the theory. Rather, a proper application of the scientific method discredited the steady state theory. Recall that any successful scientific theory must meet three criteria: continuity, clarity, and predictability. The steady state theory just narrowly passes the first two tests. It does flow from two previous principles, the phenomenon of the expanding universe and the conservation of matter principle. As we have seen, Bondi certainly stretches each principle a bit, but nevertheless, he does take both into account. The theory, therefore, exhibits the necessary continuity. As for its clarity, the primary point of confusion is the mechanism behind the continuous creation of matter. However, since the observation of the creative process is impossible, we might forgive this shortcoming. This is especially true because the steady state theory shares this particular problem with the big bang theory.

What cannot be forgiven, however, is the steady state's failure to produce accurate predictions. The steady state theory predicts that the universe should look the same from all positions in space and at all points in time. Therefore, there should be no past epoch during which anything out of the ordinary happened. The first evidence that this prediction was inaccurate came in the 1950s and '60s when data compiled by several radio telescopes suggested that in the distant past the universe looked a lot different than it does today. These results were still open to a variety of interpretations and might eventually have been explained within the cosmology of the steady state theory were it not for the discovery of the cosmic background radiation in 1965 by Arno Penzias and Robert Wilson. The discovery of the cosmic background radiation demonstrated that in the distant past the universe was much denser and, therefore, much hotter than it is today. This discovery spelled the end of the steady state theory, at least in its original form.[69]

Quasi–Steady State Cosmology

Recently Fred Hoyle, formerly of Cambridge University in Great Britain joined forces with Jayant V. Narlikar of the Inter-University Center for Autonomy in Pune, India and Geoffrey Burbidge of the University of California at San Diego to propose an updated version of the steady state theory. According to Hoyle, Narlikar, and Burbidge, this new version of the steady state theory, called **quasi-steady state cosmology (QSSC)**, is compatible with the same evidence that supports the big bang and, in fact, can explain some of the evidence that does not square with big bang cosmology.[70]

Most importantly, the new theory is compatible with the existence of the cosmic background radiation, the crucial piece of evidence that spelled the end of the original version of steady state cosmology. Moreover, QSSC explains not only the abundance of hydrogen and helium in the universe but also the expansion of the universe without resorting to the single primordial big bang inflationary epoch.[71] Perhaps even more interesting from our perspective is that Hoyle, Narlikar, and Burbidge predict that the creation events proposed by their theory may in the future be detectable by laser interferometer gravity wave detectors, which are now in the planning stages.[72]

Instead of one single big bang, quasi-steady state cosmology postulates a cyclical series of mini–big bangs that occur throughout the history of the universe at various intervals. These mini–big bangs are caused by a phenomenon labeled the **creation field** or the **C-field**. According to Hoyle, Burbidge, and Narlikar, the existence of the C-field is predicted by certain alterations in Einstein's relativity equations dealing with space-time singularities and strong gravitational fields. The C-field exists only in certain areas of the universe. These areas are characterized by a high density of matter. Typical candidates for the location of a C-field would be in the heart of a single galaxy or in regions of the enormous extragalactic clusters discussed above. Hoyle, Burbidge, and Narlikar label these regions **creation centers**.[73]

The C-field is activated by the intense concentration of matter at a creation center. Creation centers are intrinsically unstable, and, therefore, produce an enormous burst of explosive energy. This explosive burst of energy acts like a mini–big bang creating matter and activating the expansion of the universe. This sudden, short-lived explosive event is followed by a long period of gradual expansion. As the universe expands, the strength of the C-field diminishes and the expansion gradually weakens. Eventually, conditions again favor the appearance of new creation centers until, once again, the C-field is activated. This new burst of energy creates matter and activates the expansion cycle of the universe once again.[74]

How does QSSC differ from the original steady state theory? One chief difference is that in the original steady state theory proposed by Hoyle, Bondi, and Gold the creation of matter took place throughout the universe at a more or less constant rate. Another difference is that in the original theory when a creation event did occur, it produced a relatively small amount of matter. In contrast, according to the QSSC, creation events do not occur evenly throughout the universe at a more or less steady rate. Instead, they are confined to highly dense areas of matter called creation centers. These creation centers are capable of activating the C-field in an explosive burst of energy. Moreover, instead of each creation event producing a small amount of matter, the explosive creation events activated by the creation field produce large amounts of matter in a relatively short period of time.[75]

Perhaps the major advantage that QSSC has over the traditional steady state theory is that it makes verifiable predictions, something that the traditional theory failed to do. According to Hoyle, Narlikar, and Burbidge, the matter-creation events precipitated by the C-field will produce gravity waves. These gravity waves may be perceptible in the near future by laser interferometer gravity observatory (LIGO) detectors. However, Hoyle, Narlikar, and Burbidge also point out that such detection may be years in the future and may also depend on whether the gravity waves produced by the creation event vary enough to be observable.[76]

Plasma Cosmology

The third, and perhaps most exotic, alternative explanation for the origin of the universe is the theory known as plasma cosmology. **Plasma** is a gaslike state of matter made of charged particles. The idea of plasma cosmology was first introduced by Hannes Alfven of the Swedish Royal Institute of Technology. Alfven was involved in the study of plasma dynamics for more than twenty-five years. He used his work in plasma physics to explain the formation of stars and galaxies.[77] His work in the relationship between plasma and magnetic fields won the Nobel Prize in physics in 1970. Alfven's **plasma cosmology** theory promotes two ideas that are antithetical to big bang cosmology. First, plasma cosmology, like QSSC, is based on the assumption that the universe is eternal. Second, unlike both the big bang and QSSC, plasma cosmology operates on the premise that electromagnetism, rather than gravity, is the dominant force in the universe.[78] Accordingly, Alfven's theory represents neither an extension of the steady state theory nor a method of reconciling the big bang with the growing body of conflicting evidence. Instead, plasma cosmology represents a third alternative.[79]

Alfven also believes that any investigation into the origin of the universe should be based solidly upon the application of experimental results that have been demonstrated in the laboratory. He sees no reason to assume that the laws of nature would operate differently outside the laboratory, especially when those laws are applied to questions of cosmology.[80] Plasma is the basic component of most of the observable mass in the universe. Stars, for example, are basically plasmas held together by the influence of gravity. Plasmas are also thought to exist at the core of the galaxy and as a vast sea of charged particles in which galactic superclusters may be embedded. These areas of charged particles may be affected by colossal blankets of electromagnetic radiation that converge and intersect throughout the universe.[81] According to Alfven's theory of plasma cosmology, the initial state of the **metagalaxy**, a term that he uses for our local region of space, was a homogeneous substance called **ambiplasma**, which is a mixture of koinomatter and antimatter. **Koinomatter** is the term that Alfven uses for ordinary matter,[82] the matter that we are familiar with in our everyday lives. In Chapter 2, we examined the subatomic structure of this ordinary matter. The electron, for instance, is an example of ordinary matter, which has a negative charge. **Antimatter** refers to matter made of atoms that consist of particles that have charges that are the opposite of those found in koinomatter. The antielectron or positron, for example, is the antimatter counterpart of the electron. The antielectron has the same mass as an electron but has a positive rather than a negative charge.[83]

Ordinarily protons and electrons are bound together by mutual attraction, but in this case the gravitational clustering of the particles causes the heavier protons and antiprotons to move together while the lighter electrons and antielectrons mix together. As these mixtures flow through the electromagnetic fields that crisscross the universe, they create a current that separates the koinomatter from the antimatter. The current causes the koinomatter to flow one way and the antimatter to flow another way. This results in a pair of koinomatter antimatter clouds.[84]

Why don't the koinomatter antimatter clouds annihilate each other? According to Alfven, the clouds do interface and in the area of the interface they create a hot zone of annihilation. However, this zone of annihilation creates a low-density

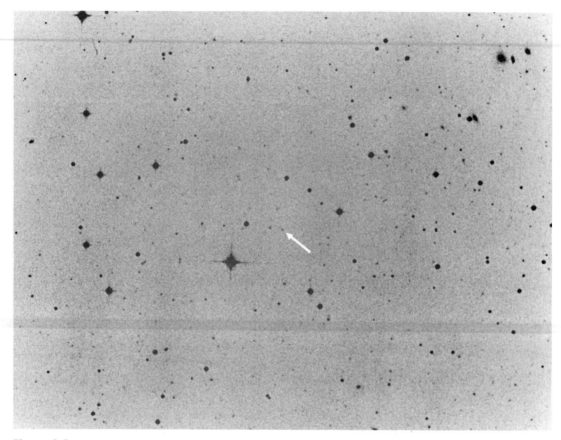

Figure 3.2 A Quasar.
A photograph of the sky, showing quasar Q0051-279. At the time of its discovery in late 1987, this quasar was the most distant object known. The quasar was initially indistinguishable from other stars in previous photographs and was only noticed after photographs taken with different filters showed it to have an unusual color. Some experts believe that the enormously strong energy output associated with quasars might be explained by matter-antimatter annihilation.
Photo Credit: Royal Observatory, Edinborough, U.K. Schmidt Telescope Unit, plate 8715204

plasma buffer that also keeps the clouds from completely engulfing one another. For awhile, clouds of koinomatter will join with other clouds of koinomatter while clouds of antimatter will join with similar clouds of antimatter. The koinomatter–antimatter clouds will continue to repel each other by riding the plasma buffer.[85] However, the gravitational contractions caused by the growing clouds will also increase the density of the clouds. The gravitational contractions will also increase the velocity of the collisions between the koinomatter and the antimatter clouds.[86]

Eventually, the plasma buffer between the koinomatter and antimatter clouds no longer provides the necessary protection. The clouds collide, and an enormous burst of energy from the koinomatter-antimatter contact causes the abrupt expansion that we associate with our region of the universe today.[87] From that point on, the universe follows the outward movement that we associate with the familiar idea of the expanding universe. The difference is, of course, that we do not have to trace the origin point of the universe to a single singularity that operated under conditions that no longer exist in the universe today.[88]

Plasma cosmology is not without its problems. One of the primary problems is that we have yet to detect regions of antimatter within the universe. Interestingly enough, Alfven points out that there is no reason, based on observational evidence alone, to rule out the conclusion that antimatter is the basic ingredient of half of the galaxies within the universe. In fact, observational evidence also does not exclude the existence of antimatter within our own galaxy.[89] Other experts believe that the enormously strong energy output associated with quasars may be explained by koinomatter-antimatter annihilation.[90] If Alfven's theory turns out to be correct, he will have provided cosmologists with a theory of the universe that does not require the type of creation event inspired by the big bang. When that happens a new chapter will have opened in humanity's quest to unlock the secrets of the universe.

3-4 THE NONSCIENTIST AND THE ORIGIN OF THE UNIVERSE

Despite all that has been said thus far about the origin of the universe and the need to solve this ultimate whodunit, many nonscientists remain unconvinced of the need to spend valuable dollars on what often turns out to be a rather expensive proposition. The past is irrevocably etched in the fabric of the universe and cannot be changed, altered, prevented, or even mildly refashioned to suit our needs. So what is the point of spending billions of dollars to unravel the mystery of the origin of the universe? To answer these questions, we pause now and examine some of the more immediate applications of the search for the origin of the universe.

Unlocking the Theory of Everything

Recall that any examination of the origin of the universe is also an examination of the subatomic universe. During the early stages of the universe, at least as the early universe is envisioned under the big bang theory, the forces of electromagnetism, the weak nuclear force, and the strong nuclear force were united as one **grand unified force**. Moreover, even earlier than that, all of the forces, including gravity, were joined together as one.

If physicists can unravel the nature of this unification, they will have reached into the heart of the universe and explained how the four forces of the universe are all manifestations of one single force. This ultimate explanation of how the universe works is commonly called the **theory of everything**.[91] Once physicists have deduced the theory of everything, instead of being astonished by the force of gravity and bewildered by the strong nuclear force, they will see both of these forces as clearly as they now see the electromagnetic force. The theory of everything may allow us to manipulate these forces just as easily as we manipulate electromagnetism today. In his novel *The Genesis Machine*, James P. Hogan describes the moment at which several scientists finally unravel the secrets of the theory of everything. They realize, almost at once, the marvelous implications of their discovery.

> It'd revolutionize the whole business of transportation. Just imagine—if you could move big loads effortlessly anywhere ... all over the world. Why bother building bridges and things when you can simply float things across rivers on a *g*-beam? Who needs roads and rails? They're only ways of cutting down friction, and this way there'd be no friction—only inertia.
>
> You'd be able to move a ten-ton block of stone around with the push of your hand.[92]

However, this is only the beginning of their conjecture. As the discussion continues and their imaginations begin to heat up, they start to realize that they have, indeed, unlocked the secrets of the universe and provided untold benefits for all humanity.

> "And what about earth-moving?" he said. "You could move mountains maybe—literally. Resculpt the whole planet ..."
>
> "Move mountains? Resculpt planets?" Morelli's voice rose to a resonant crescendo as he threw the vision out to infinity. "Think big.... Move planets! Resculpt the Solar System! Do you know there's an asteroid out there that's reckoned to contain enough iron to meet the world's needs at today's rate for the next twenty thousand years? ... Overpopulation problems? Break up another planet and park the bits in orbit around the Sun here, where it's nice and warm; that'll keep us going for a while. How do you break a planet up? Answer: gravitic engineering!"[93]

Do all of these predictions sound just a bit too fantastic? Perhaps, but then again how fantastic did human flight seem two hundred years ago, let alone space flight? What about the way that electricity drives our civilization today? How fantastic would that seem to medieval Europeans? Add telephones, televisions, fax machines, hand-held calculators, microwave ovens, and desktop computers to the list, and it is easy to see how much can be accomplished in a few short generations.

Particle Accelerators as the Hope for the Future

Despite recent setbacks in particle physics, the experimental tools exist today for the attempt to unravel first the grand unified theory that seeks to join the strong nuclear force with the electroweak force, and then the theory of everything, that will attempt to unify gravity with the other forces. These experimental tools are the enormous **particle accelerators**, which attempt to mimic the early moments of the universe. Particle accelerators first came into prominence during the Great Depression of the thirties. At that time, they were so small that some of them could be held in the palm of your hand. A particle accelerator of this size would be extremely small by today's standards.[94] The Stanford Linear Collider in California, for example, is 2 miles in length. Moreover, the Fermilab Tevatron in Illinois has a diameter of 1 mile, while the CERN electron-positron collider in Switzerland has a diameter of 19 miles.[95]

These measurements, however, pale into insignificance when compared with the Superconducting Super Collider (SSC) which was supposed to be built at Waxahachie Texas in the mid-1990s. Had its funding not been cut by Congress in 1993, the SSC would have had a 54-mile circumference.[96] The size and power of the SSC would have allowed it to generate the high energies that are needed for detecting the existence of the Higgs field that holds the key to the electroweak force.[97]

Just how do these particle accelerators work? The objective of the accelerators is to explore the early moments of the creation of the universe as closely as possible. To do this, they accelerate subatomic particles such as electrons or protons and smash them either into a stationary barrier or into one another.[98] The resulting flash of energy does not, as one might think, smash the particles into bits. Rather, the high-energy collisions actually create new particles, some of which no longer exist in nature but only in the high-energy arena of the particle accelerators. The new particles are short lived, but they can be studied by examining the particles that these short-lived particles themselves emit. With the aid of very precise detectors, scientists can learn an enormous amount about the structure of matter, which, in turn, increases their understanding of the early universe.[99] The hope is that someday they may give us the key to unlocking the theory of everything.

The Theory of Everything: Myth or Reality?

It is only fair to point out that not everyone in the scientific community is optimistic that such a time will come. In fact, some very prominent names feel that the end of physics is in sight and that the theory of everything is an unattainable dream. Chief among these individuals is Sheldon Glashow, head of the physics department at Harvard University, who believes that much of the work in particle physics today has passed from the area of the experimental verification and slipped into the realm of aesthetics and philosophy.[100] Glashow is especially critical of superstring theory, which some physicists believe may be the long-hoped-for theory of everything.[101]

According to Edward Witten, professor of physics at the Institute for Advanced Study in Princeton, **superstring theory** involves a new image of the subatomic universe, which replaces the idea that subatomic particles are one-dimensional points with the notion of subatomic entities as multidimensional strings.[102] The vibration of these multidimensional strings can account for the existence of matter and energy in the universe. One of the implications of superstring theory is that it not only accounts for the existence of gravity, that mystifying fourth force, but also actually predicts the existence of gravity.[103]

Glashow, however, has long been critical of physicists who spend their time, effort, and energy researching superstring theory because, according to Glashow, there is little hope that such a theory can ever be experimentally verified.[104] Glashow does not deny that such an approach is very appealing. He freely admits that it would be very gratifying to be able to explain the ultimate structure and operation of the universe from pure thought alone. However, he is realistic enough to remind us that we have been led astray on more than one occasion by theories that have no experimental verification. And it appears that he is willing, even eager, to abandon superstring theory for just this reason.[105]

In his recent work, *Dreams of a Final Theory*, Steven Weinberg of the University of Texas at Austin has also expressed some lingering doubts about the discovery of a final theory. Weinberg is hopeful that such a theory of everything will one day be uncovered. In fact, most of his book is written with the fundamental assumption that the discovery of such a theory is inevitable.[106] However, Weinberg is honest enough to admit that such a theory may be either nonexistent or inherently undiscoverable.[107] Weinberg also points out that "(t)here is no guarantee that progress in other fields of science will be assisted directly by anything new that is discovered about the elementary particles."[108] Moreover, throughout Weinberg's *Dreams of a Final Theory* there is very little mention of any practical payoffs that might result from the discovery of the theory of everything.[109] However, to be fair to Weinberg, he does not see the failure of scientists to link discoveries in particle physics to other fields of science as a limitation inherent in nature. Instead, he traces any failure to make any such connection to "human limitations and human interests."[110] More importantly, Weinberg is much more interested in discovering how nature works as an end in itself rather than as a means to an end.[111] It may be this sentiment that leads Weinberg to remind us that the ultimate discovery of the final theory will not mean the end of science as some writers have predicted.[112] Instead, the unlocking of that final theory will mean a change in direction opening new scientific vistas.[113]

In fact, warnings that we are approaching the end of science sound faintly reminiscent of an ill-fated forecast issued by Lord Kelvin at the end of the nineteenth century when he predicted that the end of physics was at hand. Nor was Lord Kelvin alone in his mistaken belief.[114] At about the same time, John Trowbridge, the head of the physics department at Harvard, advised students to avoid a career in physics because no important discoveries were left to them.[115] Both of these predictions proved to be badly off the mark. In fact, as we now know, at the turn of the century physics was entering an age of some of the most fantastic discoveries in the history of science, not the least of which were the theory of relativity and quantum mechanics.

It is also helpful to remember the advice issued by James Trefil and Robert Hazen of George Mason University, who in their book, *The Sciences: An Integrated Approach*, point out that present-day physicists can no more hope to predict accurately all of the practical outcomes that could result from a unified theory than their nineteenth-century counterparts could predict the spinoffs from their research into the character of electricity.[116] In fact, asking particle physicists to predict all of the potential payoffs from their research into the theory of everything is a little like asking James Clerk Maxwell to use the equations in his theory of electromagnetism to foresee track lighting or to imagine Times Square on New Year's Eve.

On the other hand, David Lindley, the senior editor of *Science* magazine and author of *The End of Physics: The Myth of a Unified Theory*, points out that unraveling a theory that unites the four forces has some inherent difficulties that may not be overcome for decades, if ever.[117] "Gravity," the most anomalous of the four fundamental forces, he tells us, "is more complicated than electromagnetism."[118] What makes it more complicated is that the force exerted by gravity feeds back into any equations attempting to unravel the interaction among objects.[119] In effect, any attempt to explain gravity results in an irreconcilable catch-22. As Lindley explains:

> (w)hen two bodies are pulled apart against their gravitational attraction, energy must be expended, and if they come together energy is released; but energy . . . is equivalent to mass, and mass is subject to gravity. Therefore, the energy involved in a gravitational interaction between bodies is itself subject to gravity. Gravity, if you like, gravitates.[120]

Still, other difficult problems, which in the past seemed to have no solution, have been resolved. We need only recall Johannes Kepler's resolution of the conflict between the motion of Mars and the predictions of the Copernican system, as discussed in Chapter 1, to realize that other apparently "unsolvable" mysteries have been unraveled in the past. In this regard it is helpful to note the words of Michio Kaku, professor of theoretical physics at the City College of the City University of New York, who in his recent book *Hyperspace* reminds us that:

> (I)t took about 70 years, between the work of Faraday and Maxwell to the work of Edison and his co-workers, to exploit the electromagnetic force for practical purposes. Yet modern civilization depends crucially on the harnessing of this force. The nuclear force was discovered near the turn of the century, and 80 years later we still do not have the means to harness it successfully with fusion reactors. The next leap, to harness the power of the unified field theory, requires a much greater jump in our technology, but one that will probably have vastly more important implications.[121]

It would be wise for us to heed these words and to remember these facts whenever we are tempted to question the need for pure research such as that conducted by the physicists who are attempting to unravel the theory of everything. This does not imply that such scientists should not be accountable for their expenditures of public money. It should, however, remind those of us who vote on such matters that the most profitable results of research are often those that are the most difficult to foresee.

THE REBIRTH OF THE COSMOLOGICAL CONSTANT

As if to underline this lack of foreseeability and to bring our discussion full circle, let us return to Einstein's cosmological constant. Recall that Einstein added the cosmological constant to his theory of relativity in order to avoid the unwelcome implication that the universe was expanding. The net effect of the cosmological constant was to add a repulsive force to the universe. This repulsive force would counteract the attractive force of gravity, creating a balance between the two forces, thus allowing the universe to exist in a steady state. Recall also that, after Hubble's work led to the theory of the expanding universe, Einstein abandoned the cosmological constant, calling it the greatest blunder of his career. Now it seem that Einstein may have been right after all, but in ways that even he could not foresee.

Over the last few years two teams of astronomers began a long-term project studying distant supernovae with the object of determining the extent to which gravity was curbing the expansion of the universe. The results of their observations, which were announced in 1998, revealed that the expansive rate of the universe was not slowing at all. In fact, it was accelerating.[122] The implications of this discovery are significant. For one thing, it indicates that Einstein's cosmological constant may in fact exist.[123] For another, instead of counteracting gravity to maintain a steady state, the cosmological constant is actually counteracting gravity with such force that the universe is expanding faster today than in the past.[124] If this is the case, then the orthodox theory of the expanding universe based on an early inflationary epoch must be either revised significantly or abandoned altogether.[125]

The fact that no one knows what the cosmological constant really is has not prevented a great deal of speculation on its true nature. Some astronomers and physicists have suggested that the cosmological constant must be a unique form of energy which somehow pervades all of space, independent of time and position, and which exists separate from matter and radiation.[126] Some astronomers and physicists have suggested that the cosmological constant must be a unique form of energy which somehow pervades all of space, while remaining independent of time and position, and which counteracts gravity, while maintaining an existence separate from matter and radiation.[127]

CONCLUSION

Whether the present state of cosmology represents a period of revolutionary or normal science may be, at least at the present time, an unanswerable question. We are probably much too close to the problem to make a proper judgment. Nevertheless, it is interesting and helpful to speculate about such matters, if only because in doing so, for a moment at least, we assume the position of objective outsiders. We need to maintain this objectivity as we enter the next realm of our study because this area is even more fantastic than the question of the origin of the universe. The topic we are about to investigate in the next chapter involves scientific discoveries that defy common sense and lie far outside the bounds of everyday experience—the world of quantum physics.

Review

3-1. If we apply Kuhn's hypothesis to the development of the big bang model, the original established paradigm would have been the belief in the static, unchanging universe. The observations carried out by Slipher and Hubble would represent attempts within normal science to further explore the extent of this paradigm. The observations of Slipher and Hubble produced an anomaly. Recall that an anomaly occurs when an observation of some sort appears that challenges the accepted paradigm. Sometimes an anomaly can be explained within the context of the accepted paradigm. However, at times an anomaly is so troublesome that it can lead to a crisis within the accepted paradigm. The nonscientist might conclude that the expanding universe has become a paradigm. Nevertheless, it may still be somewhat premature to characterize the big bang and the expanding universe as parts of a new paradigm.

3-2. Two perplexing problems have persisted in relation to the expanding universe and the big bang model. The first of these has been present since Hubble first made his original observations of the red shift phenomenon back in the 1920s. This problem concerns the age of the universe. The second problem is a relatively new one and concerns the mounting evidence that some of the galaxies within the universe are not moving as they should move in order to be in harmony with the expanding nature of the universe. This is called the mystery of galactic movement.

3-3. Several new theories have appeared in recent years that pose a direct challenge to the reign of the big bang. The three chief challengers to the standard big bang cosmology model are the steady state theory, quasi-steady state cosmology (QSSC), and plasma cosmology.

3-4. One crucial reason for continuing research into the origin of the universe is to seek a solution to the mystery of the theory of everything. If physicists can unlock the nature of the unified forces, they will have reached into the heart of the universe and unlocked the mystery of the theory of everything, which will allow them to manipulate these forces just as easily as they manipulate electromagnetism today.

Understanding Key Terms

absolute magnitude (luminosity)	Fornax cluster	particle accelerator
	galactic cluster	particle physics
ambiplasma	grand unified theory	plasma
antimatter	Great Attractor	plasma cosmology
apparent magnitude	Great Wall	quasi-steady state cosmology
Cepheid variable star	Hydra-Centaurus supercluster	streaming motion
Coma cluster	koinomatter	supercluster
creation center	light year	superstring theory
creation field (C-field)	Local Group	theory of everything
	metagalaxy	Virgo cluster

Review Questions

1. Has the expanding universe theory ushered in a new paradigm?
2. How did Freedman and her team measure the age of the universe?
3. What alternative explanation for the red shift phenomenon has been suggested by Tifft?
4. What problems in origin research were caused by the discovery of "the streaming motion"?
5. What is the Great Attractor? What is the Great Wall?

6. What are the implications of recent measurements made by the Cosmic Background Explorer (COBE)?
7. What are the theories of the steady state universe and quasi-steady state cosmology?
8. What are the elements of the plasma cosmology explanation for the origin of the universe?
9. What are some of the practical applications of origin research?
10. What are reasons for and against continuing to search for the theory of everything?

Discussion Questions

1. The first part of this chapter focuses on an examination of the question of whether or not the expanding universe and the big bang theory have ushered in a new paradigm replacing the old idea of an eternal and unchanging universe. Setting aside for a moment the question of which of the two approaches you favor, do you believe that there is enough evidence to conclude that a new paradigm based on the expanding universe and the big bang has been ushered into the modern age? Why or why not?
2. Now address the question of which of the three theories, the big bang, quasi-steady state cosmology, or plasma cosmology you prefer. Explain your preference.
3. Hannes Alfven states that he believes that any investigation into the origin of the universe should be based solidly upon the application of experimental results that have been demonstrated in the laboratory. He sees no reason to assume that the laws of nature would operate differently outside the laboratory, especially when those laws are applied to questions of cosmology. Yet, much of cosmology is based upon attempting to uncover new laws of nature that are peculiar to cosmology rather than attempting to apply existing laws to the cosmological questions of the universe. Which approach do you believe would be the most productive? Explain.
4. What is the point of spending billions of dollars to unravel the mystery of the origin of the universe? Do you support the philosophical answer to this question, or are you more inclined to support the practical side of the issue? Explain.
5. In 1951, Pope Pius XII announced in an address before the Pontifical Academy of Sciences that the big bang theory supported the Biblical account of creation. Do you believe that the Pope's announcement helped or hindered the big bang theory? Do you think that the Pope, or any other religious leader, should make pronouncements like this? Why or why not? What about politicians? Should the president announce his preference for the big bang theory, the quasi-steady state cosmology theory, or the plasma cosmology theory? Why or why not?

ANALYZING STAR TREK

Background

The following episode from *Star Trek: The Next Generation* reflects some of the issues that are presented in this chapter. The episode has been carefully chosen to represent several of the most interesting aspects of the chapter. When answering the questions at the end of the episode, you should express your opinions as clearly and openly as possible. You may also want to discuss your answers with others and compare and contrast those answers. Above all, you should be less concerned with the "right" answer and more with explaining your position as thoroughly as possible.

Viewing Assignment—*Star Trek: The Next Generation,* "Datalore"

In this episode the Starship Enterprise visits a planet in the Omicron Theta star system, where the crew of the USS Tripoli had discovered Data twenty-six years earlier. While exploring the planet, the away team from the Enterprise finds the laboratory of Data's creator, Dr. Noonian Soong. In the laboratory they also find a disassembled android that looks exactly like Data. The android is reconstructed and reactivated. He turns out to be Lore, Data's prototype. Lore fills in some of the information missing about Data's origin, including the fact that Lore was the first successful android using the positronic brain invented by Dr. Soong. The problem with Lore was that the other colonists feared and envied him. Consequently, Dr. Soong was convinced by the colonists to deactivate and disassemble Lore in favor of developing Data who, while closely resembling Lore, lacks his emotional capacity. Data is, thus, seen as being in some ways inferior to Lore.

Thoughts, Notions, and Speculations

1. Dr. Noonian Soong's development of the positronic brain makes it possible for him to create a sentient android, represented by both Data and Lore. The invention of the positronic brain is clearly one of the most extraordinary breakthroughs in the history of scientific research. Would you characterize such a breakthrough as something that would happen during a period of normal science or of revolutionary science? Explain. What type of new paradigm might be ushered in by the creation of a sentient android? Explain.
2. Should nonscientists have any input into shaping the direction of scientific research? Or should this task be left to the scientists? Should nonscientists have the authority to cancel research projects like Dr. Soong's positronic brain or the Superconducting Super Collider (SSC) in Waxahachie, Texas? Explain. If nonscientists should not have the power to promote or cancel such projects, with whom should that power lie? Explain.

3. Imagine that you are Dr. Soong. What arguments could you make to your fellow colonists to allay their fears about Lore? What arguments could you make in favor of a responsible but uncontrolled research program? Now imagine that you are in charge of the now defunct SSC project in Texas. What arguments could you make in favor of refinancing the SSC? Explain.

NOTES

1. Roy Rockwood, *By Air Express to Venus*, 103.
2. Albert Einstein, *Relativity*, 152.
3. John Barrow and Joseph Silk, *The Left Hand of Creation*, 8; Evry Schatzman, *Our Expanding Universe*, 32–33. (*Note*: A light year is the distance that light, moving at 186,000 miles per second [300,000 kilometers per second], travels within one year.)
4. Paul A. LaViolette, *Beyond the Big Bang*, 256–57.
5. Barrow and Silk, 8–9; Alan Lightman, *Time for the Stars: Astronomy in the 1990's*, 80; Joseph Silk, *The Big Bang*, 56–57.
6. Einstein, *Relativity*, 152; Albert Einstein, *The Meaning of Relativity*, 103–109, 111.
7. LaViolette, 258–59; Schatzman, 32–33.
8. Einstein, *Relativity*, 152–53; LaViolette, 258; Lightman, 78–79; Schatzman, 33.
9. Barrow and Silk, 8–9.
10. Thomas S. Kuhn, *The Structure of Scientific Revolutions*, 10–11.
11. Ibid, 52–53.
12. Ibid, 80–84.
13. Barrow and Silk, 8–9.
14. Paul Davies, *God and the New Physics*, 20; Willem B. Drees, *Beyond the Big Bang*, 23; Schatzman, 47; George Smoot and Keay Davidson, *Wrinkles in Time*, 80. (It is interesting to note that this position was reiterated to some degree by Pope John Paul II at a cosmology conference held at the Vatican in Rome in 1981. Although not officially endorsing the big bang theory, the Pope did acknowledge the theory, while cautioning cosmologists to remember that science alone will never answer the question of how the universe began. Such questions are best left to metaphysics. LaViolette, 281–82.)
15. Sam Flamsteed, "Crisis in the Cosmos," 68–69.
16. Michael D. Lemonick, *The Light at the Edge of the Universe*, 186; Michael Riordan and David Schramm, *The Shadows of Creation*, 230.
17. Flamsteed, 68.
18. Ken Croswell, "A Milestone in Fornax," 43–44; Flamsteed, 68.
19. William K. Hartmann, *The Cosmic Voyage through Time and Space*, 469. (*Note*: Some astronomical texts distinguish between galactic groups and galactic clusters. The distinction is based on size. Using this distinction, a group would be a collection of galaxies measuring up to 13 million light years across. In contrast, a galactic cluster would measure up to 60 million light years across. See Trinh Xuan Thuan, *The Birth of the Universe: The Big Bang and After*, 55–58.)
20. Flamsteed, 68–69.
21. Flamsteed, 68–69; Hartmann, 294–95.
22. Croswell, 44–45; Flamsteed, 68–69; Hartmann, 294–95.
23. Croswell, 44; Flamsteed, 69.
24. Flamsteed, 69.
25. Ibid.
26. Croswell, 43–44; Flamsteed, 72.
27. Flamsteed, 69–72. (*Note*: NGC stands for New General Catalog. The numerals represent the catalog number assigned to that galaxy. Hartmann, 477.)
28. Flamsteed, 69–72.
29. Croswell, 42–44.
30. Ibid., 44.
31. Ibid., 43–47.
32. Davies, 12–13; Hartmann, 419–21; LaViolette, 256–57.
33. Barrow and Silk, 8–9; Hartmann, 419–21; George Constable, ed., *The Cosmos*, 61–62.
34. Dava Sobel, "Man Stops Universe, Maybe," 20–21.
35. Ibid., 20.
36. Ibid.
37. Ibid., 21.
38. Ibid., 20–21.
39. Ibid., 21.

40. John Boslough, *Masters of Time*, 25.
41. Ibid., 26.
42. Hartmann, 381; Thuan, 55–58.
43. Boslough, 25; Flamsteed, 74.
44. Boslough, 26–27; Riordan and Schramm, 148–51; Thuan, 55–58.
45. Boslough, 26–27; Riordan and Schramm, 148–51.
46. Boslough, 28.
47. Ibid.
48. Ibid., 32–33; Riordan and Schramm, 138–41; Michael Rowan-Robinson, *Ripples in the Cosmos*, 152–53. (Rowan-Robinson gives the length of the Great Wall as 300 million light years.)
49. Boslough, 32–33; Riordan and Schramm, 138–41; Smoot and Davidson, 162–63; Thuan, 58–59.
50. Boslough, 46–47; Corey S. Powell, "The Golden Age of Cosmology," 19.
51. Boslough, 42; Powell, 17; Smoot and Davidson, 234–37.
52. Boslough, 46–47; Riordan and Schramm, 154–58.
53. Boslough, 47; Thuan, 71.
54. Boslough, 46–47; Powell, 17–18.
55. Powell, 17–18; Smoot and Davidson, 272, 286.
56. Marcia Bartusiak, *Through a Universe Darkly*, 281; Powell, 17–18.
57. Hermann Bondi, *Cosmology*, 140.
58. Smoot and Davidson, 68–70.
59. Bondi, 11, 141.
60. Ibid., 141–42.
61. Ibid., 142.
62. Ibid., 143–44.
63. Ibid., 143.
64. Ibid.
65. Ibid., 143–44.
66. Ibid., 144.
67. Ibid., 150–51.
68. Ibid., 144.
69. Stephen Hawking, *A Brief History of Time*, 47–48; Silk, 5–6.
70. F. Hoyle, G. Burbidge, and J. V. Narlikar, "A Quasi-Steady State Cosmological Model with Creation of Matter," 437–38; Ron Cowen, "New Challenge to the Big Bang?"
71. Hoyle, Burbidge, Narlikar, 437–38; Cowen, 236–37.
72. Hoyle, Burbidge, Narlikar, 437, 442–43.
73. Ibid., 437–38; Cowen, 236–37.
74. Hoyle, Burbidge, Narlikar, 437–38, 441: Cowen, 236–37.
75. Hoyle, Burbidge, Narlikar, 441.
76. Ibid., 437, 442–43.
77. Hannes Alfven and Per Carlqvist, "Interstellar Clouds and the Formation of Stars," 487–509. See also Hannes Alfven, *Structure and Evolutionary History of the Solar System* and Carl-Gunne Fälthammar, "Hannes Alfven," 118–19.
78. Eric J. Lerner, *The Big Bang Never Happened*, 39–42; Jeff Kanipe, "Beyond the Big Bang," in *The New Cosmos*, 132–33.
79. Lerner, 218.
80. Hannes Alfven, "Plasma Physics Applied to Cosmology," 28.
81. Knipe, 132–33.
82. Hannes Alfven, *Cosmic Plasma*, 102; Hannes Alfven, "Cosmology and Recent Developments in Plasma Physics," 162; Alfven, "Plasma Physics," 30; (*Note*: The term koinomatter comes from the Greek word "koinos" which means "common." Alfven, *Cosmic Plasma*, 98.)
83. Alan Isaacs, ed. *A Dictionary of Physics*, 15; Sybil Parker, ed., *McGraw-Hill Concise Encyclopedia of Science and Technology*, 126.
84. Alfven, *Cosmic Plasma*, 102–4; Alfven, "Cosmology," 162; Alfven, "Plasma Physics," 31, 33; Lerner, 216.
85. Alfven, *Cosmic Plasma*, 102–4, 138–39; Alfven, "Cosmology," 162; Alfven, "Plasma Physics," 33; Lerner, 217.
86. Alfven, "Plasma Physics," 30; Lerner, 217.
87. Alfven, "Plasma Physics;" 30; Lerner, 217.
88. Lerner, 217–18.
89. Alfven, "Cosmology," 162; Alfven, "Plasma Physics," 30–31. (*Note*: Alfven also makes an extensive argument for the existence of antimatter in his text, *Cosmic Plasma*. See *Cosmic Plasma*, 136–38 for the details.)
90. Alfven, "Plasma Physics," 30–32.
91. Isaacs, 425.
92. James P. Hogan, *The Genesis Machine*, 119.
93. Ibid., 120–21.
94. Harris, *The Creation of the Universe* (videocassette). Note: For an in-depth look at the history, operation, and role of particle accelerators see Leon M. Lederman and David N. Schramm, *From Quarks to the Cosmos: Tools of Discovery*, 82–127. Also, for a sociological study of the day-to-day operation of a particle accelerator see Sharon Traweek, *Beamtimes and Lifetimes*.
95. Robert M. Hazen and James Trefil, *Science Matters: Achieving Scientific Literacy*, 130–31.
96. James Trefil, *From Atoms to Quarks*, 223.
97. Daniel Kelves, *The Physicists*, xvi.
98. Isaac Asimov, *Atom: Journey across the Subatomic Cosmos*, 215–19; Hazen and Trefil, *Science Matters*, 130; Robert H. March, *Physics for Poets*, 240–41.

99. Gordon Kane, *The Particle Garden*, 72, 76; Lederman and Schramm, 85–127.
100. Paul Ginsparg and Sheldon Glashow, "Desperately Seeking Superstrings?" 7. See also: Glashow's Q&A in *Superstrings: A Theory of Everything?*, 180–91. See also John Horgan, *The End of Science*, 63.
101. Ginsparg and Glashow, 7; Horgan, 63.
102. Edward Witten, "Reflections on the Fate of Spacetime," 24–26; see also Witten's extensive Q&A in *Superstrings: A Theory of Everything?* 92–106.
103. Witten, "Reflections," 25–26. Unfortunately, string theory has not proven to be the panacea that many physicists had hoped it would be. In fact, in recent years string theory has given way to a more advanced theory generally referred to as membrane theory or more simply as M-theory. According to the basic tenets of M-theory, subatomic entities should be envisioned neither as particles, without dimension, nor as strings, with length only, but as membranes, displaying both length and width. One advantage of M-theory is that it may be subject to the type of experimental proof to which string theory seems impervious. See Michael J. Duff, "The Theory Formerly Known as Strings," *Scientific American*, February 1998, 64–69.
104. Ginsparg and Glashow, 7–8.
105. Ibid.
106. Steven Weinberg, *Dreams of a Final Theory*, 235. It might also be helpful to compare and contrast the views offered by Weinberg in his Q&A in *Superstrings*, 211–28.
107. Weinberg, 233–35.
108. Ibid., 45.
109. Ibid., 3–18; 230–40; 262–75.
110. Ibid., 45–46. It is also important to note that Weinberg devotes an entire chapter of *Dreams of a Final Theory* to the pros and cons of the building of the Superconducting Super Collider (SSC).
111. Ibid., 46.
112. Ibid., 239; See for example, John Horgan, *The End of Science*.
113. Weinberg, 239.
114. Bryan Appleyard, *Understanding the Present*, 110; David Bohm and B. J. Hiley, *The Undivided Universe*, 321; Weinberg, 13–14.
115. Appleyard, 110.
116. James Trefil and Robert Hazen, *The Sciences*, 314. (*Note:* James Trefil also emphasizes this point in his book *From Atoms to Quarks*, 224–25.)
117. David Lindley, *The End of Physics*, 217.
118. Ibid.
119. Ibid.
120. Ibid.
121. Michio Kaku, *Hyperspace: A Scientific Odyssey through Parallel Universes, Time Warps, and the 10th Dimension*, 189.
122. James Glanz, "Cosmic Motion Revealed," 2156–57.
123. Martin A. Bucher and David N. Spergel, "Inflation in a Low-Density Universe," 65.
124. Glanz, 2157; Craig J. Hogan, Robert P. Kirshner, and Nicolas B. Suntzeff, "Surveying Space-Time with Supernovae," 51.
125. Bucher and Spergel, 65.
126. Lawrence M. Krauss, "Cosmological Antigravity," 55.
127. Ibid.

Unit III   The Mystery of Quantum Physics

Chapter 4
The Emerging Quantum Universe

COMMENTARY: THE QUEST FOR THE HOLY GRAIL

Since the Middle Ages, the legend of the Holy Grail has captivated the imagination of many people. The Holy Grail is at times identified as the cup used by Christ at the Last Supper and at other times as the chalice used by Joseph of Arimathea to catch Christ's blood as Joseph helped cleanse the body after the Crucifixion. The story of the Grail is told in many ways. However, one of the most popular versions relates how the Grail was taken to England by Joseph of Arimathea during the first century A.D. After the death of Joseph, the Grail was lost. Since that time it has been the subject of many legends. The most well-known involves Sir Galahad, the most virtuous knight of King Arthur's Round Table. The Grail is supposed to possess miraculous powers, such as that of healing injuries and making unproductive land fertile again. However, the most amazing power ascribed to the Holy Grail is the ability to generate a complete transformation of the universe, that would solve all problems and cure all the ills of the entire world.[1] Consequently, the Grail has been the subject of many quests, both real and imaginary. Today, physicists have their own version of the quest for the Holy Grail. This quest is generally referred to as the search for the theory of everything. The object of the quest is to create a theory that will allow us to understand how the four fundamental forces of the universe are actually manifestations of a single unified force. Such an understanding may open the door to the ultimate mysteries of the universe and may help us manipulate the universe in ways unimaginable to us today. The quest, thus, has its practical side. Unraveling the theory of everything may help us to harness the power of the fundamental forces in ways undreamed of today. It may allow us to create matter by manipulating the subatomic constituents of that matter. Or it may help us break the hold that gravity has on us, allowing us, in effect, to turn gravity on and off as easily as we turn electric current on and off today. Of course, not all physicists are convinced that we will ever unravel the theory of everything or, if we do, that the discovery will allow us to do the wonderful things that some people have predicted. Even so, we may still wish to pursue the elusive theory of everything because there is also a philosophical side to the search. This is because, in one sense, the search for the theory of everything is a search for an understanding of the very essence of the universe. These ideas will serve as the focus of our study in the next four chapters. We attempt not only to understand the nature of quantum theory but also its philosophical and practical aspects.

CHAPTER OUTCOMES

After reading this chapter, the reader should be able to accomplish the following:

1. Outline the ancient Greek contribution to the development of atomic theory.
2. Discuss the interplay between Newton's science and his theology.
3. Detail the contributions of chemistry to our understanding of atomic structure.
4. State the basis of Maxwell's theory of electromagnetism.
5. Identify the basics of Thomson's plum pudding model of the atom.
6. Pinpoint the essence of Rutherford's planetary model of the atom.
7. Explain Max Planck's contribution to atomic theory.
8. Relate Bohr's understanding of atomic structure.
9. Clarify the nature of de Broglie's wave theory in relation to atomic structure.
10. Recognize the fundamental implications of Schrödinger's equation.

4-1   WHY STUDY QUANTUM THEORY?

A study conducted under the auspices of the National Science Foundation, entitled *The Public Understanding of Science and Technology*, came to some startling conclusions about the state of science literacy in this country at the beginning of the 1990s. The study concludes that, despite the fact that Americans are citizens of a scientific and technological culture, the vast majority of them, 93.1 percent to be exact, are scientifically illiterate. The study also defined scientific literacy to include an understanding of the scientific process, an understanding of scientific terms and concepts, and an understanding of the impact that science and technology has on society.[2]

Our goal in this text has been to improve the nonscientist's understanding of these three areas of knowledge. We now turn our attention to one of the most critically important scientific theories of the modern age. This is the theory of **quantum physics**, the study of the laws of nature as they affect matter and energy at the atomic and subatomic levels. Although our everyday speech is laced with colorful metaphors originating in the jargon of quantum physics, most of us have only a passing awareness of the theory behind quantum physics, and fewer still have an appreciation for its philosophical and practical implications. As Stephen Hawking points out in *A Brief History of Time*, more than fifty years after the formulation of the uncertainty principle, a key concept within the world of quantum theory, the extraordinary consequences of that principle have yet to be recognized by a substantial number of philosophers.[3]

Hawking later points out that the practical applications of quantum theory have also been lost on most nonscientists, despite the fact that many modern devices have been made possible by our understanding of quantum principles. Quantum theory, Hawking notes, "has been an outstandingly successful theory and underlies nearly all of modern science and technology. It governs the behavior of transistors and integrated circuits, which are the essential components of electronic devices such as televisions and computers, and is also the basis of modern chemistry and biology."[4]

The inexperience of the average citizen with quantum physics and the reluctance of many philosophers to struggle with the intricacies of quantum theory may be, if not forgivable, at least understandable. However, what is not as easily understood or forgiven is the outright animosity that sometimes exists between philosophers and physicists. Physicists tend to see the work of the philosophers as largely irrelevant to the pursuit of scientific truth. The scientists point out that philosophers cannot agree even among themselves on the premises from which they launch their complex arguments, and they have made little, if any, progress, despite centuries of debate.[5] Many philosophers, on the other hand, see physicists as unsophisticated mechanics who are remarkably shortsighted in their approach to the mysteries of quantum theory. The philosophers point out that the scientists only tinker with the surface solutions to mechanical problems and do not delve into the philosophical depths of their discoveries.[6]

This tension between philosophers and scientists is made somewhat more ironic because the two fields of knowledge were once united as a single discipline. From the earliest work of the ancient Greek philosophers until the seventeenth-century work of Galileo, Kepler, and Newton, science and philosophy operated as a single discipline, the object of which was to gain a more complete understanding of the universe. In fact, science did not become a completely separate discipline until sometime in the nineteenth century. As Helge Kragh explains in *An Introduction to the Historiography of Science*:

> The word *scientist*, in English, is only 150 years old. Before then the profession of scientists did not really exist, which is reflected in the variety of names given to those concerned with discovering the secrets of nature: savant, natural philosopher, man of science, virtuoso, cultivator of science, and so on. It was not until the middle of the 19th century that it was felt necessary in England, for practical reasons, to find a name for the professional man of science who emerged as a social phenomenon at that time.[7]

As a starting point for our examination of quantum theory, we, therefore, begin with a look at the ancient view of the structure of the universe and trace the gradual development of atomic theory from the Greeks to the advent of quantum physics at the beginning of the twentieth century. We then examine the development of quantum theory and attempt to fit together some of the more puzzling aspects of the theory.

4-2 THE HISTORICAL DEVELOPMENT OF QUANTUM THEORY

When we explore the early ideas about the structure of the universe, we find that science and philosophy are not at odds with one another. Instead, the two disciplines blend and overlap in an interrelated web of crosscurrents that reveal a holistic view of reality. In the wake of discoveries made by Galileo, Kepler, and Newton, among others in the seventeenth century, science began to take a more detached and mechanical view of the universe. As science headed in the direction of developing a mechanical picture of the universe, philosophy was left to deal with deeper issues involving the meaning of existence. However, as we will see, the advent of quantum theory may lead us back to a point where science and philosophy once again work together. Our historical study will begin with an introduction to the Greek view of atomic theory and continue with a look at the developments of atomic theory under Newton, Maxwell, Thomson, and Rutherford. This will lead us to an examination of atomic structure as seen by Niels Bohr and Louis de Broglie.

The Ancient View of Reality

What are the basic components of the universe? This is the deceptively simple question that forms the focus of this chapter and the next. Since the ancient Greeks were largely responsible for the first Western forays into philosophy, it is natural to find their imprint on the search for the ultimate structure of the universe. One of the first Greek philosophers to address this issue was Thales of Miletus (circa 624–546 B.C.), who concluded that all things are somehow made of water.[8] There is little in the record to explain exactly why Thales came to this conclusion, although there is some evidence that he may have borrowed the idea from the Egyptian and the Babylonian civilizations, both of which had cosmologies based on the idea that the Earth floated on water.[9] Still, it is possible to believe that the Earth rests on some sort of infinite expanse of water without necessarily concluding that everything is also made of water. On the other hand, there is evidence that Thales may have based his idea on the fact that all life is dependent on water for sustenance.[10]

Another Milesian, Anaximander (circa 610–546 B.C.), believed that the most basic material of the universe was a primeval substance he called **apeiron**. This conclusion was based on his observation that each of the four most basic elements of the universe (fire, air, water, and earth) are incompatible with one another. Therefore, Anaximander concluded that there must be a fifth substance, the apeiron, that served as the source of the other four.[11] The work of Anaximander later led his student, Anaximenes (circa 585–528 B.C.), to a more sophisticated understanding of the structure of matter. Anaximenes insisted that it was unwise and unnecessary to explain the nature of change by resorting to an unobservable, unverifiable process. He preferred, instead, to deal with observable phenomena.[12] He stated that of the four basic elements, air was the most fundamental and actually made up the other three.[13] Anaximenes chose air as the prime substance because all other substances are somehow a condensation or rarification of air. More importantly, this transformation process from air into other substances could be verified by observation. This is seen, for example, when air condenses into clouds and then into rain.[14] Anaximenes, thus, is at least partially responsible for introducing into Western thought the priority of observation over unfounded conjecture.[15]

There were, of course, dissenters to the idea that some unchanging prime matter served as the source of all existence. Some of this opposition came from Heraclitus (circa 540–480 B.C.), who promoted the idea that the universe was in a constant state of change.[16] It was Heraclitus who said that it is impossible to step into the same river twice.[17] His use of the river analogy was based on the fact that variations in a river are obvious, even to the most casual observer. As the river flows downstream, its condition changes. For instance, the number of fish in the immediate area may vary, the temperature may rise or fall, the silt on the bottom may be churned up, and so on. As a result, after the second, the third, or fourth stepping into the stream, the original river has vanished forever.

According to Heraclitus, things seem steadfast and unfaltering only because they exist in a delicate state of equilibrium between opposing tendencies.[18] One example often attributed to Heraclitus is the image of the tension that is generated within a strung bow. The strung bow seems to be a stable entity. However, the stability is created by the taut string pulling on the arched bow. The entire system is held in balance by that tension. This stability, however, is only a temporary condition, created by the tension between the two opposing forces: the power of the string as it pulls on the bow and the force of the bow trying to straighten itself. If the string would snap, the tension would be relaxed, and the bow would abruptly straighten.[19] Still, according to Heraclitus, this tension between opposing forces is a necessary condition of the universe. Without that tension, the system of the entire world would collapse.[20]

Parmenides (circa 515–430 B.C.) presented a picture of the universe that was quite different from the one proposed by Heraclitus. The centerpiece of the philosophy of Parmenides can be found within his statement "it is."[21] Many scholars have attempted to rephrase this statement in a variety of ways, but few have succeeded in doing a better job than Parmenides himself. What Parmenides seems to mean by this deceptively simple statement is that the universe possesses a unified, infinite, indivisible, motionless, unchanging wholeness that cannot not exist.[22] Or, as one author states, "What is, is, and cannot not-be."[23] According to Parmenides, since what is, is, it cannot experience "any process of becoming or perishing, change or movement."[24] Therefore, everything that exists, exists in the present and cannot perish. Nor is it possible for anything new to be created.[25] The fact that this position defies common sense did not seem to bother Parmenides.

Zeno (circa 490–440 B.C.), Parmenides's most famous pupil, set down a series of arguments designed to defend the position proposed by Parmenides. These arguments, which later came to be known as **Zeno's Paradoxes**, were fashioned to demonstrate that the common-sense view of motion and space could be shown to be just as absurd as the position proposed by Parmenides.[26] One of Zeno's Paradoxes suggests that it is impossible for a runner to cover the distance between two points A and B because before the runner at A can cross the entire distance between A and B, he must cross half the distance to B. However, to reach half the distance between A and B, the runner must cross half the distance to the halfway point, and so on, ad infinitum. Since the runner must cross an infinite number of halfway points, he can never reach the second point.[27]

Of course, in real life runners actually do move from point A to point B, so what was Zeno trying to prove? His goal was not to attack the world as it exists but to show that any counterargument to Parmenides could also be reduced to absurdity.[28] Therefore, the fact that Zeno's Paradoxes and Parmenides's belief in an unchanging universe both appeared to defy common sense bothered neither of these philosophers. The idea of an unchanging universe did, however, bother later philosophers, some of whom attempted to solve the problems presented by the theories of Parmenides. This attempt contributed to the birth of atomism.[29]

The Birth of Atomic Theory

Two philosophers who attempted to reconcile the conflicting views of Heraclitus and Parmenides, as well as to clarify some of the more difficult ideas within Parmenides's philosophy, were Leucippus (circa 460–390 B.C.) and Democritus (circa 460–360 B.C.). Their contribution remains with us today, for what they imagined was a world composed of infinitely small, indivisible particles called atoms [Greek for "that which cannot come apart" or "the uncuttable (bodies)"]. The **atom** became the universal name for the smallest particle of matter.[30] According to Leucippus and Democritus, the atoms were the unchanging part of the universe. What passed for our common-sense notion of change was simply a rearrangement of these atoms.[31] Leucippus and Democritus also believed that the atoms existed in many different sizes and shapes and, although the atoms did not blend with one another (like the ingredients of a cake), they did join physically to create large-scale objects. This joining was facilitated by the various atomic shapes. For instance, atoms with convex sides would fit together with atoms possessing concave sides, gradually building a macroscopic object.[32]

Not everyone in the ancient world accepted the idea that reality consists of tiny, invisible particles. Years before Leucippus and Democritus presented their ideas about the atomic structure of the universe, Anaxagoras (circa 500–428 B.C.) argued that matter in the universe consisted of different substances that could be subdivided indefinitely. According to this view, there is no "smallest" piece of anything, because every piece can be subdivided again, no matter how small. This was known as the **principle of infinite divisibility**.[33] In some quarters, atomic theory was seen as essentially atheistic. This position grew from the ideas of Lucretius (98–55 B.C.), a Roman poet, who noted that not even the power of the Divine could make something from nothing.[34] This view, of course, was not compatible with Christianity, a religion that held that the universe was created by God out of nothing.

The Theory of Transubstantiation

An interesting interplay between philosophy on the one hand, and theology on the other, can be seen in another source of the Christian opposition to atomism. This opposition lies within the doctrine of the Eucharist. According to Christian beliefs, during the celebration of the Eucharist, the priest miraculously transforms bread and wine into the body and the blood of Christ. Although this belief can be traced to the time of St. Paul, by the Middle Ages it had become deeply rooted in Christian doctrine.[35] Atomism, with its focus on the unchanging nature of the atoms, was, therefore, seen as an attack on this belief. After all, if the atoms were unchanging, then the transformation of the bread and the wine into the body and blood would be impossible. Thus, to many medieval churchmen, the science of the Greeks was seen as incompatible with the theology of the Church.[36]

However, this incompatibility was not the final word on the subject. Not to be deterred, the great medieval Christian theologian Thomas Aquinas took another part of Greek science and, with a clever twist, proved that the bread and wine could, indeed, be transformed into the body and the blood of Christ. The principle that Aquinas borrowed was Aristotle's distinction between substance and accidents. The **substance** of an object is its metaphysical "essence," that which makes it what it truly is. The **accidents** are merely its outward properties, and would include, for instance, the object's color, texture, aroma, dimensions, and so on.[37] Aquinas held that the substance of the bread and the wine changed during the miracle of the Eucharist, whereas the properties remained the same. Thus, while the bread and the wine continue to look like bread and wine, they now have the substance of the body and the blood of Christ. This change of the *substance* of the bread and wine into the *substance* of the body and blood of Christ gives the doctrine the name **transubstantiation**.[38]

Aquinas's application of Greek scientific principles to transubstantiation demonstrates the easy interplay that occurred between the scientific ideas of Aristotle and the theological conclusions of Aquinas. Rather than shy away from or ignore the apparent incompatibility of science and theology, Aquinas sought to embrace their commonality. He did not see the two fields of knowledge as in opposition. If some ideas appeared incompatible, others might be embraced that would show the two areas of knowledge to be harmonious.

Church opposition to atomism continued for quite some time. In fact, Galileo's belief in atomism was just one more reason that the church fathers of the seventeenth century opposed his teachings so vigorously.[39]

4-3 THE DEVELOPMENT OF ATOMIC THEORY

The acceptance of atomic theory diminished significantly after Lucretius, but the basic concept never went away. The Scientific Revolution that blossomed in the sixteenth and seventeenth centuries saw atomism once again come to the forefront. Many of the key figures in the Scientific Revolution, including Sir Isaac Newton, held to the basic tenets of the atomic theory. If Newton's ideas gave the Western mind a basic foundation for atomic theory, the chemists, as we shall see here, were to provide a working framework for that theory.

Newtonian Atomic Theory

Sir Isaac Newton (1642–1727) is generally credited as the man responsible for the predictable, deterministic paradigm that is such an integral part of the modern world view. It is undisputed that Newton was an atomist. In fact, the picture that he and his contemporaries had of the atom was not much more advanced than the one suggested by Democritus and Leucippus. Newton believed that the universe at its most fundamental level consisted of hard, indivisible particles that moved in relation to one another within absolute space. He also believed that these particles probably interacted with one another by the power of forces that science did not yet comprehend. Although Newton's mathematical proofs suggested the existence of particles and forces, he could not explain what caused those forces. Nevertheless, Newton did not hesitate to speculate that such forces may reside within a spiritual dimension which inhabits all material bodies.[40] Understandably, the idea that a spirit inhabited all matter gave some scientists cause to criticize this aspect of Newton's work.[41]

The fact that Newton used an incomprehensible spirit to explain the operation of forces gives us another interesting picture of the interplay between science and theology. Although Newton was clearly a scientist in the sense we use that word today, his science was still immersed in the theology of his day. Therefore, as Newton explained his ideas, he had no qualms about referring to God and God's role in the operation of the universe. In fact, Newton may have been unable to separate his belief in God from an understanding of his own theories, as this passage from *Opticks* demonstrates:

> All these things being consider'd, it seems probable to me, that God in the Beginning form'd Matter in solid, massy, hard, impenetrable, moveable Particles, of such Sizes and Figures, and with such other Properties, and in such Proportion to Space, as most conduced to the End for which he form'd them; and that these primitive Particles being Solids, are incomparably harder than any porous Bodies compounded of them; even so very hard, as never to wear or break in pieces; no ordinary Power being able to divide what God himself made one in the first Creation. . . . And therefore, that Nature may be lasting, the Changes of corporeal Things are to be placed only in the various Separations and new Associations and Motions of these permanent Particles; compound Bodies being apt to break, not in the midst of solid Particles, but where those Particles are laid together, and only touch in a few Points.[42]

What is even more interesting about Newton's preoccupation with theology is that it permitted him to see the impermanence of his own work. Newton saw God as the author of natural law, and he believed that God had the power to change natural law at will or to fashion different natural laws for different parts of the universe. The following passage from *Opticks* makes Newton's position on this issue quite clear:

> God is able to create Particles of Matter of several Sizes and Figures, and in several Proportions to Space and perhaps of different Densities and Forces, and thereby to vary the Laws of Nature, and make World of several sorts in several Parts of the Universe. At least, I see nothing of Contradiction in all this.[43]

Despite Newton's belief in God's power to vary natural law from one part of the universe to another, his conclusions about the universe contributed to the deterministic position that many people still hold today in their understanding of the universe. One of the most convincing, early supporters of the deterministic universe was Pierre Simon, Marquis de Laplace, a French mathematician, astronomer, and physicist, who stated that the universe as it exists today is simply the result of the causes which preceded it and is, in turn, the cause of whatever state comes in the future. Moreover, it is possible, at least in principle, to uncover all of the laws of nature so as to be able to determine the present state of the universe and with that knowledge, to predict every event in the future of the universe.[44] The deterministic position presented a comfortable and reliable universe that seemed as solid as rock. However, as physicists and chemists began to probe further into the mysteries of atomic theory, the predictable universe of the determinists began to unravel in the face of its greatest challenge, that of quantum physics.

The Contributions of Chemistry

Despite Newton's attempt to incorporate the forces of nature into atomic theory, just exactly how the atoms "held on to one another" to create the larger structures in the universe was not yet understood. Many people might have been willing to ignore this, however, were it not for discoveries in chemistry that demonstrated that atoms also somehow linked themselves together to form molecules. The man usually credited with this discovery is John Dalton, a British chemist, who suggested in 1808 that many of the mysteries surrounding chemical activity might be solved if this activity were explained in terms of atomism. Dalton said that this chemical activity might involve the interaction of the different atoms within different elements.[45]

These ideas eventually prompted Dimitri Mendeleev, a Russian scientist, to construct a systematic chart detailing the relationships among the elements. The result was, of course, the famous **periodic table of elements**, a fixture that hangs in the front of most high school and college chemistry labs in the world.[46] In addition, the work of the British physicist James Clerk Maxwell owes much of its inspiration to the theory of atomism. Maxwell began by applying his understanding of atomism to the attributes and the behavior of gases. Maxwell succeeded in explaining this behavior mathematically from an atomic point of view. In fact, he eventually used the movement of atoms and molecules to explain how temperature affects solids and liquids, as well as gases.[47]

Despite these advances many scientists remained unconvinced of the existence of atoms. Some positivists, for instance, sought to discredit the theory of atomism because atoms could not be observed directly. According to this view, any phenomenon that could not be observed was not worthy of science. Other scientists saw atomic theory as a convenient way of predicting the results of chemical reactions but insisted that the use of atoms was merely a "convenient fiction." This approach bears an odd similarity to the insistence of some medieval astronomers who, at the time of Copernicus, saw the Polish astronomer's suggestion that the Sun is at the center of the planetary system as a "convenient fiction." In that case it took years of debate, dozens of observations, and the arrest of Galileo to turn the tide in favor of the heliocentric system. In the case of the atomic theory, it remained for the physicists to complete the picture of the atom that we have today.[48]

4-4 THE DEBATE ON ATOMIC STRUCTURE

Even among those scientists who believed in the existence of the atom, there was disagreement about the structure of the atom. Many supporters of atomism believed that the atoms would be indivisible. However, this view gradually faded as more and more evidence was gathered that supported the view that the atom had a definitive structure. One of the most important pieces of evidence came when J. J. Thomson discovered the electron. This discovery led to the two dominant pictures of the atom, the plum pudding model and the solar system or planetary model.[49]

Maxwell's Theory of Electromagnetism

The discovery of the electron is best seen against the background of the work of James Clerk Maxwell. In 1871, Maxwell devised a set of equations that demonstrated that electricity and magnetism are actually two manifestations of the same entity, the **electromagnetic field**.[50] Maxwell's theory of electromagnetism along with the nature of the electromagnetic field will be explored at length in Chapter 6. For now, it is sufficient to see that Maxwell's conclusions concerning the unity of electricity and magnetism were significant for several reasons. Among these reasons is the fact that it led to the discovery that light is an electromagnetic vibration. Thus, Maxwell's equations showed that light is also a manifestation of the electromagnetic field.[51]

Another ramification of the unification of electricity, magnetism, and light was the fact that it helped open the door to one of the enduring mysteries of physics. Maxwell's theory demonstrated that light was simply an electromagnetic wave. If, however, light is a wave, then like a sound wave or a sea wave, it must travel in some sort of medium. It could not simply travel by itself. This difficulty led to Maxwell's introduction of the **luminiferous ether** into his theory to explain the movement of light as a wave. Maxwell's idea of the ether was not entirely original. Others before him had postulated its existence, but no one had yet demonstrated that the ether actually existed. Nevertheless, Maxwell believed that proof for the physical existence of the ether was close at hand.[52] In Maxwell's own words:

To fill all space with a new medium whenever any new phenomenon is to be explained is by no means philosophical, but if the study of two different branches of science has independently suggested the idea of a medium, and if the properties which must be attributed to the medium in order to account for the electromagnetic phenomena are of the same kind as those which we attribute to the luminiferous medium in order to account for the phenomena of light, the evidence for the physical existence of the medium will be considerably strengthened.[53]

Although Maxwell was confident that science was on the threshold of a discovery that would provide the needed physical evidence for the existence of the ether, he was still willing to admit that there were those who did not agree with his point of view. He was also aware that its addition to the world of Newtonian physics introduced a tension that was uncomfortable to many physicists. What he had no way of knowing was that the issue would not be resolved entirely until Einstein proposed his theory of relativity in the twentieth century.[54]

Still, the idea of the luminiferous ether was so enticing that it inspired the imaginations of many writers of popular fiction. One author, Roy Rockwood, writing in 1910, penned a novel entitled *Through Space to Mars*, which relates the story of a journey to the red planet. The space travelers' ship is powered by an engine known as the Etherium motor. The Etherium motor sends "wireless waves" through the aft end of the ship to push the ship forward to its rendezvous with Mars.[55] Like many similar ideas, the luminiferous ether remained in the popular imagination far beyond the time that it was shown not to exist. We look at Einstein's solution to this dilemma in Chapter 6. In the meantime, however, we must return to the mystery of the structure of the atom, the solution to which was helped along by J. J. Thomson's detection of the electron.

Thomson's Plum Pudding Model

Like many physicists at the time, J. J. Thomson, director of the Cavendish Laboratory at Cambridge University, was puzzled by the nature of **cathode rays**, a recently discovered kind of radiation. Some physicists, mostly in Germany, believed that cathode rays were a new kind of radiation that was very similar to light. A competing group of physicists saw cathode rays as particles. Thomson believed the better claim lay in the particle theory because under that theory, the movement of the particles could be predicted. In contrast, according to the wave theory, the cathode rays were supposed to travel through the ether, a substance which had yet to be detected.[56]

In an attempt to solve the problem, Thomson constructed an experiment that allowed him to test how electric and magnetic fields affected the mysterious rays. The way that the electric and magnetic fields affected the rays would allow him to determine the actual nature of the cathode rays. As a result of his experimental work Thomson concluded that cathode rays were negatively charged particles moving at a speed less than that of light.[57] In Thomson's own words:

> As the cathode rays carry a charge of negative electricity, are deflected by an electrostatic force as if they were negatively electrified, and are acted on by a magnetic force in just the way in which this force would act on a negatively electrified body moving along the path of these rays, I can see no escape from the conclusion that they are charges of negative electricity carried by particles of matter.[58]

Thomson preferred to call the newly discovered particles *corpuscles*.[59] However, several years earlier their existence had been predicted by George Johnstone Stoney, an Irish physicist, who had named them *electrons*. This was the name that prevailed.[60] Thomson's discovery led him to the conclusion that the electron was actually part of the structure of the atom and from this intuitive leap came his picture of the atom. This picture is known variously as the **plum pudding model** and the **raisin bread model**. In this model, the electrons are the negatively charged *plums* or the *raisins* that float in the positively charged pudding or bread.[61]

Rutherford's Solar System Model

There was only one problem with the plum pudding model of the atom. It did not work. This problem was not immediately obvious, however. In fact, there was an enormous amount of evidence in support of the Thomson model. Still, it did not explain everything, and so many people continued to argue about the basic structure of the atom. The man responsible for the ultimate demise of the plum pudding model was Ernest Rutherford, who, during his long and distinguished career, was associated with McGill University in Montreal, the University of Manchester in England, and the Cavendish Laboratory at Cambridge University. Ironically, Rutherford did not set out to disprove the plum pudding model. On the contrary, he actually constructed an experiment that he believed would substantiate Thomson's claims.[62]

Rutherford's road toward the destruction of the raisin bread atom and the construction of his own atomic model actually began when he discovered the nature of alpha particles. In 1906, while at McGill University, Rutherford established that **alpha particles** were positively charged particles that were many times more massive than electrons.[63] Immediately, Rutherford realized that he could construct an experiment that used the alpha particles to probe the structure of the atom. As the first step in the experiment, Rutherford placed a radiation source in a lead box and allowed a flow of alpha particles to collide with a piece of gold foil. The flow of alpha particles through the gold foil was, in turn, recorded on a fluorescent screen.[64]

• 116 • Chapter 4 The Emerging Quantum Universe

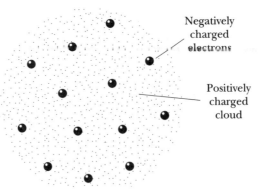

Figure 4.1 Thomson's Plum Pudding Model of the Atom.
After his discovery of the electron late in the nineteenth century, Joseph John Thomson, a British physicist and director of the Cavendish Laboratory at Cambridge University, visualized the atom as a group of negative electrons embedded in a positive cloud. The name of this model was suggested by the fact that the electrons resemble the plums in such a pudding.

Rutherford expected the alpha particles to fly through the gold foil, like a skyrocket through a snowstorm. This did, in fact, occur most of the time. However, for about every 1000 alpha particles, one ricocheted at a significant angle and for about every 10,000 particles, one rebounded backward like a baseball off a backstop.[65] This puzzled Rutherford, who, after further experimentation, decided that these results could only be explained if some sort of massive structure hovered within the atom. The model that Rutherford eventually decided upon was the solar system model. This is the model that most people recognize as the atom. At the center of the solar system model is the nucleus of the atom. The nucleus is many times more massive than an electron. The electrons, in turn, exist outside the nucleus in a spheroid space surrounding the massive nucleus. Rutherford could not explain exactly where the electrons were in his model. Nor could he explain how the structure of the atom remained intact without collapsing in on itself. However, Rutherford did determine that the nucleus almost certainly carried a positive charge because it was this positive charge that deflected some of the positive alpha particles. Eventually, this positively charged particle in the nucleus came to be known as the proton.[66] Rutherford also suggested that the nucleus might contain another uncharged particle. This particle, the neutron, was discovered by the British physicist, James Chadwick in 1932.[67]

Constructivism vs. Positivism Revisited

Before moving on to quantum theory, it will be enlightening to look momentarily at some of the social and cultural implications of Rutherford's discovery. As we saw in Chapter 1, there is a growing movement in society today called constructivism, which argues that scientists are not as objective as they claim to be. Instead, the argument goes, scientists, like everyone else, are subject to cultural and social forces that shape the conclusions they reach as they interpret data gathered during experiments. Science is, therefore, charged with the same kind of subjectivity that is characteristic of ethics, literature, and art. Constructivism seeks to displace positivism, the traditional approach to science. Positivism holds that natural law exists as an objective set of rules that can be discovered by the proper application of the scientific method, despite the cultural and social prejudice harbored by a particular experimenter.

At first look, the story of Rutherford's aborted attempt to confirm Thomson's plum pudding model would seem to support positivism. Recall that Rutherford actually approached the problem of atomic structure with the preconceived notion that Thomson's plum pudding model was an accurate representation of the atom.[68] Despite this culturally induced bias, Rutherford and his colleagues could not ignore the experimental results when the atom deflected a proper number of alpha particles. Rutherford could have ignored the results, or he could have passed them off as coincidence or experimental error. He did not. Instead he followed the results to their natural conclusions, which led to the demise of Thomson's model of the atom.

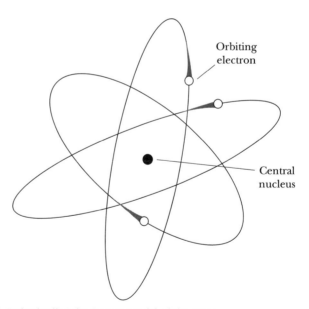

Figure 4.2 Rutherford's Solar System Model of the Atom.

After Ernest Rutherford of McGill University in Montreal, the University of Manchester in England, and the Cavendish Laboratory at Cambridge University, observed alpha particles deflected by some sort of structure within the atom, he devised the solar system model of the atom. In the solar system model, the positive charge was no longer located within a cloud. Instead, it was visualized as a centrally located mass, much like the Sun at the center of the solar system. In this model the electrons were analogous to the planets.

This episode seems to support positivism. A closer look at Rutherford's work, however, might alter this conclusion. Consider, for example, that, when it came to his own model of the atom, Rutherford was not quite as objective as he might have been in his evaluation. Apparently, Rutherford knew that his model did not explain where the electrons were, how they moved, and how the structure of the atom held itself together. Yet, he made no attempt to solve the puzzle. He was, after all, an experimentalist, not a theoretician. His self-image as experimentalist clearly shaped his point of view and prevented him from proceeding any further.[69] Perhaps even more revealing is the fact that, when Niels Bohr suggested a solution to the problems that plagued the solar system model, Rutherford was at first reluctant to accept Bohr's solution.[70]

Rutherford's position is made even more ironic since he had received a similar rebuke at the hands of his mentor, Thomson, when Rutherford presented the solar system alternative to Thomson's plum pudding model.[71] Clearly, these biases were shaped by each man's personality and by his role as an experimentalist. Just how much of this bias was culturally induced is impossible to say. Nevertheless, it does point out that outside influences can shape a scientist's conclusions about experimentally produced results. Moreover, these conclusions are not necessarily objective, even if they are accurate. Perhaps the most unequivocal observation we can make about this episode is that, even if a scientist's interpretation of data is correct, it does not necessarily prove that the interpretation was free of cultural, social, and personal bias.

4-5 THE ADVENT OF QUANTUM THEORY

Like Thomson's atom before it, Rutherford's atom had its problems, chief among which was the instability of the structure proposed by the British physicist.[72] As the electrons orbited the nucleus of the atom in the planetary model, those electrons lose energy by emitting light.[73] As the electrons lose energy, they should spiral downward and crash into the nucleus, much like a tether ball, once hit, would lose energy, wind down, and spiral into the central pole. Since most atoms remain intact and the universe has not collapsed into itself, the self-destructive dance of the planetary atom does not take place, despite these predictions. The problem was to explain this stability within what should be an unstable atom.[74] The solution was offered by Niels Bohr. When Bohr went to work with Rutherford in 1912 at the University of Manchester, he forged a lasting partnership that was destined to become one of the most legendary in the history of Western science and that would eventually lead to a solution to the instability of the planetary atomic model. Ironically, however, this solution came from an unexpected source, the work of a University of Berlin physicist named Max Planck.[75]

Max Planck and the Birth of the Quanta

Planck was working on the problem of black body radiation when he discovered something that would change the world of physics forever. **Black body radiation** is the name given to the electromagnetic radiation released as an object heats up. The wavelength of the energy released depends upon the temperature of the object; the hotter the object, the shorter the wavelength. At lower temperatures, bodies radiate infrared energy. Infrared radiation can be felt but not seen. For example, place your hand over a burner on an electric stove that has just been turned on, and you will soon feel this infrared radiation. As the temperature of the object increases, the object will begin to glow. Thus, as the temperature of the burner on the stove increases, the burner will eventually glow red. This is because the wavelengths of most of the energy released at that point have passed into the visible spectrum. If the temperature of the object is permitted to increase, the object's glow will move to orange, to yellow, to white, and to blue, and into the invisible range of ultraviolet light, X-rays, and gamma rays.[76]

The problem was that the laws of classical physics, according to Maxwell's equations, predicted that hot objects should radiate only slightly at the longer wavelengths, corresponding to the infrared side of the spectrum but very strongly at the shorter wavelengths, corresponding to the ultraviolet side of the spectrum. Theoretically, then, what should happen when an object heats up is a sudden intense blast of ultraviolet radiation. This is not, of course, what actually happens. Instead, as we noted above, a gradual rise in the object's temperature corresponds to a gradual change in wavelengths. This is what we observe as the slowly heating object changes from infrared radiation, to a red glow, to an orange glow, and so on, as the temperature rises. A hot object that has a relatively steady temperature will radiate most of its energy at the same wavelength as long as the temperature remains relatively constant. The Sun, for example, which maintains a relatively steady temperature of 5800°C, radiates yellow light. This is because most, but not all, of its energy is being released at the visible end of the spectrum, specifically at the level of yellow light.[77]

The question that remained unanswered was this: Why didn't the energy radiated by hot bodies correlate with Maxwell's equations? It remained for Max Planck to solve the riddle. He did so in 1900 when he proposed a new picture of energy. Maxwell's equations were postulated on the assumption that energy traveled in continuous waves. In fact, the idea that energy moved as a wave had provided a fairly accurate picture of energy for many years, at least until the problem of black body radiation appeared. In what Planck himself described as an act of desperation, he proposed that energy was discharged, not as a continuous wave but in individualized parcels. He labeled these individualized parcels *quanta* and, as a result, unwittingly gave a name to one of the most revolutionary concepts in modern science, quantum physics.[78]

Interestingly enough, even Planck resisted the idea that energy actually traveled in these packets. Instead, he proposed that energy was emitted and absorbed in discrete amounts. However, after being emitted and before being absorbed, the energy would travel through space as a wave.[79]

Planck, however, would be forced to reconsider the quantized view of energy after the publication of Albert Einstein's paper explaining the photoelectric effect.[80] The **photoelectric effect**, another of the perplexing mysteries of nineteenth-century physics, occurs when light hits metal, and deflects electrons from the surface of that metal. According to Maxwell, the brighter the light hitting the metal, the faster the speed of the electrons ejected from the metal. Once again, as with black body radiation, Maxwell's equations did not account for what really happened when light hit metal. Despite Maxwell's equations, the speed of the electrons ejected from the metal seemed to be completely unrelated to the brightness of the light. Some forms of light, no matter how bright, will not deflect a single electron. For instance, red light will not eject electrons no matter how intense the brightness of that light. Instead, it appeared that the speed of the ejected electrons depended on the wavelength of the light shined on that metal. The shorter the wavelength, the greater the ejection speed of the electrons.[81]

Einstein decided that Planck's notion that energy comes in packets could be applied to all energy-related problems, not just the problem with black body radiation. Einstein proposed that the photoelectric effect could be explained if we assumed that light came in packets rather than waves. The wavelength of light actually corresponds to the energy contained in that packet of light. The shorter the wavelength, the more energy in the packet, and the more likely it would be that the beam of light would eject electrons.[82] These individual packets of light were later labeled *photons*.[83] Einstein's theory of the photoelectric effect was confirmed by experiments conducted by the American physicist, Robert Millikan at the University of Chicago in 1915.[84]

Bohr's Solution to the Unstable Atom

All this is well and good, but how does it relate to the solar system model of the atom? Bohr speculated that an electron in the solar system atom does not emit energy when it occupies a particular orbital. Instead, the electrons exist only in certain stationary quantized states within the atomic structure.[85] When the electron moves from one quantized state to another, it either discharges or absorbs energy. On the other hand, when an electron occupies a particular orbital it neither emits nor absorbs energy. This would give the solar system atom the required stability that would preserve its validity as an accurate depiction of atomic structure. In this picture, the electron would act less like a tether ball on a pole and more like a bowling ball on a staircase. It takes energy to lift the bowling ball to a higher stair. This corresponds to the absorption of energy by the electron as it moves to a quantized state of higher energy. Once it reaches

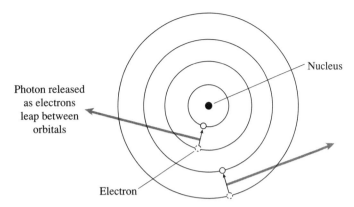

Figure 4.3 Bohr's Quantum Leap Model of the Atom.
To solve the instability inherent in Rutherford's solar system model of the atom, the Danish physicist Niels Bohr suggested that the electrons do not orbit the nucleus of an atom like the planets orbit the Sun in the solar system. Instead, applying the German physicist Max Planck's revolutionary ideas concerning the quantum nature of energy, Bohr suggested that the electrons could exist only in certain quantized states within the atomic structure. When electrons moved from one quantized state to another they either discharged or absorbed energy.

that level, the electron occupies that orbital until it absorbs or releases energy again. This position at a particular quantized state corresponds to the bowling ball at *rest* on a stair. Similarly, an electron that moves from a higher to a lower quantized state releases energy, like a falling bowling ball releases energy. Again, once it reaches that level, the electron (or the bowling ball) stays put, until it absorbs or releases energy again.[86]

The really interesting thing about this phenomena is that the electrons can exist only at certain predetermined energy levels. When an electron emits or absorbs energy, it simple stops existing in one quantized state and suddenly begins existing in another quantized state. The electron makes this *quantum leap* without crossing any space and without passing gradually between energy levels. Bohr's notion of the quantized energy levels is difficult to visualize. However, it is safe to say that these quantized energy levels are not localized in space like the orbits of the planets around the Sun. Nor is the transition from one quantized state to another a spatial phenomenon.[87]

The Structure of de Broglie's Atom

Bohr's atomic model had addressed the problem of atomic instability, and his theory had managed rather well in that regard. However, despite the advantages of his theory, it did not solve all of the mysteries. One mystery that remained untouched was how to apply Bohr's image of atomic structure to all atoms. The quantized states worked well for simple atoms. However, they did not properly account for more complex atoms that consist of numerous electrons because electrons interact with one another almost as much as they do with the nucleus.[88] Bohr's atomic theory also did not explain why the electrons do not emit energy while they occupy a particular orbital.[89] The solution to this mystery was suggested by a Frenchman, Louis Victor de Broglie, who addressed the issue in a doctoral dissertation that he presented to the faculty of the University of Paris in 1924.[90]

De Broglie's Wave Solution. De Broglie's solution to the problem posed by Bohr's atomic structure was, if nothing else, highly original. De Broglie postulated that the variations in energy levels of the electrons might correspond to changing wave patterns. The electrons might, therefore, act like waves rather than as particles. However, de Broglie was not picturing a propagating wave. Instead, he envisioned a standing or a stationary wave. Sea waves and sound waves are examples of propagating waves. **Propagating waves** travel through a medium such as water or air. The particles constituting the medium (water or air) move up and down (or right and left), crossing a midline. The highest point above the midline is called the crest, and the lowest point below the midline is called the trough. The distance between the midline and a crest or a trough is called the **amplitude** and the number of crests within a time unit to move by a specified point is designated as the wave's **frequency**. The wavelength of a wave is measured as the horizontal distance from one crest to the next. The wave itself propagates through the medium, whereas individual particles within the medium do not. To visualize this phenomenon, picture the rise and fall of a leaf in a pond.[91]

Standing waves are very much like propagating waves, but they remain stationary instead of moving. The vibrating strings of a violin or a guitar are examples of standing or stationary waves. The opposite tips of the string are fastened, while the central part of the string vibrates up and down. The immovable points at each end of the string are called **nodes**. It is possible to change the frequency of the standing wave by increasing the number of nodes. Thus, adding a node at the center of the string will change the frequency of the vibrations. The string moves up and down on each side of the center point. The resulting sound is an octave higher than the sound produced when only the two ends of the string are made immovable. The similarity between this movement of the stationary strings and Bohr's theory was evident to de Broglie.[92]

De Broglie recognized that electrons might act just like these stationary waves. Naturally, the waves would have to be circular, but as de Broglie explored further, he found that this made no difference to the analogy. His calculations eventually demonstrated that the low quantized energy states in Bohr's atom corresponded to the low frequencies, and the high quantized

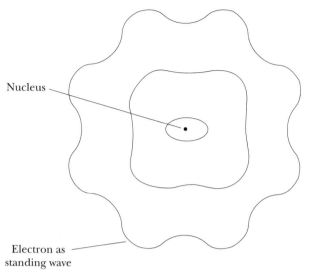

Figure 4.4 De Broglie's Standing Wave Model of the Atom.
Louis Victor de Broglie of the University of Paris suggested that the variations in energy levels of electrons might correspond to changing wave patterns. The electrons therefore, might act like waves rather than particles.

energy states acted just like the high wave frequencies.[93] Many people were grateful to de Broglie for his picture of the atom, primarily because it appeared to eliminate the disconcerting discontinuity of Bohr's model.

However, it was soon realized that both Bohr's and de Broglie's pictures were incomplete. One of the chief problems with the theory was that de Broglie could not explain how the electron, which up until this point had been assumed to be a particle, was now explained in terms of wave activity. How could the electron be both a particle and a wave? Wave–particle duality had been a familiar problem with light. Now, however, the problem had spread to the electron. This implied that wave–particle duality was a quality shared by matter, as well as the electromagnetic field. It was left to the Austrian physicist Erwin Schrödinger to provide a mathematical approach to this problem.[94]

Schrödinger's Mathematical Refinement

Schrödinger's equation is considered by many experts to be among the most critically important in the history of science, ranking with the work of Maxwell and Einstein. In fact, some experts claim that the Schrödinger equation replaces Newton's equations of motion. Others, however, are not as sure.[95] Nevertheless, it is certain that Schrödinger's equation was the final precipitating event that launched the advent of quantum mechanics. Explained simply, **Schrödinger's equation** governs "the behavior of a quantity called the 'wave function.'"[96] In his book, *The Particle Garden*, physicist Gordon Kane defines it as "the equation from quantum theory that tells how to calculate the effects of the forces on particles. It is the equivalent of Newton's second law."[97] One crucial feature of Schrödinger's equation is that it clearly supports the wave interpretation of matter.[98] In a sense it is the mathematical support needed for de Broglie's wave model of the electron within the structure of the atom. According to Heinz Pagels of the New York Academy of Sciences and Rockefeller University, one of the initial explanations for the wave model was presented by Schrödinger himself, who no longer pictured electrons as particles. Instead, he graphically described electrons as matter waves, and he insisted that they operated very much like water waves. In fact, Schrödinger even went so far as to argue that the wave phenomenon was not limited to electrons. Rather, he declared that all quantum entities were small waves, and that all existence was a part of an all-encompassing wave condition.[99] Despite Schrödinger's confidence, this did not end the controversy. As we shall see in great detail in the next chapter, the debate was only beginning.

CONCLUSION

The history of quantum theory, up to this point, has been played out within the context of classical physics. Physicists like Maxwell, Thomson, Rutherford, and Planck were educated according to classical precepts, all of which worked quite well on the macroscopic level. However, the deeper these men probed into the subatomic universe, the stranger that universe seemed to be. The world of classical physics was a comfortable, secure, predictable world. All of this was about to change. The paradigm of classical physics was about to give way to the new quantum paradigm, and the world has not been the same since.

Review

4-1. Although our everyday speech is laced with colorful metaphors originating in the jargon of quantum physics, most of us have only a passing awareness of the theory behind quantum physics, and fewer still have an appreciation for its philosophical and practical implications. The inexperience of the average citizen with quantum physics and the reluctance of many philosophers to struggle with intricacies of quantum theory may be, if not forgivable, at least understandable. However, what is not as easily understood or forgiven is the outright animosity that sometimes exists between philosophers and physicists. Physicists tend to see the work of the philosophers as largely irrelevant to the pursuit of scientific truth. Many philosophers, on the other hand, see physicists as unsophisticated mechanics who are remarkably shortsighted in their approach to the mysteries of quantum theory. This tension between philosophers and scientists is made somewhat more ironic because the two fields of knowledge were once united as a single discipline. As a starting point for our examination of quantum theory, we, therefore, begin with a look at the ancient view of the structure of the universe and trace the gradual development of atomic theory from the Greeks to the advent of quantum physics at the beginning of the twentieth century.

4-2. One of the first Greek philosophers to address the question of the structure of the universe was Thales of Miletus, who concluded that all things are somehow made of water. Another Milesian, Anaximander, believed that the most basic material of the universe was a primeval substance he called *apeiron*. The work of Anaximander later led his student, Anaximenes, to a more sophisticated understanding of the structure of matter. This idea stated that of the four basic elements air was the most fundamental, and it actually made up the other three. Heraclitus promoted the idea that the universe is in a constant state of change. Things seem steadfast and unfaltering only because they exist in a delicate state of equilibrium. Parmenides took the opposite point of view. The universe according to Parmenides exists as a unified whole, persisting in a timeless, changeless state of eternal permanence. Zeno, Parmenides's most famous pupil, set down a series of paradoxes designed to defend the position proposed by Parmenides. Two philosophers who attempted to reconcile the conflicting views of Heraclitus and Parmenides, as well as to iron out the illogical points within Parmenides's philosophy, were Democritus and Leucippus. Their contribution remains with us today, for what they imagined was a world composed of infinitely small, indivisible particles that they called *atoms*. The atom became the universal name for the smallest indivisible particle of matter.

4-3. Newton believed that at its most fundamental level, the universe consisted of hard, indivisible particles that moved in relation to one another within absolute space and that probably interacted with one another by the power of forces that science did not yet comprehend. In fact, the concept of the forces was seen by some of Newton's peers as an unnecessary addition to his theory, an addition that smacked of spiritualism. This criticism creates another interesting picture of the interplay between science and philosophy. Newton's scientific conclusions clearly contributed to the deterministic position that many people still hold today in their understanding of the universe. Newton developed a precise and consistent set of physical laws that led many people to the conclusion that the universe is stable, orderly, and systematic. Discoveries in chemistry demonstrated that atoms linked together to form molecules. The man credited with this discovery is John Dalton, a British chemist, who suggested that many of the mysteries surrounding chemical activity might be solved if this activity were explained in terms of atomism. Dalton suggested that this chemical activity might involve the interaction of the different atoms within different elements. Despite these advances many scientists remained unconvinced of the existence of atoms.

4-4. James Clerk Maxwell devised a set of equations that demonstrated that electricity and magnetism are actually two manifestations of the same entity, the electromagnetic entity. This revelation was significant because it led Maxwell to the conclusion that light is actually a manifestation of the electromagnetic field. The unification of electricity, magnetism, and light also led Maxwell to introduce his ideas of the luminiferous ether. Thomson devised a number of experiments in the 1890s that demonstrated that cathode rays were actually streams of negatively charged particles, which were eventually named electrons. Thomson's discovery led him to the conclusion that the electron was part of the structure of the atom, and from this intuitive leap came his picture of the atom. This picture is known variously as the plum pudding model or the raisin bread model. In this model, the electrons are the negatively charged "plums" or "raisins" that float in the positively charged pudding or bread. Rutherford later used alpha particles to explore the structure of the atom. To confirm the plum pudding model, Rutherford constructed an experiment in which he allowed a flow of alpha particles to collide with a thin piece of gold. He expected the alpha particles to fly through the metal. Some of the alpha particles, however, rebounded backward. This puzzled Rutherford, who decided that these results could only be explained if some sort of massive structure hovered within the atom. The model that Rutherford eventually decided upon was the solar system model.

4-5. Like Thomson's atom before it, Rutherford's atom had its problems. Chief among these was the instability of the structure proposed by the British physicist. The solution was offered by Niels Bohr, who suggested that the electrons could exist only in certain quantized states within the atomic structure. When the electron moves from one quantized state to another, it either discharges or absorbs energy. When an electron is in a particular orbital it neither emits nor absorbs energy. This gives the solar system atom the required stability that would preserve its validity as an accurate depiction of atomic structure. The electrons can exist only at certain predetermined energy levels. When an electron emits or absorbs energy, it simply stops existing in one quantized state and suddenly begins existing in another such state. The electron makes this "quantum leap" without crossing any space and without passing gradually between energy levels. Bohr's notion of the quantized energy levels is difficult to visualize. Despite the advantages of Bohr's theory, it did not solve all of the mysteries. The solution to these mysteries was suggested by a Frenchman, Louis Victor de Broglie, who postulated that the variations in energy levels of the electrons might correspond to changing wave patterns. Schrödinger's equation supports this wave interpretation of matter. It is the mathematical support needed for de Broglie's wave model of the electron.

Understanding Key Terms

accidents	luminiferous ether	proton
alpha particles	neutron	quantum
amplitude	node	quantum physics
apeiron	periodic table of elements	raisin bread model
atom	photoelectric effect	Schrödinger's equation
black body radiation	photons	solar system model (planetary model)
cathode rays	plum pudding model	standing waves
correspondence	principle of infinite divisibility	substance
electromagnetic field		transubstantiation
electron	propagating waves	Zeno's paradoxes
frequency		

Review Questions

1. What did the ancient Greeks contribute to the development of atomic theory?
2. How do Newton's science and his theology interact?
3. What were the contributions of chemistry in the development of the modern concept of atomic structure?
4. What is Maxwell's theory of electromagnetism?
5. What are the basics of Thomson's plum pudding model of the atom?
6. What is the essence of Rutherford's solar system model of the atom?
7. What did Max Planck contribute to atomic theory?
8. How did Bohr reconfigure Rutherford's understanding of atomic structure?
9. What is the nature of de Broglie's wave theory in relation to atomic structure?
10. What are the fundamental implications of Schrödinger's equation?

Discussion Questions

1. Early in this chapter we discussed the tension that exists between philosophers and physicists. As we noted then, the physicists see the work of the philosophers as irrelevant to the pursuit of science. The scientists point out that philosophers ask unanswerable questions; they cannot agree even among themselves on the premises from which they launch their complex arguments, and they have made little, if any, progress, despite centuries of debate. The philosophers see physicists as

unsophisticated mechanics who are shortsighted in their approach to the mysteries of quantum theory. The philosophers point out that the scientists only tinker with the surface solutions to mechanical problems and do not delve into the philosophical depths of their discoveries. Imagine that you have the responsibility of resolving the differences between the philosophers and the physicists. Present an argument that highlights the strengths of both groups and that demonstrates how each group can learn from the other.

2. As noted in this chapter, although Sir Isaac Newton was a scientist in the modern sense of the term, his science was still affected by theology. As Newton explained, his ideas often incorporated direct references to God and God's role in the operation of the universe. Do you believe that scientists today should follow Newton's example? Why or why not?

3. Sir Isaac Newton also saw God as the author of natural law. Moreover, he believed that God had the power to change natural law at will or to fashion different laws of nature for different parts of the universe. Do you tend to support the view that natural law may vary from one part of the universe to another? What effect would such a position have on the belief that the universe is a predictable, deterministic place? Explain your answers to both questions.

4. Do you believe that science is shaped by social or cultural forces? What do you think of the experience of Rutherford as recounted in this chapter? Does Rutherford's discovery that the plum pudding model was an inaccurate representation of atomic structure support the positivist or the constructivist point of view? Explain each of your answers.

5. One of the chief problems with Louis de Broglie's wave model of the electron is that it fails to explain how the electron, which had been assumed to be a particle, could be explained in terms of wave activity. How can the electron be both a particle and a wave? Wave-particle duality is also a problem with light, which acts as both a particle and a wave. What does this suggest to you about the nature of the quantum universe? Explain.

ANALYZING *STAR TREK*

Background The following episode from *Star Trek: The Next Generation* reflects some of the issues that are presented in this chapter. The episode has been carefully chosen to represent several of the most interesting aspects of the chapter. When answering the questions at the end of the episode, you should express your opinions as clearly and openly as possible. You may also want to discuss your answers with others and compare and contrast those answers. Above all, you should be less concerned with the "right" answer and more with explaining your position as thoroughly as possible.

Viewing Assignment— *Star Trek: The Next Generation,* "Who Watches the Watchers?"

In this episode the crew of the Starship Enterprise must come to the assistance of a science station on Mintaka III, a planet inhabited by a race of primitive people in the Bronze Age of development. The science station located on Mintaka III is having trouble with a reactor, which, among other things, powers the holographic camouflage for the study station. When the reactor shorts out, the camouflage vanishes and one of the native Mintakans is witness to the event. The Mintakan is startled when he is discovered by crew members and injures himself in a fall. The injured Mintakan is beamed back to the Enterprise and sent to sick bay where, at one point, he sees Captain Picard. Under the circumstances, the injured Mintakan assumes that Picard is an Overseer, that is, a representative of their God. When the Mintakan returns to his people, he begins to revive the religion of the Overseers based upon his "supernatural" experience. This troubles Picard, who believes that the revival of the religion will plunge the Mintakan civilization into a new era of superstition and ignorance. The head anthropologist at the science station suggests that Picard should return to the surface and give the Mintakans a set of ethical rules. The anthropologist argues that without ethical rules as a guide, the religious revival could become a spark for holy wars, an inquisition, and so on. Picard, however, is constrained by the Prime Directive, which forbids any interference with a culture such as the one on Mintakan III. Picard resolves the dilemma by beaming the female leader of the group to the Enterprise and convincing her that he is not a god but is merely another living being very similar to her. What convinces the woman that Picard is telling the truth is the death of one of the anthropologists from the science study station. She agrees to beam back to the planet to convince her people that Picard is not a god. However, the people refuse to believe her or Picard and demand that he perform a miracle by resurrecting one of their dead. The husband of the dead woman tries to prove Picard is one of the Overseers by firing an arrow at him from point blank range. Picard is, of course, injured, and this is what finally convinces the Mintakans of his mortality. As you watch the video, ask yourself whether the experience of these primitives might not be similar to the interaction between science and society discussed in this chapter.

Thoughts, Notions, and Speculations

1. Dr. Barron, the anthropologist, suggests that Captain Picard must continue the masquerade as a god and give the Mintakans a set of "divinely" inspired rules to live by. How might the scientific development of the Mintakans be helped or hindered by the development of a religion based on the revelations of Picard? As you answer this question, consider Sir Isaac Newton's sincere devotion to God. Did this devotion help or hinder his scientific work? Explain.
2. Instead of giving the Mintakans a set of religious commandments, Picard elects to bring one of their leaders on board the Enterprise to

convince her that he is not a god. The Mintakan leader who visits the Enterprise is convinced that the crew is divine until she witnesses the death of one of the anthropologists. How is her change of faith in the Overseers in the face of evidence similar to Rutherford's change of faith in Thomson's model of the atom in the face of his experimental work? Explain.

3. The Mintakan leader who has visited the Enterprise returns to her people with the idea that she will convince them of the truth of their religion. Imagine that you are guiding her in this effort and try to convince her that she will succeed if she takes the same tactic that Thomas Aquinas took when he attempted to reconcile his Christian faith with the science of Aristotle.

4. The injured Mintakan's faith in Picard's divinity is based upon personal experience. He believes that he was taken into the domain of the Overseers and saved by their miraculous intervention. In fact, he was taken on board the Enterprise and saved by a physician. Yet to him it is the same thing. In this chapter, we saw how the work of Planck, Einstein, Bohr, de Broglie, and Schrödinger is based on mathematics rather than on direct experience. Does this tell us that mathematics is more accurate than personal experience? Why or why not?

5. Although James Clerk Maxwell had no physical evidence for the existence of the luminiferous ether, he, nevertheless, had faith that such a substance did, in fact, exist. How is Maxwell's faith in the luminiferous ether similar to the Mintakan's faith in the Overseers? How is Maxwell's faith different from the Mintakans' faith in the Overseers? Explain. Also, as you answer, remember that Maxwell's faith was misplaced and he was later shown to be incorrect in his faith in the existence of the ether.

NOTES

1. John Matthews, *The Arthurian Tradition*, 65–67. Note: The era designations B.C.E. (before the common era) and C.E. (of the common era) are used throughout this book. The era designation C.E. is used only when there is a possibility of confusion.
2. Jon D. Miller, *The Public Understanding of Science and Technology in the United States, 1990*, 14–15.
3. Stephen Hawking, *A Brief History of Time*, 55.
4. Ibid., 56.
5. Michael Redhead, *From Physics to Metaphysics*, 1–2.
6. Ibid., 2. (In his book, *The Introspective Engineer*, Samuel C. Florman explores this conflict between the humanities and science from a different angle. In Chapter 6, "Image and Reflection," he examines the tension which exists between the artist and the engineer.)
7. Helge Kragh, *An Introduction to the Historiography of Science*, 25.
8. Aristotle, *Metaphysics*, 10; Aristotle, *Physics*, 4, 10; Jonathan Barnes, *The Presocratic Philosophers*, 1: 9–10; W. K. C. Guthrie, *A History of Greek Philosophy*, 1: 54–56; G. S. Kirk, J. E. Raven, and M. Schofield, *The Presocratic Philosophers*, 89–90.
9. Guthrie, 1: 58–60; Kirk, Raven, Schofield, 92–93.
10. Barnes, 1: 11; Guthrie, 1: 61–62; Kirk, Raven, Schofield, 94.
11. Aristotle, *Metaphysics*, 22; Barnes, 1: 29–30; Diane Collinson, *Fifty Major Philosophers*, 5–6; Guthrie, 1: 114, 279; Kirk, Raven, Schofield, 111–12.

12. Guthrie, 1: 120–21.
13. Aristotle, *Metaphysics*, 10; Aristotle, *Physics*, 4, 10; Barnes, 1: 43–44; Guthrie, 1: 116; Kirk, Raven, Schofield, 144–45.
14. Barnes, 1: 43–44; Guthrie, 1:119–121; Kirk, Raven, Schofield, 145–46.
15. Barnes, 1: 46–52; Collinson, 7; Dion Scott-Kakures, et al., *History of Philosophy*, 3: Guthrie, 1: 124–26.
16. Aristotle, *Metaphysics*, 19; E. L. Allen, *From Plato to Nietzsche*, 12; Barnes, 1: 65–69, Guthrie, 5; Scott-Kakures, 6.
17. Allen, 12; Barnes, 1: 66; Guthrie, 1: 441–42; Kirk, Raven, Schofield, 194, 197.
18. Guthrie, 440; Kirk, Raven, Schofield, 193; Scott-Kakures, 6.
19. Guthrie, 1: 439–40; Kirk, Raven, Schofield, 193; Scott-Kakures, 6.
20. Kirk, Raven, Schofield, 193.
21. Guthrie, 2: 13–17.
22. Ibid, 2: 14–17.
23. Ibid., 2: 16.
24. Ibid.
25. Ibid.
26. Kirk, Raven, Schofield, 277; Guthrie, 2: 87–88.
27. Aristotle, *Physics*, 123–25; Barnes, 1: 261–64, 273–75; Guthrie, 2: 91–93; Kirk, Raven, Schofield, 269–72; Julian Maria, *History of Philosophy*, 24–25.
28. Guthrie, 2: 87–88; Kirk, Raven, Schofield, 277; Maria, 25; Scott-Kakures, 10.
29. Barnes, 2: 51–52; Kirk, Raven, Schofield, 407–8; Scott-Kakures, 7–11.
30. Barnes, 2: 40–41; Guthrie 2: 389–96; Kirk, Raven, Schofield, 414–15. It is interesting to note that other terms were used, especially by Democritus, to refer to the atoms. Barnes tells us, for example, that Democritus also referred to atoms, as "thing (den)," "massy (naston)," and "being (on)." In addition, Barnes notes that Aristotle used the terms "to pleres (the full)," and "to steron (the solid)." Barnes 2: 42.
31. Aristotle, *Physics*, 47, Barnes, 2: 49–50; Scott-Kakures, 11; James Trefil, *From Atoms to Quarks*, 2.
32. Kirk, Raven, Schofield, 425–27.
33. Barnes, 2: 33–34; Guthrie, 2: 288–90; Kirk, Raven, Schofield, 365–66, 378; Scott-Kakures, 10–11.
34. J. O. Urmson and Jonathan Ree, eds., *The Concise Encyclopedia of Western Philosophy and Philosophers*, 31–32, 187.
35. Morton Scott Enslin, *Christian Beginnings*, 197.
36. Robert H. March, *Physics for Poets*, 157.
37. Aristotle, *Metaphysics*, 98–99, 121–122; Aristotle, *Physics*, 5–10, 18–19.
38. Thomas Aquinas, *Summa Theologiae*, 571–81; Higher Catechetical Institute at Nijmergen, *A New Catechism*, 342–43; Allen, 74–75.
39. Richard Tarnas, *The Passion of the Western Mind*, 260.
40. Sir Isaac Newton, *Mathematical Principles*, 300–302; 398–400; 543–47; Derek Gjertsen, *Science and Philosophy*, 168–69.
41. Peter Gibbins, *Particles and Paradoxes*, 4–5.
42. Sir Isaac Newton, *Opticks or a Treatise of the Reflections, Inflections and Colours of Light*, 400.
43. Ibid., 403–4.
44. Pierre Simon, Marquis de Laplace, *A Philosophical Essay on Probabilities*, 4.
45. Isaac Asimov, *Atom: Journey across The Subatomic Cosmos*, 10–11; Gjertsen, 173–74; Gordon Kane, *The Particle Garden: Our Universe as Understood by Particle Physicists*, 30–31; March, 159; Mendel Sachs, *Einstein versus Bohr*, 42–43; Trefil, 2–3.
46. Tony Hey and Patrick Walters, *The Quantum Universe*, 74–75; Leon M. Lederman and David N. Schramm, *From Quarks to the Cosmos*, 31–32; Kane, 30.
47. Asimov, 15; Kane, 30; Nathan Spielberg and Byron D. Anderson, *Seven Ideas that Shook the Universe*, 135–36.
48. John D. Barrow, *The World within the World*, 168; Gjertsen, 174; March, 162; Hans Christian von Baeyer, *Taming the Atom*, 15.
49. March, 168.
50. James Clerk Maxwell, *A Treatise on Electricity and Magnetism*, 2: 247–62.
51. Maxwell, 2: 431–50; Hawking, 19; Nick Herbert, *Quantum Reality Beyond the New Physics*, 5.
52. Maxwell, 2: 431; Gibbins, 5; Hawking, 19; Herbert, 5; March, 74–75.
53. Maxwell, 2: 431.
54. Gibbins, 5–6.
55. Roy Rockwood, *Through Space to Mars*, 59–60.
56. J. J. Thomson, "Cathode Rays," 293–94.
57. Ibid., 302, 315.
58. Ibid., 302.
59. Ibid., 311.
60. Barrow, 171; David Millar, ed., et al., *The Cambridge Dictionary of Scientists*, 305–6.
61. March, 168–71; Fred Alan Wolf, *Taking the Quantum Leap*, 74–75.
62. March, 174–79; Trefil, 10–11.
63. Asimov, 93–95; March, 174–75; Trefil, 10–11.
64. Hey and Walters, 38; Lederman and Schramm, 34–35; March, 174–77; Jonathan S. Wolf, *Physics*, 402–3.
65. Asimov, 94–95; Hey and Walters, 38; March, 174–76; Trefil, 10–11.

66. Asimov, 93–96; Hey and Walters, 38; Lederman and Schramm, 34–35; March, 175–77; Trefil, 10–11.
67. Barrow, 171; Lederman and Schramm, 58–59; Millar, 59.
68. March, 174.
69. Ibid., 179.
70. Daniel Kevles, *The Physicists*, 92; March, 195.
71. March, 180.
72. Niels Bohr, *Atomic Physics and Human Knowledge*, 34.
73. Saverio Pascazio, Univ. of Bari and the National Inst. of Nuclear Physics, August 22, 1996; Hey and Walters, 39.
74. Bohr, 34; Asimov, 111–12; Hawking, 59; Hey and Walters, 39; March, 171; Trefil, 13.
75. John Gribbin, *Schrödinger's Kittens and the Search for Reality*, 82; March, 179–84.
76. Asimov, 54–57; Gribbin, 82–83; Roger S. Jones, *Physics for the Rest of Us*, 138–39; March, 183; Tony Rothman, *Instant Physics: From Aristotle to Einstein and Beyond*, 160–63; J. Wolf, *Physics*, 383–84.
77. Asimov, 54–57; Gribbin, 82–83; Jones, 138–39; March, 183; Rothman, 160–63; J. Wolf, *Physics*, 383–84.
78. Asimov, 56–57; Hawking, 54; Kevles, 85; March, 184–85; Trefil, 9.
79. Gribbin, 84; Trefil, 8–9.
80. Bohr, 33; Kevles, 85–86; March, 187; Trefil, 9.
81. Asimov, 81–82; Herbert, 37–38; Jones, 141–43.
82. Asimov, 79–82; Herbert, 37–38; Jones, 141–43.
83. Herbert, 37–38; Jones, 141–43; Rothman, 166–68.
84. Gribbin, 85; Millar, 229–30.
85. Bohr, 34; Joel Davis, *Alternate Realities*, 226–27.
86. Asimov, 112–13; Davis, 227; Amit Goswami, *The Self-Aware Universe*, 25–30; Hawking, 59; Trefil, 13.
87. Bohr, 34–35; Jones, 148–49.
88. March, 200–201.
89. Davis, 227–28; F. Wolf, *Taking the Quantum Leap*, 87.
90. Goswami, 31–34; Jones, 145–46; March, 201–4; F. Wolf, *Taking the Quantum Leap*, 87–91.
91. Goswami, 32; William K. Hartmann, *The Cosmic Voyage through Time and Space*, 64.
92. Goswami, 32; F. Wolf, *Taking the Quantum Leap*, 88–89.
93. Goswami, 32–33; F. Wolf, *Taking the Quantum Leap*, 88–91.
94. Goswami, 34–35; March 212–13; F. Wolf, *Taking the Quantum Leap*, 93–94.
95. Kane, 39; Barrow, 142. Newton's three laws of motion are (1) "Every body continues in its state of rest, or of uniform motion in a straight line, unless it is compelled to change that state by forces impressed upon it;" (2) "The change of motion is proportional to the motive force impressed and is made in the direction of that force;" (3) "For every action force, there is an equal and opposite reaction force." J. Wolf, *Physics*, 64–65; for a more in-depth examination of Newton's three laws of motion see: Karl Kuhn, *Basic Physics*, 13–23. The idea here is not that Schrödinger was right and Newton wrong. Instead, we mean that Newton's laws still work when we are dealing with the macroscopic universe. By this we mean objects from the size of an atom up to the size of galactic superclusters. Schrödinger's equation, on the other hand, applies at the subatomic level; Kane, 39.
96. Barrow, 142.
97. Kane, 206.
98. Gibbins, 43.
99. Heinz R. Pagels, *The Cosmic Code: Quantum Physics as the Language of Nature*, 62.

Chapter 5
Quantum Weirdness Verified

COMMENTARY: QUANTUM PHYSICS IN WONDERLAND

A writer named Donald Cameron once penned a short story that involved the strange relationship between twin sisters named Jessica and Ursula. The story began with Jessica who was awakened by a nightmare in the early morning hours of May 20, 1945 while she was staying at her parents' home in San Diego. The nightmare concerned her married sister, Ursula, who, at that time, lived in Cleveland. As Jessica later recalled, the dream concerned some sort of explosion followed by a car chase through the city streets of an unfamiliar town. Jessica had difficulty getting back to sleep that night because of an intense pain in her left leg. Thirty minutes later, Jessica's parents received a phone call from their son-in-law in Ohio. Ursula had been involved in an automobile accident and was hospitalized with a broken leg. The next morning when Jessica learned of the incident, she did not have to be told that Ursula had broken her left leg. She already knew because of the pain she felt in her own left leg. As disconcerting as the episode was, it was not the first one that Jessica and Ursula had undergone. All through their teenage years, they had experienced similar moments of dual recognition. Until recently, an experience like that of Jessica and Ursula would have been dismissed by the scientific community as an error, an imaginary episode, or an outright hoax. Since the advent of quantum physics, however, such incidents are getting a second look from many people. Naturally, the connection between quantum phenomena and paranormal events remains tenuous at best. For that reason alone, most physicists disdain any connection between the two. Nevertheless, the uncertainties raised by the growing body of knowledge surrounding quantum phenomena have many people pondering very bizarre possibilities, some of which will be the focus of this chapter.

CHAPTER OUTCOMES

After reading this chapter, the reader should be able to accomplish the following:

1. Explain the relationship between Schrödinger's and Heisenberg's equations.
2. Identify the problem with Schrödinger's wave equation.
3. Describe Born's probability waves.
4. Relate the nature of Heisenberg's uncertainty principle.
5. State the substance of the Copenhagen Interpretation of quantum physics.
6. Explain the nature of the Bohr-Einstein debate.
7. Relate the details of Einstein's "EPR" thought experiment.
8. Clarify the essentials of Bell's Interconnectedness Theorem.
9. Identify the procedures involved in Aspect's verification experiment.
10. Define and explain nonlocality.

5-1   THE PUZZLE OF QUANTUM WEIRDNESS

Quantum physics, more than any other theory in the history of modern science, requires that we adjust our thinking about the universe. In fact, the advent of this field demands that we abandon the deterministic, predictable world of classical physics when we are speaking of events at the subatomic level. As we shall soon see, such events can be described only in terms of probabilities. Descriptions of physical events in terms of probabilities are not unknown in many fields of study. However, before the coming of quantum physics, probabilities were used because investigators were unable to know all of the factors associated with certain situations. With quantum theory, the need to speak in terms of probabilities is not due to any shortcomings on the part of the experimenter, but is, instead, due to the very nature of the quantum universe.[1]

The probabilities at the heart of the quantum universe produce some distressing consequences, among which is that any discussion of quantum events in terms of everyday life inevitably leads to confusing and contradictory images that defy common sense. This is why the term quantum weirdness is often used to describe quantum principles. **Quantum weirdness** refers to those quantum phenomena that appear to defy common experience when explained in terms of everyday life. Perhaps more troubling than the weird nature of quantum phenomena, however, is that quantum weirdness is not born of our inability to develop a more complete theory but is inherent in the nature of the quantum universe itself.[2] Fortunately, once we rise above the atomic level, the world of quantum weirdness vanishes. At that point the world again surrenders to the dictates of classical physics. However, this only adds to the mystery of quantum weirdness. After all, why does the world act with predictable regularity at the macroscopic level when it defies such predictions at the quantum level?[3] We may never know the answer to this question. However, we can at least address the opposite question. "Where does the weirdness originate?" It is to this question that we now turn.

5-2 PROBABILITIES, UNCERTAINTY, AND COMPLEMENTARITY

Schrödinger's equation was supposed to put an end to the controversial aspects of quantum theory. Certainly, in 1926, when Schrödinger's work was first making the rounds among physicists, this seemed to be the case. Schrödinger had done for de Broglie's atom what Newton had done for Kepler's elliptical orbits. Kepler had removed the problematic nature of planetary orbits within the solar system by postulating that those orbits might be elliptical rather than circular. His new assumption proved to be accurate. However, he did not explain the laws of motion and gravitation that were behind those elliptical orbits. That bit of genius was left for Newton and Einstein. Similarly, while de Broglie had postulated the existence of the wave phenomenon as an explanation for the dynamics of the atom, he had not supplied the equations that might be behind this wave phenomenon. This was left for Schrödinger, who, in a series of stunning papers in 1926, created a mathematical formulation of quantum phenomena based upon wave mechanics. As Schrödinger was soon to learn, however, this did not settle the question. In fact, it actually led, initially at least, to further quantum difficulties.[4]

Heisenberg's and Schrödinger's Equations

One of the confusing elements of Schrödinger's equation was that it did not agree with the equations of Werner Heisenberg, the German physicist who, while working on the island of Helgoland in the North Sea in 1925, had attempted to resolve the same issues that had faced Schrödinger. Heisenberg's mathematical conception, which came to be identified with the term **matrix mechanics**, had been rejected by Schrödinger. The root cause of Schrödinger's objection to Heisenberg's matrix mechanics was his argument that Heisenberg's equations predicted quantum events that could not be pictured. For this reason both Schrödinger and Einstein had rejected Heisenberg's equations as somewhat "miraculous." In typical fashion, Heisenberg had replied that there was no underlying need for this visualization. In fact, Heisenberg was convinced that Einstein and Schrödinger were in for a monumental disappointment if they insisted on being able to visualize quantum phenomena.[5] Eventually, this conflict was solved when Schrödinger himself, among others, including Paul Dirac of Cambridge University, demonstrated that the two formulations are completely equivalent.[6]

Born's Probability Waves

The second point of confusion fostered by Schrödinger's equation lay within the nature of the wave itself. The dual nature of light had been well-known for many years. Sometimes, light appeared to act as a wave, and at other times, it clearly acted as a particle.[7] Bohr's atom had emphasized the particle state of the electron. However, with the advent of de Broglie's atom and Schrödinger's wave equation, the wave property came to the forefront. Nevertheless, although Schrödinger clearly believed that a physical wave existed in reality, his formulation gave no hint as to the actual nature of this wave.[8]

Unfortunately, the idea that electrons exist as physical waves has many difficulties. Schrödinger had completed his original equations in December of 1925. As early as May 27, 1926, Hendrik Antoon Lorentz, the Dutch physicist who played a key role in the development of the theory of electromagnetism, wrote to Schrödinger and indicated a key difficulty with Schrödinger's interpretation of quantum theory. In his landmark study of quantum theory, *The Philosophy of Quantum Mechanics*, the physicist Max Jammer of Bar-Ilan University in Israel notes that Lorentz indicated that the problem with Schrödinger's wave packet is that over a very short period of time the wave would dissipate much like a water wave fans out and vanishes. For this reason Lorentz concluded that the dissipation problem makes the wave packet a highly unsuitable way to depict subatomic phenomena. Max Born, a German theoretical physicist, was inclined to agree.[9]

In fact, no one saw the problem more clearly than Born, who was working at the institute at Göttingen in 1926. Also at Göttingen at this time was James Franck, an experimentalist, whose work in electron collisions consistently demonstrated the particle nature of the electron.[10] The consistent results of Franck's experiments troubled Born, who eventually resolved the apparent contradiction by concluding that the wave that Schrödinger spoke of in his equation was not a physical wave but a **probability wave**.[11] In other words, Born suggested that the solutions to Schrödinger's equation did not demonstrate the location of the wave. Instead, these solutions, now referred to as **wave functions**, predicted only the probability of pinpointing an electron (or any subatomic entity for that matter) at a specific time and in a particular place.[12]

The fallout from Born's proposition was most startling. One possible interpretation of Born's proposition was that deterministic predictions of events at the quantum level were forever beyond our ability to foresee. Thus, we can predict quantum events, but only in a probalistic sense.[13] The only thing we could predict would be the likelihood or the probability of an event. If this were true, then the stable and deterministic universe of Newton and Laplace had vanished forever. The other interpretation was less radical but equally dissatisfying. That interpretation held that quantum theory, as it was understood at that moment in time, was not yet complete. Something about our understanding of quantum phenomena was still beyond our grasp. The stage was set for the German physicist Werner Heisenberg, who added a new twist to the dilemma.[14]

Heisenberg's Uncertainty Principle

As the problems of quantum probability and its promise of an indeterminate universe became apparent, Werner Heisenberg began his search for the nature of quantum indefiniteness. The result of that search was the legendary uncertainty principle. At its most elementary level the uncertainty principle states that an experimenter can never simultaneously pinpoint both the **momentum** and the position of a subatomic particle.[15] Because quantum phenomena are inherent in all subatomic entities, including both matter particles and force-carrying particles, we will, from this point on, use a single designation for all subatomic entities. This designation will be the **duon**. We will deviate from this term only when a particular experiment requires the use of specially designated subatomic particles. For instance, when we discuss Aspect's experiment, we use the term *photon* because Aspect used photons in his experiment.

According to the uncertainty principle, as the experimenter measures the momentum of the duon, he or she disturbs its location in an unpredictable way. As a result, the knowledge gained about the momentum of the duon will destroy any knowledge about the precise position of that duon. The same is true if the experimenter tries to ascertain the position of the duon; this will sacrifice any precise knowledge of the duon's momentum. In summary, the position can be known but not the momentum, or the momentum can be known but not the position.[16]

The uncertainty principle is the first of many examples of quantum weirdness—those quantum phenomena that appear to defy common experience when explained in terms of everyday life. However, the strange nature of quantum phenomena is "weird" only in relation to the expected classical norm, not in relation to internal quantum theory. Therefore, although the uncertainty principle would seem to be quite strange and frightening if, for example, it were applied, on the macroscopic level, to an air-traffic controller trying to pinpoint the location of an incoming jet airliner, it is a perfectly acceptable principle within the confines of quantum physics.[17]

Complementarity and Quantum Weirdness

The uncertainty principle and wave-particle duality led Niels Bohr to formulate the complementarity principle, which, in turn, led to a revolutionary change in the way physicists look at the nature of the universe and the nature of scientific experimentation. For all its importance, however, there is some ambiguity about both the concept itself and the term used to describe it. Max Jammer points out, for example, that the term *complementarity* is used somewhat indiscriminately by several writers, including Bohr, to refer to a variety of different concepts.[18] Despite this difficulty, we can construct a reasonably accurate model of Bohr's original notion of *complementarity* by following his logic in a step-by-step pattern. As a prerequisite, we must first make intellectual peace with the fact that quantum events do not occur like macroscopic events. This difference necessitates abandoning many familiar ideas about physical relationships whenever the subject turns to quantum phenomena. Chief among the ideas requiring revision is the notion of the validity of physical scientific theories. To be successful, a physical theory must meet two requirements. First, the system that is the subject of the theory must be explained in terms of causality. This requires that experiments conducted to validate the theory must be isolated so that they can be conducted without interference.[19] Second, the physical theory must be able to "explain all phenomena as relations between objects existing in space and time."[20] Let us look at each of these requirements and see the problems that arise in the quantum universe.

The Problem of Causality. To be successful, a physical theory must first be explainable in terms of causality. This means that any experiment carried out to verify the causal relationships within the theory must be performed in isolation and without interference. We must be able to perform the experiment and measure the results without interfering in the operation of the experiment. If we interfere with the experiment, we add an outside cause that creates effects that are unrelated to the natural situation we are testing. We therefore fail to verify the theory.[21] For example, if we want to know why a particular baseball player hits home runs, we must observe his batting style without interfering with his technique. If we manage to observe his batting style without interfering, we will learn a lot about his technique, and we may be able to use that knowledge to help other batters or perhaps even to verify a physical theory about baseball. If, on the other hand, we decide to shine a bright light into the batter's face just as he is about to swing the bat, we will interfere with the experimental setup and cause things that are not normally a part of the batter's environment as he faces the pitcher. The sudden illumination will distract the batter and cause him to swing the bat in a way that is not his usual manner of attacking a pitched baseball. We would eventually conclude that to study a home run hitter's batting technique, we must observe the batter without interfering with his natural batting style.

The same is true of scientific experiments aimed at measuring causal relationships. Any accurate determination of causal relationships, at least in the macroscopic world, requires isolating the system under study and observing that system without interfering with it.[22] Now the problem with attempts to verify quantum theory is that, at the quantum level, phenomena *cannot* be observed without interference. No instrumentation is small enough, nor can ever be small enough, for observation without interference. Consequently, whenever we attempt to construct an experiment that will allow us to observe a quantum system, we will inevitably interfere with that system.[23] This does not mean that we cannot make measurements of quantum systems. Rather, it means that we must make those measurements in light of the uncertainty principle, which tells us that we can measure either the momentum or the position of a duon but never both. Nevertheless, this puts a serious limit on the first of the two key requirements necessary for validating quantum theory, that is, the ability to determine causality.[24]

The Problem of Space and Time. The second requirement of a physical theory fares little better in the quantum universe. This requirement states that a physical theory must explain events as relationships among physical things that exist in time and in space.[25] The requirement has slowly been relaxed since the beginning of the twentieth century. A case in point is James Clerk Maxwell's development of electromagnetic theory. As we saw in the last chapter, Maxwell attempted to explain the electromagnetic field by adding the luminiferous ether to his theory. In 1887, Michelson and Morley demonstrated experimentally that there was no physical evidence for the existence of the ether, and in 1905 Einstein demonstrated mathematically that there was no need to postulate the theoretical existence of the ether.[26] Despite this, the equations developed by Maxwell still accurately represent the electromagnetic field.[27] This seems to indicate that when we deal with quantum phenomena, the requirement that any successful theory must involve objects that can be coordinated within space and time is not as stringent as it is when dealing at the macroscopic level. In fact, when causal relationships at the quantum level are explained by mathematical equations, we lose the ability to visualize those phenomena in space and time.[28]

Bohr's Solution: The Complementarity Principle. Bohr seized upon the problems inherent in these two requirements, saw the difficulty of trying to maintain them in light of quantum phenomena, and developed his principle of complementarity. According to Bohr, if we are dealing in classical physics, causal relationships can be expressed in the language of time and space. However, at the quantum level, if we attempt to explain a quantum phenomenon by fashioning experiments within time and space, we will interfere with that phenomenon. Therefore, although we will have a description of that phenomenon in the language of time and space, we must admit to the indeterminate nature of this description because of the uncertainty principle. If, on the other hand, we explain causal relationships with mathematical equations, we sacrifice a physical description of the phenomenon in time and space. The mathematical model may approximate what might be observed in time and space, but such approximations apply equally well to waves and to particles and are, thus, impossible to visualize. Bohr's **complementarity principle**, which is clearly an either-or proposition, can, therefore, be summarized in the following way: *Either* we describe phenomena in the context of time and space, in which case, we must deal with the uncertainty principle, *or* we describe causal relationships mathematically, in which case, we lose the ability to visualize those relationships in terms of time and space.[29]

The principle of complementarity had far-reaching effects on the interpretation of quantum theory. For one thing, in emphasizing the difficulty of separating the observer and the experimental apparatus from the phenomenon under observation, complementarity calls into question our ability to distinguish between the subjective and objective sides of the universe. Such questions had been the staple of philosophy for centuries, but now it appeared that the quiet musings of the philosophers regarding our ability to distinguish between objective and subjective reality might have to be taken seriously in the physical sciences.[30] It is the complementarity principle along with its recognition of the problems inherent in attempting to reconcile the tension between causality, on the one hand, and space-time coordination, on the other, that provides fertile ground for one of the most controversial sides of quantum theory, the Copenhagen Interpretation.[31]

5-3 THE COPENHAGEN INTERPRETATION

The Copenhagen Interpretation received its name because it was created by Niels Bohr while he was working at the Danish Institute of Theoretical Physics, which was established at the University of Copenhagen in the 1920s. At the outset it is best to point out that the Copenhagen Interpretation cannot be narrowed to a simple, easy-to-phrase statement. It can also be said that the Copenhagen Interpretation is neither consistent nor static. Instead, the term *Copenhagen Interpretation* can be used as an umbrella phrase that covers a wide variety of complex views on quantum theory.[32] Max Jammer emphasizes this complexity when he notes that the Copenhagen Interpretation cannot be narrowed to one group of concepts but is, instead, a general heading under which a number of different ideas can be linked together.[33] Perhaps the most accurate statement that can be made about the Copenhagen Interpretation is that the common denominator of which Jammer speaks is Niels Bohr.[34] Despite the variations, there are some underlying ideas that form the foundation of the Copenhagen Interpretation of quantum theory. These ideas form the basis of our discussion.

The Essentials of the Copenhagen Interpretation

The **Copenhagen Interpretation** recognizes that the deterministic picture of the universe that works so well at the macroscopic level for classical physics does not work for the world at the quantum level. This conclusion is derived from two very critical aspects of quantum physics. First, the universe at the quantum level is predictable only in a statistical sense. It is impossible to predict the movement or the position of a single duon. Instead, we must always speak in terms of statistical distributions.[35] The implications of the statistical nature of quantum phenomena should not be taken lightly. The fact that we can speak about the quantum universe only in terms of probabilities means that we can never really know the nature of quantum phenomena. Instead, we can only describe what we know *about* that phenomena. Niels Bohr put it much more succinctly, when he said, "It is wrong to think that the task of physics is to find out how nature is. Physics concerns what we can say about nature."[36] Moreover, the more we try to say about nature and the closer we examine these statistical descriptions, the more we are convinced that we really don't know all that much about quantum phenomena.[37] Heinz Pagels, former executive director of the New York Academy of Sciences, explains this situation by noting that since statistical distributions are the only way of representing quantum events, it is impossible to visualize accurately what such measurements might mean for a solitary duon. In fact, the idea of a solitary duon may be nothing more than a convenient fiction constructed to give us something to relate to as we attempt to picture quantum events.[38]

The second aspect of quantum physics that led to the Copenhagen Interpretation was the discovery that the physical state of quantum reality cannot be described without also specifying the observational technique that is used to conduct the observation.[39] If the apparatus is constructed to observe the particle property of the duons, that is what will be observed. If the apparatus is constructed to observe the wave property, then that is what the observer will see.[40] This is simply a restatement of complementarity and a reassertion of the uncertainty principle. The uncertainty principle states that measurements of duons will always be complementary, that is, we can determine a duon's position or momentum but never both at the same time. In other words, the position, that is, the particle state of the duon, can be determined, at least in a statistical sense but not at the same time as the momentum, that is, the wave state of the duon.[41]

Bohr stresses that there will always be an unavoidable interaction between quantum phenomena and the instruments used to measure those phenomena. If the apparatus is constructed to observe particles (position), particles will be observed. Bohr, and his followers, explained this phenomenon as "the collapse of the wave function."[42] By this they meant that constructing the apparatus to observe the position of the duon will cause the duon to collapse into the area where the particle is detected.[43] However, it is not correct to assume that the duon exists in a predetermined state that is changed by the measurement. Rather, the duon does not have a definite state until it is forced by the measurement to assume that state.[44] Does this mean that the reality of the subatomic universe is somehow created by the interaction of the observer and the observed? Some commentators seem to believe so.[45] For example, John Barrow, professor of astronomy at the University of Sussex, states in his work *The World within the World*:

> (T)here is no deep reality for us to discover in the traditional sense, only a description of it. The reality that we observe is determined by the act of observation. It really exists when it has been measured—it is not an illusion—but there is no sense in which we can say that it exists in the absence of an act of observation. We must recognize that "things" like photons and neutrons cannot be "real" in the same way that we think that chairs and tables are real. They are more like shadows: arising from a combination of light and the observer's situation. Shadows are real enough, but they do not exist in the sense that a book does.[46]

It is unlikely that Bohr meant to go quite this far. Still it is important to remember that Bohr was a philosopher as well as a physicist. In this regard, he is unlike many physicists who, as we have seen, are not only indifferent to philosophical interpretations of physics but also openly hostile to any attempt to draw philosophic conclusions from quantum theory.[47] However, Bohr's philosophical beliefs should not be interpreted as mystical or spiritual. He does not go so far as to support the view that the subatomic universe is the result of some sort of mystical creation event.[48] In fact, a close reading of Bohr's work reveals that he repeatedly disavowed such an interpretation of quantum theory. At one point Bohr writes:

> In this context, one sometimes speaks of "disturbance of phenomena by observation" or "creation of physical attributes to atomic objects by measurements." Such phrases, however, are apt to cause confusion, since words like phenomena and observation, just as attributes and measurements, are here used in a way incompatible with common language and practical definition. On the lines of objective description, it is indeed more appropriate to

use the word phenomenon to refer only to observations obtained under circumstances whose description includes an account of the whole experimental arrangement.[49]

On the other hand, Bohr does *not* support the position that quantum theory is an incomplete picture of reality. Some theorists who are opposed to Bohr's interpretation of quantum theory believe that the difficulty in arriving at a simultaneous measurement of position and momentum and the fact that there is an unavoidable interaction between quantum phenomena and the measuring instruments does not mean that quantum events are indeterminate. Rather, these critics argue that the difficulties in quantum measurements arise because we still have only an incomplete understanding of quantum theory. A more complete picture will, therefore, allow us to understand what is actually happening when measurements are made at the quantum level.[50] Still other theorists would say that uncertainty is caused by the interference created when we try to detect the position or the momentum of duons by using other duons. What happens is that the duons used for observation collide with the duons being observed, causing a shift in the position and the momentum of the observed duons.[51]

According to the Copenhagen Interpretation, neither of these explanations is sufficient. Instead, the Copenhagen Interpretation holds that the difficulties inherent in our attempt to measure position and momentum are not a result of our lack of knowledge or the practical difficulties involved in operating at the quantum level. Rather, these difficulties are an integral part of the underlying nature of reality. Moreover, as noted, the interaction between the measuring devices and the quantum phenomena does not disturb the preexisting state of the duon. Instead, the duon has no definite state until it is measured. At the point of measurement, the duon assumes the state that the experiment has been contrived to measure.[52]

This brings us back to the fundamental concept promoted by the Copenhagen Interpretation. This concept reveals that the deterministic picture of the universe that works so well at the macroscopic level does not work for the world at the quantum level. The deterministic world dreamt of by Laplace and Newton cannot be realized at the quantum level. Our ability to know what is happening at the quantum level is forever limited to a statistical prediction of probabilities. Moreover, the unavoidable interaction between measuring apparatus and quantum phenomena forever limits our knowledge about the position and the momentum at the quantum level.[53]

According to Bohr, the lesson to be learned from the interaction of the measuring instruments and quantum phenomena is that our ability to know certain aspects of the universe is limited. This realization should not be seen as an attempt to negate the body of knowledge built up by classical physics. It should be considered an added dimension that allows us to see the universe from a new perspective. As such, it was a step forward that opened new worlds of possibilities.⁵⁴ Before moving on to Schrödinger's and Einstein's opposition to some of the ideas implied by quantum theory, there is one important consequence of Bohr's Copenhagen Interpretation that we should pause to consider. Bohr does not deny the existence of subatomic reality. We have established this beyond doubt. However, he does point out, albeit in a very subtle manner, that, as a consequence of quantum theory, the universe must be viewed as an *interconnected whole*.⁵⁵ Bohr puts it this way:

> (T)he fundamental difference with respect to the analysis of phenomena in classical and in quantum physics is that in the former the interaction between the objects and the measuring instruments may be neglected or compensated for, while in the latter this interaction forms an integral part of the phenomena. The essential wholeness of a proper quantum phenomenon finds indeed logical expression in the circumstance that any attempt at its well-defined subdivision would require a change in the experimental arrangement incompatible with the appearance of the phenomenon itself.⁵⁶

This view of the universe at the subatomic level as a unified whole will prove to be one of the most critical conclusions of quantum theory. It will lead to the concept of nonlocality and to the interconnected nature of the entire universe. These ideas are explored later in this chapter. For the moment, however, it is useful to look at the leading opponents to Bohr's views on quantum theory. Although many scientists followed Bohr's lead and accepted the disturbing universe of the Copenhagen Interpretation, quite a few did not. Moreover, those who found themselves uncomfortable with Bohr's interpretation were in good company, because among their number was no less a pair of illustrious personages than Edwin Schrödinger and Albert Einstein.

The Mystery of Schrödinger's Cat

It is relatively safe to say that Schrödinger had severe reservations about the Copenhagen Interpretation. Interestingly enough, some people have found his position in this regard somewhat contradictory. This is because Schrödinger was attracted to Eastern religion and was apparently fascinated by the Vedantic belief that the universe exists as an undivided whole, an idea sometimes referred to as the **doctrine of identity**.⁵⁷ Schrödinger represented the belief in this way:

> Briefly stated, it is the view that all of us living beings belong together in as much as we are all in reality sides or aspects of one single being, which may perhaps in western terminology be called God while in the Upanishads its name is Brahman. A comparison in Hinduism is of the many almost identical images which a many-faceted diamond makes of some *one* object such as the sun.⁵⁸

This Copenhagen view of the universe at the subatomic level as a unified whole would seem to be in line with Schrödinger's beliefs. Such a belief would seem to make him a likely candidate for conversion to the Copenhagen point of view. Schrödinger, however, did not convert. In fact, he went on to create one of the most enduring symbols of the ongoing quantum debate. This symbol has come to be known as the mystery of Schrödinger's cat.[59]

The enigma of **Schrödinger's cat** was intended to demonstrate the limitations of Bohr's position in regard to quantum theory. In this imaginary situation a cat is placed in a box, along with a vial of poison gas, a sensitive photocell apparatus, and a radioactive element. At random intervals the radioactive element undergoes decay. **Radioactive decay** occurs spontaneously when the nucleus of an atom disintegrates due to the unstable nature of its atomic structure. When the radioactive element undergoes decay, it emits **radiation**. In this experiment, the radiation emitted will hit the photocell, activating a circuit in the apparatus that will open the vial and release the poison. The cat is killed when the poison gas is released. Everyday experience would dictate that, whether or not an observer looks into the box, the element has either decayed or it has not and, therefore, the cat is either dead or alive. However, radioactive decay is a quantum event. Therefore, according to Schrödinger's equation the state of the entire quantum arrangement can only be discussed in terms of probabilities, that is, in terms of its wave function. During the hour set aside for the experiment, there is a fifty percent chance that the radioactive element will undergo decay. Therefore, at all times there is an equal probability that the radioactive element is decayed and not decayed, and, as long as no observer looks in the box, that the cat is simultaneously alive and dead.[60]

Moreover, the cat will exist in this live/dead state until the box is opened and the observer, by his act of observing, collapses the wave function into either the live or the dead state.[61] Such a situation is, of course, flatly impossible. A cat is not a subatomic entity. It is a macroscopic object. It cannot be both alive and dead. The existence of a "live-dead cat" would imply a macroscopic superposition of states which is not possible in the macroscopic universe.[62] Since such a situation is impossible, Schrödinger concludes that Bohr's interpretation of quantum theory represents an incomplete view of reality.

Einstein's Interpretation of Quantum Weirdness

If Schrödinger was unhappy with Bohr's Copenhagen Interpretation of quantum theory, Einstein was appalled. Einstein saw Bohr's explanation of quantum weirdness as a fragmentary, misleading, and inaccurate attempt to resolve the essential problems of quantum theory. Recall that Bohr saw subatomic uncertainty as a fundamental standard of the universe. The laws of classical Newtonian physics might apply to archery, automobiles, and avalanches, but when it comes to quantum phenomena, the rules that apply are more closely allied with statistical

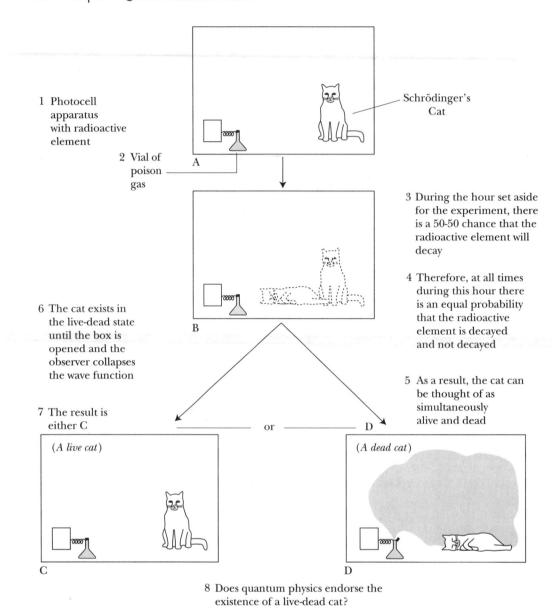

Figure 5.1 The Mystery of Schrödinger's Cat.

The mystery of Schrödinger's cat was designed to underline the fallacy of Bohr's position in regard to quantum theory. In this thought experiment a cat is placed in a box, along with a vial of poison gas, a sensitive photocell apparatus, and a radioactive element. At random intervals the radioactive element undergoes decay (Figure 5-1 A). Radioactive decay occurs spontaneously when the nucleus of an atom disintegrates because of the unstable nature of its atomic structure. When the radioactive element undergoes decay, it emits radiation. In this experiment the radiation emitted will hit the photocell, activating a circuit in the apparatus that will open the vial and release the poison. The cat is killed when the poison gas is released. Everyday experience would dictate that, whether or not an observer looks into the box, the element has either decayed or it hasn't decayed and, therefore, the cat is either dead or alive. However, radioactive decay is a quantum event. Therefore, according to Schrödinger's equation, the probabilities and complementary observations. In other words, to Bohr the underlying reality of the universe was uncertain and, therefore, ultimately unknowable, at least in a deterministic sense associated with classical physics. Einstein was never satisfied with this approach. He believed that if quantum theory failed to reveal all the answers associated with the subatomic universe, it was the theory and not nature that was at fault. To Einstein, the underlying reality of the universe was solid, unchanging, and ultimately deterministic. Quantum theory indicated otherwise only because it was an incomplete theory.[63]

The Essential Focus of the Debate

This disagreement between Einstein's classical approach and Bohr's Copenhagen Interpretation is known as the **Bohr-Einstein debate**. It is important to note that Einstein's quarrel with Bohr was not one of degree. It is not as if they agreed on the general conclusions but differed on the details. In fact, the direct opposite is true. The debate focused on the very nature of reality. Much of the confusion and difficulty surrounding the Bohr-Einstein debate can be eliminated by focusing on Bohr's explanation of the heart of the debate. According to Bohr, "from the very beginning the main point under debate has been the attitude to take to the departure from customary principles of natural philosophy characteristic of the novel development of physics which was initiated in the first year of this century by Planck's discovery of the universal quantum of action."[64]

Bohr's summary of the problem accurately pinpoints the central cause of the problem. Discoveries in quantum theory since the beginning of the twentieth century signaled an abrupt departure from the traditional, classical interpretation of the laws of physics. We have seen this departure from classical norms in the problems created by the uncertainty principle, the solution suggested by the complementarity principle, the revelation of the statistical nature of quantum phenomena, and the discovery that the physical state of quantum reality cannot be described without also specifying the observational technique that is used to conduct the observation.

Basically there are two ways to interpret this departure from classical norms. One position we have already examined. This is Bohr's belief, as expressed in

---

state of the entire quantum arrangement can only be discussed in terms of probabilities, that is, in terms of its wave function. During the hour set aside for the experiment, there is a fifty percent chance that the radioactive element will undergo decay. Therefore, at all times during the hour the probability that the radioactive element is decayed equals the probability that it is not decayed. Thus, as long as no observer looks in the box, it is as likely that the cat is dead as that it is alive (Figure 5-1 B). Moreover, the cat will exist in this live/dead state until the box is opened and the observer, by his act of observing, collapses the wave function into either the live or the dead state (Figure 5-1 C and Figure 5-1 D). Such a situation is, of course, flatly impossible. A cat is not a subatomic entity; it is a macroscopic object. It cannot be both alive and dead. The existence of a "live-dead cat" would imply a macroscopic superposition of states which is not possible in the macroscopic universe. Since such a situation is impossible, Schrödinger concludes that Bohr's interpretation of quantum theory represents an incomplete view of reality.

the Copenhagen Interpretation, that quantum theory represents an accurate representation of the world of the subatomic universe. This interpretation requires the need to come to terms with Bohr's conclusion that the quantum universe exists in an indefinite state until a measurement compels it to assume the state that the experiment has been contrived to measure.[65] The other way to interpret this departure from classical norms is to view it as a momentary situation that will be resolved when more is known about quantum theory. At that point, the classical norms will be reestablished and will apply equally well to macroscopic and subatomic phenomena. This latter position is the one that was taken by Einstein.[66] Bohr explains the problem this way:

> The question at issue has been whether the renunciation of a causal mode of description of atomic processes involved in the endeavours to cope with the situation should be regarded as a temporary departure from ideals to be ultimately revived or whether we are faced with an irrevocable step towards obtaining the proper harmony between analysis and synthesis of physical phenomena.[67]

It is also important to recall that the debate is not, like many aspects of modern folklore, a myth. Historically, the debate began in October of 1927 in Brussels at the Solvay conference on quantum physics. Daniel Kevles, Koepfli Professor of Humanities at CalTech, describes the Solvay conference in the following way:

> Einstein remained convinced that it was possible to obtain a model of reality, a theory that represented "things themselves and not merely the probability of their occurrence." Now, at Brussels, Einstein pressed his objections to quantum mechanics. At breakfast he would happily pose an experiment that, he was sure, violated Heisenberg's uncertainty principle, then trundle down the street to the conference, absorbed in discussion with Bohr. By evening, having pondered the experiment all day, Bohr would triumphantly demonstrate to Einstein how the experiment did not violate but reaffirmed the principle of uncertainty. Every morning Einstein would challenge Bohr with a new experiment; every evening Bohr and quantum mechanics would win the day.[68]

This lively back-and-forth exchange between the two physicists continued for the rest of their lives. Neither was willing to surrender to the other, and neither ever did, despite years of intense debate.

5-4 THE EINSTEIN–PODOLSKY–ROSEN THOUGHT EXPERIMENT

Einstein did not believe that he had an interpretation for quantum events that would replace the Copenhagen Interpretation nor did he necessarily think that such an interpretation was on the horizon. However, he did believe that reality should follow deterministic principles even if those principles were at the present unknown. As a result, his approach to the debate with Bohr was to continually challenge Bohr's position with a variety of cleverly constructed but highly controversial thought experiments.

The Nature of the Einstein–Podolsky–Rosen Thought Experiment

A **thought experiment** (*gedanken* experiment) is one that is visualized in the mind of a scientist rather than one that is actually performed. Generally, thought experiments cannot be carried out because of certain practical limitations.[69] Both Bohr and Einstein had a predilection toward the creation of these thought experiments because such experiments helped relate the mathematical abstractions inherent in quantum theory to the physical universe.[70] It is crucial to understand, however, that many thought experiments are eventually carried out once technology catches up with the physicists' imaginations.[71] The most famous of Einstein's thought experiments appeared in an article that the physicist coauthored with Boris Podolsky and Nathan Rosen, while the three of them were at the Institute for Advanced Research at Princeton. The article, "Can Quantum-Mechanical Description of Physical Reality Be Considered Complete?" which was published in the May 15, 1935 issue of *Physical Review*, introduced a thought experiment that purported to demonstrate the incomplete nature of quantum theory.[72]

The experiment, which came to be known as **EPR** after Einstein, Podolsky, and Rosen, challenged the very heart of the Copenhagen Interpretation. Before examining its details, it will be helpful to recall once more the essence of the Bohr–Einstein debate. Einstein maintained that reality has an objective, deterministic existence that is unaffected by the observer. Therefore, a duon, which is a subatomic particle, has a definite position and momentum, regardless of whether or not an observer measures that duon. Bohr says that reality is both indefinite and inherently unknowable. Thus, a duon has neither position nor momentum but instead exists in an indefinite state until a measurement forces the particle to assume one state or the other to conform to that measurement.

In the EPR article Einstein creates a thought experiment that he believes demonstrates that the subatomic universe has a definite objective existence, and that quantum theory as it existed at that time, was an incomplete theory. In the thought experiment Einstein constructs a quantum mechanical state in which two duons are at fixed distances apart. In addition, the two duons have a definite total momentum. Suppose now that we measure the position of the first duon. If we know the position of the first duon, we also know the position of the second duon because it is located at a fixed position in relation to the first duon. Suppose we then measure the momentum of the first duon. We now know the momentum of the second duon because the total momentum of the system is fixed. We, therefore, know *both* the position *and* the momentum of the second duon without having disturbed that second duon. Moreover, we have gained this knowledge with absolutely no interaction with the second duon. More importantly, in determining the position and the momentum of the second duon, we have contradicted the Heisenberg uncertainty principle. We have simultaneously established the position and the momentum of the second duon, something the uncertainty principle clearly states is not possible.[73] Einstein's conclusion is that the subatomic universe is, therefore, definite, and quantum theory of that time did not describe physical reality as completely as its proponents, notably Bohr, claimed.[74]

Not to be outdone, Bohr penned a response to the EPR thought experiment. Bohr's response appeared in the October 15, 1935 issue of *Physical Review*, under the identical title that Einstein, Podolsky, and Rosen had used in their earlier article.[75] In that article and many others to follow, Bohr attempts to discredit the EPR argument by noting that the very formulation of the thought experiment devised by Einstein contained an essential ambiguity when applied to quantum phenomena.[76] Moreover, Bohr disputes Einstein's conclusion that since quantum mechanics cannot predict the physical reality of a quantum system without disturbing the system, the theory is incomplete. The ambiguity that Bohr pinpoints is found in the notion that the system is not disturbed in any way. Bohr claims that quantum theory can predict the physical reality of a physical system without disturbing the system as long as the term *disturbance* is defined according to quantum, rather than classical, rules.[77]

In constructing his rebuttal to EPR, Bohr freely admits that the way Einstein created his thought experiment seems to show that the system has not been disturbed.[78] However, what Einstein overlooks, Bohr notes, is the overall unity of the entire experimental system. According to Bohr, in order to grasp the implications of the thought experiment, we must keep in mind that all parts of the system, including the first duon, the measuring apparatus, and the second duon, are inextricably bound up within the same unified system. In other words, the two duons are as much a part of the system as the measuring apparatus, and vice versa. Bohr points out that we cannot relate what actually happens to either duon unless we measure that duon. If we measure a single duon, as Einstein suggests, that is the only accurate measurement we have. Bohr admits that we have some sort of measurement for the second duon and that we have obtained that measurement indirectly as a result of our measurement of the first one. However, Bohr argues that this second measurement is a number and nothing more. It does not tell us the actual state of the second duon. To determine its actual state, we must measure that duon. However, that measurement is a different experiment that will yield different results. It will tell us the momentum (or position, depending on what we are measuring) of the second duon, and it will give us a number that purports to give us the momentum (or position) of the first duon. This number, however, will not represent the actual momentum (or position) of the first duon because the measurement of the second duon has disturbed the entire system, thus rendering the momentum (or position) of the first duon indefinite.[79]

Bohr concludes that the measurement of one duon really tells us nothing about the momentum (or position) of the other, or perhaps more precisely, the measurement of the one duon affects the state of the other duon, thus making the quantum state of the second duon indefinite.[80] Einstein labeled Bohr's idea, that the measurement of one duon leads to the indefinite state of the second duon, as spooky action-at-a-distance.[81] A more precise label for this state of affairs is nonlocality. **Nonlocality** refers to the fact that the measurement of the first duon affects the state of the second duon despite the fact that the measurement of the first duon is a nonlocal, that is, a distant, event in relation to the second duon.[82] Einstein found the idea of nonlocal effects to be unacceptable because the existence of nonlocal effects would eliminate objective causal reality from the quantum universe.[83] Once Einstein had eliminated nonlocal effects from consideration, he was compelled to concede that quantum theory was incomplete. This meant that physicists would have to search for a hypothetical **superquantum theory** that, once understood, would reveal any hidden variables that would be needed to forecast the results of any quantum measurement. In fact, once these local hidden variables were in hand, physicists would be able to figure out by direct measurement the simultaneous position and momentum of any duon.[84]

Bohr, however, saw no need for a superquantum theory. He believed that quantum theory, as it existed at the time, was complete. There was nothing ambiguous or contradictory about it. However, to accept this conclusion, Einstein would have to concede the existence of nonlocal effects. In fact, Bohr went so far as to state that such an interpretation of the EPR thought experiment would eliminate the problems of which Einstein complained. According to Bohr, if we simply allow that the duons and the apparatus measuring those duons are all part of the same system that acts in a nonlocal way to produce an indefinite state, the problems disappear.[85] In an article entitled "Discussion with Einstein," Bohr concludes:

> This description, as appears from the preceding discussion, may be characterized as a rational utilization of all possibilities of unambiguous interpretation of measurements, compatible with the finite and uncontrollable interaction between the objects and the measuring instruments in the field of quantum theory. In fact, it is only the mutual exclusion of any two experimental procedures, permitting the unambiguous definition of complementary physical quantities, which provides room for new physical laws, the coexistence of which might at first sight appear irreconcilable with the basic principles of science.[86]

Thus, quantum theory can be seen as complete only when we admit to the indivisible nature of any quantum system that leads to nonlocal interactions.

Still, even this says it poorly. Bohr does not seem to conceive of a quantum system as "interacting," despite his use of the term "interaction." Nor does he seem to see the parts of the system as "parts."[87] In fact, later in the same article, Bohr admits to the essential difficulty of trying to talk about any quantum phenomena "where no sharp distinction can be made between the behaviour of the objects themselves and their interaction with the measuring instruments."[88] Apparently, then, he has no problem with the nonlocal effects that trouble Einstein because Bohr sees any quantum system as a unified whole.

The Continuing Debate

In the ongoing debate between Einstein and Bohr, neither could understand the stubbornness of the other. For Einstein's part, he could not see why Bohr insisted on returning humanity to the center ring of the universe by giving observers the power to determine the nature of quantum reality simply by means of a measurement of that quantum system. Einstein was much more comfortable with an objective picture of the universe existing "out there" rather than one that exists in an indefinite state. Because Einstein began with this premise, he had to accept one of two conclusions. Either quantum theory is complete and quantum reality involved nonlocal events, or quantum theory is incomplete and could not account for certain local hidden variables that surely must exist.[89] As we have seen, Einstein could not accept the idea of nonlocal events. This was why he refers to such events as spooky action-at-a-distance. Therefore, he is forced to conclude that quantum theory is incomplete.[90] Bohr, in contrast, saw little difference between the role of the measurement in quantum theory and the observer's role in Einstein's own theory of relativity. Moreover, Bohr was profoundly puzzled by Einstein's stubborn refusal to see that quantum theory is simply a variation of Einstein's own position in other theoretical areas.[91] Einstein did not relent and neither did Bohr, and so the debate remained unsettled for many years. It was not until John Bell created his interconnectedness theorem in 1964 that a possible method for settling the debate was at hand.

5-5 EXPERIMENTAL VERIFICATION OF QUANTUM WEIRDNESS

John Bell, a theoretical physicist working at CERN, the European Organization for Nuclear Research, attacked the problem of EPR in 1964 and, by using mathematical logic, demonstrated the limits that would exist in the interaction between two widely separated duons. Although Bell was "deeply impressed" by Einstein's insistence that quantum theory is an incomplete view of reality, Bell's approach to solving the paradox of EPR does not appear to suffer from any bias in Einstein's favor.[92] In fact, in an article entitled "On the Einstein-Podolsky-Rosen Paradox," Bell notes:

> The paradox of Einstein, Podolsky, and Rosen was advanced as an argument that quantum mechanics could not be a complete theory but should be supplemented by additional variables. These additional variables were to restore to the theory causality and locality. In this note that idea will be formulated mathematically and shown to be incompatible with the statistical predictions of quantum mechanics.[93]

It is clear, then, that whatever Bell may have thought as he first attacked the EPR paradox, his final conclusions remain an objective interpretation of the proposition.

Bell's Interconnectedness Theorem

Recall that Einstein's unwillingness to admit to the possibility of nonlocal effects led him to conclude that the present version of quantum theory is incomplete and that a superquantum theory must exist. Once this hypothetical superquantum theory is unraveled, physicists will be able to determine the existence of local hidden variables that will help them to determine the results of any quantum measurement without referring to nonlocal effects.[94] **Bell's interconnectedness theorem** begins with the assumptions that for any version of the EPR experiment, local hidden variables do exist.[95] With this assumption in mind, Bell developed a theorem that allowed him to determine a limit to the interaction between duons in the measurement of a quantum system, given the existence of those local hidden variables. This limit is referred to as **Bell's inequality**. If the limit is not surpassed, then local hidden variables rule. However, if the interaction continues beyond Bell's limit, this constitutes a violation of Bell's inequality. Such a violation cannot be explained by local hidden variables. Rather, it could be explained if we admit either to the indefinite nature of the quantum universe or to the existence of nonlocal effects.[96] The violation of Bell's inequality would demonstrate that Bohr was correct in his belief that nonlocality exists and the spooky action-at-a-distance Einstein so abhorred is a reality.[97] Thus, Bell's mathematical proof would provide a way to determine an actual winner of the Bohr–Einstein debate.[98]

Unfortunately for Bell, the technology needed to test the essential truth of his theorem was not in place in the 1960s. The proof would, in fact, have to wait a few more years. However, as early as 1972, John Clauser, a physicist at Berkeley, constructed an experiment using photons to test Bell's inequality. Much to the surprise of Clauser, who apparently believed in the existence of local hidden variables, the experiment succeeded in violating Bell's inequality.[99] Bohr's faith in the completeness of quantum theory was on the road to vindication. However, several difficulties were later pinpointed in Clauser's work. Consequently, although there were other experiments conducted after Clauser's work, it was not until 1982 at the University of Paris, that Alaine

Aspect and his colleagues Jean Dalibard and Gerard Roger devised an experiment capable of testing Bell's interconnectedness theorem to the satisfaction of most physicists.[100]

The Aspect Verification Experiment

Aspect and his team of experimenters used a laser to excite calcium atoms to the energy level necessary for their polarization experiment. As a result, correlated pairs of photons were released from the atoms. The polarization of the photons was neutral.[101] The photons were then sent in opposite directions toward two separate acoustooptical switches. The switches could alter the photon stream from transmission to reflection states every ten billionth of a second. This would determine the polarizer toward which the photons would travel. One polarizer blocked the photons while the other transmitted them according to a set of exact probabilities.[102]

The experimenters then used photomultipliers to ascertain what happened to the photons and an electronic coincidence monitor to tally any synchronous activity. The measuring devices were so accurate that they could catch the photons before they had moved more than a few yards from the source. The most crucial step in the test was to figure out the degree of correlation of the polarization of the photon pairs. The results showed the type of correlation that violated the Bell limit. Aspect, Dalibard, and Roger had, at long last, named the winner of the Bohr–Einstein debate. And the winner was clearly Bohr.[103] The crucial facet of Aspect's work was the ability of the experimenters to measure the results of the experiment in such a way that any possibility of communication between the photons was eliminated. This was accomplished by placing the devices used for measuring far enough apart so that any chance of a causal connection is removed. The actual distance in this experiment was 15 meters. Aspect refers to this partition as an "Einstein separability." The establishment of the **Einstein separability** means that the correlation between the photons cannot be caused by any type of "local" event.[104]

A team of physicists working at the University of Geneva took the concept of Einstein separability one step further when in 1997 they conducted an experiment that confirmed nonlocality at a distance of seven miles. The center of the experiment was located in Geneva, where photons were shone into a potassium-niobate crystal, splitting them into pairs of photons. Each photon in a pair possessed less energy than the original photon but the total energy of the pair equaled that of the original. The Geneva team then transmitted the stream of photon pairs on the fiber-optic telephone lines that run between Geneva and the towns of Bellevue and Bernex. The total distance between the origin point and the detectors in the two outlying towns was seven miles. As expected, the experimenters found a degree of correlation between the photon pairs that violated the Bell limit and confirmed nonlocality.[105]

The implications of this may not be immediately clear, so let us take a moment to review the relationship between Bell's theorem and the Aspect experiment. Since the polarizations of the photons correlate instantly, despite the distances involved, it would appear that Bell's inequality has been violated. Recall that this violation cannot be explained by local hidden variables. However, such a violation could be explained if we admit to the existence of nonlocal effects.[106] Clearly, the violation of Bell's inequality inherent in the results of the Aspect experiment demonstrates that nonlocality exists and therefore that "spooky action-at-a-distance" is a reality.

5-6   THE IMPLICATIONS OF NONLOCALITY

How can the high degree of correlation between the photons in the Aspect experiment be explained? Do the photons somehow communicate with one another, despite the fact that such a feat would require a faster-than-light signal? Do the results of the Aspect experiment demonstrate that the quantum universe does not exist as an independent, objective reality? Or has Aspect simply added a new twist to a debate that will eventually be explained by the elusive superquantum theory that Einstein sought? Naturally, there are no definitive answers to these questions. However, several physicists have created theories that attempt to explain the implications of nonlocality. Although it is always unwise to generalize about such things, we can say that there are two schools of thought in this regard. For the sake of simplicity and clarity, if not for complete accuracy, we will label these schools of thought the *quantum school* and the *classical school*.

The Quantum School of Nonlocality and Unity

The quantum school would, in general, hold that at least one of the two assumptions about the nature of quantum reality will have to be sacrificed to satisfy the predictions of Bell's theorem and the results of the Aspect experiment. In other words, to satisfy these results we must admit either to the indefinite nature of the quantum universe or to the existence of nonlocal effects. Most individuals would opt to preserve a definite existence of the quantum universe by admitting to the reality of nonlocal effects, despite the fact that such a position means dealing with the paradox of the faster-than-light correlation suggested by the results of the Aspect experiment.[107] The acceptance of the existence of nonlocality, however, introduces its own set of troubling implications, which several physicists and philosophers have attempted to explain.

The physicist David Bohm, for example, author of one of the most popular textbooks on quantum theory, suggested that the acceptance of nonlocality means that at the subatomic level at least, the entire universe is a single unified reality.[108] Michael Talbot explains Bohm's position in relation to Aspect's experiment by noting that the apparent distinctiveness of the photons in Aspect's experiment is largely illusionary. The photons are not separate entities communicating with one another via faster-than-light signals. Instead, they are interconnected, as all parts of the universe are connected, at a deeper, more elementary level of existence. But even that explanation is poor because, at that deeper level, the "parts" of the universe really are not "parts" at all, but are an outward manifestation of a single cosmic entity.[109]

This image of the universe introduces an entirely new outlook on reality. No longer can nature be seen as a haphazard gathering of separate physical entities. Instead, the universe is characterized by a unified wholeness. Some commentators have attempted to explain Bohm's idea of a unified universe by using images such as "an interrelated web of forces" or "a vast blanket of interconnected energy." Bohm, however, suggests that we should resist the image of the universe as some sort of interconnected web. Instead, he prefers to see all entities within the universe carrying within them the inherent "image" of all other entities within the universe. The entire universe is thus "enfolded" within each entity in the universe. The objects that we encounter in the macroscopic universe are a temporary unfolding of this enfolded universe, but it is the enfolded universe that is the ultimate "stuff" of existence. Bohm refers to the enfolded nature of the universe as the **implicate order**, and he labels the continuous unfolding of the enfolded universe as the **holomovement**.[110]

Nor is Bohm alone in his belief that the universe is a unified whole. Nick Herbert, a physicist and author of *Quantum Reality: Beyond the New Physics*, appears to hold a position that is remarkably similar to the one held by Bohm. Herbert underscores the significance of Bell's theorem and its relationship to the results of the Aspect experiment. In *Quantum Reality*, he emphasizes, for example, that Aspect's experiment verified the wholeness of a universe that appears to be dominated by local, causally connected events but is actually supported by an imperceptible, underlying, interconnected reality.[111] Herbert lays out a vision of the universe that is remarkably similar to the implicate order and the holomovement of Bohm. Unlike many commentators, in his explanation Herbert is careful to point out the connection between Bell's theorem, the Aspect experiment, and our everyday experience. He begins by reminding us that the quantum universe is not located "out there" somewhere. Rather, everything in the universe, including each of us, is a quantum system. Moreover, all quantum systems have always interacted with each other. In this way all parts of the universe are connected with one another. However, the nature of this connectedness should not be visualized as a vast blanket or an enormous web, but as an enfoldment of each quantum system within all other quantum systems.

Although the absolute nature of this enfoldment may remain forever beyond our comprehension, Bell's work has revealed that it is not a fabrication or a convenient metaphor. Instead, it is the essence of reality. As such, according to Herbert, we cannot deny its existence.[112] In *The Tao of Physics*, Fritjof Capra, who has done high energy physics research at the University of Paris, the University of California, and Stanford University, says it another way. In explaining the implications of nonlocality, Capra writes that the activity of matter and its being are inseparable.[113] In this view the entire universe becomes an interwoven tapestry of interconnected dynamic relationships.[114]

The Classical School of Locality and Reality

Of course, not all physicists are eager to make this leap. In fact, despite the results of Aspect's experiment, many still prefer the predictable deterministic universe of Newton and Einstein. Some attribute the difficulties of quantum theory to the inadequacies of language. Chief among the antinonlocality school of thought is David Klyshko of Moscow State University. In a 1994 paper entitled "Quantum Optics: Quantum, Classical, and Metaphysical Aspects," Klyshko demonstrates his disdain for all interpretations of quantum physics that attempt to see quantum phenomena in metaphysical terms.[115]

Klyshko admits that in the formative years of quantum theory there may have been some need for metaphysical language. However, according to Klyshko, that time has passed, and the requirement for metaphysical language has passed along with it. Unfortunately, from Klyshko's perspective, a mythical thesaurus of metaphysical terms associated with quantum theory has endured in the minds of some physicists. To solve this problem, Klyshko calls for the development of a standard set of guidelines that will "red flag" metaphysical terms and distinguish them from properly used scientific language.[116] According to Klyshko, properly used scientific language includes the language of quantum theory itself, that of classical theory and that of semiclassical theory, a position which sees the atom in quantum terms but which looks at field theory from a classical perspective.[117] The problem of language is made even worse because metaphysical expressions are often interspersed within explanations that use the other, more legitimate, languages.[118] Despite all his talk about terms and terminology, Klyshko's real objection is not merely that metaphysical language is unscientific, but that such language also represents an interpretation of quantum theory that he and others in the classical school find untenable. In short, Klyshko objects to any interpretation that cannot be verified according to the accepted scientific method.[119]

He is especially critical of the concept of nonlocality, which he labels as an unjustified and deceptive term and which he argues can be used only in a limited symbolic sense. Again, it is not the language that Klyshko objects to as much as the concept that the language represents. He states flatly that there is no such thing as nonlocality in quantum theory. Therefore, the only legitimate use of the term should be to represent symbolically the fact that classical and quantum predictions cannot be spoken of in the same language.[120]

Other physicists, notably Heinz Pagels, have had similar negative reactions to the implications of nonlocality. Pagels, in fact, appears to deny the existence of nonlocality altogether, at least as an accurate reflection of reality. He does not deny the results of experiments like those performed by Aspect. In fact, he readily admits that such experiments appear to demonstrate the existence of nonlocality.[121] However, he also notes that all attempts to verify the existence of Bell's inequality change the initial state of the experiment and, thus, change the results. Even if the results tell us that Bell's inequality has been violated (and they will tell us this), the fact that the initial state of the experiment has been changed means that we are forever prevented from pointing to the violation of Bell's inequality as proof of nonlocality. The only thing we have demonstrated is something we knew all along. Any experimental endeavor to measure duons interferes with those duons and changes the results of the experiment.[122] Therefore, Pagels concludes that a violation of Bell's inequality cannot be seen as a verification of nonlocal effects.[123]

Another authoritative voice within the classical school is the Nobel Prize–winning physicist Murray Gell-Mann, who in his recent study *The Quark and the Jaguar*, concludes that physical effects cannot propagate from one duon to another, despite the claims of those commentators who support nonlocality. In fact, Gell-Mann goes as far as to label such claims "flapdoodle."[124] He argues convincingly that such interpretations would violate the fact that nothing in the known universe can travel faster than light. He is even more critical of those commentators who claim that the phenomenon of nonlocality might be responsible for paranormal events. Such a conclusion, Gell-Mann insists, would not only upset the ideas of classical physics but also those of quantum physics. In fact, such claims would amount to a total rejection of the known laws of nature.[125]

Other physicists, who have had reactions not quite as severe as Klyshko's, Pagels, and Gell-Mann's, nevertheless, become very conservative in their descriptions of the results of Aspect's work. Instead of claiming that the universe displays a unified nature, these physicists cautiously describe the entangled state of the photons as "correlated" rather than "connected." Similarly, instead of describing the relationships as nonlocal, some physicists prefer to talk about local hidden variables that will forever remain undetected by even the most sensitive of instruments. There are also those physicists who choose to talk of hidden dimensions that permit local events to appear nonlocal. Finally, some physicists act as if the whole issue will go away if they simply choose to ignore it.[126]

5-7 THE CHILDREN OF THE COPENHAGEN INTERPRETATION

The Copenhagen Interpretation of quantum theory has earned a reputation as the standard theory of quantum physics. As such, it occupies the same position of orthodoxy that the theory of the big bang occupies in cosmology or that the theory of evolution occupies in biology. This by no means indicates, however, that it is unchallenged. The best that can be said is that, like the theory of the big bang, it has yet to be dislodged from its position of prominence. As previously recounted, according to the Copenhagen Interpretation complementarity, wave-particle duality, and uncertainty cannot be explained away as problems caused by defects in the measuring apparatus or by an incomplete understanding of the theory. Rather, these three principles are an integral part of the underlying nature of reality. In other words, reality is inherently unknowable.[127]

In the Copenhagen Interpretation it takes an *act of observation* to change what the equations report as a probability within a quantum system to what the observer sees as an actuality. Thus, there is always an inherent gap between our knowledge of the duon and the duon itself. If the Copenhagen Interpretation is taken to its logical extremes, must we conclude that there is no objective reality at the quantum level? Probably not, according to Niels Bohr. In fact, on many occasions Bohr expressly discourages any unwarranted mystical interpretation of quantum phenomena.[128] This is not to say that the Copenhagen Interpretation is without radical implications. On the contrary, it suggests a very real limit to our ability to understand the true nature of quantum events.

In *The Arrow of Time*, Peter Coveney and Roger Highfield point out the truly radical nature of Bohr's position when they note that, under the Copenhagen Interpretation, any attempt to describe the subatomic universe is limited by the weaknesses of language used as an interpretation of sense data. According to this point of view the universe is both classical, at the level of the measurement itself, and quantum, at the level of the duon being measured. This means that the universe that we perceive at the macroscopic level is real but still rests on the "unreal" subatomic universe. Since this limit cannot be avoided, physics will never be able to grasp the fundamental nature of quantum events.[129] As noted, there are many physicists who regard this position as unsatisfactory. Some of these physicists are in the Klyshko/Pagels/Gell-Mann camp and therefore are certain that any gaps in our knowledge about the subatomic realm are caused by the incomplete nature of the theory rather than by something inherent in the subatomic universe. Others, however, have moved beyond the Copenhagen Interpretation into realms that make even the bizarre implications of Bohr's position seem conservative by comparison.[130]

The Princeton Interpretation

The **Princeton Interpretation** is a radical point of view that takes quantum theory one step beyond the Copenhagen Interpretation and places the responsibility for quantum reality squarely within the mind of the conscious observer. In his book *Physics for Poets*, Robert March suggests the label "Princeton Interpretation" for this point of view because its foremost disciples, Eugene Wigner, John von Neumann, and John A. Wheeler, worked at Princeton for many years.[131] Nick Herbert, who prefers the simpler label "Copenhagen Interpretation, Part II," explains that at the heart of this position is the belief that reality is *created* by the observer.[132] This means that, although the subatomic universe really does exist, subatomic events are nonexistent until observed. When someone is present to observe a subatomic phenomenon, that phenomenon materializes. Otherwise, in the absence of an observer, the phenomenon does not take place.[133]

The essential difference between the Copenhagen Interpretation and the Princeton Interpretation is that in the former there are no elementary phenomena without acts of observation, while in the latter the acts of observation actually create the elementary phenomena. Thus, the Princeton Interpretation of quantum reality gives the act of observation a central role in the creation of the quantum phenomena. In fact, it seems as if the entire universe would not exist if there were no observer events to give rise to that universe.[134] It is clear then that from Wheeler's point of view the existence of the phenomenal universe may depend upon the existence of observers.[135]

In his book *At Home in the Universe*, Wheeler dramatizes this concept with the story of an imaginary conversation between the universe and an observer. The universe boasts that it provides both space and time for the observer to inhabit. The observer is thus reduced to a role of infinite insignificance. The observer, however, responds by pointing out that the universe is simply a conglomerate of individual phenomena. These phenomena would be meaningless, however, without acts of observation to register their existence.[136] Taking all of this into account, it is but one small step to the belief that the universe exists to provide an environment for conscious observers.[137] Wheeler says it better, however, when he notes, "nothing is more astonishing about quantum mechanics than its allowing one to consider seriously on quite other grounds the same view that the universe would be nothing without observership as surely as a motor would be dead without electricity."[138]

The Subquantal Reality Theory

The most curious aspect of Wheeler's point of view is the unexpected feedback loop hidden within the observer-created universe. The loop works like this: the physical universe becomes real only because conscious observers exist to view that universe; however, those sentient observers are created by the physical universe that their observation is responsible for bringing into existence.[139] Which came first, the observer or the universe? E. H. Walker of the Aberdeen Research and Development Center and Nick Herbert of the CORE Physics Technologium in Boulder Creek, California would have no problem with this question. They would answer that quantum processes are responsible for the emergence of human consciousness. This theory is put forth by Walker and Herbert in their article "Hidden Variables: Where Physics and the Paranormal Meet."[140]

Walker and Herbert begin their article by reviewing Bell's theorem and pointing out that Bell's inequality requires that any variables that would be a part of a "subquantal reality" have to be of the nonlocal variety.[141] According to Walker and Herbert, this **subquantal reality** exists beneath the quantum level and is responsible for all quantum phenomena. Although the nature of this subquantal reality cannot be explained at the present time, its mechanism will be uncovered at some time in the future.[142]

The idea of a subquantal reality that is responsible for quantum events is not in itself remarkable. In fact, it springs from the idea previously suggested by Herbert that everything in the universe is a quantum system and that all quantum systems have in the past interacted with each other.[143] In this way all parts of the universe are connected to one another by way of this subquantal reality. However, in "Hidden Variables: Where Physics and the Paranormal Meet," Walker and Herbert take the idea a few steps further. They suggest that this subquantal reality is the essential source of consciousness from which emerges the sentient nature of human intelligence. They go on to propose that the interaction of the subquantal level of thought and the physical universe may be responsible for some rather astonishing paraphysical phenomena.[144] Moreover, Walker and Herbert also briefly outline several experimental attempts to verify their theory.[145]

If, in fact, such experiments do verify the theories proposed by Walker and Herbert and if human thought is, therefore, subject to nonlocal subatomic phenomenon, then experiences like those of Jessica and Ursula recounted in the opening commentary are actually instances of nonlocality at work. The nonlocal events that affect the thoughts of Ursula are linked at the subquantal level with Jessica's thoughts. Thus, it is not surprising that Jessica awakens at the precise moment that Ursula has her accident. Nor is it surprising that Jessica "feels" the pain that Ursula experiences even though they are many miles apart. On the subquantal level at least, the separation between the two sisters is virtually nonexistent. Or to put it somewhat playfully, ESP may be verified by EPR.

Before we get carried away with this rather fanciful interpretation, however, we should note that Walker and Herbert admit that theirs is a minority position and that most physicists staunchly resist the type of link between quantum events and paraphysical phenomena that they suggest.[146] In fact, the connection that Walker and Herbert propose may be little more than fanciful conjecture. After all, the nonlocal interaction between widely separated, albeit entangled, subatomic particles is still a random, unpredictable interaction. Because of the random nature of this interrelationship the nonlocal connection between the widely separated duons cannot be used to send signals back and forth. Thus, the apparent superluminal aspect of nonlocality may not be the explanation for ESP that it at first appears to be.[147]

The Many-Worlds Theory

Another intriguing way to address some of the problems associated with quantum physics is offered by the Many-Worlds Theory of Hugh Everett and Bryce DeWitt. Everett of Princeton University and, more recently, DeWitt of the University of Texas, were working independently on a solution to the Uncertainty Principle of quantum physics when they first suggested the **many-worlds theory**, which proposes that every time a quantum measurement is made by someone, somewhere, a new parallel universe comes into existence.[148] The operation of the human mind as a sentient physical network is tied directly to quantum phenomena. This means that every time a person makes a decision of any type, that person is involved in a quantum transition that creates a new parallel universe. Every person in the universe inhabits each of these universes. Moreover, every one of these separate universes has a history that is unique to that universe.[149]

The story does not end here, however. Each of these newly created universes branches off, creating another universe every time a decision is made within that new universe, and so on ad infinitum.[150] The many-worlds theory could also solve the dilemma of Schrödinger's cat. Recall that in Schrödinger's thought experiment, the cat in the box exists in a live/dead state until the box is opened and the observer, by his act of observing, collapses the wave function into either the live or the dead state. According to Everett and DeWitt, the wave function does not collapse but exists in two universes, one in which the cat is alive and one in which the cat is dead. In the many-worlds theory, the wave function never collapses to a single state, but continues to exist in a multiplicity of universes.[151]

As might be expected, the many-worlds approach to quantum physics is not accepted by most physicists. However, it does seem to be a favorite among some of the more outspoken representatives of that field. In *The Physics of Immortality*, for instance, Frank Tipler states rather emphatically that the conclusions of the many-worlds theory are the inevitable result of the mathematics of quantum theory. In effect, Tipler concludes that the many-worlds interpretation is the only accurate interpretation of quantum theory.[152] Other experts, notably Paul Davies and Amit Goswami, are not quite as certain about this as Tipler. In *The Mind of God*, for example, Davies says that the many-worlds theory is inadequate because the existence of the parallel worlds created by these quantum transitions can never be verified.[153] Similarly, Goswami notes in *The Self-Aware Universe* that it is somewhat futile to speculate on the presence of parallel universes that do not interact with one another and whose existence, therefore, can never be verified.[154]

As the foregoing comments suggest, we must resist the impulse to take the implications of quantum theory too far. As we have seen, quantum theory, even in its most extreme forms, does not deny the existence of the real world. Neither Bohr, Wheeler, Everett, nor DeWitt would say that the subatomic universe does not exist. Nor would they argue that quantum events are not predictable. They would simply note that our knowledge of that predictability is limited to a set of probabilities rather than a set of absolute certainties.

CONCLUSION

The various facets of quantum weirdness explored in this chapter indicate that the world of quantum phenomena is one of the strangest fields encountered in modern science. The discoveries of quantum theory place a high premium on the interaction between the subject of an observation and the observer of that subject. Moreover, those same discoveries suggest that the apparent separateness of the universe is an illusion. Such discoveries cannot help but have an effect on other fields of knowledge. Ideas such as the uncertainty principle, the principle of complementarity, the Copenhagen Interpretation of quantum theory, and the phenomenon of nonlocality have reached beyond the boundaries of physics and, rightfully or wrongfully, for better or worse, have infiltrated philosophy, literature, law, and ethics. Or is it the other way around? Have the uncertainties of the modern world affected physicists, whose work in quantum physics simply mirrors the disintegration of the modern world? Certainly this is a possibility. However, before we surrender to this somewhat pessimistic point of view, let us explore another important theory of modern science, the theory of relativity. Perhaps the answers provided by relativity will be more down to Earth than those presented by quantum physics. But, then again, maybe not.

Review

5-1. In this chapter, we explored the growth of quantum theory as it reached maturity in the twentieth century. This included a look at Heisenberg's uncertainty principle, the Copenhagen Interpretation, the Bohr-Einstein debate with regard to quantum theory, the Einstein-Podolsky-Rosen thought experiment, Bell's theorem, Aspect's University of Paris experiment, and the mystery of nonlocality.

5-2. Heisenberg's matrix mechanics had been originally rejected by Schrödinger because Heisenberg's equations predicted quantum events that could not be visualized. Eventually, this conflict was resolved when Schrödinger and Dirac demonstrated that the two sets of equations consistently reached matching conclusions. The second point of confusion fostered by Schrödinger's equation lay within the nature of the wave itself. Max Born eventually resolved the apparent contradiction by concluding that the wave that Schrödinger spoke of in his equation was not a physical wave but a probability wave. When the problems of quantum probability became apparent, Werner Heisenberg began his search for the nature of quantum indefiniteness. The result of that search was Heisenberg's uncertainty principle. The uncertainty principle states that an experimenter can never pinpoint both the momentum and the position of a subatomic particle (a duon). The phenomenon of wave-particle duality reveals that duons can sometimes act as particles and sometimes as waves. Bohr's complementarity principle, which is clearly an either-or proposition, can, therefore, be summarized in the following way: *Either* we describe phenomena in the context of time and space, in which case, we must deal with the uncertainty principle, *or* we describe causal relationships mathematically, in which case, we lose the ability to visualize those relationships in terms of time and space.

5-3. Bohr's explanation of the uncertainty principle, wave-particle duality, and the complementarity principle is called the Copenhagen Interpretation. According to the Copenhagen Interpretation, complementarity, wave-particle duality, and uncertainty cannot be explained away as problems caused by defects in the measuring apparatus or an incomplete understanding of the theory. Rather, the three principles are an integral part of the underlying nature of reality. Einstein saw Bohr's attempt to explain quantum weirdness as a misleading attempt to resolve the essential problems of quantum theory. He believed that, if quantum theory failed to reveal all the answers associated with the subatomic universe, then it was the theory and not nature that was at fault. To Einstein, the underlying reality of the universe was solid and unchanging. Quantum theory indicated otherwise only because it was an incomplete theory.

5-4. Einstein's approach to the debate with Bohr was to challenge Bohr's position continually with a variety of cleverly constructed but highly controversial thought experiments. A thought experiment is one that cannot actually be carried out because of the practical limitations that exist at the time the experiment is first imagined. The most famous of Einstein's thought experiments appeared in an article that the physicist coauthored with Boris Podolsky and Nathan Rosen. The article introduced a thought experiment that purported to demonstrate the incomplete nature of quantum theory. The experiment, which came to be known as EPR, after Einstein, Podolsky, and Rosen, challenged the very heart of the Copenhagen Interpretation. In the EPR article Einstein created a thought experiment that he believed demonstrated that the subatomic universe has a definite objective existence and that quantum theory, as it existed at the time, was an incomplete theory. Not to be outdone, Bohr penned a response to the EPR thought experiment, in which Bohr attempted to discredit the EPR argument by noting that the very formulation of the thought experiment devised by Einstein contained an essential ambiguity when applied to quantum phenomena. Bohr concluded that the measurement of one duon really tells us nothing about the momentum (or position) of the other duon, or perhaps more precisely, the measurement of the one duon affects the state of the other duon, thus making the quantum state of the second duon indefinite. Einstein labeled Bohr's idea as spooky action-at-a-distance. Einstein did not relent and neither did Bohr, and so the debate remained unsettled for many years. It was not until John Bell created his interconnectedness theorem in 1964 that a possible method for settling the debate was at hand.

5-5. John Bell, a theoretical physicist working at CERN, the European Organization for Nuclear Research, attacked the problem of EPR in 1964 and, by using mathematical logic, demonstrated the limits that would exist in the interaction between two widely separated duons. Bell's theorem begins with the assumptions that the quantum universe has a definite, objective existence and that local hidden variables do exist. With these assumptions in mind, Bell developed a theorem that allowed him to determine a limit to the interaction between duons in the measurement of a quantum system, given the existence of those local hidden variables. This limit is referred to as Bell's inequality. Alaine Aspect and his team of experimenters at the University of Paris created an experiment to test Bell's inequality. Aspect and his team used a laser to excite calcium atoms to the energy level necessary for their polarization experiment. As a result, correlated pairs of photons were released from the atoms. The polarization of the photons was neutral. The photons were then sent in opposite directions toward two separate acoustooptical switches. The switches could alter the photon stream from transmission to reflection states every ten billionth of a second. This would determine the polarizer toward which the photons would travel. One polarizer blocked the photons while the other transmitted them according to a set of exact probabilities. The experimenters then used photomultipliers to ascertain what happened to the photons and an electronic coincidence monitor to tally any synchronous activity. The measuring devices were so accurate that they could catch the photons before they had moved more than a few yards from the source. The most crucial step in the test was to figure out the degree of correlation of the polarization of the photon pairs. The results showed the type of correlation that violated the Bell limit. Aspect and his colleagues had, at long last, named the winner of the Bohr–Einstein debate, and the winner was clearly Bohr.

5-6. How can the incredibly high degree of correlation between the photons in the Aspect experiment be explained? There are two schools of thought in this regard—the quantum school and the classical. The quantum school would, in general, hold that at least one of the two assumptions about the nature of quantum reality will have to be sacrificed to satisfy the predictions of Bell's theorem and the results of the Aspect experiment. Thus, to satisfy these results we must admit either to the indefinite nature of the quantum universe or to the existence of nonlocal effects. Most individuals would opt to preserve a definite existence of the quantum universe by admitting to the reality of nonlocal effects, despite the fact that such a position means dealing with the paradox of the faster-than-light correlation suggested by the results of the Aspect experiment. Of course, not all physicists are eager to make this leap. In fact, despite the results of Aspect's experiment, the classical school of locality and reality still prefers the predictable deterministic universe of Newton and Einstein. Some attribute the difficulties of quantum theory to the inadequacies of language. Others deny the existence of nonlocality altogether. Still, the ultimate scientific and philosophical ramifications of nonlocality are as yet unknown.

5-7. The Copenhagen Interpretation of quantum theory is the standard theory of quantum physics. This by no means indicates, however, that it is unchallenged. The best that can be said is that it has yet to be dislodged from its position of prominence. The Copenhagen Interpretation, complementarity, wave-particle duality, and uncertainty cannot be explained away as problems caused by defects in the measuring apparatus or an incomplete understanding of the theory. Rather, these three principles are an integral part of the underlying nature of reality. In other words, reality is inherently unknowable. Some theories have moved beyond the Copenhagen Interpretation into realms that make even the bizarre implications of Bohr's position seem conservative. The Princeton Interpretation, for example, is a radical point of view that takes quantum theory "one step beyond" the Copenhagen Interpretation and places the responsibility for quantum reality squarely within the mind of the conscious observer. According to the theory of subquantal reality, all parts of the universe are connected to one another by way of a subquantal reality. Moreover, this subquantal reality is the essential source of consciousness from which emerges the sentient nature of human intelligence. The Many-Worlds Theory proposes that every time a quantum measurement is made by someone, somewhere, a new parallel universe comes into existence. As noted above, the operation of the human mind as a sentient physical network is tied directly to quantum phenomena. What this means is that every time a person makes a decision of any type, that person is involved in a quantum transition that creates a new parallel universe. Every person in the universe inhabits each of these universes. Moreover, every one of these separate universes has a history that is unique to that universe.

Understanding Key Terms

Bell's inequality	EPR	quantum weirdness
Bell's interconnectedness theorem	holomovement	radiation
	implicate order	radioactive decay
	many-world theory	Schrödinger's cat
Bohr-Einstein debate	matrix mechanics	subquantal reality
complementarity principle	momentum	superquantum theory
Copenhagen Interpretation	nonlocality	thought experiment (gedanken experiment)
	Princeton Interpretation	uncertainty principle
duon		
Einstein separability	probability wave	wave function

Review Questions

1. What is the relationship between Schrödinger's and Heisenberg's equations?
2. What are the problems with Schrödinger's wave equation?
3. What is Born's probability wave solution?
4. What is the nature of Heisenberg's uncertainty principle?
5. What is the substance of the Copenhagen Interpretation of quantum physics?
6. What is the nature of the Bohr-Einstein debate?
7. What are the details of Einstein's EPR thought experiment?
8. What are the essentials of Bell's interconnectedness theorem?
9. What are the procedures involved in Aspect's verification experiment?
10. What is nonlocality?

Discussion Questions

1. Both Einstein and Bohr were steadfast in their beliefs about quantum theory. Neither one would give in to the other. Einstein held firmly to his belief that we would one day find a superquantum theory that would explain quantum phenomena; Bohr was equally convinced of the indefinite state of the subatomic universe. Which position do you find more convincing? Explain.
2. As noted in this chapter, it is important to remember that Bohr was a philosopher as well as a physicist. In this regard, he is unlike many physicists who, as we have seen, are not only indifferent to philosophic interpretations of physics but are also openly hostile to any attempt to draw philosophic conclusions from quantum theory. Do you think that it is appropriate for scientists to draw philosophical conclusions from their theoretical work? Or would you agree with David Klyshko, who seems to believe that there is no place for philosophy in physics? Explain.
3. A thought experiment is one that cannot actually be carried out because of the practical limitations that exist at the time the experiment is first imagined. Many physicists have a predilection toward the creation of these thought experiments because such experiments help relate the mathematical abstractions inherent in quantum theory to the physical universe. What do you think about the validity and usefulness of such thought experiments? Explain your answer. Now ask yourself about the usefulness of Einstein's EPR thought experiment. Did you see the usefulness in that experiment? What about Schrödinger's cat experiment? Did that thought experiment help you see some of the problems with the Copenhagen Interpretation? Explain.
4. Walker and Herbert suggest that a subquantal reality may be the essential source of consciousness from which emerges the sentient nature of

human intelligence. They also propose that the interaction of the subquantal level of thought and the physical universe may be responsible for some rather astonishing paraphysical phenomena. Do you believe in this hypothesis? Explain.
5. The physicist David Bohm has suggested that the acceptance of nonlocality means that at the subatomic level at least, the entire universe is enfolded and unfolded through a phenomenon he labels the holomovement. What do you think of Bohm's proposition? Explain.

ANALYZING *STAR TREK*

Background

The following episode from *Star Trek: The Next Generation* reflects some of the issues that are presented in this chapter. The episode has been carefully chosen to represent several of the most interesting aspects of the chapter. When answering the questions at the end of the episode, you should express your opinions as clearly and openly as possible. You may also want to discuss your answers with others and compare and contrast those answers. Above all, you should be less concerned with the "right" answer and more with explaining your position as thoroughly as possible.

Viewing Assignment—*Star Trek: The Next Generation*, "Sarek"

In this episode, the Starship Enterprise has been commissioned to transport Ambassador Sarek to a crucial diplomatic meeting. At this meeting, Sarek, a Vulcan, is to negotiate with the Legarans to settle the terms of their entry into the United Federation of Planets. Sarek has boarded the Enterprise with his wife and two staff members. Almost immediately after Sarek boards the Enterprise, crew members begin to experience a series of inexplicable random acts of violence. Eventually, Dr. Crusher and Counselor Troi determine that the cause of the violence is Sarek, who is suffering from Bendii syndrome. Apparently, Bendii Syndrome affects Vulcans in the later stages of old age. One of the primary symptoms of the disease is an inability to control emotions. This is especially critical in Vulcans because they are telepathic. Therefore, the loss of emotional control is not localized but affects others, even those located a great distance from Sarek. Those individuals who are affected by this nonlocal interference also lose emotional control—hence the sudden appearance of incidents of random violence on board the starship. Captain Picard informs Sarek of the findings of Dr. Crusher and Counselor Troi. A confrontation between the two men convinces Picard that Sarek has indeed lost emotional control and that this loss of control threatens the success of Sarek's mission. He, therefore, resolves to cancel the negotiations. However, at the suggestion of Sarek's wife, Picard reconsiders and decides instead to undergo a mind meld with

Sarek. The mind meld will allow Picard to serve as a receiver of Sarek's emotions, thus keeping Sarek under control and preventing the spread of violent flareups elsewhere in the vicinity by localizing the emotional responses in Picard. While Sarek conducts the negotiations, Picard undergoes a series of emotional outbursts. However, by undergoing the mind meld, Picard has managed to buy enough time for Sarek, who completes the mission.

Thoughts, Notions, and Speculations

1. How could the phenomenon of nonlocality be used to explain the influence that Sarek's emotional breakdown has on other crew members of the Enterprise? Explain.
2. How might each of these physicists respond to the use of nonlocality to explain the nonlocal loss of emotional control experienced by the crew of the Enterprise due to Sarek's condition: Albert Einstein, Niels Bohr, David Bohm, David Klyshko, Fritjof Capra, Murray Gell-Mann, John A. Wheeler, and Heinz Pagels? Explain your response in each case.
3. E. H. Walker and Nick Herbert suggest that subquantal activity is the essential source of consciousness from which emerges the sentient nature of human, and presumably Vulcan, intelligence. They also propose that the interaction of the quantum level of thought and the physical universe may be responsible for some rather astonishing paraphysical phenomena. Could this explanation be behind the effects of the Bendii syndrome? Could it not also explain the Vulcan ability to mind meld? Explain your response to each question.
4. David Bohm suggests that we should resist the image of the universe as some sort of interconnected web. Instead, he prefers to see all entities within the universe carrying within them the inherent "image" of all other entities within the universe. The entire universe is thus "enfolded" within each entity in the universe. The objects that we encounter in the macroscopic universe are a temporary unfolding of this enfolded universe, but it is the enfolded universe that is the ultimate "stuff" of existence. Bohm refers to the enfolded nature of the universe as the implicate order, and he labels the continuous unfolding of the enfolded universe as the holomovement. Use the concepts of the implicate order and the holomovement to defend the proposition that the type of experiences that Picard undergoes while mind melded to Sarek are actually possible.
5 Now reverse your position. This time use the work of David Klyshko, Heinze Pagels, and Murray Gell-Mann to dismiss such a proposition as meaningless. Then explain whether you personally believe the position stated in questions 4 or 5. Explain.

NOTES

1. Paul Davies, *The Cosmic Blueprint*, 165–66.
2. David Lindley, *Where Does the Weirdness Go?* xii.
3. Ibid.
4. Max Jammer, *The Philosophy of Quantum Mechanics*, 31–33; Jeremy Bernstein, *Cranks, Quarks, and the Cosmos*, 54–55; Peter Coveney and Roger Highfield, *The Arrow of Time*, 118–19; Daniel J. Kevles, *The Physicists*, 164.
5. Bernstein, 55; Coveney and Highfield, 118–19; Kevles, 163–64.
6. Bernstein, 55; Coveney and Highfield, 119; Kevles, 164; Robert H. March, *Physics for Poets*, 218; Saverio Pascazio, Univ. of Bari and the National Inst. of Nuclear Physics, August 22, 1996; Fred Alan Wolf, *Taking the Quantum Leap*, 108.
7. John D. Barrow, *The World within the World*, 132.
8. Bernstein, 55.
9. Max Born, *Atomic Physics*, 92 and 95; Jammer, 31.
10. Jammer, 39; Kevles, 165; March, 214.
11. Born, *Atomic Physics*, 95; Jammer, 38–39.
12. Born, *Atomic Physics*, 95; Bernstein, 55–56; Kevles, 165; March, 214–16.
13. Saverio Pascazio.
14. Coveney and Highfield, 121; March, 216–17. Recall that Laplace had very strong convictions about the predictable nature of the universe. It was Laplace who said, "Given for one instant an intelligence which could comprehend all the forces by which nature is animated and the respective situation of the beings who compose it—an intelligence sufficiently vast to submit these data to analysis—it would embrace in the same formula the movements of the greatest bodies of the universe and those of the lightest atom; for it, nothing would be uncertain and the future, as the past, would be present to its eyes." See Pierre Simon, Marquis de Laplace, *A Philosophical Essay on Probabilities*, 4.
15. Werner Heisenberg, *The Physical Principles of Quantum Theory*, 20; Jammer, 58, 61–63. Jammer points out that Heisenberg actually used a number of different terms for his concept of uncertainty. In fact, in Heisenberg's original 1927 paper Jammer could locate only three uses of the term "uncertainty" (Unsicherheit), two of which were in the postscript. This is especially interesting since Niels Bohr's influence is evident in the writing of the postscript. Jammer points out that Heisenberg preferred the terms inexactness or impreciseness (Ungenauigkeit) and precision or degree of precision (Genauigkeit). Another term which appears only twice is "indeterminacy" (Unbestimmtheit). Jammer, 61.
16. Heisenberg, 20–30; Jammer, 38, 61–63; Paul Davies, *God and the New Physics*, 102; March, 218–19.
17. Robert M. Hazen and James Trefil, *Science Matters*, 66–69.
18. Jammer, 89.
19. Heisenberg, 62–63.
20. Ibid., 63.
21. Niels Bohr, *Atomic Physics and Human Knowledge*, 39; Heisenberg, 62–63.
22. Heisenberg, 62–63; Jammer, 86–87, 91.
23. Bohr, 39–40, 90; Heisenberg, 63; Jammer, 86–87, 91.
24. Bohr, 39–40; Heisenberg, 62–64.
25. Heisenberg, 63.
26. Stephen Hawking, *A Brief History of Time*, 19–20.
27. Heisenberg, 63.
28. Ibid., 65.
29. Bohr, 39–41; Heisenberg, 62–65.
30. Heisenberg, 65.
31. Jammer, 87.
32. Peter Gibbins, *Particles and Paradoxes*, 47; Jammer, 87.
33. Jammer, 87.
34. Gibbins, 47–48.
35. Bohr, 40, 90; Roger S. Jones, *Physics for the Rest of Us*, 180; Heinz R. Pagels, *The Cosmic Code: Quantum Physics as the Language of Nature*, 76.
36. Aage Peterson, "The Philosophy of Niels Bohr," in *Niels Bohr: A Centenary Volume*, eds. A. P. French and P. J. Kennedy, 305.
37. Jones, 180; March, 226.
38. Pagels, 76.
39. Bohr, 40–41, 74, 90; Pagels, 76–77.
40. Bohr, 40–41, 74, 90; Pagels, 74–75.
41. Bohr, 40–41; John Gribbin, *Schrödinger's Kittens and the Search for Reality: Solving the Quantum Mysteries*, 16.
42. Gribbin, 10; Lindley, 61.
43. Gibbins, 11–12; Gribbin, 10.
44. Lindley, 70.
45. Nick Herbert. *Quantum Reality*, 17–18.
46. Barrow, 150.
47. Aage, 299; Pagels, 75.
48. Bohr, 73, 91.
49. Ibid., 73.
50. Ibid., 88.
51. Gribbin, 17; M. Y. Han, *The Probable Universe*, 63; Hazen and Trefil, 66–68.

52. Lindley, 70.
53. Bohr, 88–91.
54. Ibid., 91.
55. Ibid., 72.
56. Ibid.
57. Erwin Schrödinger, *My View of the World*, 21, 95, 106.
58. Ibid., 95.
59. March, 227–28.
60. Barrow, 152; Coveney and Highfield, 130–31; March, 227–29; Wolf, 189–91.
61. Danah Zohar, *The Quantum Self*, 39–40.
62. Pascazio.
63. Davies, *God and the New Physics*, 102.
64. Bohr, 32–33.
65. Ibid., 39–40.
66. Ibid., 33.
67. Ibid.
68. Kevles, 167.
69. March, 108.
70. Gibbins, 49.
71. Ibid., 23.
72. Albert Einstein, Boris Podolsky, and Nathan Rosen, "Can Quantum-Mechanical Description of Physical Reality Be Considered Complete?" eds. A. P. French and P. J. Kennedy, 146; Bohr, 58–59; Herbert, 201; Barrow, 146–47; see also Chapters 6 and 7 of Max Jammer, *The Philosophy of Quantum Mechanics* for an in-depth study of the entire EPR controversy.
73. Pascazio.
74. Bohr, 59; Jones, 184.
75. Niels Bohr, "Can Quantum-Mechanical Description of Physical Reality Be Considered Complete?" 146; Bohr, *Atomic Physics*, 59.
76. Bohr, *Atomic Physics*, 60; see also the abstract at the opening of the Niels Bohr article, "Can Quantum-Mechanical Description of Physical Reality Be Considered Complete?," 146.
77. Bohr, *Atomic Physics*, 60.
78. Ibid., 60–61.
79. Bohr, *Atomic Physics*, 60–61; Jones, 184–85.
80. Bohr, *Atomic Physics*, 60–61; Jones, 184–85.
81. Jones, 185; Pagels, 141; Lindley, 3–8.
82. Herbert, 214; Gibbins, 116.
83. Jones, 186; Pagels, 141–42.
84. Gibbins, 118; Pagels, 142.
85. Bohr, *Atomic Physics*, 61.
86. Ibid.
87. Bohr, *Atomic Physics*, 61; Jones, 185–86.
88. Bohr, *Atomic Physics*, 61.
89. Pagels, 141.
90. Ibid.
91. Herbert, 201.
92. Ibid., 211–12.
93. J. S. Bell, "On the Einstein-Podolsky-Rosen Paradox," 14.
94. Bell, 14; Gibbins, 118; Pagels, 142.
95. Bell, 14–15; Herbert, 212; Pagels, 142–43.
96. Davies, *God and the New Physics*, 105–106; Herbert, 222; Pagels, 142–43; Michael Talbot, *Beyond the Quantum*, 31–32.
97. Barrow, 147; Talbot, 31–32.
98. Talbot, 31–32.
99. Herbert, 225–26; Talbot, 32.
100. Herbert, 226; Talbot, 32.
101. Alaine Aspect, "Alaine Aspect," 40–41; Talbot, 32–33; P. C. W. Davies and J. R. Brown, eds., "The Strange World of the Quantum," 17–19; Herbert, 226–27.
102. Talbot, 32–33; Davies and Brown, "The Strange World of the Quantum," 17–19; Herbert, 226–27.
103. Aspect, 41–43; Talbot, 32–33; Davies and Brown, 17–19; Herbert, 226–27.
104. Aspect, 41–42.
105. Robert Pool, "Score One (More) for the Spooks," 53.
106. Aspect, 42–44; Herbert, 226–27; Talbot, 31–32.
107. J. C. Polkinghorne, *The Quantum World*, 76.
108. David Bohm, *Unfolding Meaning*, 7.
109. Michael Talbot, *Mysticism and the New Physics*, 146.
110. Bohm, 12–13.
111. Herbert, 227.
112. Ibid., 223.
113. Fritjof Capra, *The Tao of Physics*, 192.
114. Ibid., 81.
115. David Klyshko, "Quantum Optics: Quantum, Classical, and Metaphysical Aspects," 13–27.
116. Ibid., 14–15.
117. Ibid.
118. Ibid., 15.
119. Ibid.
120. Ibid.
121. Pagels, 150–51.
122. Ibid.
123. Ibid.
124. Murray Gell-Mann, *The Quark and the Jaguar*, 172–73.
125. Ibid., 173.
126. Talbot, *Beyond the Quantum,* 36–39.
127. Barrow, 150.
128. Bohr, 91.

129. Coveney and Highfield, 124.
130. Herbert, 16–29.
131. March, 228.
132. Herbert, 17–18.
133. Ibid.
134. John A. Wheeler, *At Home in the Universe*, 292.
135. Ibid.
136. Ibid., 128.
137. March, 228.
138. Wheeler, 39.
139. Paul Davies, *The Mind of God*, 224–25.
140. E. H. Walker and Nick Herbert, "Hidden Variables," 245–55; See also the following articles included in *Future Science*: William Tiller, "The Positive and Negative Space/Time Frames as Conjugate Systems," 257–79; J. H. M. Walker, "The Convergence of Physics and Psychology," 289–308.
141. Walker and Herbert, 248.
142. Ibid., 248.
143. Herbert, 223.
144. Ibid.
145. Ibid., 253–55.
146. Ibid., 250.
147. Barrow, 148; Alastair Rae, *Quantum Physics Illusion or Reality*, 46–47, 72–74.
148. Davies, *God and the New Physics*, 116; Amit Goswami, *The Self-Aware Universe*, 81; Frank J. Tipler, *The Physics of Immortality: Modern Cosmology, God and the Resurrection of the Dead*, 167–69; Zohar, 32–33.
149. Davies, *God and the New Physics*, 116; Paul Halpern, *Cosmic Wormholes*, 191–93; Tipler, 168–69.
150. Davies, *God and the New Physics*, 116; Halpern, 193; Tipler, 168–69.
151. Tipler, 168–69.
152. Ibid., 169.
153. Davies, 190.
154. Goswami, 81.

Unit IV The Theory of Relativity

Chapter 6
The Special Theory of Relativity

COMMENTARY: THE ASSASSINATION OF ADOLF HITLER

In 1991, John Byrne, a highly innovative writer of speculative fiction, penned "The Man Who Made Tomorrow," an interesting fable that included several perplexing time-travel paradoxes.[1] One of the most intriguing paradoxes in the story concerned a mission into the past to assassinate Adolf Hitler. The time traveler in Byrne's narrative plans his assassination attempt for a date well in advance of the advent of World War II. His objective is, of course, to eliminate Hitler before he has a chance to start the war and, as a result, to avoid the war altogether. Up to this point in the narrative, Byrne's plot is not unlike many others. Various writers have hatched time-travel stories in which the time traveler attempts to alter the past. In many of these stories, however, the attempted alteration of the timeline does not take place, and the history of the "real" world is preserved. Byrne, however, has a different outcome in mind. In his story, the time-traveling assassin succeeds. Hitler is eliminated, and the Second World War is prevented. Unfortunately, the results of this time-travel mission are most unexpected. Instead of creating a future paradise, the opposite occurs. The future turns out to be much more dehumanizing and infinitely more dangerous than anything that ever happened in the "real" world. Is Byrne, in some strange and twisted way, lending his moral approval to the death and destruction caused by World War II? Certainly not. His objective is to point out that tampering with the past is, at best, a risky business and, at worst, downright dangerous. At the end of his fable Byrne is careful to note that such dangers are remote, because, as far as he is concerned, time travel is a fantasy. However, as we shall see in this chapter and the next, Byrne may be somewhat overconfident. Albert Einstein's theory of relativity does not eliminate the possibility of time travel. In fact, the theory actually suggests the possibility that time travel could one day become a reality.

CHAPTER OUTCOMES

After reading this chapter, the reader should be able to accomplish the following:

1. Discuss the nature of psychological relativity.
2. Describe relativity as understood by the philosophers of ancient Greece.
3. Discuss the relative nature of uniform motion.
4. Describe how the laws of physics operate within inertial systems.
5. Explain why Maxwell included the lumniferous ether in his theory of electromagnetism.
6. Explain how the existence of the ether was disproven.
7. Relate the conflict caused by the constant speed of light.
8. Explain Einstein's resolution of that conflict in the special theory of relativity.
9. Identify the time dilation effect.
10. Describe the phenomenon of relativity of simultaneity.

6-1   THE RELATIVE NATURE OF RELATIVITY

"All things are relative." Or so the philosophers of the popular press tell us. Certainly everyone has experienced the kernel of truth within this familiar but deceptively simple adage. Change is in the nature of things. We can hardly go through a single hour without experiencing the accuracy of this statement. Yet, the nature of relativity goes beyond the fact that change is inevitable. Relativity is manifested in a number of different ways. Perhaps the most familiar brand of relativity is psychological relativity.

Psychological Relativity

Psychological relativity demonstrates that the same experience can be seen from different perspectives. At times these perspectives are so dissimilar that a single occurrence can take on the nature of unrelated events. Consider, for example, the 1995 World Series. If you were a fan of the Cleveland Indians, you saw the World Series as a delightful vindication for forty-one years of baseball suffering. As a fan of the Atlanta Braves, you may have dreaded the thought of another lost opportunity and may have been just as happy to see your team sit this one out. Of course, as it turned out, as a Braves' fan you celebrated Atlanta's ultimate victory over Cleveland. If you were a fan of the Seattle Mariners or the Cincinnati Reds, in all likelihood, you were content to ignore the entire event since your teams had been given early tickets home. And, as a Native American, you would have viewed the entire series as the most politically incorrect sporting event in recent memory.

Although this type of relativity is very real to us, it is, nevertheless, clearly psychological in its nature and effect. It is caused by different backgrounds, experiences, nationalities, religions, and so on. Most people have also experienced **psychological time (subjective time)**, by which the nature of an experience may alter the perception of the duration of that event. If an individual is enjoying a series of events, then time passes very quickly. In contrast, if the experience is boring, painful, or unpleasant, time will drag interminably for that individual. Albert Einstein was not a stranger to the concept of psychological time. In fact, in explaining his theories, he takes great pains to explain the nature of psychological time, and to distinguish it from the changes in real time associated with special relativity:

> The experiences of an individual appear to us arranged in a series of events; in this series the single events which we remember appear to be ordered according to the criterion of "earlier" and "later," which cannot be analyzed further. There exists, therefore, for the individual, an I-time, or subjective time. This in itself is not measurable. I can, indeed, associate numbers with the events, in such a way that a greater number is associated with the later event than with an earlier one; but the nature of this association may be quite arbitrary.[2]

This common feeling of subjective time is not what Einstein had in mind, however, when he created the theory of relativity. Instead, Einstein introduced the world to a new understanding of time that is entwined within the fabric of reality itself.

The Ancient Greek Notion of Stability

Despite our familiarity with psychological relativity, common sense tells us that the Earth itself is not subject to this notion of relativity. On the contrary, common sense dictates that the Earth is stable and unmoving. If the Earth were moving, then we should certainly feel that motion. Moreover, we would observe objects toppling over, clouds rushing by, and people holding on to stable articles to prevent themselves from falling down, as the Earth moved forward. This was, in fact, the point of view adopted by many of the ancient Greeks. Although a few pre-Socratic philosophers thought that the Earth might be moving in a circle around a fiery hub of some sort, most Greeks held to the common-sense belief that the Earth was stable. The Earth's stability was assured because it resembled an enormous tree trunk with its roots firmly planted within the deepest region of the underworld.[3]

The pre-Socratic philosopher Anaximander even came up with an explanation for why the stable Earth, which stood at the absolute center of the universe, did not fall. According to Anaximander, since the Earth is at the center of the universe, it must be equidistant from all the extreme points in the universe. As a result, there is no possible way for the Earth to move closer to or farther away from any of those points. Therefore, it remains centrally located in a stable condition.[4]

Anaximander's work is also noteworthy because he proposed the idea of an absolute reality underlying the entire fabric of the universe. This concept also underlies the Greeks' apparent belief in the absolute, stable universe. Anaximander believed that the most basic material of the universe was an infinite substance he called *apeiron*. Exactly what he meant by an infinite substance is somewhat problematic. However, one interpretation would be that the apeiron is infinite or everlasting in a temporal sense. If the apeiron were not everlasting, then it would have had a beginning and that would require something that is not the apeiron from which the apeiron originated. However, if that were the case then the apeiron would not be the absolute reality underlying the universe.[5]

Another interpretation would be that the apeiron of Anaximander is spatially infinite as well. Anaximander probably did not think of the apeiron as being infinite in size. It is more likely that he visualized it as an enormous sphere surrounding and containing the entire universe. To imagine the apeiron as infinite in size would have required a knowledge of mathematics that the Greeks did not possess in Anaximander's time.[6] Another interpretation might be that Anaximander thought of the apeiron as infinite because it was inexhaustible. Logically, an inexhaustible supply of apeiron would make sense. After all, as far as Anaximander knew, all things came from the apeiron, and, because new things come into existence all the time to replace things that perish, the supply of apeiron from which those new things come must also have no quantifiable limit.[7] Aristotle would later dispute this idea, noting that things that perish do not vanish into nothingness. Instead, they are simply transformed from one state to another.[8]

More important to our present study, it is likely that Anaximander also thought of the apeiron as existing without boundaries. This belief in the indivisibility of the apeiron may have caused him some difficulty at first because the universe, as it presently exists, clearly consists of separate things delineated by boundaries. Such boundaries mark off the difference between the four primary elements of earth, water, air, and fire. According to Anaximander, however, there was no reason to give any one of these elements priority over the others. There was no reason, for example, to think, as Thales had taught, that water was the primary substance, giving rise to fire, earth, and air. It was far more logical to Anaximander that all four elements came from a common, neutral substance. This preexisting, common substance, the apeiron, originally contained those elements in equal measure.[9] Aristotle sums up Anaximander's belief in the infinite nature of the apeiron in his book, *Physics*, by noting, rather succinctly, that the apeiron has no beginning, will exist forever, and cannot be destroyed.[10]

Later in the same text when explaining Anaximander's belief in the indivisible quality of the apeiron, Aristotle says, "the infinite cannot be something that is separate from perceptible things and an independent being which is itself infinite. For if infinity is neither a magnitude nor a plurality, but is itself a primary being, not an accident of it, then it will be indivisible."[11] From this perspective, the apeiron sounds very much like the lumniferous ether that Maxwell suggested in the nineteenth century as the one absolute substance of the universe. There were, of course, those among the Greeks who opposed the idea of an absolute, infinite substance. The chief dissenter was Heraclitus, who promoted the idea that the universe is in a constant state of change. According to Heraclitus, things seem steadfast and unfaltering only because they exist in a delicate state of equilibrium between two opposing tendencies.[12]

Relativity According to Galileo and Newton

During the Middle Ages, the Greek notion of an absolute substance underlying the universe fit rather well with the Christian idea of an absolute, unchanging God ruling over the world. The unchanging God described by Thomas Aquinas in his *Summa Theologiae*, for instance, has much in common with the apeiron of Anaximander, at least in the sense that both the God of Aquinas and the apeiron of Anaximander serve as the absolute backdrop for all existence. Aquinas writes that God is indivisible, eternal, and unchangeable. In fact, he goes so far as to identify God and eternity as one and the same.[13] Nevertheless, the idea of an absolute, infinite substance pervading the universe was not to remain unchallenged. In the seventeenth century, Galileo's work went a long way toward nullifying the belief in an absolute substance and replacing it with the idea of relativity in physics. Later, Sir Isaac Newton's work went even further in helping us to understand the relative nature of motion. It was not until Maxwell's work on electromagnetism in the nineteenth century that problems began to appear. These problems, as we shall see, led Maxwell to the reintroduction of an absolute unchanging substance. In this case, that absolute substance was called the ether.

The Relativity of Uniform Motion. Galileo introduced relativity into the laws of physics when he determined that all uniform motion is relative. According to Galileo, everything in the universe is moving relative to everything else. This is referred to as **Galilean relativity**. In Galilean relativity it is, therefore, not possible to say in an absolute sense that an object is moving at X miles per hour. The only way movement can be measured accurately is relative to the movement of some other object.[14] As is true of most ideas in science, the notion of an absolute substance did not disappear quickly. In fact, Sir Isaac Newton, whose work on the laws of motion and gravity clearly supported Galilean relativity, did not welcome the idea of the relative nature of uniform motion. In fact, Newton refused to abandon the possibility that an absolute substance might exist at some level in the universe, even if that substance is undetectable from our region of space. As he notes in *Mathematical Principles*:

> It is a property of rest, that bodies really at rest do rest in respect to one another. And therefore as it is possible, that in the remote regions of the fixed stars, or perhaps far beyond them, there may be some body absolutely at rest; but impossible to know, from the position of bodies to one another in our regions, whether any of these do keep the same position to that remote body, it follows that absolute rest cannot be determined from the position of bodies in our regions.[15]

Let us consider some of the ramifications of Galilean relativity. Since the general principle of Galilean relativity confirms that nothing in the universe exists in a state of absolute rest, the movement of all things makes the movement of all other things relative. To understand this concept, we must begin with a definition of speed. The **speed** of an object is the rate at which it moves, without regard to its direction.[16] Consider, for instance, an automobile traveling on the interstate. The driver observes that her speedometer registers a speed of 65 miles per hour. This means that the car is moving so as to cover 65 miles in 1 hour of time. The speed displayed on the speedometer, however, is not absolute. On the contrary, it must be measured relative to something that is not moving. The speedometer uses the surface of the Earth as its reference point. This means that, relative to the surface of the Earth, the car is traveling at 65 miles per hour. However, this measurement is based upon the convenient fiction that the Earth is motionless. Yet we know that the Earth is not stationary. In fact, it is moving very rapidly in a variety of ways. It is rotating on its axis; it is orbiting the Sun; it is within a solar system that is revolving around the galactic center, and it is part of the Milky Way Galaxy, which is also moving in an assortment of different ways throughout the universe.[17]

However, for the sake of convenience, we ignore all of this movement and pretend that the Earth is not moving. This fiction allows us to consider the surface of the Earth as a fixed reference point. Now let us assume that the car is moving *north* with a **velocity** (speed in a specific direction) of 65 miles per hour relative to a police officer with a radar device standing at the side of the interstate. Imagine that another car moving north at 85 miles per hour is gradually gaining on the first car. Relative to our police officer with the radar gun, the second vehicle is moving at 85 miles per hour. Relative to the first vehicle, however, the second vehicle is moving at only 20 miles per hour. It is even more enlightening to imagine a vehicle moving south at 65 miles per hour toward the first vehicle that is still traveling north at 65 miles per hour. At the precise moment that the two vehicles pass each other, relative to one another, they are within the same frame of reference. This may be hard to visualize at speeds of 65 miles per hour, so imagine two slowly moving ocean liners passing within a yard of each other. At the precise moment that your porthole passes a porthole on the opposite ship, you could reach out and shake the hand of your counterpart on the other vessel just as easily as if the two of you were standing safely on the dock.[18]

Let us move the example off the surface of the Earth and on to a Boeing 747. At this point consider movement within the airliner. The flight attendant about to deliver drinks to the passengers occupies a place within the 747 and moves together with the 747. While he stands still, he occupies a position of rest relative to the 747 and everything in the airliner. However, relative to the Earth, which, for this example, we can consider at rest, he is moving with the same velocity as the plane. However, we know that the Earth is not really at rest, but is also in motion. Therefore, the actual movement of the flight attendant includes the movement of the plane and the movement of the Earth. Moreover, as he begins to deliver drinks to the passengers, his movement also includes that movement relative to the contents of the 747. Therefore, as he moves about the cabin of the 747, his movement consists of the movement of the earth, the movement of the plane relative to the earth, and his movement about the cabin of the 747 relative to the contents of the plane.[19] (Newton's examples in *Mathematical Principles* use a "ship under sail" rather than a jet airliner and a sailor rather than a flight attendant. Obviously, the principle remains the same.)

The Laws of Physics and Uniform Motion. Equally interesting is that, according to Galilean relativity, as these vehicles move at a uniform rate of speed, those individuals within the vehicles will experience no alteration in the laws of physics. This is why a flight attendant on an airliner can pour a cocktail for you while the two of you are moving at speeds in excess of 500 miles per hour.[20] In fact, if you were to pull down the plastic shade on your window, take out your calculator, and begin to work on your tax return, you would not even know that you were moving, let alone moving at more than 500 miles per hour relative to the surface of the Earth. Naturally, this presupposes that you experience no acceleration, no deceleration, and no turbulence, and that you have a very good pilot who can hold the plane on a steady course. Another way to say this is to note that the laws of physics remain the same in all **inertial systems**. Einstein said the same thing a bit more succinctly when he wrote, "every universal law of nature which is valid in relation to co-ordinate system C, must also be valid, as it stands, in relation to a co-ordinate system C^1, which is in uniform translatory motion relatively to C."[21] In other words, as long as the motion of that airplane remains uniform, you can play catch, role dice, shuffle cards, flip a coin, eat your dinner, and drink your coffee just as easily as if were on the "stationary" Earth.[22]

We could take the analogy even further. Let us suppose you have joined an aviator in the cockpit of an F-16. As the F-16 moves forward at 1000 miles per hour, your pilot fires a missile that moves away from the jet at 1000 miles per hour, relative to the moving jet. This means that, relative to an observer on the ground, the missile is moving at 2000 miles per hour. How do we arrive at this conclusion? It works like this. The missile was already moving forward at a speed of 1000 miles per hour before it was launched. Now it leaves the plane going in the same direction as the plane, and the pilot tracks its speed as 1000 miles per hour relative to the speed of the rapidly moving F-16. To arrive at the speed of the missile as it is observed from the ground, the pilot must add its speed of 1000 miles per hour relative to the F-16 to the speed it was traveling before it was launched. By adding the two speeds together the pilot arrives at an estimate of the speed of the missile relative to the observer on the ground. None of this should be surprising because it is all based on the idea that all uniform motion is relative.[23] However, a new card is about to be dealt from the bottom of the deck. That card involves the speed of light.

6-2   LIGHT AND THE MYSTERY OF THE ETHER

Despite initial resistance to the idea, the principles of relative motion began to seem like universal laws of nature. All this changed when light was introduced into the equation. With the introduction of light, the principle of relativity was threatened, and the possibility of an absolute substance within the universe was reintroduced. One of the major characters credited with introducing the problems associated with light is the Scottish mathematician and physicist, James Clerk Maxwell. Maxwell's equations in the nineteenth century led him to the development of the theory of electromagnetism, in which he demonstrated that light is a manifestation of the electromagnetic field.[24]

Electricity, Magnetism, and Light

To develop a proper picture of Maxwell's achievements, we must step back and look at the work of another nineteenth-century physicist, Michael Faraday. Faraday attempted to explain electricity and magnetism in terms of a field that envelops bodies in space. A **field** is a region encircling an object in which a force operates.[25] According to Faraday's vision, bodies in space add to the field and also react to the combined fields of all other bodies. It is likely that Faraday came up with this idea because of his familiarity with the work of the eighteenth-century physicist and Jesuit priest Ruder Boscovich. It was Boscovich who, with amazing foresight, proposed that matter and energy need no longer be seen as different entities. In fact, Boscovich went so far as to argue that atoms were really stable concentrations of force. With this idea in mind, it is not difficult to see how Faraday might conclude that bodies in space would produce a field of force and would react to the combined fields of other spatial bodies.[26]

To help him picture how a field moved between and among bodies in space, Faraday invented a way to draw the field using what he called **lines of force**. These lines of force could visually depict the direction of a force in a field as well as the strength of that force.[27] After conducting numerous experiments, Faraday noted a reciprocal relationship between electricity and magnetism.[28] What Faraday had discovered was that an electric field in motion will create a magnetic field. Moreover, the opposite proposition is also true; that is, a moving magnetic field will create an electric field.[29] As a result, Faraday realized that it no longer made sense to speak of electricity and magnetism as separate. Instead, they were manifestations of a single electromagnetic field.[30]

Maxwell picked up on Faraday's work and succeeded in expressing Faraday's unification of electricity and magnetism in exact mathematical form. As Maxwell worked with Faraday's ideas, he also noted something that Faraday had not noticed. Maxwell found that the transfer of energy within a field involves a delay that is equal to the speed of light.[31] This discovery led Maxwell to the belief that light can also be explained in terms of the electromagnetic field. Accordingly, Maxwell concluded in his landmark publication, *Treatise on Electricity and Magnetism*, that light is a manifestation of the electromagnetic field.[32] This led to an understanding that all light, not just visible light, but light of all frequencies, from radio waves to gamma rays, was a manifestation of the electromagnetic field.[33] Maxwell's equations also led him to another conclusion that would eventually be crucial to the development of the special theory of relativity. That conclusion was that in empty space, light always moved at a velocity of 3×10^8 meters per second. Again, this includes not just visible light, but light of all frequencies.[34]

An assumption that underlies Maxwell's work with electromagnetism and light is that light is propagated as a wave. Sir Isaac Newton had favored the view that light is actually a stream of particles. Nevertheless, despite Newton's endorsement of the particle theory, by the middle of the nineteenth century, experiments had demonstrated that light actually moves as a wave. In his *Treatise on Electricity and Magnetism*, Maxwell acknowledges the debate between those physicists who support the particle theory and those who support the wave theory. Maxwell, however, also declares his preference for the wave theory and, in fact, openly admits that, in writing his treatise, he plays the role of an advocate promoting the wave theory, rather than that of an impartial adjudicator judging the feasibility of both theories. Maxwell, of course, realized that the conflict between the two theories would have to be resolved at some point because, from his point of view at least, they seemed to represent opposing ideas that could not be reconciled.[35]

Each theory had its problems, of course. However, Maxwell was most concerned with the difficulties of the wave theory which stem from the fact that a light wave, like a sound wave or an ocean wave, must move through some sort of medium. The particle picture did not need such a medium because the force was either carried by the particles or somehow passed from particle to particle. Without the particles, the wave theory had to add a medium to carry the waves. Just as sound needs the air to carry sound waves, light needs a medium to carry light waves. This meant that space could not be empty but had to be filled with some sort of matter. Maxwell concluded that space was permeated with a medium called the *lumniferous ether*.[36] Fields were now seen as having an actual, physical existence and could be viewed as states of the ether.[37] The ether acted as the medium for light and as an absolute reference point by which the constant speed of light could be measured.

> When light is emitted, a certain amount of energy is expended by the luminous body, and if the light is absorbed by another body, this body becomes heated, shewing [sic] that it has received energy from without. During the interval of time after the light left the first body and before it reached the second, it must have existed as energy in the intervening space. . . . According to the theory of undulation (i.e., the wave theory), there is a material medium which fills the space between the two bodies, and it is by the action of contiguous parts of this medium that the energy is passed on, from one portion to the next, till it reaches the illuminated body.[38]

As it turned out, Maxwell's solution was incorrect. However, before we see why, we should look a little more closely at the mystery of the ether.

The Mystery of the Ether

Maxwell's addition of the ether to his theory of electromagnetism may seem to be a desperate attempt to explain an inexplicable problem. This is not exactly the case, however. In fact, the existence of the ether not only solved the wave problem, but also provided a way to address several other theoretical difficulties.

For example, it helped lead to the quantitative unraveling of several optical mysteries, such as the process of diffraction and refraction as well as the phenomenon of polarization. The ether was also used to explain the nature of electric charge and static electric fields and to show how they create the force of electricity. Perhaps even more impressive is that the nature of the ether suggested the existence of both electrons and antielectrons before either had been detected.[39] Despite the eventual usefulness of the ether concept, Maxwell seems almost apologetic when he introduces it into his *Treatise*:

> To fill all space with a new medium whenever any new phenomenon is to be explained is by no means philosophical, but if the study of two different branches of science has independently suggested the idea of a medium, and if the properties which must be attributed to the medium in order to account for electromagnetic phenomena are of the same kind as those which we attribute to the lumniferous medium in order to account for the phenomena of light, the evidence for the physical existence of the medium will be considerably strengthened.[40]

Part of Maxwell's uncertainty about the advisability of including the ether in his theory may be due to the fact that the actual composition of the ether was never explained with any degree of satisfaction. Clearly, the ether was thought to permeate all space. However, the ether was *not* something that existed simply as some sort of invisible "filling" for the vacuum of outer space. Rather, it was the fundamental, eternal essence of all the universe.[41] In fact, the atoms themselves were also made of the ether.[42] But what was the ether itself made of? Strangely enough, the prevailing image saw the ether as a substance that was both rigid and flexible at the same time. To describe the ether's existence as simultaneously rigid and flexible, Einstein used the term *quasirigid*.[43] The ether had to be flexible to allow for the passage of large, slowly moving objects like the planets. Yet it also had to be rigid to provide a medium for the light wave. The rigid nature of the ether would allow each layer of the ether to impact on each subsequent layer allowing the wave to pass from one layer to the next.[44]

The Michelson-Morely Test. The ether also provided a reference point for the speed of light. This is what Maxwell suggested when he came up against the problem presented by the unwavering speed of light. The light waves moved through the ether, but the ether itself did not move, giving Maxwell his absolute reference point. In other words, Maxwell guessed that his finding that the speed of light never varied was due to the fact that his equations were made only in relation to a single reference point, the unmoving ether. If this were the case, however, a different approach to measuring the speed of light should reveal the existence of the ether.[45]

Just such a different approach was taken by Albert Michelson and Edward Morley at the Case School of Applied Science in Cleveland in 1887. The idea behind their experiment was the assumption that, as the Earth orbits the Sun, it changes its position in relation to the ether through which it is moving.[46] Therefore, a light shining in one direction at all times will change its speed in relation to the unmoving ether as the Earth travels around the Sun. Now imagine that someone is positioned on the unmoving ether. That someone is observing the movement of the Earth and measuring the speed of the light. On January 1, the Earth is moving away from the person. At this time the velocity of the Earth should be added to the velocity of the light. Six months later, on July 1, the Earth is moving toward the person. Now the velocity of the Earth should be subtracted from the velocity of the light.[47]

Of course, the experiment does not have to be done on different days. The entire experiment can be carried out on the same day by changing the direction in which the light is aimed. At some time, the light will be moving against the ether and will be slowed down. When pointed in the opposite direction, it will be moving with the ether, and it will speed up. Actually what Michelson and Morley did was to split a single light so that the resulting beams were at right angles to each other. The beams were then reflected back by strategically placed mirrors, and the speed of the beams measured by a delicate set of instruments. If the Earth were traveling through the ether, then the two beams should be traveling at different speeds.[48]

The results should be analogous to two swimmers navigating a river. One swimmer is traveling downstream, while the other is attempting to swim across the river from one bank to the other. If all of the other variables are constant, then the swimmer traveling downstream would be moving faster than the one struggling against the current as he swam across the river. Since the light beams were traveling at right angles to each other, one beam, like the swimmer swimming downstream, should be traveling at a rate of speed different from that of the other beam. Michelson and Morley ran the experiment many times, each time aiming the beam in a variety of different directions. However, no matter what directions they chose, and no matter how many times they ran the experiment, the results were always the same. The speed of light did not vary. They could reach only one viable conclusion. The speed of light in empty space is constant, even though there is no absolute position against which to measure it.[49]

The Conflict Caused by Light

The problem with light is not that it has an absolute upper speed limit but rather that this absolute upper speed limit is the same for everyone and everything in the universe. This makes very little sense. If the ether did not exist then nothing in the universe exists in a state of absolute rest against which to measure the movement of all other bodies. As a result, the movement of all things makes the movement of all other things, including light, relative. The problem occurs because light does not operate this way. In a vacuum the speed of light is a constant 3×10^8 meters per second (300,000 kilometers per second).[50] The speed of light, which is usually represented by the letter c, is most often expressed as 186,000 miles per second. Therefore, that figure will most often be used in the following discussion and in Chapter 7. However, at times the figure may be converted to 300,000 kilometers per second, thirty billion centimeters per second, or 670 million miles per hour, depending upon the example or analogy in use at the time.[51]

The logic of the preceding argument would suggest that the velocity of light should change along with the velocity of its source. Let us try this with the F-16. As the F-16 approaches the landing field, it is moving at 1000 miles per hour. The pilot switches on the landing lights. How fast is the light moving? You might be tempted to answer that the velocity of the light is relative. That is, the speed of light will change, depending on whether the measurement is made by the pilot in the F-16 or by an observer on the ground. The pilot would say that the light is moving at 670 million miles per hour. The observer would say that light is moving at 670 million miles per hour plus the 1000 miles per hour presently clocked by the F-16 (i.e., 670,001,000 miles per hour).[52] This is not, however, what actually happens. Remember that the speed of light is constant. Einstein said it this way: "light in vacuo always has a definite velocity of propagation (independent of the state of motion of the observer or of the source of the light.)"[53]

Just as the speed of light was the same for Michelson and Morley, no matter how they directed their light beams, it will be the same for both the pilot in the F-16 and the observer on the ground. All observers will clock the speed of light in empty space at 670 million miles per hour. How is this possible? How can the speed of light appear to be absolute when there is no absolute reference point? Logically, the motion of light should obey the same laws of relative motion that everything else in the universe seems to obey. Yet, it does not. That is the mystery that Einstein set out to unravel in 1905 when he formulated his special theory of relativity.

6-3   THE SPECIAL THEORY OF RELATIVITY

Before delving into Einstein's solution to the dilemma of the light paradox, several myths about the theory of relativity should be dispelled. First, when Einstein set down his theory in 1905, it was not called the theory of relativity, let alone the special theory of relativity. The ideas that came to be known as the special theory of relativity actually appeared in *Annalen der Physik* in a 1905 paper entitled "On the Electrodynamics of Moving Bodies." The term *special*

• 186 • Chapter 6 The Special Theory of Relativity

Figure 6.1 Albert Einstein at About the Age of 26.
We are accustomed to seeing pictures of Albert Einstein as a white-haired scientist in his seventies. However, when he published his first paper on special relativity, "On the Thermodynamics of Moving Bodies" in 1905, he was only twenty-six years old.
Photo Credit: Corbis-Bettmann

theory of relativity came into use much later.[54] Second, contrary to the reverence with which it is viewed today, when it was first proposed by Einstein, the theory of relativity, and many of his other theories for that matter, were not well-received. In fact, they were widely rejected. For instance, Einstein's theories were repudiated in Germany and the Soviet Union. They were expressly rejected by an American cardinal who had the support of the Vatican, and they were labeled blasphemous by a Jewish newspaper.[55] Part of the difficulty that people had with Einstein's work was, of course, that the very nature of his discovery was revolutionary. However, a large part of the opposition was also due to the approach that Einstein chose to take. When faced with the conflict between the motion of material bodies and the propagation of light, Einstein did not attempt to adjust the data, revamp the old theory, or modify his thinking. Instead, he chose to rebuild the universe from the ground up.

Restructuring the Universe

In his text, *Relativity Visualized*, Lewis Carroll Epstein, professor of physics at the City College of San Francisco and the University of California at Berkeley, constructs a very revealing analogy that helps explain the risk that Einstein took in his attempt to restructure the universe. Epstein compares Einstein's problem to a homeowner's discovery that the front door of her house will no longer close all the way. The homeowner is faced with two choices. One choice that she has is to concentrate on the door and, by rehanging it, realign the door with the door frame. Most of us would choose this approach to the problem.[56]

However, there is a second way to solve the problem. The homeowner can leave the door alone and try to realign the rest of the house with the door. To realign the house with the door, she will have to readjust the very foundation upon which the house is built. She will have to jack up the foundation and realign the entire house with the front door. Of course, once she has succeeded in realigning the house with the front door, every other door in the house, and probably all of the windows, may be out of alignment. For this reason (as well as for practicality), the realignment of the entire house would be the incorrect approach to take in solving the problem of the misaligned front door.[57]

Nevertheless, this latter approach was the one that Einstein took. He elected to realign the entire house to fit the door. In this case, of course, the house that he chose to realign was the entire universe. The problem facing Einstein was the apparently unresolvable conflict between the principle of relative movement of material bodies and the unwavering speed of light. Einstein stated the conflict in the form of two assumptions that would serve as the basis of his special theory of relativity. The first assumption notes that

> a co-ordinate system that is moved uniformly and in a straight line relatively to an inertial system is likewise an inertial system. By the "special principle of relativity" is meant the generalization of this definition to include any natural event whatever: thus, every universal law of nature which is valid in relation to a co-ordinate system C, must also be valid, as it stands, in relation to a co-ordinate system C', which is in uniform translatory motion relatively to C.[58]

Einstein's second assumption was that the speed of light in a vacuum is constant, regardless of the source of the light or the position of a person observing the light. This assumption about the nature of the speed of light was supported by the work of Maxwell.[59]

It seemed as if one or the other assumption must be sacrificed. Either the principle of relativity had to be abandoned or the constant speed of light in empty space had to be disproven. The easiest approach would be to retain relativity and abandon the speed of light. Developments in physics, however, had demonstrated that this was not the proper course to follow. Other developments had shown that relativity could not be abandoned either. At this point, Einstein intervened and wrote his history-making paper, "On the Electrodynamics of Moving Bodies."[60]

The Resolution of the Conflict

Sherlock Holmes once remarked that once all other possibilities have been eliminated, whatever is left, no matter how improbable, must be the truth. In a very real sense, this was the course of action that Einstein took in resolving the conflict between the motion of material bodies and the propagation of light. Einstein believed that the speed of light in a vacuum would always be measured as 186,000 miles per second by everyone, regardless of the velocity at which they are moving. However, if the speed of light in empty space is the same for all observers, and, at the same time, the principle of Galilean relativity cannot be abandoned, then the resolution of the conflict must certainly lie elsewhere.[61]

The solution to the dilemma lay in the very nature of space and time. Einstein concluded, that if the speed of light does not change and the principle of relativity is to remain intact, then the spatial and temporal coordinates measuring objects moving at the speed of light must be different for different observers moving at different speeds. If an experimenter on the Moon turns on a searchlight, that experimenter will measure the speed of light at 186,000 miles per second. If a second experimenter on board a starship chases the searchlight at 100,000 miles per second, he will not measure the speed of light at 86,000 miles per second, as ordinary logic would dictate, but at 186,000 miles per second.[62] This is simply a restatement of the problem explored earlier in this chapter.

Because the speed of light remains the same for each experimenter, Einstein concluded that the factors used to determine the speed, that is, time and space, had to change for different experimenters based on each experimenter's motion. In order for both the first and the second experimenters to measure light at 186,000 miles per second, then the miles and the seconds, that is the space and the time, must change for each observer. In summary, Einstein decided that in order to maintain both relativity and the constant speed of light, Newton's concepts of absolute space and absolute time had to be discarded. Thus, in one fell swoop, Einstein eliminated two of the most fundamental assumptions of the Newtonian universe. Both time and space had become elastic and changeable.[63]

6-4 THE ALTERATION OF SOME BASIC ASSUMPTIONS

In deciding that time and space are elastic and changeable, Einstein revolutionized our way of looking at reality. Even today, almost a century after his historic paper, most people in the Western world have not yet come to grips with the truth of his theory of relativity, let alone the philosophical implications of this revolutionary way of looking at the universe. As was noted, one of the most significant implications of Einstein's revolution is the concept of elastic space-time. This elasticity has led to the abandonment of several long-standing assumptions about the nature of the universe. The abandonment of these assumptions is based upon Einstein's use of the Lorentz transformation.[64]

The Lorentz Transformation

Recall that Einstein decided that Newton's concepts of absolute space and absolute time had to be discarded to maintain both relativity and the constant speed of light. Newton saw space and time as absolute. Einstein replaced this idea with the notion that space and time change in relation to the movement of the observer. Einstein used this notion to develop a set of equations based upon the **Lorentz transformation**, which explains how the relationships among space, time, and mass are altered by relative motion. The Lorentz transformation is named for the Dutch physicist Hendrick Lorentz who, before Einstein, devised the transformation to explain the problems of the absolute speed of light in relation to the ether. Einstein, of course, in his 1905 paper, concluded that the ether was no longer needed to explain the phenomenon. Nevertheless, the transformation still bears Lorentz's name.[65]

Einstein concluded that the movement from one inertial system to another will be controlled by the Lorentz transformation. This means that an observer in an inertial system that is seen as stationary will detect four changes in the contents of a moving inertial system. First, the length of an object will shrink in the direction that the object is moving. Another way to say this is to note that the space that the object occupies will contract in the direction of the moving object. This is known as the **Lorentz contraction**. Second, the mass of the moving object will increase. Third, the faster an object travels, the slower time will move for that object relative to a stationary observer. This is known as **time dilation** or the **time dilation effect**.[66] Finally, the experience of a simultaneous present moment for the moving object and a stationary observer will vanish. This last conclusion is known as the **relativity of simultaneity**.[67]

As a consequence of these conclusions we can no longer legitimately see time and space as separate entities. If we take this step by step, we should be able to see why time and space must be joined together. First, as noted, the faster an object moves, the slower time will move for that object, relative to a stationary observer. Also, as the object increases its speed, the space that it occupies contracts. Putting these conclusions together we can say that the speed of an object will cause its space to contract and its time to expand. In a sense, we can, therefore, conclude that space is changed into time. The two entities, once thought to be independent, are not. Instead, we must see them as inextricably intertwined into one entity known as space-time or the **space-time continuum**.[68] Therefore, the geometry of space-time requires that we plot not only the three coordinates of space, but also a fourth coordinate for time.[69] In the next chapter we explore the effects of gravity on space-time. For now, however, we narrow our focus to time dilation and the relativity of simultaneity.

The Time Dilation Effect

In our study of time we are concerned with the third transformation, the time dilation effect, and the fourth transformation, the relativity of simultaneity. First, consider the time dilation effect—that the faster an object moves, the slower time moves for that object relative to a stationary observer.[70] At normal cruising speeds in everyday life, such alterations in time go unnoticed. Even if we take one of the fastest forms of human travel, the space shuttle, we find that when the shuttle orbits the Earth at five miles per second, its chronometer runs only one ten-millionth of a percent slower than a similar timepiece back at the Houston space center.[71]

However, once a moving object approaches the speed of light, the effects become much more readily apparent. A clock on board a starship traveling close to the speed of light will run more slowly relative to clocks moving at slower speeds. For the following examples, we will use the Earth as a reference point. If an astronaut boards a starship and travels at a rate that approaches light speed, a clock on the starship will run more slowly relative to a clock on the Earth. A mathematical formula called the **Lorentz factor** can tell us exactly how slowly the starship clock will run relative to the earthbound clock.[72]

For the sake of simplicity in the following examples, we will bypass the mathematical aspects and focus on the results of the Lorentz factor.[73] Suppose that we synchronize four timepieces at 6:00 A.M. The first timepiece stays on Earth. The other three timepieces will be loaded on starships, each one of which will travel at increasingly more rapid speeds. The second timepiece will be loaded on a starship that travels at a rate of 0.25 light speed, or $0.25c$. (As noted earlier, the speed of light is represented by the letter c. The third timepiece will travel at $0.5c$, and the fourth at $0.75c$. The Lorentz factor tells us that, when the timepiece back on Earth reads 12 noon, the timepiece traveling at $0.25c$ will read 11:48 A.M., whereas the one traveling at $0.5c$ will read 11:12 A.M., and the one moving at $0.75c$ will read 9:58 A.M. This is the time dilation effect in action. Time dilates, that is lengthens or expands for the moving clocks relative to the stationary clock back on the Earth.[74] (See Table 6.1 for mathematical verification of these figures.)

TABLE 6.1 • Time Dilation Verification

The final formula that Einstein used to measure time dilation is:

$$t = \frac{t^1}{\sqrt{1 - \frac{v^2}{c^2}}}$$

where t = time in the frame of reference considered stationary
t^1 = time in the moving frame of reference
v = speed of the moving frame of reference
c = light speed

We will consider the time elapsed on the third starship in the example used in the text. The clock on the third starship reads 9:58 A.M. This means that the time that has elapsed on that starship is 3 hours and 58 minutes or 3.96 hours. The starship was traveling at a velocity of .75c. With these two figures we can calculate the time that elapsed on the earthbound clock using the formula:

$$t = \frac{t^1}{\sqrt{1 - \frac{v^2}{c^2}}}$$

$$t = \frac{3.96 \text{ hours}}{\sqrt{1 - \frac{(.75c)^2}{c^2}}} = \frac{3.96 \text{ hours}}{\sqrt{1 - \frac{.5625c^2}{c^2}}} = \frac{3.96 \text{ hours}}{\sqrt{1 - .5625}} =$$

$$\frac{3.96 \text{ hours}}{\sqrt{.4375}} = \frac{3.96 \text{ hours}}{.661437828} = 5.98695725 \text{ hours}$$

If we round this to 6 hours, the time on the earthbound clock reads 12 noon.

Source: Vincent Icke, *The Force of Symmetry* (Cambridge: Cambridge University Press, 1995), 115–17.

We can illustrate this proposition by using a thought experiment involving an imaginary device referred to as a **Lorentz light clock**, which consists of two parallel mirrors, one of which is fastened to the ceiling of a room, while the other is located on the floor. Imagine that a light ray flashes from the top mirror to the bottom one. Each time that the ray hits the bottom mirror, it registers a tick, and each time it hits the top mirror it registers a tock. Each round trip, therefore, corresponds to an interval of time. Taking two such clocks, we allow one to remain on Earth and place the other in a starship that is moving relative to the stationary Earth. An observer on the Earth watching the earthbound Lorentz light clock will see the light ray bounce up and down vertically between the two mirrors. However, if that same observer could watch the bouncing light ray in the Lorentz clock on a starship, she would note that the light ray does not move vertically. Instead, from her perspective it moves along a slanted line. This means that the distance traveled by the slanted light ray on the moving clock is greater than the distance traveled by the vertical light ray on the stationary clock.[75]

Despite this difference, the speed of the light rays is always the same whether the observer watches the moving clock or the stationary one. This is because the speed of light remains constant, regardless of the motion of an observer. Since the speed remains the same, then the time that it takes for the light ray to flash from the top mirror to the bottom one will be different for the moving clock, as measured by the Earth observer. The observer on the Earth will measure fewer ticks and tocks for the moving clock when compared to the stationary clock. This would correspond to fewer time intervals for the moving clock. Time is, therefore, moving more slowly on the starship relative to time as measured on the Earth. The Lorentz factor can tell us exactly what that difference is, so that we arrive at the different times. Significantly, it does not matter whether we use a Lorentz light clock, a mechanical clock, an electric clock, an atomic clock, or a biological clock. The results are the same.[76]

Time Dilation and the Twins Paradox. The fact that time dilation also affects all clocks, including biological clocks, leads to some intriguing results. Consider, for example, the twins paradox. Let's imagine that two twins, Ashley and Megan, age thirty, have decided to take separate vacations (see Figure 6.2). Ashley has arranged to spend her vacation on a planet in a nearby star system. To reach the star system, her starship travels at 0.5c. She experiences the passage of one year on her starship, whereas her sister, Megan, who prefers to stay back on Earth, experiences the passage of 1.2 years. This difference in age—Ashley (31), Megan (31.2)—while quite real, is still somewhat negligible. On her next trip, however, which she immediately commences, Ashley travels at 0.99c. Again she experiences the passage of one year. Back on Earth, however, Megan experiences the passage of seven years. Megan is now 38.2 years old, whereas Ashley is only 32. Immediately upon returning, Ashley decides to take a trip to an even more distant star system. To make the round trip within a year, her starship must travel at the incredible velocity of 0.999c. When Ashley returns, she will be one year older. On the other hand, since Megan has experienced the passage of twenty-two years, she is now 60.2 years old.[77] (See Table 6.2 for mathematical verification of these figures.)

The Twins Paradox and Future Time Travel. It is important to recall that, since the laws of physics remain the same for all inertial systems, each twin is entitled to think that she is the one who is not moving. Therefore, each twin will be unable to perceive any change in the passage of her own time. As far as each twin is concerned, the laws of physics operate the same, and no difference in the passage of time is noted. The entire situation seems normal. Suppose the traveling twin is required to pull an eight-hour duty shift each day of her journey on the starship. She must get out of bed, freshen up, have breakfast, and report for duty each day at 0800 hours. She has lunch at 1200 hours

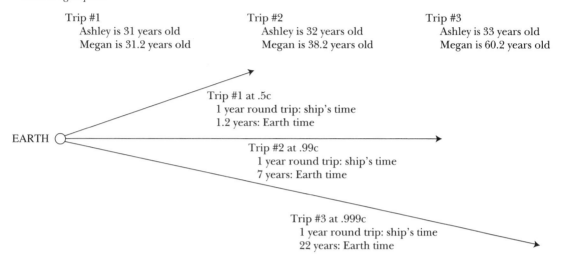

Figure 6.2 Time Dilation and the Twins Paradox.
The twin's paradox demonstrates the effects of time dilation. In this thought experiment, two twins, Ashley and Megan, age thirty, have decided to take separate vacations. Ashley has arranged to spend her vacation on a planet in a nearby star system (trip #1). To reach the star system, her starship travels at .5c. She experiences the passage of one year on her starship while her sister, Megan, who stays on Earth, experiences the passage of 1.2 years. This difference, although quite real, is still somewhat negligible. On her next trip (trip #2), however, Ashley decides to travel at .99c. Again she experiences the passage of one year. Back on Earth, however, Megan experiences the passage of seven years. Megan is now 38.2 years old whereas Ashley is only 32. Finally, Ashley decides to take a trip to an even more distant star system (trip #3). To make the round trip within a year, her starship must travel at the incredible velocity of .999c. When Ashley returns, she will be one year older. In contrast, since Megan has experienced the passage of twenty-two years, she is now 60.2 years old.

and goes off duty at 1700 hours. By 2200 hours, she is back in her quarters and fast asleep in anticipation of her next duty shift. During this twenty-four hour day, as measured by the ship's chronometer, she experiences the passage of time just as she would have on Earth. There is no sensation of speed nor any feeling of movement in fast motion. The only way that the traveling twin will know that there has been a difference between her time and time as experienced by her sister is to return to Earth.[78]

Remember that the laws of physics operate as usual for the traveling twin as long as she is in a state of uniform motion. So the laws of physics will operate for her in the same way that they operate on the "stationary" Earth. Therefore, while the traveling twin is moving at a uniform rate of speed, she will see her place in the universe as stationary. From the traveling twin's perspective, the

TABLE 6.2 • Mathematical Verification of the Twins Paradox

Using the Lorentz factor:

$$t = \frac{t^1}{\sqrt{1 - \frac{v^2}{c^2}}}$$

The time dilation effect for each trip described would look like this:

Before the first trip both Ashley and Megan are 30 years old.

Trip #1 Ashley travels at .5c for one year on the starship. The question before us is: How much time has passed for Megan on Earth?

$$t = \frac{1 \text{ year}}{\sqrt{1 - \frac{(.5c)^2}{c^2}}} = \frac{1 \text{ year}}{\sqrt{1 - \frac{.25c^2}{c^2}}} = \frac{1 \text{ year}}{\sqrt{1 - .25}} = \frac{1 \text{ year}}{\sqrt{.75}} = \frac{1}{.866} = 1.1547$$

For convenience we will round 1.1547 to 1.2. Therefore after Trip #1

> Ashley is 31
> Megan is 31.2

Trip #2 Ashley travels at .99c for one year on the starship. How much time has passed for Megan on Earth?

$$t = \frac{1 \text{ year}}{\sqrt{1 - \frac{(.99c)^2}{c^2}}} = \frac{1 \text{ year}}{\sqrt{1 - \frac{.9801c^2}{c^2}}} = \frac{1 \text{ year}}{\sqrt{1 - .9801}} = \frac{1 \text{ year}}{\sqrt{0.0199}} = \frac{1}{0.1407} = 7.1$$

For convenience we will round 7.1 to 7. Therefore after Trip #2

> Ashley is 32
> Megan is 38.2

Trip #3 Ashley travels at .999c for one year on the starship. How much time has passed for Megan on Earth?

$$t = \frac{1 \text{ year}}{\sqrt{1 - \frac{(.999c)^2}{c^2}}} = \frac{1 \text{ year}}{\sqrt{1 - \frac{.998001c^2}{c^2}}} = \frac{1 \text{ year}}{\sqrt{1 - .998001}} = \frac{1 \text{ year}}{\sqrt{0.001999}} = \frac{1}{.044710178} = 22.4$$

For convenience we will round 22.4 to 22. Therefore after Trip #3

> Ashley is 33
> Megan is 60.2

Source: Vincent Icke, *The Force of Symmetry* (Cambridge: Cambridge University Press, 1995), 115–17.

Earth would seem to be in motion. The same effect can be seen in a moving car. The passenger in the car is entitled to see his position as stationary, while the scenery can be seen moving past the car. Now, if the traveling twin could look back at the twin on the Earth, she would see that the earthbound twin's clock is running more slowly than the clock on the starship. But if this is the case, why does the traveling twin's biological clock move more slowly than that of the Earth-bound twin? Shouldn't each twin be able to declare that she is the younger of the two?[79]

This apparent contradiction lies at the heart of the twins paradox. The problem was first presented by an English astrophysicist named Herbert Dingle, who argued that, because the traveling twin would see the Earth moving relative to her apparently stationary position, and since she would see the clock of the earthbound twin moving more slowly than her clock, the problem was not one of actual time distortion but one of contrasting perspectives.[80] Dingle's argument is persuasive. After all, he is correct that each twin will see the other twin's clock moving more slowly than her own. What Dingle failed to consider, however, is that the traveling twin is the one who is really in motion relative to the stationary starting point, that is the Earth. The traveling twin has, therefore, experienced not only a constant state of motion, but also acceleration. She had to accelerate to leave the Earth and, on the return trip, she must accelerate again. It is at these points of acceleration that most of the time on Earth passes by, relative to the time on the starship.[81] It is the traveling twin who is younger because it is she who experiences the effects of the acceleration on her body. The role of acceleration was not immediately obvious in Einstein's formulation of the special theory of relativity. This fact explains the source of the original paradox. It was only when the general theory of relativity was formulated that the effects of acceleration could be factored into the problem.[82] As we shall see in the next chapter, with the advent of the general theory of relativity, Einstein formulated the principle of equivalence, which states that acceleration and gravity are equivalent.[83]

All of this would be just so much speculation if there were no experimental proof to support Einstein's equations. Fortunately, such experimental proof does exist. One such experiment was performed using the particle accelerator at the European Organization for Nuclear Research (CERN) at Geneva in 1966. The experiment in question involved the acceleration of **muons**, which are subatomic particles that are created when cosmic rays hit molecules located high in the Earth's atmosphere. Muons are especially useful for time dilation experiments because they undergo radioactive decay at an exceedingly rapid pace.[84] At CERN, muons were created by high-energy collisions in the particle accelerator. These muons were then magnetically directed to travel on circular paths until they reached speeds in excess of $0.997c$. As a result of this, the life expectancy of the muons was measured to have increased by eight to twelve times their normal life spans. These figures were well in line with those predicted by Einstein's relativity equations, thus confirming the validity of the time-dilation effect.[85]

The Relativity of Simultaneity

The fourth transformation, the relativity of simultaneity, has undermined another long-standing assumption about the nature of time. Prior to the advent of the theory of relativity, most people held firmly to the belief that there was a universal present moment for the entire universe. The theory of relativity has demonstrated that there is no such thing. Consequently, we must abandon the belief in the existence of universal simultaneous events and replace that belief with a new assumption, the relativity of simultaneity. The relativity of simultaneity, which is sometimes referred to as the nonuniversal nature of "nowness," holds that events that appear to be simultaneous in one inertial frame of reference may not be simultaneous in another inertial frame of reference. Moreover, since everyone is entitled to consider their frame of reference as stationary, there is no such thing as an absolute frame of reference, and, therefore, no way to label objectively any set of events as either simultaneous or sequential.[86]

Einstein explains the relativity of simultaneity by conjuring up the image of a train moving down a railroad track. The train is traveling at a uniform speed. To the passengers, the train is their inertial system, and they consider all events occurring in time from that perspective. Naturally, anything that occurs on the embankment that parallels the track also occurs at some point in relation to the train. A person standing on the embankment would also experience these events. However, do these events occur simultaneously when viewed from the train and from the embankment? Suppose, for example, that lightning strikes the track behind the train and in front of the train simultaneously at points A and B, respectively, while the train is moving toward point B. Will the passengers and an observer on the track see the lightning hit the track simultaneously? The answer is, "no."[87]

The difference is due to the motion of the train and the constant nature of the speed of light. The lightning strikes are considered to be simultaneous in relation to the midpoint M on the track, that is the point M which is equally distant from points A and B, and the point M' opposite M on the embankment. Therefore, the light from the lightning flashes will meet at M simultaneously, and also simultaneously at M'. An observer on the embankment standing at M' will see the flashes at exactly the same time. What about a passenger sitting on the train at M? If the train is not moving, the passenger at point M and the observer at point M' on the embankment are in the same inertial system, and they see the lightning flashes at the same time. However, if the train is moving toward point B, the passenger at M will not remain at the same position as the observer at M' on the embankment but will travel away from A and will, therefore, be slightly ahead of the light traveling from the lightning strike at A. Conversely, he is moving toward point B and is, thus, traveling to meet the light from the strike at point B.[88]

If the speed of the train could be added to and subtracted from the speed of the light as it approaches the train from points A and B, then the passenger on the train would see the lightning strikes as simultaneous. However, as we saw when we discussed the F-16 example on page 182, this is not what happens. Since the speed of light is always the same, the speed of the train cannot be added to the speed of light as the light travels from point A, nor can the speed of the train be subtracted from the speed of light as the light travels from point B toward the train. Therefore, the movement of the train toward B means that the passenger will see the lightning strike at B first, followed by the lightning strike at point A.[89] In other words, for the passenger, the events are not simultaneous.[90]

Another way to understand this is to visualize a passenger on a moving starship (system A) and an observer who is considered stationary (system B) relative to the moving starship. The passenger on the starship intends to synchronize two timepieces, one located in the nose of the starship, the other located in the tail. The synchronization will be accomplished by activating each timepiece by shining a light beam from the center of the starship simultaneously forward to the nose and backward to the tail. The passenger does so and synchronizes each clock. The clocks will not, however, be synchronized from the perspective of the observer in system B. The passenger in system A sees the two clocks as synchronized because, from her perspective, she is entitled to see her inertial system as unmoving. However, the observer in system B will see the starship in motion. Therefore, from his point of view, the timepiece in the nose of the starship is moving away from the light beam while the timepiece in the tail is moving to meet the light beam. As a result, from the perspective of system B, the timepiece in the tail will reach the light beam before the timepiece in the nose. The same result would *not* occur if the passenger in system A uses bullets to activate the timepieces. This is because even though the bullet traveling to the nose of the starship must travel further than the bullet moving toward the tail, the speed of the bullet moving forward must be added to the speed of the starship, whereas the speed of the bullet moving to the tail must be subtracted from the speed of the starship. Therefore, from the perspective of system B the bullets arrive simultaneously. This does not happen when light beams are used because the speed of the light beams is unaffected by the speed of the starship. Both the passenger in the starship (system A) and the observer (system B) will see the light move at the same speed.[91]

Einstein, therefore, concluded that incidents that are seen to happen simultaneously in one frame of reference may not appear simultaneous in other frames of reference. Because it makes no sense to ask which frame of reference is "correct," there is no way to conclude whether the events are simultaneous or sequential.[92] This principle applies to all similar situations. Since all inertial systems have their own time frame, there is no way to determine the time of an event unless we also know what inertial system we are asking about. Otherwise, the timing is all relative. This is the principle of the relativity of simultaneity. Einstein summarized it this way:

> Now before the advent of the theory of relativity it had always tacitly been assumed in physics that the statement of time had an absolute significance,

i.e. that it is independent of the state of motion of the body of reference. But we have just seen that this assumption is incompatible with the most natural definition of simultaneity; if we discard this assumption, then the conflict between the law of propagation of light *in vacuo* and the principle of relativity . . . disappears.[93]

Bergson's Objections

As might have been expected, the relativity of simultaneity played havoc with the established ideas of some of the most influential Western philosophers of Einstein's day. One such philosopher was the French scholar, Henri Bergson, who argued that all time exists as a single universal moment of simultaneity. According to Bergson, time's multiplicity, which in *Time and Free Will* Bergson appears to identify with duration, is given reality only by the perception of a conscious mind. Bergson believed that the only thing that exists is the present moment. Time's multiplicity, that is, time's duration, on the other hand, is a perception of the conscious mind, which can recall past states of simultaneity and, although Bergson never addresses this directly, presumably can anticipate future states of simultaneity. Given this line of thought, Bergson concludes that the only thing that really exists is the constant state of universal simultaneity for the entire universe.[94]

Bergson illustrates this point by referring to the oscillations of the pendulum of a clock. According to Bergson, the only oscillation that exists is the present one. Moreover, the present oscillation exists with universal simultaneity throughout the entire universe. The conscious mind of an observer, however, recalls the previous series of oscillations and anticipates future oscillations, and, thus, imposes upon time a duration and a multiplicity that in reality does not exist.[95] When Einstein introduced the principle of relativity of simultaneity, this clearly challenged Bergson's idea of universal time. The proposition that two events that are seen as simultaneous by one observer would be seen as sequential by other observers undermined the foundation upon which Bergson had based his ideas. These ideas were further contradicted by the time-dilation effect which, as we have seen, demonstrates that the time measured by a moving clock slows when compared to the time measured by a clock in a "stationary" frame of reference. Bergson, of course, did not sit idly by and allow Einstein's ideas to go unopposed. As might be expected, Bergson attacked Einstein's theory when the physicist visited Paris in 1922; the battle between the two scholars was carried on for decades afterward, if not by Einstein and Bergson personally, then by advocates for each side.[96] It is not surprising that Bergson's opposition to Einstein's theory sounds similar to Herbert Dingle's objection to the same. Both Bergson, the philosopher, and Dingle, the scientist, argued that time dilation and the relativity of simultaneity can be blamed on different perspectives, not on different realities in time.[97]

Bergson's opposition to Einstein's principle of the relativity of simultaneity continued for many years. Moreover, Bergson's influence in this regard has continued to be felt to the present day. For instance, the American physicist Mendel Sachs, of the State University of New York at Buffalo, has argued that the slowing of time as measured by a moving clock relative to time as measured by a stationary clock is a matter of perception only. Sachs states, rather convincingly, that once the moving clock returns to the same frame of reference as the stationary clock, the two clocks will be synchronized once again. Similarly, when the traveling twin of the twins paradox returns to Earth, she will be the same age as the twin who remained behind. Sachs concludes that the apparent slowing of any timepiece on board the moving starship would result from different perspectives and would not change the rate at which the two twins age in relation to one another. Sachs concludes by stating that the idea that the twins undergo some sort of **asymmetric aging** is nonsense. Asymmetric aging is nothing more than an illusion caused by the relative movement of the two twins but which can have no real effect on the actual physical aging of their respective bodies.[98]

The influence of philosophers like Bergson and scientists like Dingle and Sachs continues to affect our interpretation of the theory of relativity. The implications of universal simultaneous time and the nonphysical effects of relativity, for instance, have had a profound impact on philosophical concepts such as determinism and free will. These ideas, however, are reserved for Chapter 8. In the next chapter, we turn to the general theory of relativity to see how Einstein developed that theory in order to deal with some of the puzzles raised by the special theory of relativity.

CONCLUSION

There is a hidden significance to the time-dilation effect and the relativity of simultaneity that cannot be ignored. Time dilation and the relativity of simultaneity mean, not only that time is elastic, but also that it is malleable. Thus, a sufficiently advanced civilization in a highly developed technological state could conceivably be able to manipulate time to such an extent that time travel into the future and perhaps even into the past may be possible. Certainly, as we shall see in the next chapter, the time dilation effect allows for the *possibility* of time travel into the future, and we will explore some of the techniques that perhaps might be used to accomplish such a feat. On the other hand, we must admit that the operative word here is "possibility." The best that we can say is that nothing in the theory of relativity completely forbids travel forward and backward in time. Moreover, some of the obstacles preventing time travel, not the least of which is our inability to travel at light speed, are so difficult to overcome that we may as well conclude that no one will ever break the time barrier. Still, because relativity does not flatly eliminate time travel, and because several physicists have constructed theoretical time-travel techniques, we should keep an open mind on the subject as we explore these possibilities. In addition, the idea of time travel cannot be divorced from the creation of time-travel paradoxes and the moral responsibility that will accompany each of these paradoxes.

Review

6-1. Psychological relativity demonstrates that the same experience can be seen from different perspectives. At times, these perspectives are so dissimilar that a single occurrence can take on the nature of unrelated events. In ancient Greece, Anaximander proposed the idea of an absolute reality underlying the entire fabric of the universe. This also underlines the Greeks' apparent belief in the absolute, stable universe. During the Middle Ages, the notion of relativity took a back seat to the Christian idea of an absolute, unchanging God ruling over the universe. The unchanging God described by Thomas Aquinas in his *Summa Theologiae* has much in common with the apeiron of Anaximander, at least in the sense that both the God of Aquinas and the apeiron of Anaximander serve as the unchanging, indivisible, infinite, absolute backdrop for all existence. Galileo determined that all uniform motion is relative. Newton's work on the laws of motion and gravity clearly supports Galilean relativity. Equally interesting is that the laws of physics remain the same in all inertial systems.

6-2. James Clerk Maxwell's calculations led him to the conclusion that in empty space, light always moves at a velocity of 3×10^8 meters per second. This, of course, appeared to defy not only common sense but also the laws of gravity and motion as established by Galileo and Newton. There were only two ways to escape this paradox. Either Maxwell was mistaken or some sort of absolute ether existed throughout the universe, acting as the medium for light and as an absolute reference point. The ether was seen as the fundamental, eternal essence of all the universe, pervading all matter. Michelson and Morley, however, ran a series of experiments in Cleveland that demonstrated that the ether did not exist. But then, how can the speed of light appear to be absolute when there is no absolute reference point? That is the mystery that Einstein set out to unravel in 1905 when he formulated his special theory of relativity.

6-3. The solution to the dilemma lay in the very nature of space and time. Einstein concluded that if the speed of light does not change but the principle of relativity is to remain intact, then the spatial and temporal coordinates measuring objects moving at the speed of light must change accordingly. Consequently, Einstein decided, in order to maintain both relativity and the constant speed of light, Newton's concepts of absolute space and absolute time had to be discarded. Thus, Einstein eliminated two of the most fundamental assumptions of the Newtonian universe, the notion of absolute space and the belief in absolute time.

6-4. Einstein concluded that an observer who is considered stationary in relation to a moving object will detect the following four transformations in the moving object. First, the length of an object will shrink in the direction that the object is moving. Another way to say this is to note that the space that the object occupies will contract in the direction of the moving object. This is known as the Lorentz contraction. Second, the mass of the moving object will increase. Third, the faster an object travels, the slower time will move for that object relative to a stationary observer. This is known as time dilation or the time-dilation effect. Finally, the experience of a simultaneous present moment for a moving object and a stationary observer will vanish. This last conclusion is known as the relativity of simultaneity.

Understanding Key Terms

asymmetric aging
field
Galilean relativity
inertial system
lines of force
Lorentz contraction
Lorentz factor

Lorentz light clock
Lorentz transformation
lumniferous ether
muons
principle of equivalence
psychological relativity

psychological time
relativity of simultaneity
speed
subjective time
time dilation (time-dilation effect)

Review Questions

1. What is the nature of psychological relativity?
2. How was relativity understood by the philosophers of ancient Greece?
3. What is the relative nature of uniform motion?
4. How do the laws of physics operate within inertial systems?
5. Why did Maxwell include the lumniferous ether in his theory of electromagnetism?
6. How was the existence of the ether disproven?
7. What conflict was caused by the constant speed of light?
8. How did Einstein resolve that conflict in the special theory of relativity?
9. What is the time-dilation effect?
10. What is the phenomenon of relativity of simultaneity?

Discussion Questions

1. During the Middle Ages, the notion of relativity took a back seat to the Christian idea of an absolute, unchanging God ruling over the universe.

The unchanging God described by Thomas Aquinas in his *Summa Theologiae* has much in common with the apeiron of Anaximander, at least in the sense that both serve as the unchanging, indivisible, infinite, absolute backdrop for all existence. Do you agree with the assessment that the God of Aquinas is comparable to the apeiron of Anaximander? Why or why not?
2. James Clerk Maxwell's calculations led him to the conclusion that in empty space, light always moves at a velocity of 3×10^8 meters per second. This, of course, appeared to defy the laws of gravity and motion as established by Galileo and Newton. There were only two ways to escape this paradox. Either Maxwell was mistaken or some sort of absolute ether existed throughout the universe, acting as the medium for light and as an absolute reference point. Had you been a contemporary of Maxwell, which position would you have taken? Explain.
3. If you and your hypothetical twin had the opportunity to engage in a journey similar to the one explained in the twins' paradox, would you choose to be the traveling twin or the twin that remains on Earth? Explain your choice.
4. What are some of the philosophical ramifications of time dilation? Explain.
5. What are some of the philosophical ramifications of the principle of relativity of simultaneity? Explain.

ANALYZING *STAR TREK*

Background

The following episode from *Star Trek: The Next Generation* reflects some of the issues that are presented in this chapter. The episode has been carefully chosen to represent several of the most interesting aspects of the chapter. When answering the questions at the end of the episode, you should express your opinions as clearly and openly as possible. You may also want to discuss your answers with others and compare and contrast those answers. Above all, you should be less concerned with the "right" answer and more with explaining your position as thoroughly as possible.

Viewing Assignment—*Star Trek: The Next Generation*, "Yesterday's Enterprise"

In this episode the crew of the Enterprise encounters a temporal rift. Before the crew can react properly, a disabled starship appears out of the temporal rift. At the very second that the damaged ship appears, the present moment changes, and the Enterprise is transformed into a battle cruiser in the midst of a twenty-year war with the Klingon Empire. The crew suddenly realizes that the disabled ship is a starship that was reported lost twenty-two years in the past. The ship from the past also bears the name Enterprise. Its appearance from the temporal rift has altered the present timeline. As a consequence of the change in timelines, Tasha Yar has replaced Worf as the officer in charge of security. Eventually, the crew of the present Enterprise learns that the Enterprise from the past had been answering an emergency signal from a Klingon outpost that had been under attack by four Romulan battle cruisers. In the battle that followed, a volley of photon torpedoes and phasers combined in an enormous energy blast, creating the temporal rift. When the Enterprise from the past drifted into the rift and disappeared, history changed. Had the Enterprise from the past attempted the rescue of the Klingon outpost, it would have been a show of honor and courage that would have favorably impressed the Klingons and sealed the peace treaty that the Federation was negotiating with the Klingons at the time. Guinan is the only person on either ship who senses that something is wrong. She convinces Captain Picard that the Enterprise from the past must return to its own time and continue its attempted rescue of the Klingon outpost. Picard, in turn, convinces Captain Rachel Garrett of the past Enterprise to return to that ship's "present" (or is it the past?—actually, it's our future) to continue the battle with the Romulans, even though such a return will almost certainly mean the deaths of all crew members. Meanwhile, Tasha Yar learns from Guinan that in the other timeline she has died a meaningless death. Consequently, she asks Captain Picard for permission to join the crew of the Enterprise from the past as it returns to its own time to battle the Romulans. Picard agrees. When the past Enterprise reenters the temporal rift, the original timeline is instantly restored.

Thoughts, Notions, and Speculations

1. Relativity of simultaneity tells us that all inertial systems have their own time frame, and there is no way to determine the time of an event unless we also know what inertial system we are asking about. How, then, is it possible for the crew of the present Enterprise to know about the disappearance of the past Enterprise before it actually happens? Which event actually occurred in the past in relation to the present Enterprise—the attempted rescue of the Klingon outpost or the disappearance of the past Enterprise? Explain.
2. If we ignore the idea that the Enterprise from the past traveled through a temporal rift, we can reconfigure the episode to demonstrate what would happen if the ships in Starfleet were subject to the time-dilation effect. Suppose that the past Enterprise were on a one-year journey traveling at $0.999c$, relative to the stationary Earth. If this were the case, then the past Enterprise dropping out of warp and meeting a present Enterprise twenty-two years in the future would be a common event experienced by all starships. Ships from the past would be forever encountering ships from the present (or is it the future?) and vice versa. Relate this experience to the twins paradox. Why does this not happen in the Star Trek universe?

3. Why doesn't the appearance of the past Enterprise violate the laws of cause and effect? In contrast, how does the return of the past Enterprise to its own time violate the laws of cause and effect? Explain.
4. In this episode Tasha Yar is about twenty-six years old. When she travels twenty-two years into the past, she will, therefore, coexist with her own four-year-old self. Construct an argument that says that such a paradox makes time travel into the past unlikely.
5. Using what you have learned about the special theory of relativity, explain how the crew of the past Enterprise could rejoin the crew of the present Enterprise, after returning to the former crew's own time through the temporal rift.

NOTES

1. John Byrne, "The Man Who Made Tomorrow."
2. Albert Einstein, *The Meaning of Relativity*, 1.
3. G. S. Kirk, J. E. Raven, and M. Schofield, *The Presocratic Philosophers*: 9–10, 134, 343.
4. Kirk, Raven, Schofield, 133–34.
5. W. K. C. Guthrie, *A History of Greek Philosophy*, 1: 84.
6. Ibid., 1: 84–85.
7. Ibid.
8. Ibid.
9. Ibid., 1: 85–86.
10. Aristotle, *Physics*, 47.
11. Ibid., 49.
12. Kirk, Raven, and Schofield, 194–97.
13. Thomas Aquinas, *Summa Theologiae*, 23.
14. Sir Isaac Newton, *Mathematical Principles*, 6–9; Jonathan S. Wolf, *Physics*, 429; Lewis Carroll Epstein, *Relativity Visualized*, 1–3.
15. Newton, 8–9.
16. Alan Isaacs, ed., *A Dictionary of Physics*, 399; Robert H. March, *Physics for Poets*, 12.
17. Epstein, 1–3; Wolf, 63; See also Chapters 2 and 3 of this book.
18. Barry Chapman, *Reverse Time Travel*, 34–35; Epstein, 1–3; Wolf, 63, 429.
19. Newton, 7. Newton uses a ship under sail and a sailor to illustrate these principles.
20. Alan J. Friedman and Carol C. Donley, *Einstein as Myth and Muse*, 51; Newton, 9; Tony Rothman, *Instant Physics*, 116–18; Wolf, 63.
21. Albert Einstein, *Out of My Later Years*, 56.
22. Newton, 9; Rothman, 116–18.
23. Albert Einstein, *Relativity*, 19–20; Epstein, 21–25; Chapman, 34–35; Rothman, 116–18.
24. James Clerk Maxwell, *A Treatise on Electricity and Magnetism*, 2: 431.
25. Isaacs, 144; Wolf, 500. For Einstein's explanation of the concept of field see Albert Einstein, *Ideas and Opinions*, 366–72.
26. March, 69–72.
27. March, 71–72; Rothman, 98–99.
28. March, 72–73; Rothman, 98–101, Peter Coveney and Roger Highfield, *The Arrow of Time*, 59.
29. March, 73; Coveney and Highfield, 59.
30. March, 73–74; Coveney and Highfield, 59; Stephen Hawking, *The Illustrated A Brief History of Time*, 29; Rothman, 111.
31. March, 74; Maxwell 2: 431.
32. March, 75; Maxwell 2: 431; Rothman, 111.
33. Mendel Sachs, *Relativity in Our Time*, 23.
34. Rothman, 117; Sachs, 23, 27.
35. Maxwell, 1: x–xi.
36. Jeremy Bernstein, *Cranks, Quarks, and the Cosmos*, 23; Epstein 8–9; March, 74–75.
37. Albert Einstein, *Ideas and Opinions*, 368.
38. Maxwell, 2: 432.
39. Epstein, 9, 18–19.
40. Maxwell, 2: 431.
41. Epstein, 11. For Einstein's explanation of the ether see Albert Einstein, *Ideas and Opinions*, 369.
42. Epstein, 10.
43. Albert Einstein, *Sidelights on Relativity*, 6.
44. Epstein, 17–18.
45. Bernstein, 22–24; Hawking, 30–31; Rothman, 116–21; Bertrand Russell, *The ABC of Relativity*, 35–39.
46. Rothman, 118–19.

47. Ibid.
48. The Michelson-Morley experiment is the stuff of which legends are made. For several versions of the story see the following sources: David Cassidy, *Einstein and Our World*, 24–25; Hawking, 30–31; Russell, 35–46.
49. Cassidy, 24–25; Hawking, 30–32; Russell, 35–37.
50. Einstein, *Relativity*, 21–24; Rothman, 117–19; Sachs, 27.
51. Einstein, *Relativity*, 21; Epstein, 22; Rothman, 117; Sachs, 27–28.
52. Einstein, *Relativity*: 21–24; Epstein, 21–25.
53. Einstein, *Out of My Later Years*, 56.
54. Nigel Calder, *Einstein's Universe*, 180; Remy Lestienne, *The Children of Time*, 60.
55. Calder, 230–31.
56. Epstein, 24–25.
57. Ibid., 24–25.
58. Einstein, *Out of My Later Years*, 56.
59. Ibid.
60. Einstein, *Relativity*, 22–23.
61. Einstein, *Out of My Later Years*, 56: Einstein, *Relativity*, 22–23; Lestienne, 58–59; Rothman, 120–21.
62. Einstein, *Relativity*, 34–35; Paul Davies, *About Time*, 52–53; Rothman, 120–24; Trinh Xuan Thuan, *The Secret Melody*, 68–69.
63. Einstein, *Relativity*, 34–35; Davies, *About Time*, 52–53; Rothman, 120–24; Thuan, 68–69.
64. Nick Herbert, *Faster Than Light: Superluminal Loopholes in Physics*, 25–26.
65. Herbert, 25–26; Lestienne, 59–60; Sachs, 29.
66. Einstein, *Relativity*, 40–42, 52; Herbert, 25–26.
67. Einstein, *Out of My Later Years*, 104–105; Herbert, 25–26; Lestienne, 59–60.
68. Isaacs, 395; Thuan, 69.
69. Isaacs, 395; John W. Macvey, *Time Travel*, 50; Thuan, 69.
70. Einstein, *Relativity*, 41–42; John Gribbin, *Time-Warps*, 76–81; Macvey, 50–51; Barry Parker, *Cosmic Time Travel*, 30–31.
71. Lawrence M. Krauss, *The Physics of Star Trek*, 21.
72. Chapman, 37–41; March, 118–20; Vincent Icke, *The Force of Symmetry*, 115–17.
73. Icke, 116 (for the math).
74. Ibid., 116–17; Chapman, 41.
75. Icke, 117; March, 118–20; Clifford M. Will, *Was Einstein Right?*, 253–55.
76. Icke, 117; March, 118–20; Will, 253–55.
77. Gribbin, 79–81; Parker, 34–35; Thuan, 68–69.
78. Eric Chaisson, *Relatively Speaking*, 73–74; Davies, *About Time*, 59–65; Thuan, 68–69.
79. Coveney and Highfield, 81–82; March, 126–27; Calder, 153–54.
80. Calder, 155. For an in-depth look at Dingle's interpretation of the special theory of relativity see Herbert Dingle, "Scientific and Philosophical Implications of the Special Theory of Relativity," 537–54.
81. Calder, 156; Chapman, 44–45; Lestienne, 78–80; March, 126–27; Parker, 85–86.
82. Calder, 156; Parker, 35, 85–86.
83. Chapman, 70–71; Coveney and Highfield, 84–85.
84. Chaisson, 85–88; Coveney and Highfield, 81; Will, 255.
85. Chaisson, 85–87; Coveney and Highfield, 81; Will, 255. Each source reports a different figure for the increased lifespan of the muons. Chaisson reports that the lifespan of the muons increased by a factor of eight; Coveney and Highfield report a factor of nine, and Will, a factor of twelve.
86. Einstein, *Relativity*, 29–31; Herbert, 28–29; Thuan, 70–71.
87. Einstein, *Relativity*, 29–30.
88. Ibid.
89. Epstein, 36–67.
90. Einstein, *Relativity*, 29–30.
91. Richard P. Feynman, *Six Not-So-Easy Pieces*, 64; March, 111–12; Will, 251–53; Einstein, *Relativity*, 29–31; Epstein, 36.
92. March, 111–12; Will, 251–53; Einstein, *Relativity*, 29–31; Epstein, 36.
93. Einstein, *Relativity*, 31.
94. Henri Bergson, *Time and Free Will*, 120–21. See also Gilles Deleuze, *Bergsonism*, 78–80.
95. Bergson, 107–9.
96. Deleuze, 79–85; Lestienne, 81.
97. Deleuze, 84–85; Lestienne, 80–81.
98. Sachs, 71–78.

Chapter 7
Past and Future Time Travel

COMMENTARY: THE PREMATURE DEATH OF JOHN F. KENNEDY

Over the years, the film industry has addressed the dramatic implications of time travel in a variety of different films. Some films, such as *Back to the Future* and *Time Bandits*, have exploited the humorous aspects of time travel. Others, like *The Time Machine* and *The Philadelphia Experiment*, have taken a more serious approach. One such film, *Running against Time*, focused on several perplexing time-travel paradoxes. One of the most intriguing of these paradoxes concerned a mission into the past to prevent the assassination of President John F. Kennedy. David Rhodes, the time traveler in the movie, plans to arrive on the roof of the Texas Book Depository shortly before noon on November 22, 1963. His primary objective is, of course, to stop Lee Harvey Oswald from firing the fateful shots that kill President Kennedy. His secondary objective, however, is to prevent the war in Vietnam. Rhodes, a history professor at a prominent university, is convinced that had Kennedy lived, he would have withdrawn American forces from Southeast Asia in 1963, thereby stopping the war before it began. Unfortunately, Rhodes fails to stop the assassination. Instead, circumstances conspire to make the authorities believe that Rhodes, and not Oswald, killed the President. What follows is a complex series of events during which Rhodes and his allies from the future attempt to restore history in relation to the assassination, while still preventing the war in Vietnam. Unfortunately, the results of this tampering with history are disastrous. Instead of creating a future that avoids the Vietnam War, the opposite occurs. Events quickly get out of hand. The war escalates to unprecedented heights, eventually leading to a nuclear confrontation. Thankfully, at the end of the film, the universe is restored to order, and the original time line is returned. However, gradually Rhodes learns that his time-traveling adventures have made some permanent, if somewhat disconcerting, changes in the present. What would happen if someone actually succeeded in altering the past? How would such an alteration impact the present? Would the present cease to exist, or would it simply be altered in subtle and unnoticeable ways? Would we who exist in the "present" even realize that changes occurred within the "past" of our particular timeline? These are only a few of the bewildering issues that serve as the focus of this chapter.

CHAPTER OUTCOMES

After reading this chapter, the reader should be able to accomplish the following:

1. Describe how speed affects travel into the future.
2. Identify the features of the general theory of relativity.
3. Explain how gravity affects travel into the future.
4. Outline the essential nature of a Kerr black-hole time machine.
5. Relate the essential ingredients of Gott's time-loop theory.
6. Identify the characteristics of a Morris/Thorne wormhole.
7. Distinguish between contradictory-event paradoxes and information paradoxes.
8. Discuss the nature of the grandmother paradox and the conservation paradox.
9. Clarify the nature of the paradox of simultaneous temporal existence.
10. State the theory behind the possible creation of alternate time lines.

7-1   TRAVEL INTO THE FUTURE

According to Einstein's theory of relativity, one way to create a time distortion is to increase the speed of an object relative to another inertial system. The faster an object travels, the slower time moves for that object in relation to other inertial systems. At normal cruising speeds in everyday life such alterations in time are minuscule. However, once the speed of an object approaches the speed of light (i.e., 186,000 miles per second) the effects become noticeable. Because time slows as the light barrier is approached, a time traveler can slow his present moment, relative to another inertial system, by nearing that light barrier. For the sake of convenience, we will make Earth the stationary inertial system and we will place our time traveler in a starship capable of traveling at speeds that approach the speed of light. As the time traveler accelerates his starship to nearly the speed of light, his time frame slows, while a stream of successive present moments pass by rather quickly on Earth.[1] When the time traveler slows his starship and returns to Earth, the movement of time within both inertial systems will match once again. However, during the trip, time on board the starship will have expanded relative to time on the earth. Therefore, the time traveler will not have aged at the same rate as the people on Earth. In a sense, then, the traveler has passed into the future of the Earth.[2]

Shortcomings of Time-Dilation Time Travel

Of course, the time-dilation travel into the future is not without its shortcomings. Chief among these is the fact that each trip is a one-way journey. Because time dilation reconfigures the forward movement of time in one inertial system relative to another, time-dilation travel does not give the time traveler access to the past. This is simply another way of saying that the time traveler will never be able to return to his "present" because it would exist only in the past. The other shortcoming is the light barrier itself. Objects may approach light speed, but, at least based on our current understanding of the laws of physics, they will never be able to equal or exceed it.[3] To understand why the universe has an upper speed limit of 186,000 miles per second, we recall that energy is required to move an object. This energy can be employed in two ways. Either it will be used to make the object go more rapidly, or it will be transformed into mass, which makes the object more massive. When an object moves at a relatively slow speed, most of the energy is applied to the movement, and the mass does not increase any appreciable amount. For instance, an object moving at only 10 percent light speed, which at 18,600 miles per second is actually rather fast, will have an increase in mass that amounts to approximately 0.5 percent. As an object approaches light speed, the increase in mass becomes much more appreciable, so that an object moving at 90 percent light speed will see a 50 percent increase in mass. Moreover, when that object reaches light speed, its mass stretches infinitely. Infinite mass would require infinite energy to move the object faster, and that just cannot be obtained. Consequently, objects are limited to a maximum speed of 186,000 miles per second.[4]

Faster than light (FTL) travel is thus not possible, based on our current understanding of the laws of physics. For the sake of argument, let us assume for a moment that at some time in the future FTL propulsion systems become commonplace. Even if this did come to pass, the time-dilation effect would create a number of frustrating consequences. Space exploration and colonization on a galactic level are not practical without FTL or its equivalent. Yet, FTL makes the administration of any galaxywide exploration and colonization effort impractical. Consider, for example, the galactic federation created for the *Star Trek* universe. The organized administration of the United Federation of Planets is supposedly made possible because warp-drive engines allow starships to travel faster than light. Without FTL travel, it would take dozens, even hundreds, of years for Federation starships to travel from one star system to another. As a result, space colonies, star bases, alien civilizations, and remote outposts like Deep Space Nine, would soon find themselves isolated from one another for decades, perhaps even centuries. FTL eliminates this problem and allows the Federation to run about as efficiently as any Earth-side government.

At least that is the theory. But would such a system actually function smoothly? Probably not. At warp factor 1 the Starship Enterprise is traveling at light speed, that is, at 186,000 miles per second. Warp speed is calculated using a unit of measurement known as the cochrane. The cochrane, which is named for the inventor of warp drive, Zephram Cochrane, measures the distortion of space created by the continuum distortion propulsion (CDP) system or, as it is usually referred to, the warp drive. Each cochrane is equal to one unit of light speed. However, the distortion of space-time caused by CDP is not **linear** (i.e., its graph is not that of a straight line). Instead, it advances according to a complex formula that begins with warp 1 equal to light speed, but which progresses so that warp 2 is equal to ten times light speed, warp 3 is equal to 39 times light speed, warp 4 is equal to 102 times light speed, and so on.[5] Translated into miles per second, at warp factor 3, the ship has reached 7,245,000 miles per second. At warp factor 6, the ship exceeds 72,912,000 miles per second, and at warp factor 9, the ship's top speed, the Enterprise travels at 281,976,000 miles per second.[6]

Now add to this the fact that the other starships in Starfleet are racing around at various degrees of warp speed, and the problem of holding a coherent galactic federation together should become evident. The various speeds of these ships would render any attempt to manufacture a universal time frame throughout the Federation an exercise in futility. Each time a ship drops out of warp, the crew must ask, "What year is it back on Earth?" Even assuming there were a way to answer the question, each crew would get a different answer at each stop. The time-dilation effect would reduce the Federation to a mismatched mess of eternally shifting time frames.[7]

Of course, none of this mismatched time shifting ever occurs in *Star Trek* because the writers have created several ingenious plot devices to circumvent the problem of light speed and time dilation. Part of their solution lies in the fact that a starship does not actually move faster than light. Instead the ship's warp drive engines somehow "warp" space allowing the ship to move from one point in the universe to another without crossing the space in between. On the other hand, when a starship uses its impulse engines, it is actually moving in real time. Therefore, time-dilation effects must be taken into consideration. For this reason, when starships use their impulse engines they cannot exceed 0.25c. This speed limit keeps the time-dilation effect at manageable levels. Finally, Federation scientists have invented a method of adjusting the ship's internal-computer clock system so that starship time corresponds to "real present" time as calculated by Starfleet headquarters on Earth. To coordinate a starship's chronometer with real present time on Earth, the crew can simply access one of the many timebase beacons located throughout Federation space.[8]

The General Theory of Relativity

Speed, however, is not the only way to distort, twist, or dilate time. Gravity can have the same effect on time. To understand how gravity affects time, we must discuss Einstein's general theory of relativity. Recall that the special theory of relativity deals with inertial systems that travel at uniform speed in a straight line. This narrow view of movement in the universe was sufficient for a while, but it could not stand alone forever. Gradually, it became clear to Einstein that, to make his relativity theory complete, he would also have to include the effects that occur when an object is not moving in a uniform, steady manner. In other words, Einstein sought a way to include acceleration into his theory of relativity.[9] **Acceleration** is the rate at which a moving object increases its velocity.[10] Uniform (or constant) acceleration can be determined by dividing an alteration in velocity, by the time it takes to execute that alteration.[11]

The Problem Facing Einstein. It seemed to Einstein that if a body is accelerating, there must be some absolute standard used to measure that acceleration. In other words, the question before Einstein at this point was whether relativity applied to nonuniformly moving bodies in the same way that it applied to uniformly moving bodies. The answer led Einstein to the formulation of the general theory of relativity.[12] Before we get to the answer, however, let us look at the problem once more. Einstein framed the difficulty in this way:

> When by the special theory of relativity I had arrived at the equivalence of all so-called inertial systems for the formulation of natural laws (1905), the question whether there was not further equivalence of coordinate systems followed naturally, to say the least of it. To put it in another way, if only a relative meaning can be attached to the concept of velocity, ought we nevertheless to persevere in treating acceleration as an absolute concept?[13]

In attempting to answer this question, Einstein received a helping hand from the work of Ernst Mach, an Austrian philosopher-physicist, who disagreed with Newton's belief in the existence of absolute space.[14] Recall that, although

Newton incorporated Galilean relativity into his laws of gravity and motion, he still held stubbornly to the belief that somewhere in the universe, perhaps beyond the region of the fixed stars, some absolute, unmoving body existed.[15] Mach thought otherwise. He believed that relativity could be preserved throughout the universe. To demonstrate his position, Mach compared uniformly moving bodies to accelerating bodies. According to Mach, objects moving at a uniform rate of speed would not experience any forces affecting them, whereas those that are accelerating would experience the effects of forces.[16]

The Principle of Equivalence. Let us say, for example, that you are cruising down the interstate at 65 miles per hour. At this uniform rate of speed, you would feel no forces pushing or pulling you. However, when you accelerate to pass a slowly moving truck, you feel the acceleration as you are pushed backward in the seat of the car. If you drive to the end of the road and over the edge of a cliff, you, the car, and all its contents will fall to Earth under the acceleration of gravity. However, while you are in free fall, you will feel no gravitational force. In effect, you will be weightless. Suppose, further, that as the car plunges downward, it impacts on the nose cone of a rocket heading upward. As the rocket accelerates, you, the car, and its contents weigh more than they did on the surface of Earth. The acceleration of the rocket creates a gravitational force of its own. With these ideas in mind, Einstein began to see the relationship between gravity and acceleration. Making this connection led Einstein to the formulation of his principle of equivalence, which states that gravitational fields and acceleration cannot be distinguished. They are identical.[17] Einstein explains his discovery of the principle of equivalence in this way:

> The principle of the equality of inertial and gravitational mass could now be formulated quite clearly as follows: In a homogeneous gravitational field all motions take place in the same way as in the absence of a gravitational field in relation to a uniformly accelerated coordinate system. If this principle held good for any events whatever (the "principle of equivalence"), this was an indication that the principle of relativity needed to be extended to coordinate systems in non-uniform motion with respect to each other, if we were to reach a natural theory of the gravitational fields.[18]

Another way to explain the principle of equivalence is to note that bodies react to acceleration in the identical way that they react to a gravitational field. Therefore, according to the principle of equivalence, an individual who is within an accelerating system that is free of gravity will feel effects that are indistinguishable from the effects felt by another individual within a gravitational field.[19] In fact, a blindfolded person in an elevator in a gravity-free environment would not be able to tell, in principle at least, whether the force holding him to the floor results from gravity or acceleration.[20]

Einstein's Theory of Gravity. The principle of equivalence led Einstein to conclude that in order to apply the theory of relativity to accelerating bodies, he would have to develop a theory of gravity that was different from Newton's theory. Gravity as visualized by Newton acted as a force between two bodies. Einstein, in a series of papers between 1908 and 1915, suggested that gravity does not operate as Newton imagined. Instead, according to Einstein, gravity is actually a distortion in the geometry of space-time. Notice the phrasing here. Einstein did not say that gravity *causes* a distortion in the geometry of space-time. Rather, he said that gravity *is* a distortion in space-time. The source of this distortion is an object's mass. We could, therefore, say that mass causes space-time to curve or warp. Or, if we look at it from the opposite perspective, space is curved in the presence of massive objects. For example, within the solar system, the most massive object is the Sun. Therefore, the Sun's mass causes the greatest curvature in space-time in the solar system. From this perspective, the planets are not drawn into orbit around the Sun by a force emanating from the Sun, but, instead, they follow trajectories in curved space caused by the Sun's mass.[21]

One way to picture this is to visualize an enormous, tightly drawn elastic sheet. In the absence of a massive object the sheet is pulled taut. However, if a massive object, such as a bowling ball, were placed in the center of the elastic sheet, the sheet would sag under its influence. This sag or indentation in the elastic sheet is roughly analogous to the curve in space-time caused by the presence of a massive object. The more massive the object, the deeper the indentation in the elastic sheet. Similarly, the more massive the celestial object, the more pronounced the curve in space-time.[22] If we were to roll a marble on to the elastic sheet, it would follow a trajectory determined by the indentation made by the bowling ball. The original position and the initial speed of the marble would determine its ultimate resting point. For example, the marble might not have enough speed to escape the indentation, in which case it would smash into the bowling ball. On the other hand, if it were moving rapidly enough, it might roll into and out of the indentation; or at a lesser speed, it might travel in an elliptical path around the bowling ball. This latter alternative is roughly analogous to the elliptical orbits of the planets around the most massive object in the solar system, the Sun.[23]

Predictions Made by the General Theory of Relativity. With this revolutionary view on the nature of gravity in mind, Einstein went on to predict several remarkable consequences, one of which was that even light would be affected by the distortion of space-time. This prediction was supposedly confirmed by the British astrophysicist Sir Arthur Eddington in 1919, while in Africa on a special expedition, when he measured star positions during a solar eclipse. When he compared those measured positions to the positions of the stars when the Sun was located elsewhere, he found that the presence of the Sun deflected the starlight by a measurement that coincided with the predictions made by Einstein's theory.[24]

Another prediction made by Einstein's general theory of relativity involves the relationship between time and gravity. According to the general theory, time will move more slowly under the influence of a strong gravitational source.[25] Einstein goes on to explain that the effects of gravity on time are similar to the effects produced by speed. A clock will slow down in a strong gravity well just as it will as it approaches light speed. In fact, Einstein asserts that time could even approach a near standstill if the gravitational field were powerful enough.[26] This should not be surprising. As we have seen, the principle of equivalence states that acceleration and gravity are equivalent. If accelerating clocks slow down in relation to stationary ones, then clocks under the influence of a strong gravitational field should slow down as well.[27]

What all of this implies is that the relationship between time and gravity can provide us with other methods of time travel. For example, because there is no universal present moment in relative time and because time slows in a strong gravity well, our time traveler can slow his present moment by nearing a strong gravitational field. When the traveler exits the strong gravitational field and returns to the system from which he left, he will find that the present moment in that part of the universe, say the present moment near the planet Earth, would become his present moment even though it was in the distant future when he originally left the system for his trip into the strong gravitational field.[28]

Black Holes and Time Travel into the Future

Where would our fictional time traveler find a gravity well strong enough to dilate time? The most promising area would be in the vicinity of a black hole. To understand how a black hole would provide the necessary gravitational influence to dilate time, we must first explore two new ideas: the concept of escape velocity and the nature of an object's gravitational radius. **Escape velocity** is the lowest speed necessary for an object to move out of the gravitational influence of a massive astronomical body. For example, if an astronaut wants to leave Earth, his ship must reach a speed of 6.8 miles per second. We, therefore, say that the escape velocity of Earth is 6.8 miles per second.[29]

Deep within astronomical bodies like the Sun and the Earth, there is a point at which the escape velocity of that astronomical body reaches the speed of light. The distance from the center of the body to the point at which the escape velocity reaches light speed is called the **Schwarzschild radius**. The Schwarzschild radius is named for Karl Schwarzschild, a German astronomer, who first calculated this radius in 1916.[30] To understand how he calculated the gravitational radius, we must examine Einstein's gravitational field equations. The purpose of

Einstein's **gravitational field equations** is to determine the relationship between the power of a gravitational field and the matter and energy that produced that field.[31] The equations can be applied to determine several factors. First, they can determine the magnitude of a gravitational field generated by a particular arrangement of matter and energy. Second, the equations will reveal the degree to which time is dilated by a gravitational field. Schwarzschild's application of Einstein's equations reveals that time dilation can be determined by a formula based upon the distance from the center of an astronomical body. The use of this formula can also give us the Schwarzschild radius of an astronomical body.[32]

As we have already seen, the Schwarzschild radius gives us a region located a certain distance from the center of a massive body whose escape velocity is the speed of light. This occurs because at the Schwarzschild radius the gravitational field is so enormous that it robs the photons of their energy, making them powerless to resist the gravitational pull. At any place closer to the body than the Schwarzschild radius, the escape velocity, therefore, will be greater than the speed of light. In essence, this means that the gravitational pull is so strong that not even light can escape from the event horizon. More importantly for our study, at the Schwarzschild radius, time dilation is infinite.[33] Usually, the Schwarzschild radius can be disregarded because most astronomical bodies have a Schwarzschild radius that is much smaller than their actual radius. The Schwarzschild radius of the Sun, for example, is approximately one mile, although the Sun's actual radius is about 435,000 miles. Therefore, under these circumstances, the Schwarzschild radius and the time-dilation effect would be of no consequence.[34]

Schwarzschild Black Holes and Time Travel. However, if the Schwarzschild radius were somehow moved to the exterior of an astronomical body, the time-dilation effect could not be ignored. For the Schwarzschild radius to move outside an astronomical body like a star, that star would have to retain its mass but shrink to a very small size. This is what happens in the creation of a black hole. As we saw in Chapter 2, black holes are actually the final stage in the life cycle of massive stars. During most of a star's life, it will maintain a state of equilibrium as the radiation produced by fusion is counteracted by gravity. When the star's hydrogen has been nearly depleted, the star will call upon other sources of fuel to maintain its existence. For instance, it will burn away the hydrogen in its outer layers, which will lead to a temporary expansion of the star. Once that hydrogen is gone, the star will burn helium and, after that, other available elements.[35]

Eventually, however, the nuclear fuel is depleted; the state of equilibrium will be overcome as gravity becomes the dominant force and the star collapses in upon itself. The ultimate fate of a star will depend on its mass. With stars that are extremely massive—stars whose mass is more than five times that of the

Sun—the collapse will be so intense that the star will contract in upon itself, forming an infinitely dense point called a singularity.[36] At the singularity the usual laws of physics based upon the general theory of relativity seem inapplicable, the rules of cause and effect no longer seem to apply, and space-time appears to reach a point of infinite curvature.[37] Of course, exactly what happens at the singularity is still the subject of intense debate. The theoretical physicist Kip Thorne of Caltech, for example, supports the notion that at the singularity the rules of quantum gravity take over and space-time is once again separated into space and time. Time flashes out of existence and space is converted into what Thorne calls a type of "quantum foam."[38] For our purposes here it is not critical that we settle this debate. What is important, however, is that we recall that, once the star has collapsed to this state, its Schwarzschild radius will exist far outside the singularity. When the collapsed star reaches this state, it is referred to as a black hole.[39] This type of black hole is also called a **Schwarzschild black hole**, a label given to any black hole that is a perfect sphere and is not rotating.[40]

If we were to diagram the structure of a black hole, we would find the singularity at its center. Surrounding the singularity would be a black sphere. The surface of this sphere is called the **event horizon** of the black hole. The distance between the event horizon and the singularity is equal to the Schwarzschild radius, which, as we have seen, now exists outside the star. Also, because the distance between the event horizon and the singularity is equal to the Schwarzschild radius, the gravitational field surrounding the singularity would be so powerful that the escape velocity at the event horizon would be the speed of light. Consequently, at the event horizon not even light would be able to escape the gravitational influence of the singularity.[41] This is why the sphere is black and why the entire structure is referred to as a black hole.[42] In fact, the only reason that the sphere would be visible at all is that it would block the light emanating from stars located beyond it.[43] Anything that crosses the event horizon will be unable to escape because at the event horizon, the escape velocity is the speed of light.[44] Remember also that this type of black hole is of the nonrotating variety.[45]

Now we can return to the possibility of using a black hole for time travel into the future. First, recall that, at the Schwarzschild radius, time dilation is infinite. Recall also that in the case of a black hole the Schwarzschild radius is located outside the star and is equal to the radius of the event horizon. Because of this, a daring and resourceful time traveler might move her ship close enough to the event horizon to take advantage of the time dilation that occurs there. In this way, her clock would slow relative to clocks in other selected frames of reference. To see this process in more detail, let us return to our time-traveling twins from the preceding chapter. Let us assume that both Ashley and Megan are on a starship that is located in the vicinity of a black hole. Megan takes a shuttle craft and, using her impulse engines, moves the shuttle close to the event horizon of the black hole. Ashley remains on the starship and observes Megan's progress. Ashley notes that Megan's clock is running more slowly than the clock on the starship. From Ashley's perspective, the closer that Megan's shuttle comes to the event horizon of the black hole, the more slowly Megan's clock appears to run. Although Megan's clock will never stop completely, from Ashley's point of view, it will come very, very close to stopping.[46]

If Megan could look back at the starship, she would see a different set of circumstances. From her perspective, the clocks on the starship would speed up, while the clock on board her shuttle craft would be running at a normal rate. In fact, the closer that Megan gets to the event horizon of the black hole, the faster the clocks on the starship appear to move relative to the normal movement of her clock. Since time is passing rapidly on the starship, if Megan could return, she would have traveled into the future as measured by the clock on her shuttle, contrasted with the clock on the starship.[47]

Unfortunately, although such trips sound good in theory, from a physical point of view they may be impossible. Remember that a strong gravitational field will warp not only time but also space. In Section 7-2, we will explore how the gravitational warping of both space and time beyond the event horizon may provide the time traveler with a bridge to the past. For now, however, we are concerned only with the effects that the gravitational field will have on the space taken up by the time traveler's shuttle craft. As the shuttle craft approaches the event horizon, the forward end of the shuttle will feel a greater gravitational pull than the aft portion. In short order, the shuttle and its passenger would be ripped apart by that enormous gravitational pull.[48]

Quasars, Active Galaxies, and Primordial Black Holes

Thus far, we have discussed only black holes that result from the collapse of a single massive star. Black holes, however, come in other varieties. For instance, supermassive black holes are believed to be at the center of, and to produce energy within, quasars. Similarly, active galaxies contain powerful, although somewhat smaller, massive black holes. Finally, some theorists believe that primordial black holes were created under the extreme conditions that existed during the first few moments of the expanding universe.

Supermassive Black Holes and Quasars. Quasars are highly luminous deep-space objects that probably lie at the centers of distant galaxies. Since quasars look like distant stars, they were first labeled "*quasi-stellar* objects." From this initial designation, came the shortened term *quasar*. A black hole at the center of a quasar would likely be equal to one billion solar masses.[49] Moreover, some astronomers speculate that the most powerful quasars may harbor black holes at their centers that are equal to one trillion solar masses.[50] It is likely that such incredibly massive black holes were actually built up from many smaller ones that, over the course of a galaxy's lifetime, fell together due to their mutual gravitational attraction. Such a supermassive black hole would destroy all stars that wander into its gravitational field. The destruction of these stars by the supermassive black hole is probably what produces the enormous energy output emanating from individual quasars.[51] Of course, there is some question as to the source of energy for these quasars. If quasars do swallow nearby stars, once those stars have been consumed, the quasar should extinguish itself. Recent observations made by astronomers using the Hubble Space Telescope and by others using an Earth bound telescope at the Dominion Astrophysical Observatory in Canada, may provide a solution to this puzzle. These observations indicate that most quasars are associated with galactic collisions. Such collisions would place a large number of stars in the vicinity of the quasar thus providing a rich source of energy. Galactic collisions as a primary energy source would also account for the high number of quasars in the early universe. At an earlier epoch, when the universe was more densely packed, galactic collisions would have been more common thus producing more quasars than at other times.[52]

The enormous gravitational field produced by the supermassive black hole at the heart of a quasar might at first seem to set up ideal conditions for time travel. Unfortunately, this is not the case. For one thing, the activity in the region of a supermassive black hole is likely to be quite violent. As stars fall under the influence of the black hole's enormous gravitational field, they will be torn apart and gasses from these disintegrating stars will spiral into the black hole, forming an **accretion disc** around its central region. These gases will also heat up, releasing high-energy particles from this area.[53] Radiation from virtually every point on the spectrum, including gamma rays, X-rays, ultraviolet light, visible light, infrared light, and radio waves, would also be emitted from the quasar.[54] If these conditions are not enough to discourage a potential time traveler, the location of the nearest quasar probably would. The light from quasars is so red-shifted that the quasars appear to be among the most distant objects in the entire universe—making them virtually unreachable.[55]

Massive Black Holes and Active Galaxies. Because quasars are so distant, they are impractical as a source of the intense gravity needed for time travel; therefore, we should continue our search for other types of massive black holes. Some astronomers are convinced that massive black holes exist at the center of **active galaxies**, which are so named because they are a source of intense radiation.[56] The radiation emanating from active galaxies is now thought to result from the destruction of stars caused by the strong gravitational pull of centrally located massive black holes. Compared with the black holes that are supposed to lurk at the center of quasars, the black holes associated with active galaxies

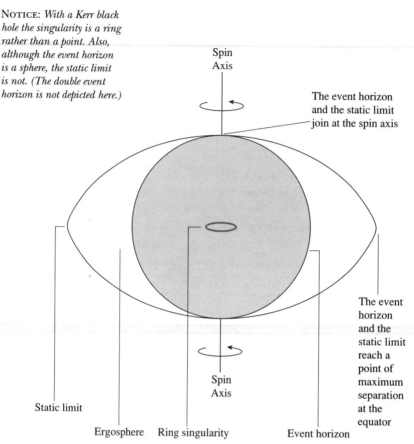

Figure 7.1 A Rotating Black Hole: One Possible Avenue to the Past.

A rotating or spinning black hole is referred to as a Kerr black hole. Unlike the singularity within a stationary Schwarzschild black hole, the singularity at the heart of a rotating Kerr black hole is not a single point. Instead, the singularity is shaped like a ring. A rotating singularity also displays several features that do not exist with the nonrotating variety. Since the singularity is spinning, the equator bulges outward much like the equator of the Earth bulges because of its rotation. In the case of the spinning singularity, however, this rotation creates an outer boundary, called the static limit, which has its greatest separation at the equator but touches the event horizon at each pole. Between the static limit and the event horizon there is an area called the ergosphere. The term "static limit" is used to describe this boundary because once anything, including a time traveler's starship, passes that boundary, it will not be able to stop. Kerr black holes also have two event horizons. Within the outer event horizon of a Kerr black hole there exists a space-time inversion zone where the roles of space and time are reversed. However, beyond the inner event horizon space and time resume their normal roles. Once within the inner event horizon, a time traveler can pass through the center of the ring singularly without being crushed by the gravitational pull. While she remains in this region, close to the ring singularity, but safely away from its crushing power, she will be able to travel backward in time. The direction of her orbit, the number of times she circles the axis of rotation, and the rotational speed of the black hole, will determine how far backward her clock moves. When she exits this region, she would again enter the space-time inversion zone between the inner and the outer event horizons. She could then move beyond the outer event horizon and eventually into normal space once again.

are somewhat smaller, and as a result, considerably less powerful. Nevertheless, these black holes are neither weak nor powerless. They would still have the ability to tear stars apart and in that process to produce enormous energy outbursts.[57]

If this activity sounds similar to what happens at the heart of a quasar, that should be expected. In fact, some astronomers are convinced that quasars and active galaxies are actually the same phenomenon at different evolutionary stages.[58] Unfortunately, the violent nature of the environment surrounding a black hole at the center of an active galaxy makes it an unlikely candidate as a gravitational source for time travel.[59] In addition, although not as distant as quasars, active galaxies are far enough away from our present location in the universe to make any trip to such a galaxy highly impractical.

Primordial Black Holes. The last type of black hole, whose existence has been suggested by the British physicist Stephen Hawking, is the **primordial black hole**, which may have been created in the high-density, high-temperature environment of the first moments of the expanding universe.[60] Primordial black holes are also known as **mini–black holes** because, when first formed 15 to 20 billion years ago, they probably had masses of about 500 billion kilograms each. This is approximately equivalent to the mass of a good-sized mountain.[61] However, during their lifetimes primordial black holes evaporate. As they evaporate, they lose mass, release radiation, and experience a rise in temperature. Eventually, when a mini-black hole's mass reaches about 20 micrograms, it explodes with the power of a 10 million megaton hydrogen bomb.[62] Some theorists believe that the only way to detect the existence of a mini-black hole is to observe its gravitational influence on other bodies.[63] Unfortunately, at that stage the usefulness of a primordial black hole as a vehicle for time travel would be gone forever.

Neutron Stars and Future Time Travel. Since, despite their enormous gravitational power, black holes seem to be highly impractical as a way to travel into the future, it might be advisable for our time traveler to search for other gravitational sources for time-travel purposes. Other astronomical bodies might generate gravitational fields strong enough to help the courageous time traveler journey into the future without facing the dangers associated with black hole time travel. One cosmic body that might provide such a gravitational field is a neutron star, which is a second cousin of a black hole. Both of these astronomical bodies are actually stars that have used up their nuclear fuel and have collapsed in upon themselves. The difference between the two is their mass. A neutron star results from a star which is much less massive than one that ends as a black hole.[64] Less mass also means that a weaker gravitational field surrounds a neutron star. This is not to say, however, that the gravitational field of a neutron star will be weak. On the contrary, compared with the gravitational effects that we experience in our everyday lives here on Earth, the gravitational effects experienced by the time traveler would be spectacular.[65] As Paul Davies explains:

> [A] spinning neutron star [is] an object so compact that its gravitational field is a billion times stronger than Earth's. The effect on time is dramatic. Time at the surface of a typical neutron star is slowed by about 20 percent relative to Earth time.... Time is warped so drastically on a neutron star because the star, while possessing the mass of the sun or more, is nevertheless compressed to a radius of just a few kilometers. The stronger the gravity at the surface of an object, the more time is slowed, or stretched.[66]

Could the gravity well of the neutron star be used to suspend the time flow of a time traveler parked in orbit around that star? Perhaps. On the other hand, despite the powerful gravitational pull of a neutron star, it still does not compare with that of a black hole.[67] That is why, when we want to discuss travel into the past we must return to the most powerful gravitational source of all—a black hole.

7-2 TRAVEL INTO THE PAST

By traveling into the future the time traveler is really just accelerating a trip she would take anyway. After all, if she did nothing, the future would come to her. Not so with retrograde time travel. To travel into the past, the time-traveling adventurer will have to bend, twist, and reconfigure the traditional laws of the universe. The consequences of this bending and twisting may threaten the very fabric of space-time itself. However, before the time traveler deals with the paradoxical consequences of retrograde time travel, we must first determine whether such a journey is possible in the first place.

Rotating Black Hole Time Machines

The work of Roy Kerr, a physicist and mathematician from New Zealand, indicates that retrograde time travel may be possible. To understand the nature of the time-travel tunnels which would be possible according to Kerr's equations, we must return to the type of black hole that results from the collapse of a single massive star. As we have seen, the powerful gravitational field created by a Schwarzschild black hole pulls everything that comes too close to the event horizon toward the singularity at its center.[68] As was noted, the radius of the event horizon is the same as the gravitational radius of the collapsed star. The gravitational radius is the radius associated with any object wherein the gravitational field is so strong that the escape velocity from that object reaches the speed of light.[69]

As we have seen, in astronomical bodies like the Sun or the Earth, the gravitational radius is hidden deep within the object and can, therefore, be ignored. However, when an astronomical body has collapsed to the degree of a singularity, the gravitational radius will exceed the object. This will place the gravitational radius outside the object, creating a strong gravitational field in the vicinity of the collapsed star. Earlier in this chapter, we saw how this strong gravitational field might be used by a time traveler to move into the future.[70] However, we also saw the difficulties associated with this sort of journey. One principle difficulty is that the gravitational field will have an impact not only on time, but also on space. Consequently, the forward part of the shuttle craft will experience a much more powerful gravitational pull than the rear. This gravitational pull would destroy the shuttle craft and the time traveler along with it.[71] However, if somehow this gravitational pull could be avoided, the time traveler would find that beyond the event horizon exists a region in which there is a strange reversal in the customary relationship between space and time.[72] If the traveler could enter this region, she would find herself crossing into that region in which the space-time continuum reverses its customary order. Outside the event horizon she would be able to manipulate space but not time. Within the backward realm beyond the event horizon, she will soon find that she can maneuver through time but not through space.[73]

Can this domain of time-space reversal be used as a method of retrograde time travel? The answer to this question depends on the nature of the black hole. As was noted, if the black hole that the time traveler approaches is a Schwarzschild black hole, which is nonrotating, then the answer is probably "no." Because a nonrotating black hole would exert enormous gravitational pressure on the time traveler, she would undoubtedly be crushed. She might be confused temporarily because of her ability to manipulate time within the zone of time-space reversal. However, the enormous gravitational power of the collapsed mass would pull her into the infinitely dense singularity. Once she crosses the event horizon, perhaps even before she arrives at the event horizon, she will have reached a point of no return and will not be able to pull free of the enormous gravity well.[74]

Rotating Kerr Black Holes. Suppose, for a moment, that we could find a black hole that is not stationary. Much like any astronomical body—for example, the Earth, such a black hole would rotate on its axis. A rotating or spinning black hole is referred to as a **Kerr black hole** (see Figure 7.1). With a Kerr black hole the possibility of a successful time trip increases. Unlike the singularity within a stationary Schwarzschild black hole, the singularity at the heart of a rotating (Kerr) black hole is not a single point, but is, instead, shaped like a ring. The ringlike configuration of the singularity would allow a time traveler to journey to the center of the black hole without reaching a place where the gravity is so powerful that it causes an infinite curvature in space-time and crushes the time traveler out of existence.[75]

A rotating singularity also displays several features that do not exist with the nonrotating variety. Since the singularity is spinning, the equator bulges outward, much like the equator of the Earth bulges due to its rotation. In the case of the spinning singularity, however, this rotation creates an outer boundary called the **static limit**, which has its greatest separation at the equator, but which touches the event horizon at each pole. Between the static limit and the event horizon there is an area called the **ergosphere**.[76]

This brings us to another peculiarity of the Kerr black hole—the fact that it has a double event horizon. Between the **outer event horizon** of a Kerr black hole and the **inner event horizon**, there exists a space-time inversion zone where the roles of space and time are reversed, as was the case within the event horizon of a Schwarzschild black hole. However, *within* the **inner event horizon** we would find that space and time have resumed their regular roles.[77] Once within the inner event horizon, the time traveler can pass by the ring singularity without being crushed by the gravitational pull. While remaining in this region, close to the ring singularity, but safely away from its crushing power, our time traveler will be able to travel backward in time.[78] The direction of the orbit, the number of times she circles the axis of rotation, and the rotational speed of the black hole, will determine how far backward the clock moves.[79] When the time traveler exits this region, she would again enter the space-time inversion zone between the inner and outer event horizons. She could then move beyond the outer event horizon and eventually into normal space once again.[80]

Dangers within a Kerr Black Hole. One way to use a Kerr black hole as an interstellar gateway is to link it to a second black hole in a distant area of space creating a Kerr tunnel. While such a trip sounds good on paper, it is not without its problems and risks. For instance, Kerr's equations indicate that a trip through the Kerr tunnel would be a one-way journey. This means that our time traveler may be unable to leave the tunnel once she reaches the second black hole. In fact, when she attempts to exit the outer event horizon of the distant black hole she may find that she is trapped and cannot escape the one-way barrier represented by the outer event horizon any more than she could escape from the outer event horizon when she first entered the first Kerr black hole. She may have traveled to a distant part of the universe to no avail, because she cannot move beyond the event horizon. Her only hope might be to duplicate her original journey. However, once she has accomplished this feat, she would simply end up in another distant black hole.[81]

Would our traveler be doomed to repeat such futile journeys over and over like some sort of black hole–hopping Sisyphus? Such would appear to be her fate unless, at the end of her journey she finds, not a duplicate black hole, but the exact opposite, a white hole. A **white hole** is a theoretical region in space that may exist at the other end of a black hole. Since a black hole pulls in matter and radiation due to its enormous gravitational field, a white hole would serve as that region at the opposite end of the black hole where the matter and radiation is ejected.[82] When our time traveler appears within the event horizon of the white hole, she would find herself instantly ejected from it, just as all matter and radiation are ejected. In fact, the event horizon would actually operate as an anti–event horizon. As such, it would be impossible to cross back into the white hole. Doing so would be like trying to swim up a waterfall or reenter a one-way turnstile. Consequently, even though our time traveler is no longer trapped within a black hole, she has still made a one-way trip and cannot use the white hole to return to her own time or her own region of the universe.[83]

Do white holes actually exist in our universe? No one knows for certain. It is interesting to note, however, that regions of intense radiation have been observed within our universe. As we saw earlier in this chapter, quasars are one such energy source. Some theorists have proposed that the intense energy produced by quasars is not caused by supermassive black holes at their centers, as others have suggested. One alternative explanation for the intense energy output of quasars is that they are actually distant white holes. Although there is no concrete proof for this proposition, it nevertheless cannot be completely ignored. Unfortunately, even if quasars are white holes, this is of little help to our time traveler. Quasars are among the most distant objects in our universe.[84] Consequently, upon emerging from the white hole, a trip, which is irreversible, the time traveler would be in a vastly distant region of the universe and would, therefore, be faced with a journey home that is so long that the original trip would not be worthwhile. However, this is not the worst fate that might befall the time traveler. It is also possible that the white holes in our universe are actually connected with black holes in other universes. As a result, when the time traveler exits the white hole at the end of the journey, she might not be in the past of the original universe but, instead, in an entirely different universe.[85]

As disheartening as this scenario may appear, it is still not the worst outcome of a journey into a Kerr black hole. Another, far more ominous consequence suggests that, once the time traveler has passed the outer event horizon of a Kerr black hole, she will have surrendered all hope of returning to the home universe. This interpretation is offered by the theoretical physicist Fred Alan Wolf, who argues that at the precise moment that the time traveler passes the outer event horizon of a Kerr black hole, the universe left behind instantly ages to infinity and, therefore, no longer exists. In this scenario, the time traveler's home universe has vanished.[86] Spatial dislocation, however, is not the only complication posed by a Kerr black hole. Another, more deadly crisis occurs because Kerr black holes are inherently unstable. Thus, the entrance of any object into a Kerr black hole can disrupt the delicate balance within its structure and cause the entire thing to collapse. The entrance of the time traveler's ship itself would probably provide enough of a disruptive influence to cause such a collapse. The compression would crush the time traveler out of existence. These problems make it highly unlikely that a Kerr black hole can be used for time travel in the near future.[87]

Kerr-Newman Black Holes. However, this is not the end of the story for rotating singularities. There are two other possibilities involving Kerr black holes. One such possibility results from a series of equations proposed by Ezra Newman of the University of Pittsburgh. Newman's work changes the nature of a rotating Kerr black hole by introducing the possibility that such a black hole might be electrically charged, rather than electrically neutral, as we have thus far assumed.[88] The advantage of a charged **Kerr-Newman black hole** is that the region allowing time travel exists not only within the ring singularity but also beyond the ring into a region referred to as the **ring envelope**. For this reason, the time traveler need not maneuver the timeship into the central region of the ring singularity. Instead, the ring envelope that extends beyond the ring singularity can be accessed. Such a trip would be less risky than one within a normal Kerr black hole since the time traveler can avoid the powerful gravitational field that lurks within the ring singularity itself.[89]

However, the Kerr-Newman black hole does not eliminate all time-travel problems associated with rotating black holes. For instance, the problem of one-way travel that is associated with an ordinary Kerr black hole does not evaporate with the Kerr-Newman black hole. It is, therefore, possible that our time traveler will end up once again in a spatially distant region of the home universe.[90] Additionally, there is little evidence to support the supposition that charged rotating black holes actually exist within the physical universe. A Kerr-Newman black hole may, therefore, be nothing more than a hypothetical mathematical construct.[91]

Super-Extreme Kerr Objects. One final option remains available for time travel using a rotating black hole. This option lies with an astronomical object called a **super-extreme Kerr object (SEKO)**. To understand the nature of a SEKO, it is advisable to step back for a moment and reexamine a normal Kerr black hole. Recall that the rotation of the singularity at the heart of a Kerr black hole produces two event horizons, the outer and the inner event horizons. If the spin of the rotating black hole were allowed to increase so that the object's angular momentum is equal to its mass, the two event horizons would slide toward one another, until they would merge. This would produce what is known as an **extreme Kerr object**. If the spin were allowed to increase further, so that the object's angular momentum exceeded its mass, the merged event horizons would disappear completely, and along with them, much of the danger associated with black hole time travel. Once the event horizon has vanished, the internal ring singularity would no longer be hidden, and the time traveler could use the region surrounding the ring to travel backward in time, by orbiting the ring axis of rotation in an appropriate manner.[92] In this situation, the time traveler would no longer face the risk of ending up in an alien universe or in some spatially distant region of the home universe. The chance of being trapped by the strong gravity well of an ordinary Kerr black hole also disappears with a SEKO.[93]

Unfortunately, for the prospect of time travel, SEKOs are not without their obstacles. The chief complication is that such extreme objects are, like Kerr-Newman black holes, probably nothing more than mathematical constructs and, therefore, may not exist within the physical universe. The reason for this difficulty is that at the heart of *any* black hole lies a singularity. Whether we are speaking of a stationary or a rotating black hole is of no consequence. All black holes possess a singularity, the structure of which is so extreme that all space-time ceases to exist in its vicinity and the laws of physics as we understand them break down completely. Such an alarming set of circumstances is troubling enough philosophically when the singularity is hidden from the rest of the universe beyond the event horizon of a black hole. In a SEKO, however, the event horizon is gone and the singularity is exposed to the rest of the universe. This is why it is referred to as a **naked singularity**. Such a situation is untenable to most physicists and, as a result, most believe that naked singularities have no real physical existence.[94] Because all Kerr objects, including Kerr black holes, Kerr-Newman black holes, and SEKOs have difficulties that make their use as time machines impractical, we explore two other methods of retrograde time travel.

Gott's Temporal Distortion Twist Theory

In an article in *Physical Review Letters*, J. Richard Gott of Princeton University proposed another method of retrograde time travel called the GOTT/Closed Time Curve/Cosmic String/Temporal Distortion Twist (GOTT/CTC/CS/TDT). To embark on a retrograde time travel mission using a GOTT/CTC/CS/TDT, three primary elements are needed: (1) a starship, (2) two cosmic strings traveling at each other from opposite directions at nearly the speed of light, and (3) two more cosmic strings traveling at each other from opposite directions at nearly light speed for the return trip[95] (see Figure 7.2).

A **cosmic string (CS)** is a slender, elongated strand of superenergy remaining as a by-product of the big bang. Cosmic strings are so massive (perhaps more than a thousand trillion tons for every inch) that they easily distort space-time within their vicinity. Despite the massive nature of these strings, one of

• 226 • Chapter 7 Past and Future Time Travel

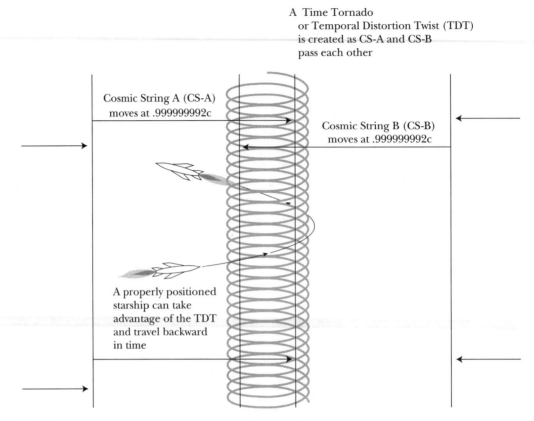

Figure 7.2 GOTT's/Closed Time Curve/Cosmic String/Temporal Distortion Twist.

To embark on a time-travel trip into the past using a GOTT/CTC/CS/TDT, as proposed by J. Richard Gott of Princeton University, three elements are needed: (1) a starship, (2) two cosmic strings traveling at each other from opposite directions at 0.999999992c, where c is light speed, and (3) two more cosmic strings traveling at each other from opposite directions at 0.999999992c for the return trip. A cosmic string (CS) is a slender, elongated strand of superenergy remaining as a by-product of the big bang. Cosmic strings are so massive (perhaps more than a thousand trillion tons for every inch) that they easily distort space-time within their vicinity. The first step in Gott's time-travel procedure is to locate or, if the space-faring civilization is sufficiently advanced, manufacture two cosmic strings. The second step is to maneuver these two cosmic strings so that they are racing toward each other at 0.999999992c. The next step is to accelerate the starship toward one CS at a predetermined, calculated speed. When these two massive, elongated strands of energy "zip" by each other, they bend and curl time, much like two opposing winds bend and curl the air as they rush by each other, creating a tornado. This "time tornado," or temporal distortion twist (TDT), will send a properly positioned starship back into its own past, so that it may actually arrive back in time before it has left on its mission.

them is not enough. Two are needed to distort space-time sufficiently to allow time travel into the past. Consequently, the first step in Gott's time-travel procedure is to locate or, if your space faring civilization is advanced enough, manufacture two cosmic strings. The second step is to maneuver these two cosmic strings so that they are racing toward each other at nearly the speed of light. The next step is to accelerate the starship toward one CS at a predetermined, calculated speed. When these two massive, elongated strands of energy zip by each other, they bend and curl time, much like two opposing winds bend and curl the air as they rush by each other, creating a tornado. This time tornado or temporal distortion twist (TDT) will send a properly positioned starship back into its own past so that it may actually arrive back in time before it has left on its mission.[96] Once again, however, the elements needed to create a GOTT/CTC/CS/TDT are clearly beyond our present technological, not to mention financial, capabilities. For this reason, let us take a look at one final technique for travel into the past.

Morris and Thorne's Traversable Wormholes

Kip Thorne and Michael Morris of Caltech have also contributed a theoretical study that demonstrates the feasibility of retrograde time travel. Morris and Thorne believe that cosmic wormholes may be used to travel in time. The concept of the wormhole was first introduced in the 1950s by the physicist John Wheeler of Princeton. Wheeler proposed that **wormholes** or tunnels might exist in space that could theoretically connect distant regions of the universe, even regions that were billions of miles apart.[97] A wormhole would allow the crew of a starship to traverse interstellar, perhaps even intergalactic, distances instantaneously without bothering actually to cross the real space in between.[98]

The work of Morris and Thorne has also produced a way to use wormholes to travel back in time. To accomplish this feat it would be necessary to take one end of the wormhole and accelerate it to nearly the speed of light. This could be done by placing the entrance of the tunnel near a massive spatial body such as a meteoroid, an asteroid, a moon, or even a small planet (see Figure 7.3). The goal is to allow the mouth of the wormhole to fall under the gravitational influence of that cosmic body. Once the opening is captured by the gravitational field, the cosmic body could be accelerated to nearly the speed of light. The mouth of the tunnel would be pulled along for the ride. After a carefully calculated trip at a speed near the light barrier, the mouth of the wormhole would be returned to a site in the vicinity of the other opening. This near-light-speed journey would subject the accelerated entrance to the time dilation effect, thus creating a time variance between the accelerated and stationary wormhole entrances. The crew of a starship could take advantage of this time variance to travel backward in time. The first step would be to journey in real space from the stationary to the accelerated entrance. If the crew of a starship were to enter the mouth of the accelerated tunnel and then exit from the stationary tunnel, the ship would have flown back in time.[99] In his work *Cosmic Time Travel*, Barry Parker explains the operational details of this time tunnel in space:

> To understand how this is possible, you have to consider clocks inside and outside the mouths. All clocks inside the wormhole will record the same

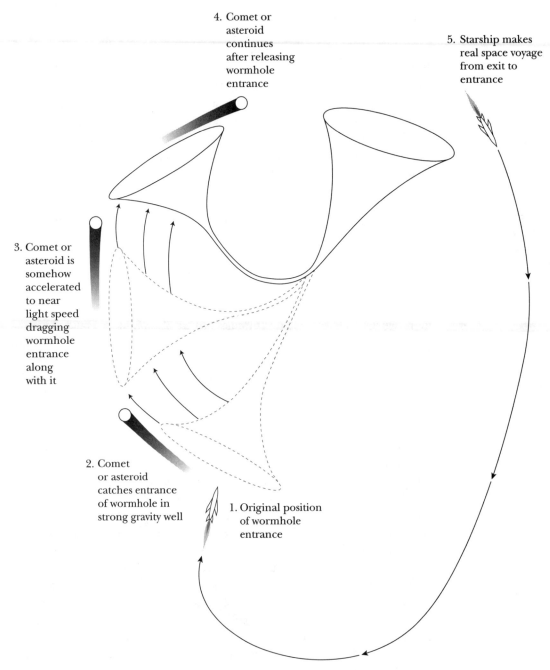

Figure 7.3 A Morris/Thorne Accelerated Wormhole.
  Kip Thorne and Michael Morris of the California Institute of Technology believe that cosmic wormholes may be used to travel in time. A wormhole is a hypothetical tunnel in space connecting distant regions of the universe. A wormhole would allow the crew of a starship to cross interstellar, perhaps even intergalactic distances instantaneously without actually crossing the real space in between. Morris and Thorne have produced a way to use wormholes to travel back in time. To accomplish this feat, it would be necessary to locate first the mouth of a relatively stable

7-2 Travel into the Past • 229

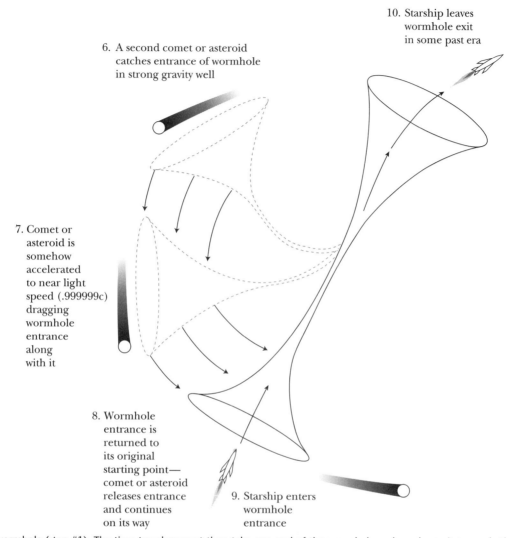

10. Starship leaves wormhole exit in some past era

6. A second comet or asteroid catches entrance of wormhole in strong gravity well

7. Comet or asteroid is somehow accelerated to near light speed (.999999c) dragging wormhole entrance along with it

8. Wormhole entrance is returned to its original starting point—comet or asteroid releases entrance and continues on its way

9. Starship enters wormhole entrance

wormhole (step #1). The time travelers must then take one end of the wormhole and accelerate it to nearly the speed of light. This could be done by placing the entrance of the tunnel near a massive spatial body, such as a meteoroid, an asteroid, a comet, a moon, or even a small planet (step #2). The goal is to allow the mouth of the wormhole to fall under the gravitational influence of that cosmic body. Once the opening is captured by the gravitational field, the cosmic body could be accelerated to nearly the speed of light. The mouth of the tunnel would be pulled along for the ride (steps #3 and 4). Another way to force one mouth of the wormhole to move is to charge the mouth of the wormhole—with a positive charge, for example—and cause it to fall under the influence of a negatively charged body that would attract the positively charged mouth. Whichever technique is used, after a carefully calculated trip at a speed near the light barrier, the mouth of the wormhole would be returned to a site in the vicinity of the other opening (steps #6, 7, and 8). This near-light journey would subject the accelerated entrance to the time dilation effect, thus creating a time variance between the accelerated and stationary entrances. The crew of a starship could take advantage of this time variance to travel backward in time. The first step would be to journey in real space from the stationary to the accelerated entrance (step #5). If the crew of a starship were to enter the mouth of the accelerated tunnel (step #9) and then exit from the stationary tunnel (step #10), the ship would have flown back in time.

time, regardless of the motion of one of the ends. The reason is that the two ends are not connected by ordinary space. As we saw earlier, a wormhole can reach half way across the universe, but it doesn't thread its way through the space between the two points.

Furthermore, the clock just outside the stationary mouth has not moved relative to the one just inside and will therefore show the same time. The clock just inside the mouth that has moved, on the other hand, will show an earlier time, in the same way that a moving twin, when he returns, will be younger.

Therefore, if you go in through the mouth that has moved and come out through the stationary mouth, you will go back in time. You will exit at the second mouth before you entered the first one. In the same way, if you go in through the stationary mouth and exit through the mouth that has moved, you will go into the future.[100]

The Morris/Thorne Traversable Wormhole is not without its shortcomings, however. Chief among these is the fact that a traveler can never go back in time to a period predating the discovery or the construction of the wormhole. This is because the time and the date of the wormhole discovery or construction will be the earliest time that can be registered on the clock at the stationary entrance.[101] Another shortcoming is that wormholes, if they exist at all, would be extraordinarily erratic. They snap open, narrow rather quickly, and then seal themselves up again. Some theorists believe that they may open and close at regular, if not predictable, intervals.[102] Others believe that each wormhole is allowed but a single opening in time. Whichever the case, some means of extending the life of a wormhole would be necessary before it would become feasible to use it as a permanent means of retrograde time travel. Thorne and Morris have faced this problem by postulating the existence of an antigravity force that would prop the wormhole open long enough to allow the time traveler to find her way through.[103]

7-3   TIME-TRAVEL PARADOXES

Time travel, especially retrograde time travel, challenges our usual experience with the laws of nature. In fact, the many variables and the exotic ingredients necessary to construct a time machine capable of traveling into the past suggest that such trips are highly unlikely. On the other hand, from what we have seen here about the general theory of relativity and the various theoretical approaches to retrograde time travel, such trips into the past are possible at least in principle, even if they are extremely unlikely in reality. Moreover, the very fact that relativity does not eliminate the possibility of time travel suggests that we ought to think about the potential consequences of such trips through time. As we do this, we will soon find out that time travel, especially retrograde time travel, can cause some serious paradoxes within the space-time continuum.

Realistically, it is highly unlikely that anyone alive today will have to face such paradoxes. Still, exploring some of these paradoxes will give us a better understanding of the intricate relationships that exist among the physical laws of the universe. Moreover, addressing these paradoxes will also allow us to discuss, in a relatively nonthreatening way, whether scientists should pursue all the possible products of their research. Another way to state this question is to ask: "Must scientists do everything that their research tells them they *can* do?" In relation to the question of time travel, there are two major types of paradoxes that can occur due to retrograde time travel. These are contradictory-event paradoxes and information paradoxes. As we shall see, both types can cause serious problems for the stability of the space-time continuum. Scientists must evaluate the risks, weigh them against the benefits, and decide whether the gains that can be realized by time travel are worth the hazards involved in tampering with the space-time continuum.[104]

Contradictory-Event Paradoxes

Contradictory-event paradoxes occur when incompatible incidents converge within the same time line. Such paradoxes are inherent in the inconsistency of the events themselves rather than in our understanding of those events.[105] The three most famous contradictory time-travel paradoxes are the grandmother paradox, the conservation paradox, and the simultaneous temporal-existence paradox. All three of these disturbing puzzles have been the subject of much debate. They have also formed the basis for the development of several films, a number of television dramas, and dozens of science fiction novels.

The Grandmother Paradox. According to the **grandmother paradox**, alternately known as the **grandfather** or **grandparent paradox**, the retrograde time traveler visits the past and meets her grandmother. Through a series of misadventures, the time traveler prevents her grandmother from meeting her grandfather before the birth of the time traveler's mother. As a result, the time traveler's mother is never born. Since the time traveler's mother does not exist, it is impossible for the time traveler to exist. Herein lies the convergence of incompatible events. If the time traveler's mother never existed and the time traveler never existed, then how can she travel back in time to prevent her mother's birth? Either she never existed and, therefore, could not travel in time or she has traveled in time but cannot stop the liaison between grandmother and grandfather. Yet, that appears to have happened. How is this possible?

In reality, a solution to any contradictory paradox is elusive only if the solution is limited by **Boolean logic**, that is an either-or approach. If we were limited to Boolean logic, we would conclude that only one of the two incompatible events can exist in real time. Either the time traveler exists and can stop her mother's birth, or she does not exist and, therefore, cannot travel back in time to prevent that birth. In contrast, a non-Boolean approach would allow many more solutions to the quandary. For instance, once her mother's birth is prevented, the time traveler may instantaneously cease to exist. In effect, when the time traveler prevents her own birth, she has committed a bizarre time-distorted, retroactive suicide. This is the result that occurs in a movie entitled *The Philadelphia Experiment: Part II*. In this film the time traveler's intervention results in the premature death of his father. In a grotesque form of retribution the forces of the space-time continuum reach out from a time vortex and reclaim the time traveler, scattering his atoms through the vast expanse of the eternal time stream.

Michael J. Fox, playing the time-traveling Marty McFly, has much better luck in *Back to the Future*, a light-hearted tale of a time-travel mishap that distorts the space-time continuum between 1955 and 1985. In this case, McFly's mother, a teenager in 1955, becomes enamored with the young time traveler and, consequently, snubs McFly's father. As a result, McFly's parents never marry, and his entire family begins a slow fade from history. This gradual fade is graphically represented by the disappearance of McFly's siblings from a photograph he has conveniently stashed in his wallet. Fortunately for McFly, his disappearance from the time stream occurs much more slowly than the vanishing act of the time traveler in *The Philadelphia Experiment*. This allows McFly the extra time to correct events and restore the history of the years 1955–85 to its proper sequence.

The disappearance of the time traveler from the time stream is only one solution to the grandmother paradox. There are others. For instance, it is possible that the time traveler cannot alter history no matter what she knows or how hard she tries. In fact, her presence in the past may actually be what is needed to insure that her grandmother and grandfather meet one another. Unknown to her the time trip into the past is actually a part of history. Another possibility is that our stalwart time traveler arrives in the past in some sort of incorporeal form. In this disembodied form she can see and hear the events as they occur but can never interact with them.

A clever variation of the grandmother paradox was created by the American science fiction writer Murray Leinster in a short story entitled "Dear Charles." In this strange tale of role reversal, the grandfather becomes the time traveler who moves several centuries into the future in an attempt to prevent his great (fifty-two times)-grandson, Charles, from marrying his, that is Charles's, fiancée, a girl named Ginny, who is ultimately destined to be Charles's great-great-etc.-grandmother. The danger inherent in this situation arises from the possibility that Charles, now his great-great-etc.-grandfather's rival, will somehow prevent his ancestor from winning the affections of Ginny. If this occurs, the space-time continuum will be caught in a repeating **time loop** from which there is no escape. The time loop would work like this. If Charles stops Ginny from returning to the past with his ancestor, Charles will not be born and he will not be present to prevent his ancestor from marrying Ginny, which means that Charles's ancestor will marry Ginny and Charles will be born and will stop his ancestor from marrying Ginny, which means that Charles will not be born, and will not exist to marry Ginny or to stop his ancestor from marrying Ginny, and so on, ad infinitum. Fortunately, Charles's ancestor concocts a way to signal Ginny from the past. As a result, Ginny is forewarned and prevents Charles from instituting the repeating time loop.[106]

The Conservation Paradox. While the grandmother paradox is troubling, it is not quite as baffling as the second contradictory-event paradox, the conservation paradox. The **conservation paradox** suggests that time travel may be impossible because it would violate the First Law of Thermodynamics. According to the **First Law of Thermodynamics**, "energy will always be conserved in a physical process, even though it may be converted from one form to another."[107] The sudden and inexplicable appearance of a time traveler in the past would violate the law of the conservation of energy. Energy can be converted from one form to another, but it can neither be created nor destroyed. Moreover, according to Einstein's famous equation $E=mc^2$, the mass of a material object is actually a measurement of the energy that object contains.[108] Given this immutable law, how can the mass of the time traveler disappear from the present and suddenly appear in the past? The disappearance from the present would represent the destruction of energy in the present, and the materialization in the past would represent the creation of energy in the past from nothing. If the total amount of energy cannot be changed within the closed system of the universe, what accounts for the energetic subtraction of the time traveler from the present and his energetic addition to the past?

In his novel *The Dark Tower*, the famous British writer C. S. Lewis pinpoints the essential nature of this problem with unerring accuracy. His main character, Dr. Orfieu, explains that the particles that make up the body of a time traveler will still exist in whatever time period the time traveler visits. However, those particles are likely to be scattered all over the Earth. As a result, a time traveler moving into the past or the future, might reach an era in which the traveler's body does not exist. Yet, the particles making up the body will exist, scattered haphazardly all over the Earth. Therefore, a potential time traveler cannot enter a different time era because, when arriving in that time period, the particles that make up her body will be someplace else. Orfieu concludes that, since the same particles cannot coexist in the same time and place, time travel, at least time travel by anything physical, like a living person, is impossible.[109]

The solution to the puzzle suggested by Lewis may lie within the confines of a radical new theory about the nature of mass and energy. This new theory, known as the zero-point field (ZPF) theory, suggests that the mass of an object is not, as Einstein says, the measure of its energy content. According to ZPF, the most fundamental components of the universe are not mass and energy, at least not as different entities. Rather, the fundamental components of the universe are electric charge and the electromagnetic field. This electromagnetic field, the ZPF, interacts with these massless electric charges to create masslike apparitions that are really unified clusters of charges embedded in the electromagnetic field. What this means from a practical point of view is that the chair on which you are seated actually has no mass. Rather, it is a cluster of massless electric charges interacting with the ZPF to give the appearance of mass.[110]

It is crucial to see that ZPF theory represents a fundamental alteration in our understanding of Einstein's most famous equation. The traditional understanding of $E=mc^2$ is that mass and energy are equivalent; that is, the mass of an object is equivalent to its energy content. However, according to ZPF theory, it is no longer accurate to assert that mass is equivalent to energy. Rather, we must say that mass actually *is* energy. The ZPF theory holds that the electromagnetic field gives rise to masslike particles. Because mass is an illusion that represents a reconfiguration of ZPF energy as it interacts with electric charge, Einstein's equation tells us the amount of energy needed to create these masslike apparitions. Thus, particles and antiparticles that appear to "snap" into existence do not really come from "nothing." Instead, those particles were there all along, existing in a form that had yet to interact with the electromagnetic field.[111]

In the same way, the appearance of the time traveler in the past does not represent the creation of mass from nothing. Rather, her appearance in the past represents the reconfiguration of massless electric charges as they interact with the ZPF. The time traveler, like the particles and antiparticles that appear to "snap" into existence, does not really come from "nothing." Instead, she was there all along, existing in a form that had yet to react with the electromagnetic field. It is, of course, an enormous leap of faith from the spontaneous and short-lived creation of subatomic particles and antiparticles to the spontaneous appearance of a full-grown, long-lived time traveler from the future. Nevertheless, the ZPF theory, at the very least, promises a fertile area for speculation on a solution to the conservation paradox.

Significantly, the idea that the conservation law can be preserved in time travel does not depend solely upon the ZPF theory. Although the ZPF theory has several advantages as an explanation for the apparent creation of mass from nothing, not the least of which is the absence of any need for a creation event, certain conventional theories can also be called upon to explain this phenomenon. In an article entitled "Recent Work on Time Travel," for example, John Earman, professor of philosophy of science at the University of Pittsburgh, has suggested that the conservation law is preserved if we assume that, at the departure point in the present, the time traveler's mass is transformed into energy, while at the arrival point in the past, an equivalent amount of energy is changed back into mass, representing the reconfiguration of the time traveler. Unfortunately, Earman does not explain this process in any detail.[112] Nevertheless, his endorsement of a possible solution to the conservation paradox is, at the very least, a sign of encouragement for any would-be time travelers.

The Simultaneous-Temporal-Existence Paradox. Let us assume for a moment that the conservation paradox can be solved by the principles of the ZPF theory. Does this solution eliminate all of the difficulties associated with time travel? Unfortunately, the answer to this question is "probably not." In fact, we still face the most puzzling enigma of all. A **simultaneous-temporal-existence paradox** occurs when the time traveler journeys into the past and meets a younger version of him- or herself in that past era. Which of the two people is the "real" one, the time traveler or her past self? The following illustration goes a long way toward illuminating the logical complications associated with an event paradox in which a time traveler meets herself in the past. Twenty students in a philosophy of science class were asked to unravel the following puzzle. "Suppose that twenty years from now, you enter a time machine and journey from the future into the past. While walking around your old neighborhood, you meet yourself. The two of you are now face to face. Which one is the *real* you?" Strangely, every student in the class answered that the future self was the *real* person. They were so convinced of this conclusion that they defended their position for thirty minutes. After the discussion ended, they were asked to consider a deceptively simple follow-up question. "Suppose your future self walked in the door right now. According to your own arguments, she is the *real* you. If that is true, then who are you?" Despite the staunch conviction with which the students had presented their first responses, most of them were now thoroughly confused.

The object of this case history is to point out the problems that arise when time travelers zip from the future into their own past. As was true with the grandmother paradox, as long as the focus is on a Boolean approach, only two solutions to the paradox present themselves. However, with non-Boolean logic, the explanations behind the enigma are endless. One answer is proposed in *Running against Time*, in which the principle character, David Rhodes, returns to 1963 to prevent the assassination of President John F. Kennedy. In that year, however, Rhodes coexists with his twelve-year-old self. Unfortunately, according to the jargon used in the film, the laws of physics prevent two bodies from sharing the same "life force" in the same time period. Unexpectedly, the younger Rhodes is the one who suffers. He falls victim to a mysterious affliction and is near death. Only the prompt return of the older Rhodes to the future saves his childhood self.

Another far less pleasant solution to the duplicate dilemma is presented in the film *Timecop*, in which the evil antagonist travels into the past to alter history for his own personal profit. For most of the film, his time-twisting plan succeeds. Only at the climax does he fall victim to the vengeance of the time stream. Having been warned repeatedly that the same matter cannot occupy the same space in the same time period, the villain scrupulously avoids running into his past self for most of the film. In the final moments, however, the infamous time traveler is tricked into direct physical contact with his past self. When these past and future selves collide, the space-time continuum erupts, and the two mirror-image villains are stretched and pulled "into infinity" as they rapidly annihilate one another.

A highly original solution to the dilemma of dual existence within the same time stream was suggested by the popular television program *Quantum Leap*. In this show, the time-traveling Dr. Sam Beckett can leap through time only within the years that form the boundaries of his own lifetime. His physical self, however, cannot occupy the same space-time continuum. As a result, every time he leaps from one year to another, his consciousness must occupy the physical body of a person who actually belongs within that time line. The First Law of Thermodynamics is preserved because the consciousness of the person whose body the good doctor inhabits is trapped in some sort of "metaphysical waiting room" while Dr. Beckett goes about correcting history. Of course, any scientific explanation for how all this happens is conveniently left out of the dialogue of the show.

Information Paradoxes

An **information paradox**, which is sometimes referred to as a **knowledge paradox**, occurs when knowledge is brought from the future to make past events occur that are based upon that future knowledge. In other words, the events in the past are made to happen because of information brought back from the future. A true information paradox occurs only if the knowledge brought from the future is used to cause itself.[113] An information paradox would occur, for instance, if the time traveler would journey back to 1920 with a first edition of *The Great Gatsby*, 5 years before it was actually published. If she were to seek out F. Scott Fitzgerald and allow him to copy the novel word for word so that he could reproduce the manuscript, she would have created an information paradox. The heart of the paradox is found in the fact that the process of putting the ideas and words together to create the novel cannot be generated by the novel itself.

> Knowledge [information] paradoxes violate the principle that knowledge can come into existence only as the result of problem-solving processes, such as biological evolution or human thought. Time travel appears to allow knowledge to flow from the future to the past and back, in a self-consistent loop, without anyone or anything ever having to grapple with the corresponding problems. What is philosophically objectionable here is not that knowledge-bearing artifacts are carried into the past—it is the "free lunch" element. The knowledge required to invent the artifacts must not be supplied by the artifacts themselves.[114]

Fitzgerald cannot use a novel he has not yet written to write that same novel. This would require the novel to exist before it has been written. Another example of this type of paradox appears in a science fiction story entitled *Time Breakers: Mind Out of Time*. In that story, one of the characters, Hamilton Shaw, writes out a copy of *Hamlet*, which he takes with him as he travels to sixteenth-century England. At that point, Shaw gives Shakespeare the manuscript and suggests that the playwright transcribe the manuscript and produce the play. Shakespeare does so, creating a clear information paradox. Such a set of contradictory circumstances seems irreconcilable. The logic of the situation seems to demand that nobody ever actually wrote *Hamlet*.[115] However, once again, the paradox remains mystifying only if the solution is forced to conform to Boolean logic. Boolean logic requires either that *Hamlet* was written before it was shown to Shakespeare, or that it was not. Both conclusions cannot exist simultaneously within the strict confines of Boolean thought. Following a non-Boolean approach, however, a number of explanations can be constructed for the apparent paradox. One of these solutions is the existence of alternate time lines.

7-4 ALTERNATE TIME LINES

Information paradoxes, like the Fitzgerald enigma, may be disconcerting, but they are not inherently dangerous. In contrast, the implications of time-travel twists caused by contradictory events, like the grandmother paradox, are not only difficult to understand but also threaten the very existence of the space-time continuum. Despite this, some theorists believe that they may have found a solution to these baffling dilemmas. The most intriguing of these theories is the possible existence of alternate time lines.

Parallel Universes and Alternate Time Lines

An extremely clever explanation for all of these paradoxes can be found within the parallel-universe theory of Hugh Everett of Princeton and Bryce DeWitt of the University of Texas. When they created their parallel-universe theory, Everett and DeWitt were not working with time travel but with several problems associated with quantum physics. What they suggested, as a solution to the problem of the uncertainty principle, was the existence of parallel universes. The **parallel universe theory**, which is also called the **many-worlds theory** and the **many-universes theory**, postulates the creation of a new universe every time a quantum measurement is made. Every time a person makes a decision of any type, he is involved in a quantum transition which creates a new universe. Each person inhabits all of these universes. Moreover, each universe has its own unique history that is separate from the history of all other universes. For example, if a young boy is trying to decide whether to attend church or play hooky by going to a baseball game, he actually does both. In one universe he dutifully goes to church under the watchful eyes of his proud parents, while in the other he trots off for an enjoyable morning at the ball park. From that point on, each universe develops its own separate historical time line. To complicate things further, each of these universes will branch off into other universes, all of which will formulate their own histories and so on ad infinitum.[116] With a little imagination the parallel universe theory can be applied to time travel. Instead of containing a single time stream, the universe holds an extradimensional timelike river with an infinite number of branching tributaries. Every time a time traveler makes a trip into the past, the traveler creates an alternate universe complete with its own alternate history. To distinguish these alternate universes from the parallel universes that coexist within our same time frame, these time-related parallel universes have been called **alternate time lines**.

The Everett/DeWitt alternate time line theory resolves many of the time-related paradoxes discussed in this chapter. For example, the grandmother paradox is avoided because the alternate time line which the time traveler visits includes many events that do not occur in the original time line. As a result, the time traveler's grandmother fails to meet her grandfather in the alternate time, but the time traveler's original time line remains undisturbed. Similarly, the paradox of simultaneous existence also disappears. A time traveler can never actually meet herself. At best, she can run into a duplicate self in another time line. The energy-conservation paradox is also avoided because the time traveler is supposed to exist within the alternate time line just as much as the original version of the time traveler. Finally, information paradoxes are avoided because the knowledge brought into one time line was actually created in another time line. Thus, the copy of *The Great Gatsby* that the time traveler takes into the past was actually written by another Fitzgerald in another time line.

Science Fiction and Alternate Universes

Science fiction folklore is replete with examples of alternate time lines. As far back as 1934, David Daniels imagined the existence of such alternate universes in a science fiction story that appeared in *Wonder Stories*. As the theoretical physicist Paul Halpern recounts in *Cosmic Wormholes*, Daniels manages to erase all contradictory events by switching to an alternate universe at precisely the correct moment. Moreover, every time that the time traveler moves into the past, a duplicate time line is created. Curiously, one time line is labeled "conventional history" whereas all of the others are referred to as "alternate realities." How it is possible to know which time line represents conventional history and which represent alternate realities is never made clear.[117] Since the appearance of Daniels's short story, many tales have explored this clash between conventional history and altered realities. In *The Proteus Operation*, by James Hogan, for example, several time lines intersect, including one in which Hitler won the Second World War.[118] In *The Gallatin Divergence*, by L. Neil Smith, the North

American Confederacy has replaced the United States of America.[119] Finally, in *At the Narrow Passage*, by Richard C. Meredith, the United States is supplanted by the British North American Colonies, which, in 1971, are still fighting the First World War.[120]

Many stories of alternate time lines stretch back even further into the receding mists of time. Several tales, including *Dinosaur Nexus*, by Lee Grimes, and *First Frontier*, by Diane Carey and James Kirkland, explore alternate time lines in which the dinosaurs are not destroyed but instead evolve into the dominant species on Earth.[121] *First Frontier* is especially notable because it represents one of the many *Star Trek* tales that explore the existence of alternate time lines. One of the most popular *Star Trek* episodes of the original series, "The City on the Edge of Forever," is based upon an alternate history in which the United States' entry into World War II is delayed by the intervention of a pacifist who should have died within the universe of conventional history. Another episode, "Mirror, Mirror," contemplates an alternate universe in which the benevolent United Federation of Planets is replaced by a ruthless totalitarian empire. These stories represent only a sampling of the many science fiction tales based upon the concept of alternate universes.

Perhaps the most extreme approach to the alternate–time line scenario was created by Orson Scott Carr in his novel *Pastwatch: The Redemption of Christopher Columbus*. In this novel Carr postulates that a single trip into the past would completely obliterate the time line from which the time traveler originated. Every person who ever lived, every event that had occurred, every instant of time after the point in time at which the time traveler arrives in the past would be gone. In fact, according to Carr, those people and those events never existed in the first place. Naturally, there is a hidden paradox buried within this line of reasoning. If the time traveler's time line no longer exists, how could the time traveler, who is a product of that time line, exist? Carr devises a clever, albeit somewhat complex, escape from this paradox. He insists, first of all, that time and causation are not the same thing, although we tend to think of them in that way. Time exists as a series of separate moments that follow one another in a discernible pattern. Causes appear to precede effects only because of our perspective. We say that an event that precedes another event caused that second one.[122]

However, with the construction of a time machine, it would be possible to take the latest moment in a series of separate moments and place that moment earlier in the time stream. Once that happens, all points in time leading from that past event, in this case the appearance of the time traveler, will be different from the first line of events. Since contradictory events cannot occupy the same moment in time, the old time line vanishes and the new one takes its place, playing itself out moment after moment. The first time line does not cease to exist from a causal point of view. The original time line still led to the displaced moment. However, the original time line ceases to exist in the sense that it flows from the series of moments that will lead from the displaced moment. The time traveler occupies a unique position. Since she comes from the previous series of moments, representing her time line, she will possess a memory of that time line. She will also accumulate a new set of memories as she lives in the new time line. However, she cannot return to her time line because it no longer exists.[123]

CONCLUSION

The ideas explored in this chapter suggest that time travel into both the future and the past are feasible, albeit unlikely. Although it is true that all of the time-travel methods contemplated here have required the application of highly advanced technology, none of them is a scientific impossibility. Stephen Hawking has suggested that time travel will never become a reality because the present has not been swamped with curious tourists from the future. Hawking makes a valid point. On the other hand, any civilization advanced enough to develop time travel would probably be aware of the problems associated with any expedition into the past. Consequently, such a civilization would most likely demand that time travelers educate themselves so as to blend into any time period they wished to visit. Moreover, all time travelers would probably be bound by a noninterference directive that would require those time travelers to avoid disturbing events and people in the past. More importantly, even if Hawking is correct, his argument need not discourage us from imagining the wonderful opportunity that retrograde time travel presents—a chance to relive history.

Review

7-1. One way to create a time distortion is to increase the speed of an object relative to other objects in the universe. The faster an object travels, the slower time moves for that object in relation to other objects moving at slower speeds. Speed, however, is not the only way to distort, twist, or dilate time. Time dilation can also be created by changes in gravity. The most promising source of a strong gravity well is in the vicinity of a black hole. Another astronomical body that generates a gravitational field that may be strong enough to help time travelers journey into the future is a neutron star.

7-2. The work of Roy Kerr indicates that retrograde time travel may be possible. A time traveler could theoretically navigate her ship through the region of a rotating singularity without being crushed, and end up in the past. J. Richard Gott has proposed another method of retrograde time travel, called the GOTT/Closed Time Curve/Cosmic String/Temporal Distortion Twist (GOTT/CTC/CS/TDT). Kip

Thorne and Michael Morris of Caltech have also contributed a theoretical study that demonstrates the feasibility of retrograde time travel using wormholes.

7-3. There are two major types of paradoxes that can occur due to retrograde time travel: contradictory-event paradoxes and information paradoxes. Contradictory-event paradoxes occur when incompatible incidents converge within the same time line. The three most famous contradictory time-travel paradoxes are the grandmother paradox, the conservation paradox, and the simultaneous-temporal-existence paradox. Information paradoxes occur when knowledge is brought from the future to make past events occur that are based upon that future knowledge.

7-4. An extremely clever explanation for all of these paradoxes can be found within the parallel universe theory of Hugh Everett of Princeton University and Bryce DeWitt of the University of Texas. The parallel universe theory postulates the creation of a new universe every time a quantum measurement is made. Every time a time traveler makes a trip into the past, she creates an alternate time line complete with its own alternate history.

Understanding Key Terms

acceleration
accretion disk
active galaxy
alternate time line
black hole
Boolean logic
conservation paradox
contradictory event paradox
cosmic string
ergosphere
escape velocity
event horizon
extreme Kerr object
First Law of Thermodynamics
grandmother paradox
gravitational field equations

gravitational radius (Schwarzschild radius)
information paradox (knowledge paradox)
inner event horizon
Kerr black hole
Kerr-Newman black hole
linear
mini–black hole
naked singularity
neutron star
outer event horizon
parallel universe theory (many-worlds theory, many-universes theory)
primordial black hole
principle of equivalence

quasar (quasi-stellar object)
ring envelope
rotating black hole
Schwarzschild black hole
simultaneous-temporal-existence paradox
spinning black hole
static limit
stationary black hole
stationary limit
super-extreme Kerr object (SEKO)
time loop
white hole
wormhole
zero-point field (ZPF)

Review Questions

1. How does speed affect travel into the future?
2. What are the essential principles of the general theory of relativity?
3. How does gravity affect travel into the future?
4. What is the essential nature of a Kerr black hole space-time bridge?
5. What are the essential ingredients of Gott's time-loop theory?
6. What are the characteristics of a Morris/Thorne wormhole?
7. What is the difference between contradictory-event and information paradoxes?
8. What is the nature of the grandmother paradox? the conservation paradox?
9. What is the nature of the paradox of simultaneous temporal existence?
10. What is the theory behind the possible creation of alternate time lines?

Discussion Questions

1. Speculate on some of the effects of an altered time line. Would our memories be changed, or would some of us vanish because we do not exist in the new time line? What would happen to the history books? What about all references to past events that might occur in movies and television programs? What would happen to the events recorded in newspaper accounts? Explain your answers to each question.
2. If you had the ability to travel in time, which era would you decide to visit first? Explain. Which eras would you avoid? Explain. If you discovered a way to travel in time, which past events would you try to change? Would you confine yourself to events in your own life, or would you, like the character David Rhodes mentioned in the opening commentary of this chapter, try to alter all of history? Explain.
3. What paradoxes might occur if you tried to change history? Explain. Also speculate on the dangers of time travel. Would you consider time travel too dangerous to be pursued, or would you encourage the development of time travel? Explain.
4. Speculate on the effects of time dilation on any attempt to establish an extensive galactic organization, such as the type of federation that appears in *Star Trek*.
5. If you discovered a way to travel in time, would you reveal your discovery to the world, or would you keep it a secret and reserve the ability to travel in time for your own use? Explain.

ANALYZING STAR TREK

Background   The following episode from *Star Trek: The Next Generation* reflects some of the issues that are presented in this chapter. The episode has been carefully chosen to represent several of the most interesting aspects of the chapter. When answering the questions at the end of the episode, you should express your opinions as clearly and openly as possible. You may also want to discuss your answers with others and compare and contrast those answers. Above all, you should be less concerned with the "right" answer and more with explaining your position as thoroughly as possible.

Viewing Assignment— *Star Trek: The Next Generation,* **"We'll Always Have Paris"**

While fencing in one of the recreation areas of the Enterprise, Captain Picard and a young lieutenant experience a moment in which time repeats itself. Upon checking with the bridge crew, Picard learns that the time loop occurred throughout the entire ship. When the captain arrives on the bridge, he learns that the ship has received a distress call from a deep-space experimental laboratory run by Dr. Paul Manheim. Manheim is a physicist who proposed several revolutionary theories regarding time. When they were not well-received, he left Earth to establish a private, deep-space laboratory on Vandor IV. Apparently, something has gone wrong with his experiments because the space-time continuum is caught in a series of recurring time loops. When the Enterprise arrives at Vandor IV, the crew beams Dr. Manheim and his wife, Jenice, to sick bay. The doctor is caught in some sort of phase shift. He seems to drift in and out of our dimension of space-time. Eventually, it becomes clear that Picard and Jenice were romantically involved twenty-two years earlier. Meanwhile, the Enterprise continues to experience time-loop distortions. Manheim reports that his work on Vandor IV has managed to break open a rift in time to another dimension. The rift must be eliminated, or the entire space-time continuum will be disrupted. Since Data appears to be the only person unaffected by the time loops, he is assigned the task of attempting to eliminate the rift. While he is in Manheim's laboratory on Vandor, another time loop occurs in which three time-shifted versions of Data appear simultaneously. At first none of the three can figure out which is the real-time Data. However, eventually they do determine the real-time Data, and the rift is sealed.

Thoughts, Notions, and Speculations

1. When Data is originally explaining Dr. Manheim's work, he notes that one of Manheim's theories involves the relationship between time and gravity. He further comments that this theory found little support among other members of the scientific community. Given what you have learned about relativity, explain why such a statement would be inaccurate.
2. Identify the time paradoxes that occur on the Enterprise while under the influence of the time-loop distortions. Speculate on which of the time-travel techniques discussed in this chapter is closest in theory to the time-loop distortions.

3. Imagine that you have the responsibility of obtaining a grant for Dr. Manheim's research project from a funding council. How would you defend his research? Explain. Now switch sides and imagine that you are opposed to his research. What arguments would convince the council to reject his request for a grant? As part of both arguments consider whether people will welcome Manheim's discovery of time travel, or will they see it as threatening to their way of life? Explain.
4. Explain how the theory of parallel universes and alternate time lines might apply to the time-loop distortions.
5. Identify which of the time loops that occur on the Enterprise violate the First Law of Thermodynamics and which do not. Explain your decision.

NOTES

1. Albert Einstein, *Relativity: The Special and General Theory*, 41–42; John Gribbin, *Time-Warps*, 79–81; John W. Macvey, *Time Travel*, 50–51; Barry Parker, *Cosmic Time Travel*, 30–35.
2. Gribbin, *Time Warps*, 79–81; Parker, 30–35; Barry Zimmerman and David Zimmerman, *Why Nothing Can Travel Faster Than Light*, 82–83.
3. Albert Einstein, *Relativity: The Special and General Theory*, 51–52; Parker, 33; Zimmerman and Zimmerman, 80–81.
4. Albert Einstein, *Relativity: The Special and General Theory*, 51–52; Stephen Hawking, *The Illustrated A Brief History of Time*, 31–32; Parker, 34. It is critical to remember that the speed limit of 186,000 mps (300,000 kps) is the speed of light in a vacuum. In other mediums light slows down. For instance, in water light travels at 140,000 mps. Other particles, such as electrons, have been made to go faster than light in other mediums. This is what happened in the 1930s when Pavel Cerenkov, a Russian scientist, increased the speed of electrons in water to 160,000 mps. Still, the upper speed limit of the universe remains the speed of light in a vacuum, 186,000 mps. Zimmerman and Zimmerman, 83.
5. Rick Sternbach and Michael Okuda, *Star Trek: The Next Generation Technical Manual*, 54–55.
6. Ibid., 55.
7. Lawrence M. Krauss, of Case Western Reserve Univ. also tackled this problem in *The Physics of Star Trek*.
8. Sternbach and Okuda, 78. In the *Star Trek: The Next Generation* episode entitled "Cause and Effect," for example, the Enterprise must access a Federation time base beacon in order to adjust its internal chronometer. The adjustment is made necessary because the Enterprise has been caught in a temporal causation loop.
9. Einstein, *Relativity: The Special and General Theory*, 75–77; Tony Rothman, *Instant Physics: From Aristotle to Einstein, and Beyond*, 209–10.
10. Alan Isaacs, ed., *A Dictionary of Physics*, 3.
11. Karl F. Kuhn, *Basic Physics*, 2.
12. Einstein, *Relativity: The Special and General Theory*, 69–70.
13. Albert Einstein, *Ideas and Opinions*, 286.
14. Ernst Mach, *The Science of Mechanics*; for an in-depth study of Mach's influence on Einstein see Mendel Sachs, "The Mach Principle," 98–104.
15. Sir Isaac Newton, *Mathematical Principles*, 8–9.
16. Einstein, *Ideas and Opinions*, 286–87; Rothman, 210–11.
17. Einstein, *Ideas and Opinions*, 286–87; Albert Einstein, *Out of My Later Years*, 45–46; Einstein, *Relativity: The Special and General Theory*, 75–77; Paul Davies, *About Time*, 87–88; Rothman, 210–11.
18. Einstein, *Ideas and Opinions*, 287.
19. Eric Chaisson, *Relatively Speaking*, 78–79; Robert M. Hazen and James Trefil, *Science Matters*, 167.
20. Chaisson, 78–79; Alan J. Friedman and Carol C. Donley, *Einstein as Myth and Muse*, 61; Robert H. March, *Physics for Poets*, 140; Rothman, 211.

21. Chaisson, 80–83; Barry Chapman, *Reverse Time Travel*, 74–75; Peter Coveney and Roger Highfield, *The Arrow of Time*, 90–91; Davies, *About Time,* 102; Friedman and Donley, 62.
22. Chaisson, 81–83; Chapman, 74–75; Coveney and Highfield, 90–91; Paul Halpern, *Cosmic Wormholes*, 35–36.
23. Halpern, 36.
24. Albert Einstein, *The Meaning of Relativity*, 92–93; Einstein, *Relativity: The Special and General Theory*, 84–85; Hawking, 42; Rothman, 212–13. It now appears that Eddington's calculations may have resulted as they did because he knew the conclusion that he wanted to reach. Nevertheless, the predictions made by Einstein's theory have been verified by subsequent measurements. See Hawking, 42; Rothman, 212–13.
25. Einstein, *Meaning*, 91–92.
26. Macvey, 112–13; Parker, 134–35.
27. Einstein, *Meaning*, 90–92.
28. Macvey, 112–13; Parker 134–35.
29. Isaacs, 136; Halpern, 38.
30. David Millar, ed. et al., *The Cambridge Dictionary of Scientists*, 289; Davies, 106–7; Halpern, 38.
31. Davies, *About Time*, 107.
32. Ibid.
33. Ibid.
34. Davies, *About Time*, 107; Halpern, 37–38; Parker, 114–15.
35. William K. Hartmann, *The Cosmic Voyage through Space and Time*, 290–91.
36. Parker, 130–33; Colin A. Ronan, *The Natural History of the Universe from the Big Bang to the End of Time*, 89.
37. Nick Herbert, *Faster Than Light: Superluminal Loopholes in Physics*, 111–12; Parker, 130; Kip S. Thorne, *Black Holes and Time Warps*, 476–77, 557.
38. Thorne, 476–77 and 557.
39. Parker, 130–33; Ronan, 89.
40. Parker, 135.
41. Halpern, 38–39; Parker, 130–33.
42. Halpern, 39.
43. Parker, 130–33.
44. Davies, *About Time*, 116–17; John Gribbin, *Unveiling the Edge of Time*, 144; Halpern, 38–40.
45. Parker, 135.
46. Davies, *About Time*, 114–15; Parker, 134–35.
47. Davies, *About Time*, 114–15; Parker, 134–35.
48. Parker, 135.
49. Isaacs, 345; Ronan, 70. Trinh Xuan Thuan, *Secret Melody*, 175–77.
50. Ronan, 73.
51. Ronan, 71–73; Thuan, 175–77.
52. Michael Disney, "A New Look at Quasars," 55–57.
53. Hartmann, 403–6; Thuan, 175.
54. Hartmann, 405.
55. Isaacs, 345; Thuan, 176–77.
56. Ronan, 66.
57. Thuan, 177.
58. Hartmann, 405; Ronan, 67, 70.
59. Thuan, 177.
60. Hawking, 127; Isaacs, 36; Ronan, 64; Thuan, 115.
61. Hawking, 127; Thorne, 447.
62. Ronan, 64; Thorne, 447; Thuan, 115.
63. Hawking, 127.
64. Isaac Asimov, *The Collapsing Universe*, 143–47; Hartmann, 303–4.
65. Davies, *About Time*, 105.
66. Ibid.
67. In *Cosmic Wormholes: The Search for Interstellar Shortcuts*, Paul Halpern, a physicist with the Philadelphia College of Pharmacy and Science, also notes that neither a white dwarf nor a neutron star would create a gravity well powerful enough to serve as a gateway or "wormhole" for interstellar travel. Instead, Halpern suggests that only a black hole would provide the extreme conditions needed to create an interstellar wormhole. Halpern, 48–71.
68. Davies, *About Time*, 116–17.
69. Asimov, 181.
70. Parker, 134–35.
71. Parker, 135.
72. Gribbin, *Unveiling*, 118; Halpern, 153–54; Parker, 132.
73. Gribbin, *Unveiling*, 118; Halpern, 153–54; Parker, 132.
74. Halpern, 153–55.
75. Chapman, 95; Halpern, 73; Herbert, 119.
76. Asimov, 217–18; Nigel Calder, *Einstein's Universe*, 89–90; Parker, 141.
77. Fred Alan Wolf, *Parallel Universes*, 168.
78. Wolf, 168–69.
79. Halpern, 156; Herbert, 121; Gribbin, 163.
80. Halpern, 156; Herbert, 121.
81. Halpern, 75.
82. Gribbin, *Unveiling*, 237; Halpern, 77–81; Parker, 162.
83. Halpern, 81–83.
84. Isaacs, 345; Carl Sagan, *Cosmos*, 250; Thuan, 176–77.
85. Herbert, 121; Parker, 161.
86. Wolf, 168.
87. Gribbin, *Unveiling*, 173–74; Halpern, 74.
88. Gribbin, *Unveiling*, 162; Halpern, 155–56; Herbert, 121–22; Parker, 162.
89. Halpern, 156.

90. Ibid.
91. Gribbin, *Unveiling*, 162; Parker, 162. Moreover, recent calculations by Shahar Hod and Tsvi Piran of the Hebrew University, demonstrate that even without the entrance of an outside influence, the entire structure may collapse in and of itself before it has a chance to form. For details on this see Shahar Hod and Tsvi Piran, "Mass Inflation in Dynamical Gravitational Collapse of a Charged Scalar Field," 1554–57; and Meher Antia, "Lost Horizon," 6.
92. Halpern, 157; Herbert, 123–25; Parker, 155–56.
93. Halpern, 157; Herbert, 123–25.
94. Halpern, 157–58.
95. David Deutsch and Michael Lockwood, "The Quantum Physics of Time Travel," 70; Davies, *About Time*, 248; John Earman, "Recent Work on Time Travel," 279–80; J. Richard Gott, "Closed Timelike Curves Produced by Pairs of Moving Cosmic Strings," 1126; Halpern, 151–52; Ivars Peterson, "Timely Questions," 202–3. For two popular accounts of Gott's work see Michael Lemonick, "How to Go Back in Time," 74; Richard Morris, "The Perils of Time Travel," 60. For a detailed explanation of the origin and nature of cosmic strings see: Gribbin, *Unveiling*, 181–83.
96. Deutsch and Lockwood, 70; Davies, *About Time*, 248; Earman, 279–80; Gott, 1126; Peterson, 202–3; Lemonick, 74; and Morris, 60.
97. Davies, *About Time*, 245; Halpern, 2.
98. Halpern, 2.
99. Halpern, 161–62; Parker, 224–26; Pickover, *Time: A Traveler's Guide*, 198–200.
100. It is crucial to avoid taking this diagram too literally. Remember this illustration is a two-dimensional diagram showing a tunnel between two points in three-dimensional space. A pair of genuine wormhole entrances would actually create a tunnel between two points in four-dimensional space-time. Consequently, the crew of a starship approaching the opening of a wormhole would probably not see the stylized indentation depicted in the graphic. Rather, as they advanced toward the entrance they would probably see a sphere-like phenomenon. Pickover, 199.
101. Ibid., 224–26.
102. Ibid., 120.
103. Davies, *About Time*, 247.
104. Deutsch and Lockwood, 71–72.
105. Ibid., 71.
106. Murray Leinster, "Dear Charles," 28–44.
107. Coveney and Highfield, 150.
108. Mendel Sachs, *Relativity in Our Time*, 85–87.
109. C. S. Lewis, *The Dark Tower and Other Stories*, 17–18.
110. Bernard Haisch, Alfonso Rueda, and H. E. Puthoff, "Beyond E=mc²," 26–30.
111. Haisch, Rueda, and Puthoff, 26–30; see also Owen Davies, "Volatile Vacuums," 50–54.
112. Earman, 271–72.
113. Deutsch and Lockwood, 71.
114. Ibid.
115. Rachel Pollack and Chris Weston, *Time Breakers*, 10.
116. Paul Davies, *God and the New Physics*, 116; Halpern, 193; Deutsch and Lockwood, 72–74.
117. Halpern, 196.
118. James P. Hogan, *The Proteus Operation*.
119. L. Neil Smith, *The Gallatin Divergence*.
120. Richard C. Meredith, *At the Narrow Passage*.
121. Lee Grimes, *Dinosaur Nexus*; Diane Carey and James I. Kirkland, *First Frontier*.
122. Orson Scott Carr, *Pastwatch*, 215–20.
123. Ibid.

Unit V The Social and Cultural Impact of the Theory of Relativity

Chapter 8
The Philosophical Side of Relativity

COMMENTARY: RELATIVITY AND THE NATURE OF ALBERT EINSTEIN

Approximately three weeks after the public at large learned about the theory of relativity, Albert Einstein was invited to write an article about his work for the November 28, 1919 edition of the London *Times*. In that article Einstein praised the international spirit of the 1919 British expedition which had confirmed his theories. He was especially impressed by the willingness of the British scientists to expend time, money, effort, and energy to confirm the theories of a German scientist so soon after a world war which had seen Great Britain and Germany as mortal enemies. In fact, the international spirit of the British expedition went beyond even Einstein's expectations because it was actually planned and financed while the war was still in progress.[1] Even more interesting is the fact that the expedition went ahead as planned even though the British military was in desperate need of new recruits for the coming Battle of the Marne and could have, indeed almost did, conscript Sir Arthur Eddington, the leader of the expedition, despite his status as a conscientious objector.[2] At the end of the article Einstein comments that, with the verification of his theory of relativity,

"In Germany I am called a German man of science and in England I am represented as a Swiss Jew."[3] He then added, somewhat in jest, that if relativity is ever disproven, "I shall become a Swiss Jew for the Germans and a German man of science for the English!"[4] An editorial on relativity also appeared in the same edition of the *Times*. In that editorial, the *Times* not only recognized Einstein's praise of British scientific neutrality, but also suggested that his description of himself was an example of relativity. "We concede him his little jest," the *Times* reported. "But we note that, in accordance with the general tenor of his theory, Dr. Einstein does not supply an absolute description of himself."[5] Is Einstein's self-description, in fact, an expression of relativity? Or does the comment by the *Times* represent some of the common misconceptions about Einstein's theory? Has everything been proven relative, or can we still hold to some absolute truths? How has the theory of relativity impacted outside of science? Has philosophy, for instance, been rendered pointless by Einstein's theory of relativity? Or has Einstein's thinking somehow verified the ideas proposed by certain modern philosophers?

CHAPTER OUTCOMES

After reading this chapter, the reader should be able to accomplish the following:

1. Discuss the cultural impact of relativity.
2. Outline the relationship between philosophy and relativity.
3. Relate Kant's theory of space and time.
4. Explain how relativity verifies empiricism.
5. Define the nature of idealism.
6. Outline the relationship between idealism and relativity.
7. Discuss the simultaneous existence of time and the nature of the space-time cube.
8. Define the principle of the fixity of time.
9. Identify the problems of free will created by the simultaneous existence of all time.
10. Explain the role of philosophy in the modern world.

8-1   EINSTEIN'S INFLUENCE ON PHILOSOPHY

Our objective in this text has been to explore not only the scientific theories of the modern age but also the impact of those theories beyond the world of science. The theory of relativity, for example, has reconfigured our understanding of the very nature of space and time. Today, because of the theory of relativity, we know that space and time are not passive cosmic platforms upon which all events take place. Instead, we now know that space and time have no meaning in the absence of matter and energy. Space and time are, therefore, tied directly to fabric of the universe itself. As a result, it is not possible for a philosopher to speak of space and time without considering the implications of relativity theory. Thus, it is important that we examine the relationship between relativity and philosophy.[6] Of course, there are those who argue that the effects of relativity on philosophy are negligible. For instance, V. F. Lenzen, of the American Philosophical Association, argued effectively as early as 1945 in *The Philosophical Review* that it was possible for scientific concepts to be neutral in relation to philosophy. Lenzen, who was both a physicist and a philosopher, held the view that scientific concepts could be explained in terms that would not prejudice any individual philosophical system.[7]

Exactly what Lenzen meant by this statement is still open to debate over fifty years later. For instance, Andrew Paul Ushenko, a philosopher at Princeton University, has argued that Lenzen's perspective is valid only if it is taken to mean that science cannot support any one theory of knowledge to the exclusion of all other theories. On the other hand, Ushenko believes that some philosophical systems are incompatible with certain scientific discoveries. As an example of this, he points out that the philosophical system of solipsism is incompatible with the theory of relativity.[8] **Solipsism** is the belief that only the self exists. According to solipsism, all other existents exist only in the mind of the self.[9] Some experts have argued that solipsism is a logical extension of René Descartes's belief in pure rationalism. Descartes's rational philosophy found its most succinct expression in his famous statement, "I think, therefore I am." In fact, it appears that Descartes himself may have been somewhat concerned that his beliefs might lead to solipsism.[10]

Ushenko's conclusion that solipsism is incompatible with relativity results from the following argument. Relativity has demonstrated that time and space vary depending upon the observer's position and motion. The relative nature of space and time requires, by definition, at least two observers. The existence of two observers is contrary to the basic assumption of solipsism that only one observer, the self, exists. Therefore, solipsism and the theory of relativity are clearly incompatible. Ushenko is careful to point out that such an argument does not necessarily disprove solipsism. It merely shows that solipsism is incompatible with relativity. On the other hand, the argument does demonstrate that science in general, and relativity in particular, make some philosophical belief systems *less likely* to be true than others. Therefore, Ushenko concludes, contrary to Lenzen's optimistic outlook, that science is *not* neutral to philosophy.[11]

The Nature of Einstein's Influence

Having concluded that science, in general, and the theory of relativity in particular, do indeed affect philosophy, the next issue that we must explore is how that influence is propagated. One thing can be said about the theory of relativity and its author at the outset. Einstein made few direct philosophical pronouncements, and those that he did make were generally reserved to the ethical and social responsibilities of the scientist.[12] In fact, it is safe to say that, unlike some of his contemporaries, Einstein had no illusions that he could be both a physicist and a philosopher. His foremost ambition was always to explore the universe via physics, not by way of philosophy.[13] This is not to say that he failed to recognize the interaction of physics and philosophy. Nor does it mean that he did not believe that scientists had to make philosophic adjustments as the nature of the universe gradually revealed itself. On the contrary, this perspective seems to pervade his thoughts. Nevertheless, the fact remains that he made only a few direct philosophical declarations.[14] How then have his ideas managed to have such a profound effect upon philosophy?

According to Hans Reichenbach of the University of California at Los Angeles, Einstein's influence is felt beyond science because of his unwavering faith in positivism.[15] Positivism, as we have seen, is based on the belief that laws of nature exist as an objective set of rules that can be discovered by neutral scientific processes. This is the approach that we discussed at length in Chapter 1 when we looked at the scientific method. According to positivism, science can be distinguished from most other areas of knowledge because scientific truth is derived from an objective evaluation of carefully controlled observation and experimentation. Positivism, then, is the way that most practicing researchers would describe their work.[16] It is also the frame of mind that Einstein took as he approached his own work.[17]

Einstein's positivistic approach is significant because it implies that philosophy must evolve out of science rather than the other way around. According to positivism, physics comes first and philosophy second.[18] This attitude is not necessarily intended as a value judgment on the importance of one area of knowledge over the other. Rather, it is a description of the causal connection between physics and philosophy. In a sense then, philosophy can be looked at as a product of physics. The physicist reveals a new way of understanding how the physical universe operates. At that point, the philosopher of physics takes over and analyzes the implications of this new way to view the universe. The physicist, in the meantime, continues to explore the physical laws that govern the operation of the universe.[19]

In an article entitled "The Philosophical Significance of the Theory of Relativity," Reichenbach points out that physics and philosophy require not only different temperaments but also different techniques.[20] The physicist is interested in unlocking new discoveries about the universe. This approach requires that he focus on the objective application of the scientific method. Nevertheless,

Figure 8.1 Albert Einstein, in later life.
Albert Einstein did not believe that he could be both a physicist and a philosopher. Accordingly, he made few philosophical pronouncements. Instead, Einstein limited his nonscientific statements to social and ethical issues. Here he is shown testifying before the Anglo-American Commission conducting hearings in Washington in 1946 on the Palestine question.
Photo Credit: UPI/Corbis-Bettmann

Reichenbach believes that the physicist is also driven by inspiration, intuition, or, as he says, by a personal creed of some sort. Einstein, for instance, was driven to his discovery of relativity by his firm belief that the universe must be an orderly and harmonious place.[21] Accordingly, he did not criticize nor analyze the process by which he approached relativity. Instead, he conducted the process so that he could determine what those results would be. In a sense then, the physicist is guided by two complementary forces, a faith in the scientific process and a personal conviction that his intuition is correct.[22]

In contrast, according to Reichenbach, philosophers must be logical and analytical in both temperament and technique. They will be driven by neither a creed nor a vision. Instead, they will be methodical and analytical as they examine those theories that have been proposed by the physicists. Philosophers will investigate the theory itself, rather than the route that the physicist took to arrive at that theory. The philosopher is interested in whether the results are valid and consistent within themselves and whether those results can be justified within the overall context of physics. Philosophers, according to Reichenbach, will not criticize a physicist's individual creed or personal vision. In fact, philosophers clearly recognize that frequently such a vision is the driving force behind an important discovery. Philosophers, therefore, will make allowances for personal beliefs as long as those beliefs do not taint the scientific process nor bias the results. Philosophers are also interested in the logical results that flow from a new theory or a radical discovery. Accordingly, they will try to pinpoint its effects in the larger context of science and society.[23]

The Union of Physics and Philosophy

Taking all of this into consideration, Reichenbach concludes that Einstein's positivism has drawn science and philosophy together under the same roof forever. Philosophy can no longer claim to be a separate discipline which seeks an understanding via a route that is different from the one followed by science. The direction that all philosophic thought must take is carefully laid down first by the scientist.[24] Reichenbach says it this way:

> There is no separate entrance to truth for philosophers. The path of the philosopher is indicated by that of the scientist: all the philosopher can do is to analyze the results of science, to construe their meanings and stake out their validity. Theory of knowledge is analysis of science.[25]

Nor is Reichenbach alone on his conjecture. Physics professor James Trefil of George Mason University and the author of *Reading the Mind of God* has also argued that, out of all the academic disciplines, only science arrives at answers that can be tested and either proven or disproven by experimental results. Consequently, Trefil points out that scientific conclusions are verifiable, whereas philosophical ones, at least those that contradict verified scientific conclusions are not.[26] Unfortunately, neither Reichenbach nor Trefil has really settled anything. In fact, their conclusions simply displace the problem from one arena to another. According to Reichenbach, for instance, before Einstein, philosophers had free license to fashion whatever concept of reality they could imagine, regardless of how fantastic or unrealistic that concept might be. Now the philosopher must follow the lead of the scientist. However, this does not mean that the philosopher no longer has a function. Nor does it subordinate the function of the philosopher for that of the scientist. Instead, it alters the role of the philosopher from that of originator to that of interpreter. In essence, philosophers must now accept the objective validity of certain theories that have been verified by the scientists. However, the interpretation of the impact of that theory on human knowledge, or on the meaning of life, or on the ability to make ethical decisions is still open to a wide variety of philosophical perspectives. Or is it?

This question will be the focus of the rest of this chapter. Rather than argue the point abstractly, we will examine two case histories, each of which involves a different interpretation of the theory of relativity. One of these interpretations is proposed by Reichenbach himself. The other is presented by Kurt Gödel, a logician who was one of the preeminent minds of the twentieth century. As we have seen, Reichenbach believes that the theory of relativity as proposed by Einstein has eliminated any subjective interpretation of space and time.[27] In contrast, Gödel believes that relativity demonstrates that space and time are subjective concepts created by the mind to give order to the universe.[28]

If Reichenbach is correct that the philosopher must follow the scientist's lead,[29] then it follows that all philosophers should reach identical conclusions about all scientific discoveries. This, however, is not the case. As noted, Reichenbach reaches one conclusion about relativity while Gödel uses relativity to support a proposition diametrically opposed to that of Reichenbach. Reichenbach uses relativity to show that space and time have a real existence, whereas Gödel uses the same theory to show that the passage of time exists only in the mind. All of these differences are verifiable facts. Therefore, the remaining question is: Do these differences matter? As we shall see, there is no definitive answer to this question. However, there are indications that these different interpretations do actually make a profound difference. For instance, Gödel's interpretation of relativity has at least two profound implications that do not flow from Reichenbach's interpretation. Gödel's interpretation implies, first, that all time exists simultaneously and, second, that human volition must, therefore, be an illusion. Both of these ideas run counter to those of Reichenbach. For the remainder of this chapter, we explore these issues. First, we will look at Reichenbach's belief that relativity eliminates the idea that space time is created by the mind. Then we reverse our perspective and see how Gödel uses relativity to support the notion that space and time are mental concepts. Finally, we follow Gödel's ideas to their logical end as we explore the possible simultaneous existence of all time and the elimination of free will.

8-2 POSITIVISM AND RELATIVITY

Before Einstein's work in the twentieth century, one of the most convincing explanations of space and time was proposed by the eighteenth-century German philosopher Immanuel Kant. Like Einstein's theory of relativity, when Kant's theory of space and time was first developed, it was truly revolutionary. In fact, his theory was so radical that Kant himself labeled it as the new "Copernican revolution."[30] Before Kant, the basic assumption of philosophy was that the human intellect must adapt itself to the objective reality of the physical universe.

Naturally, there were problems with this approach. One difficulty was that, according to this belief, it was impossible to know whether the information processed by the mind was accurate. The intellect might, after all, misprocess information by succumbing to an illusion, by misinterpreting sense data, or by surrendering to a deliberate attempt to mislead it.[31] This concern was one of the fears expressed by Descartes, who rejected empiricism and embraced rationalism because he believed that data from our senses were frequently misleading.[32]

Kant's Ideas on Space and Time

According to Kant, the physical universe is not something that the mind must struggle to perceive either correctly or incorrectly. Instead, the physical universe must correspond to the activity of the mind. The mind, in a sense, shapes physical reality.[33] Lodged within the mind are certain a priori forms that construct reality, not according to experience but according to the pure activity of the mind. The mind places all sense experiences, that is, all **sensations**, within time and space to give them coherence. In a sense then, through the activity of the mind, the unstructured sensations become organized **perceptions**. Time and space, however, exist within the mind as necessary, a priori principles that make experience possible.[34]

According to Kant, there are also other categories within the mind that help shape experience. These categories take the perceptions that have been shaped by the intuitive forms of space and time and transform them into **conceptions** that follow certain predictable patterns. Once perceptions have become conceptions, they can be seen as universally valid and, therefore, objective.[35] These categories are quantity, quality, relation, and modality. **Quantity** embodies the concepts of unity, plurality, and totality. **Quality** embraces the notions of affirmation, negation, and limitation. **Relation** involves the ideas of substance-accidents, cause-effect, and causal reciprocity. Finally, **modality** includes the concepts of possibility, actuality, and necessity.[36]

The advantage of Kant's theory of knowledge is that it insures that a thinking subject's understanding of reality will be accurate. The thinking subject knows that his or her understanding of reality is accurate because the intuitive forms of space and time impose an organization onto all unstructured sensations, thus creating perceptions that are then further organized by the categories into logically predictable conceptions. If the forms and the categories did not impose that organizational scheme on sensations, then those sensations would be experienced in some other way. However, that way would also be determined by the forms and the categories. Thinking subjects see reality according to a predictable pattern because that pattern is imposed on reality by the intuitive forms of space and time and by the categories of quantity, quality, relation, and modality.[37] In this way, the forms and the categories make experience possible.[38]

Reviewing the process that occurs when an individual seeks to "know" something may make Kant's ideas about space and time somewhat clearer. According to Kant, the process begins when an individual picks up the sensation of a physical object—a tree, for example—through the senses. Initially, these sensations are received in an unstructured and disorganized way.[39] So, at the first instant that an individual looks at a tree, the mind of that individual may pick up an unstructured combination of sensations, including for example, colors such as green and brown, a rough texture corresponding to the bark of the tree, and so on. The mind then works to impose order on these individualized sense experiences. This organization stage involves processing these sensations via the a priori forms of space and time.[40] Thus, an individual can recognize a tree as a tree because the previously unstructured sensations have been organized into a perception by the activity of the mind.[41]

The perception is then further organized by the categories into a conception and eventually a universal experience.[42] For the individualized experience to becomes a universal experience, there must be a firm foundation in reality. When the individualized experiences of many thinking subjects are found to correspond to one another, they are said to be universal; since they are universal, they must also be objective.[43] Of course, there is a downside to Kant's theory of knowledge. Because our knowledge of the world is shaped by the intuitive concepts of space and time, we know only the appearances of things, not the things themselves. Therefore, human perceptions are always somewhat removed from the actual physical objects. Thus, the mind never really knows an object in its pure state of existence. Kant says that there is always a gap between the perception of a thing and the "thing-in-itself."[44]

The Einstein-Kant Debate

According to Reichenbach, Einstein's work with relativity has shown that Kant was wrong. Relativity has taught us that space and time do not exist within the mind, as Kant believed, but have a real existence that depends upon matter and energy. In fact, without matter and energy, there would be no time and no space. However, in saying this, we must be careful not to go too far. Even Einstein believed that space and time can still be thought of as human constructs. According to Einstein, space and time *are created* by the mind, but not in the way that Kant thought. Kant believed that space and time exist a priori in the mind and shape reality so as to give it coherence. In contrast, Einstein states that space and time may be constructed by the mind, but they still correspond to existing physical relationships. According to Reichenbach, many concepts that are structured by the mind correspond to real physical relationships, whereas others do not.[45]

For instance, the word "cousin" identifies a set of existing physical relationships that have a basis in concrete fact. In contrast, the word "sorcerer" has no such physical referent and does not describe any set of relationships outside the world of make believe. According to Reichenbach, this is the critical difference between the theories of Kant and those of Einstein. Kant's a priori concepts of space and time exist in the mind of any thinking subject, but they have no physical referents. In contrast, when Einstein speaks of space and time, he also sees them as mental constructs. However, according to Einstein's theory, these mental constructs, like the word "cousin," have real physical referents that express real physical relationships.[46]

Let us take a closer look at the theory of relativity, this time with an eye on its relationship to Einstein's positivism and its effect on philosophy. According to Einstein, the concept of space-time describes actual physical relationships among real things, such as photons and clocks. Moreover, these relationships determine certain basic characteristics of the physical universe. What is more, Einstein has consistently demonstrated the validity of these characteristics with consistent mathematical equations.[47] Perhaps even more important, other reputable individuals, among them the British astrophysicist Sir Arthur Eddington, have supported those equations with observational data.[48] This is, of course, the heart of positivism. As we have seen, positivism supports the inevitability of correct conclusions in scientific research. In this case, according to Reichenbach at least, Einstein has the right answer and Kant does not. It is in this way that Einstein's work has the indirect philosophic impact of which Reichenbach speaks. Kant's theory of space and time sounds very convincing until it is placed up against the solidly verifiable explanations of Einstein. At that point, Reichenbach says, Kant's theory is seen for what it really is, a clever and rational explanation based upon nothing more than a fanciful imagination.

Perhaps even more significant than this, however, is another conclusion that Reichenbach comes to in his article, "The Philosophical Significance of the Theory of Relativity." This is the conclusion that there are some things that science simply does better than philosophy. Recall that earlier in this chapter, we noted that philosophy can be looked at as a product of physics. According to this perspective, the physicist reveals a new way of understanding the physical universe. At that point, the philosopher takes over and interprets the implications of this new way to view the universe. The physicist, in the meantime, continues to explore the physical laws that govern the operation of the universe. The scientist neither criticizes nor analyzes either the scientific process or the results of that process. Instead, the scientist conducts the process so that those results can be determined.[49]

Reichenbach argues that if the scientific process is carried out properly, it can take the scientist to places that could not have been conceived of in the wildest dreams of the philosopher. It is precisely because the scientist follows the scientific process, at times unaware of the outcome, that the scientist can arrive at these uncharted and unexpected destinations. In the case of relativity, the destination that Einstein reached changed our understanding of reality. According to Reichenbach, the mathematical and observational approach to the concept of time as a part of relativity theory was more successful than any philosophical analysis ever could be. For instance, Reichenbach says, it is unlikely that any philosopher could have dreamt up the notion that time is intertwined with the matter and energy within the universe. It is equally unlikely that any philosopher could have come up with the idea that time is not necessarily unidirectional.[50] **Unidirectional** in this context means that time flows in one direction, from the present to the future. Before Einstein, most philosophers simply assumed that time is unidirectional and that it could not be altered nor further analyzed. Time existed. It flowed forward, and that was that. Reichenbach points out that such a conclusion is simply a polite way for the philosophers to say that they did not know how to explain time. It is true that the mind can imagine fracturing time, slowing time, and perhaps even reversing it. However, only Einstein showed that all these things were possible by mathematical equations and observational data.[51]

This is enough for Reichenbach. As far as he is concerned, this alone has demonstrated that philosophy must always follow science. According to Reichenbach, Einstein's unexpected solutions to the problems posed by relativity have demonstrated the validity of positivism beyond any doubt. Therefore, there is nothing left for the philosopher to do but pick up the trail and follow the scientist so that the philosopher can see where the scientist's discoveries lead.[52] As should be expected, not everyone in the scientific and philosophic communities is as convinced as Reichenbach of the ascendancy of positivism. In fact, Kurt Gödel uses the theory of relativity to prove an idea that is diametrically opposed to the conclusion that Reichenbach has reached.

8-3   IDEALISM AND THE SIMULTANEOUS EXISTENCE OF TIME

Kurt Gödel, who was at the Institute for Advanced Study at Princeton at the same time as Einstein, disagrees with Reichenbach's thesis that the validity of positivism can be proven by a proper application of Einstein's theory of relativity. Moreover, Gödel uses relativity to show not only that time is not unidirectional but also that time does not really "flow" in any direction. According to this perspective, all time exists simultaneously so that the past, the present, and the future are all one.

Idealism and Relativity

Before examining Gödel's proof of idealism using relativity, it will be helpful to determine first just exactly what idealism is. **Idealism** holds that all reality is simply the product of the mind. The origin of idealism can be traced to the eighteenth-century British philosopher, Bishop George Berkeley. According to Bishop Berkeley, it is not possible for the human mind to perceive physical objects.[53] All the mind can perceive are the qualities of those objects. Thus, a thinking subject may know the color, size, shape, and texture of an object, but not its substance. With this premise in mind, Berkeley concludes that the substance of physical objects and, therefore, the physical objects themselves do not exist.[54] Berkeley then denies the existence of the entire physical universe.[55]

Kant can also be considered an idealist, at least in the sense that he believed that the physical universe had to be organized by the a priori forms of space and time, which exist only in the mind. In this way Kant gave the human mind the power to shape existence. Moreover, as we have seen, his theory of knowledge also moved the thinking subject several steps away from "things-in-themselves."[56] It remained for later philosophers to develop the concepts of idealism more fully. The British philosopher Thomas Hill Green, for example, held that the entire universe existed as a set of interrelated parts, all of which are held together solely by the power of the mind. Moreover, Green believed that the difference between appearance and reality is the difference between what the inadequate human mind can perceive and what the absolute mind of the universe can create and maintain.[57] Another idealist, the British philosopher John Ellis McTaggart, takes idealism into new territory by claiming that the entire universe actually results from a union of many different minds directly linked together.[58]

How is it possible to unite idealism with relativity? At first look, the two beliefs seem as different as Shakespeare and Mickey Spillane. Nevertheless, Gödel argues that relativity requires an idealistic view of reality. Gödel's argument is set down in detail in his article, "A Remark about the Relationship between Relativity Theory and Idealistic Philosophy."[59] In this article Gödel maintains that the theory of relativity not only validates Kant's idea that space and time exist only as a priori mental constructs but also proves the claim put forth by the pre-Socratic philosopher Parmenides, who argued that the universe exists in a state of eternal "changelessness."[60] Gödel's claim seems to defy rationality. After all, Kant's ideas on space and time are clearly rooted within the mind, whereas Einstein's refer to definitive physical references. Moreover, relativity unequivocally requires a forward movement in time, whereas Parmenides and Kant vigorously deny that such movement exists.

Nevertheless, Gödel succeeds in showing how these apparently diverse philosophies may be considered compatible.

Gödel begins his argument by stating that change results from the passage of successive temporal intervals. Moreover, the passage of successive temporal intervals requires the movement of successive present moments. Time, then, is nothing more than an infinite sequence of "nows."[61] The theory of relativity, however, has clearly demonstrated that simultaneity is relative. This means that the concept of "nowness" is not universal. On the contrary, there is no universal present moment, that is, no universal "now." Since the experience of now can differ from observer to observer, a unique infinite sequence of nows is not possible. Another way to say this is to note that no single now can claim to have objective existence. Consequently, there can be no absolute passage of time and no change anywhere within the universe.[62] This means that Parmenides was correct when he noted that the universe exists in an eternal state of changeless permanence. Why, then, do we perceive the passage of time, when it is clear that such temporal movements are not possible? Gödel has an answer for this question, too. He says that our perception of time is created within the human mind. Therefore, Kant is correct, and time exists only within the human mind.

Gödel's interpretation of time must be distinguished from the interpretation offered by the French philosopher Henri Bergson. This distinction is critical because, on the surface at least, the two views sound remarkably similar in tone and texture. As we saw in Chapter 6, Bergson believes that the multiplicity of time is actually a creation of the mind which has the capacity to recall past states of simultaneity and to anticipate such future states.[63] This sounds oddly reminiscent of Gödel's and Kant's observation that time exists only within the human mind. Although the ideas of the three philosophers do, in fact, resemble one another, there is, nevertheless, a profound difference. Gödel links his belief in the ability of the mind to create time with the relativity of simultaneity and concludes that it is possible, indeed likely, that all time exists simultaneously.[64] This conclusion is based upon Gödel's interpretation of the relativity of simultaneity. According to Gödel, the relativity of simultaneity tells us that no one's now is the absolute objective now for the entire universe. Therefore, one observer's present is another observer's past, and because both exist simultaneously, all time must exist simultaneously.[65]

In contrast, Bergson says that the only thing that really exists is one constant state of simultaneity. The sensation of duration is created by the conscious mind. As we saw in Chapter 6 Bergson illustrates his concept of universal time with the image of the oscillating pendulum of a clock. In this illustration, the only oscillation of the pendulum that exists is the present oscillation. The multiplicity of successive moments arises because the conscious mind recalls the past oscillations that led to the present oscillation. The past oscillations, however, do not exist in the physical universe. That privilege is reserved for the present one.[66] In contrast, if Gödel were to interpret Bergson's pendulum illustration, he would conclude that all oscillations and all moments exist simultaneously.[67]

Bergson is not the only scholar to disagree with Gödel's notion of time. Gödel himself outlines an argument put forth by the British astrophysicist James Jeans in which Jeans claims to prove that an absolute, objective measure for the passage of time exists for the entire universe. According to Jeans, all individualized observations of time can be taken to represent different measurements of the overall passage of an objective, absolute time within the universe. If we compile each of these different measurements, we can calculate the passage of true, objective time for the universe.[68] Gödel's explanation as to how such a result can be calculated is at best unclear. This ambiguity may be due to the fact that Gödel does not support Jeans's proposition. In fact, he spends the remainder of the article demonstrating that the passage of objective time cannot exist.[69]

Gödel clearly does not believe in the passage of absolute time for the entire universe. He sees, instead, a universe in which all time exists simultaneously. Interestingly enough, Gödel supports the idea of the simultaneous existence of time by referring to the possibility of reverse time travel.[70] Such a possibility, as we have seen, is not ruled out by the theory of relativity.[71] According to Gödel, reverse time travel would not be possible unless the past, the present, and the future coexist simultaneously. Such coexistence would allow a traveler to journey from the present to the past and back to the future in much the same way that more conventional travelers traverse distances within simultaneous space. Therefore, whenever a time traveler moves to a different region of time, someone will be able to say that the traveler has moved into the past. This then removes the objective passage of time.[72] Instead, we have a universe in which all time exists simultaneously. If all time exists simultaneously, there is no reason to believe that time passes from one moment to the next. Therefore, Gödel concludes that our perception of the passage of time is a construct of the mind.[73]

The Eternal Space-Time Cube

Nor is Gödel the only one to come up with the idea that all time exists simultaneously. In fact, in recent years this idea has been proposed more than once as an explanation for several space-time paradoxes that result when time is thought of as moving in a unidirectional manner. For instance, in his book, *The Fourth Dimension: A Guided Tour of the Higher Universes*, Rudy Rucker, a professor of mathematics and computer science at San Jose State University, suggests a new way to visualize Gödel's explanation of simultaneous time. Rucker argues that, to remain consistent with the theory of relativity, space and time must be seen as wrapped up together in a single, unified reality called the **space-time cube**, or the **block universe**. Rucker urges us to think of space-time single enormous object. This enormous four-dimensional object comprises all of space-time. Three of the four dimensions are reserved for space, and one for time. To appreciate the implications of this image of the universe, we must envision ourselves standing outside of the space-time cube and observing it from a detached and objective position.[74] The concept of the space-time cube did not originate with Rucker. In fact, Rucker's idea of a space-time cube is based upon an idea first proposed by the Russian-born, German mathematician Hermann Minkowski, in 1908. Minkowski believed in the **spacialization** of time. This means that time should be seen in much the same way that space is seen. Time, Minkowski said, must be "spatialized."[75]

Rucker's detailed explanation of the space-time cube begins with the premise that each of us inhabits a universe that consists of the totality of our perceptions laid out like a long line of successive events within a four-dimensional space-time cube. In other words, each of us has a pattern of perceptions nestled comfortably within the space-time cube. This pattern of perceptions is called a **lifeline** [or **worldline** or **lifeworm**[76] (see Figure 8.2)]. It is helpful to remember that this and other illustrations in this chapter cannot possibly be drawn to scale. As you view each illustration, you should remind yourself that you are viewing only a small fragment of the eternal space-time cube. You could think of each illustration as the equivalent of examining the entire Pacific Ocean by viewing a single drop of water under a microscope. This point of view should help keep things in proper perspective. Rucker explains the concept of the worldline or the lifeworm in this way:

> My world is, in the last analysis, the sum total of my sensations. These sensations can be most naturally arranged as a pattern in four-dimensional space-time. My life is a sort of four-dimensional worm embedded in a block universe. To complain that my lifeworm is only (let us say) seventy-two years long is perhaps as foolish as it would be to complain that my body is only six feet long. Eternity is right outside of space-time. Eternity is right now.[77]

Time then, as Gödel and Rucker conclude, is not something that flows. Nor is time the measurement of change. Time is simply another dimension within the space-time cube, which is actualized by a series of perceptions that make up one's lifeline. For example, let us imagine a series of events that take place around Christmas, 2009. At 6 P.M. on December 24, Christmas Eve, Jennifer is enjoying a family get-together in her grandmother's living room. On December 26, she has breakfast at 9 A.M. with her grandmother, Louise, at their favorite restaurant. On December 28, she takes back a sweater that she received as a Christmas gift and is, therefore, at the Lone Pine Mall at noon. On December 30, she returns to her favorite restaurant and has dinner with a friend at 9 P.M. Finally, on January 1, 2010, she is back at her grandmother's celebrating the New Year at 1 A.M. with her family.

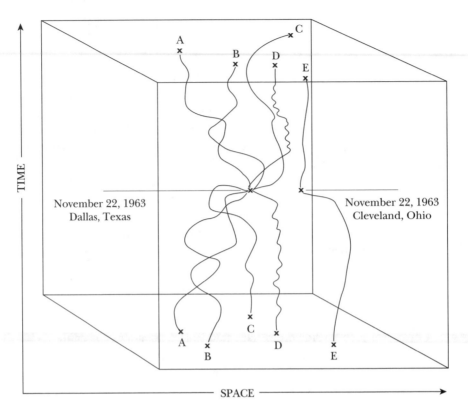

Lives A, B, C, D all converge on November 22, 1963
in Dallas, Texas
while life E does not

Figure 8.2 Lifelines within a Space-Time Cube.

The figure represents an infinitesimally minute portion of the eternal space-time cube visualized by Rudy Rucker in his work *The Fourth Dimension*. In this version of the space-time cube we have focused on five lifelines. By necessity, of course, we see only a small portion of each lifeline. Nevertheless, from this isolated view we can appreciate the nature of space-time according to Rucker's perspective. Each line represents the life of a person moving forward in time, represented by the vertical axis, and moving about in space, represented by the horizontal axis. At only one point in time are four of the lifelines (A, B, C, and D) at the same point in space. That time is November 22, 1963, and the point in space is Dallas, Texas. Only the fifth lifeline, that belonging to E, is at a different point in space at the same moment in time. At that time, E is not in Dallas like the others, but is, instead, in Cleveland.

At 2 A.M. that New Year's Day, Jennifer becomes reflective. She sits apart from other family members in her grandmother's living room thinking about recent events. She thinks about December 24 at her grandmother's, December 26 at the restaurant, December 28 at the mall, December 30 back at the restaurant, and January 1 back in her grandmother's living room. Jennifer lived through each of these events. Each one is, therefore, real to her. They exist with her as part of her lifeline of perceptions. They are much more real to her than either the mall or the restaurant as they exist at 2 A.M. on January 1, 2010. Because she sees neither the restaurant nor the mall at 2 A.M. that morning, they do not exist for her. In fact, the very existence of both the restaurant and the mall is doubtful, while the existence of the past events cannot be doubted. To see this just a bit more clearly, assume that Jennifer blacks out for the intervals between each of these events. She wakes up successively on December 24 in her grandmother's living room, on December 26 at the restaurant, on December 28 at the mall, on December 30 back at the restaurant, and on January 1, in her grandmother's living room. If this were to happen, only these events would be real to Jennifer. Space would have rearranged itself as her time moved forward (see Figure 8.3).[78]

Like Jennifer, every individual exists as a lifeline of perceptions within the eternal space-time cube. The space-time cube exists as a whole. Each lifeline exists as a finished product within the space-time cube. All of the moments of our lives are, therefore, uniformly real and equally existing at the same time. Rucker explains it this way:

> [E]ach of us is a certain space-time pattern in the block universe. Today, or the day of my birth, or the day of my death—all are equally real, all are different pieces of the block universe. I will never stop living this instant. This instant will never cease to exist; this instant has always existed.[79]

This image of the universe as an eternal space-time cube that exists now and has always existed in its entirety makes Gödel's explanation of simultaneous time easier to visualize. Recall that Gödel's concept of simultaneous time means that there is no common past, present, or future for the entire universe. In fact, if we conceive of the universe as an eternally existing space-time cube, we can see how this view is inevitable. In Figure 8.4 we see a stylized space-time cube containing the localized lifelines of Jennifer and Louise. For the sake of simplicity we have focused on several days in their lives. On the lifelines we see how they converge at three points: the family get-together on December 24, the breakfast on December 26, and the New Year's Eve celebration on January 1. If we were asked which of these events is occurring now, the answer would be that there is no absolute now. All of the points on the lifelines exist simultaneously with one another. Despite the illusion of duration that we experience in everyday life, time really does not pass. In fact, Rucker concludes, as Gödel did before him, that there is no passage of time. The entire universe, all space and all time, is simply there. As Rucker points out in *The Fourth Dimension*:

> The great advantage the block universe has over the other viewpoints is that in the block universe there is no objectively existing "Now." Nothing is moving in the block universe, and there is no need to try to find some absolute and objective meaning for the horizontal space sheet that the other . . . models depend on.

• 264 • Chapter 8 The Philosophical Side of Relativity

Spacetime Event	A Jennifer's Sensations	B Conventional View of Spacetime
The present for Jennifer: New Year's Day January 1, 2010— 2 A.M.	Living Room	Living Room Fireside Restaurant Mall Note: All three exist now regardless of Jennifer's sensations. They do not exist in the past.
Dinner at Fireside Restaurant, December 30, 2009— 9 P.M.	Fireside Restaurant	
Shopping at the mall, December 28, 2009— Noon	Mall	
Breakfast at Fireside Restaurant, December 26, 2009— 9 A.M.	Fireside Restaurant	
Family reunion in living room December 24, 2009— 6 P.M.	Living Room	

← SPACETIME →

Space-time event

Column A represents Jennifer's lifeline. What is real for Jennifer is a line of successive perceptions. So in Jennifer's present only the living room exists. All other locations exist in her past.

Column B represents our conventional view of spacetime. All three locations exist in space "now." Regardless of Jennifer's sensations. They do not exist in the past.

Figure 8.3 Space, Time, and Jennifer's Perceptions.

According to Kurt Gödel and Rudy Rucker, time is simply another dimension within the space-time cube, which is actualized by a series of perceptions that make up an individual's lifeline. In the figure we see how such a series of perceptions can be visualized. Figure 8.3 depicts a series of events that take place around Christmas, 2009. At 6 P.M. on December 24, Christmas Eve, Jennifer is enjoying a family get-together in her grandmother's living room. On

As it turns out, it is actually *impossible* to find any objective and universally acceptable definition of "all of space, taken at this instant." This follows ... from Einstein's special theory of relativity. The idea of the block universe is, thus, more than an attractive metaphysical theory. It is a well-established scientific fact.[80]

Gödel explains it this way:

(T)he decisive point is this: that for *every* possible definition of a world time one could travel into regions of the universe which are passed according to that definition. This again shows that to assume an objective lapse of time would lose every justification in these worlds. For, in whatever way one may assume time to be lapsing, there will always exist possible observers to whose experienced lapse of time no objective lapse corresponds (in particular also possible observers whose whole existence objectively would be simultaneous). But, if the experience of the lapse of time can exist without an objective lapse of time, no reason can be given why an objective lapse of time should be assumed at all.[81]

Lifelines, Light Cones, and Elsewhen

The fact that the universe may exist as an eternal, space-time cube does not mean that all parts of the cube are accessible to each of us at all times. In fact, those parts of the cube that are available to us are severely limited. The primary limiting factor is the speed of light. Nothing can travel faster than the speed of light. Therefore, our future consists of only those areas within the space-time cube that can be reached by traveling at the speed of light. On the other hand, our past consists of only those areas that we could have reached had we traveled at the speed of light in the past.[82]

Since this concept is difficult to grasp, physicists usually resort to depicting the geometry of four-dimensional space-time using two-dimensional **Minkowski diagrams**.[83] A typical Minkowski diagram is laid out like a simple graph (see Figure 8.5). The vertical axis represents time, and the horizontal axis represents

---

December 26, she has breakfast at 9 A.M. with her grandmother, Louise, at their favorite restaurant. On December 28, she takes back a sweater that she received as a Christmas gift and is, therefore, at the Lone Pine Mall at noon. On December 30, she returns to her favorite restaurant and has dinner with a friend at 9 P.M. Finally, on January 1, 2010, she is back in her grandmother's living room celebrating the New Year at 1 A.M. with her family. At 2 A.M. on New Year's Day, Jennifer becomes reflective, and she reviews the events of the past few days. She thinks about December 24 in her grandmother's living room, December 26 at the restaurant, December 28 at the mall, December 30 back at the restaurant, and January 1 back in her grandmother's living room. Jennifer lived through each of these events. Each one is, therefore, real to her. They exist with her as part of her lifeline of perceptions. They are much more real to her than either the mall or the restaurant as they exist at 2 A.M. on January 1, 2010. Since she sees neither the restaurant nor the mall at 2 A.M. on January 1, they do not exist for her. In fact, the very existence of both the restaurant and the mall is doubtful, whereas the existence of the past events cannot be doubted.

• 266 • Chapter 8 The Philosophical Side of Relativity

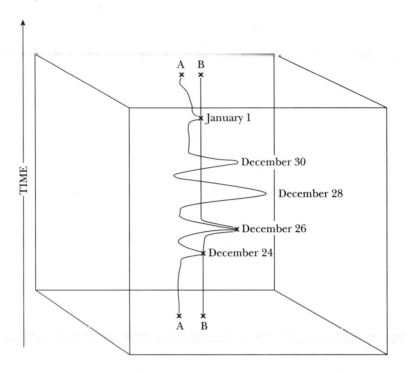

Note: Jennifer's lifeline (A) moves in space several times. Louise's (B), however, moves only once on December 26 when she and Jennifer have breakfast at the restaurant. Still Jennifer's lifeline intersects with Louise's at three separate points: on December 24 (the reunion), December 26 (the breakfast), and January 1 (the New Year's celebration).

Figure 8.4 The Lifelines of Jennifer and Louise.

In the figure, we see a stylized space-time cube containing the localized lifelines of Jennifer (A) and Louise (B). For the sake of simplicity we have focused on several days in their lives. On the lifelines we see how their lifelines converge at three points: the family get-together on December 24, the breakfast on December 26, and the New Year's celebration on January 1. Notice that Jennifer's lifeline appears more erratic than that of Louise. This is because Jennifer is moving about in space, whereas Louise moves in space only on December 26 to join Jennifer at the restaurant for breakfast. If we were asked which of these events is occurring "now," the answer would be that there is no absolute "now." All of the points on the lifelines exist simultaneously.

space. Thus, vertical movement up the graph without any corresponding horizontal movement represents a traveler standing still in space but moving in time. In contrast, a horizontal line, generally moving from left to right, represents movement in space without a corresponding movement in time. The movement of light is generally represented by two lines drawn at 45-degree angles (with either axis). The light leaving an event on the lifeline of a traveler

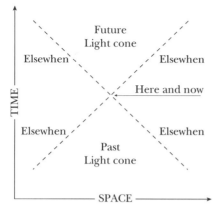

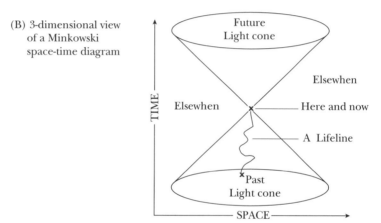

Figure 8.5 A Minkowski Space-time Diagram.
   Even though the universe exists as an eternal, space-time cube, most parts of that cube are inaccessible to us because of the speed of light. Accordingly, our future consists of only those areas within the space-time cube that can be reached by traveling at the speed of light. On the other hand, our past consists of only those areas that we could have reached had we traveled at the speed of light in the past. The geometry of this four-dimensional situation can be visualized by examining a two-dimensional Minkowski diagram. Figure 8.5 A represents a typical Minkowski diagram. Figure 8.5 B is the same figure drawn to show its three-dimensional nature. In both versions, the vertical axis is used to represent time, whereas the horizontal axis represents space. Thus, vertical movement up the graph without any corresponding horizontal movement represents a traveler standing still in space but moving in time. In contrast, on a horizontal line, generally from left to right, movement in space is represented without a corresponding movement in time. The movement of light is generally represented by lines drawn at a 45° angle with the horizontal. The light leaving an event on the lifeline of a traveler angles upward to the left and to the right of a central point, which represents the "here and now" for that traveler. This creates the future light cone of that traveler. The light that converges on that point of "here and now" from below represents the past light cone of that traveler. Everything outside the two cones exists in a space-time region known as elsewhen.

angles upward to the left and to the right of a central point which represents the "here and now" for that particular traveler. This creates the **future light cone** of that traveler. The light that converges on that point of "here and now" from below represents the **past light cone** of that traveler. Everything outside the two central triangular regions exists in a space-time region known as **elsewhen**, which is also sometimes referred to as **elsewhere**. Because "elsewhen" more properly combines the concepts of space and time, that is the term we will use here. Elsewhen, which consists of the triangular regions on the left and right, represents those areas of the space-time cube that are inaccessible.[84] However, elsewhen is inaccessible only as long as we are limited by the speed of light. If the barrier represented by the speed of light could be broken, all points within elsewhen might be accessible. Traveling at the speed of light would allow us to reach the outer limit of the future light cone.[85]

Minkowski diagrams can also clear up some of the ambiguity surrounding the twins paradox discussed in Chapters 6 and 7. Figure 8.6 represents a Minkowski diagram depicting the twins paradox. The origin of the journey begins with Ashley and Megan on Earth (point A of Figure 8.6). This moment constitutes one unified site on their converging lifelines. Remember, Ashley is the twin who decides to take a deep-space journey on a starship. The starship will be traveling at $0.999c$. Because Megan stays on Earth, she moves in time but not space; thus her lifeline simply moves vertically up the time axis. Ashley's lifeline diverges from Megan's because Ashley is moving in space (horizontally) and in time (vertically). Her lifeline can be depicted as a line leaving the one unified site which represents the last moment that she and her twin were together on Earth. Because she is traveling at $0.999c$, her lifeline will be depicted moving away from that single unified site at nearly a 45-degree angle. Ashley's ship reaches its destination (point B), turns around, and heads back to Earth, converging again with Megan's lifeline (point C).[86]

At this point we meet the most paradoxical aspect of the trip. Recall that relativity has demonstrated that the time which elapses between the two points of convergence, points A and B, will vary depending upon the speed at which each twin travels. Therefore, the lifelines of the two twins will not be the same. Ashley's lifeline has remained as straight as an arrow because she has not moved in space. In contrast, Megan's lifeline has developed a sharp turn at point B. Although Megan's lifeline appears longer on paper, the time duration of her trip is actually shorter. As a lifeline moves away from the vertical axis of time in the Minkowski diagram, space-time equations require that the duration be shortened by a special multiplication factor. The calculations demonstrate that, as measured by a clock, the longest duration between A and C is the lifeline that Megan traveled. This is why Ashley's lifeline converges with Megan's in a shorter span of time. Thus, the moment of convergence is the same for both of them, but the length of time experienced since they parted company is different.[87]

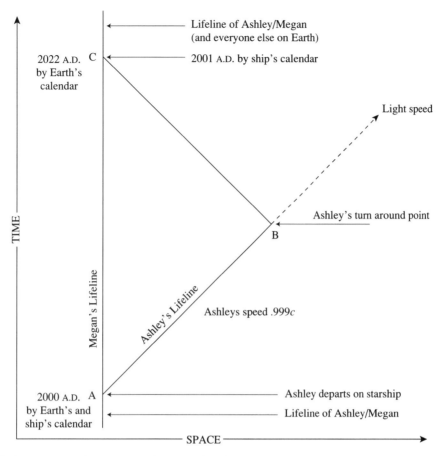

Figure 8.6 The Twins Paradox as Depicted on a Minkowski Space-Time Diagram.

The diagram depicts the twins paradox in two dimensions. The origin of the journey in the twins paradox begins with Ashley and Megan on Earth. This event constitutes a single unified point on their converging lifelines. Remember, Ashley is the twin who decides to take a deep-space journey on a starship, which travels at .999c. Since Megan stays on Earth, she moves in time but not in space; thus, her lifeline simply moves vertically up the time axis. Ashley's lifeline diverges from Megan's because Ashley is moving in space (horizontally) and in time (vertically). Her lifeline can be depicted as a line leaving a single unified point, representing the last moment that she and her twin were together on Earth point A. Since she is traveling at .999c, her lifeline will be depicted moving away from that single unified point at nearly a 45°-angle. Ashley's ship reaches its destination point B, turns around, and heads back to Earth, converging again with Megan's lifeline at point C. At this point we meet the most paradoxical aspect of the trip. Recall that relativity has demonstrated that the time that elapses between the two points of convergence, A and B, will vary, depending upon the speed at which each twin travels. Therefore, the lifelines of the two twins will not be the same. Megan's lifeline has remained as straight as an arrow because she has not moved in space. In contrast, Ashley's lifeline has developed a sharp turn at point B. While Ashley's lifeline appears longer on paper, the time duration of her trip is actually shorter. As a lifeline moves away from the vertical axis of time, space-time equations require that the duration be shortened by a special multiplication factor. The calculations demonstrate that, as measured by a clock, the longest duration between A and C is the lifeline that Megan traveled. This is why Ashley's lifeline converges with Megan's in a shorter span of time. Thus, the moment of convergence is the same for both of them, but the length of time experienced since they parted company is different.

8-4 PHILOSOPHY AND THE SPACE-TIME CUBE

It is tempting to conclude that the differences between the interpretations proposed by Reichenbach and Gödel are simply differences of degree that really have little consequence beyond the scholarly journals in which they first appeared. Unfortunately, such is not the case. In fact, the difference between the two positions profoundly affects our basic understanding of the nature and purpose of human existence. Specifically, if Gödel is correct in his conclusion regarding the simultaneous existence of all time, then we will see that the time-honored assumption that human beings have free will must be discarded. **Volition** or **free will** is the innate ability of human beings to make decisions for themselves without being controlled by outside influences. A philosophy that stands in opposition to volition is **determinism**, which holds that human decision-making is subject to forces that predetermine those decisions and compel human beings to act according to a preordained pattern.

Choosing whether volition or determinism characterizes human behavior is not without serious consequences. In fact, if determinism is accepted and volition abandoned, then concepts such as democracy, law, morality, and creativity must also be abandoned because these concepts are inexorably tied to the ability of human beings to be responsible for their own actions. To amplify the significance of Gödel's work in relation to time and volition, we devote the remainder of this chapter to the impact of simultaneous time on the concept of volition. First, we look at why, as a consequence of simultaneous time, volition must vanish. Then we explore the theological implications of the loss of volition. Finally, we seek out an alternative explanation that may rescue the idea of human volition after all.

The Principle of the Fixity of Time

If we follow Rucker's lead and attempt to visualize Gödel's thesis, that all time exists simultaneously, using the image of the space-time cube, we conclude that all the points on all lifelines within the cube exist simultaneously with one another. Like Gödel and Rucker, we also see that, despite the illusion of duration that we experience in everyday life, time really does not **pass**. In fact, there is no passage of time at all. The space-time cube exists as an eternally present, unified whole. This concept of simultaneous existence is not troubling to many of us when we consider the past. After all, it is not that difficult to see that the past exists as a tapestry of events that are fixed and immutable. This belief is known as the **principle of the fixity of the past**, or PFP.[88]

Even the retrograde-time-travel scenarios that we explored in Chapter 7 do not necessarily violate PFP. For example, we saw that there were several methods of solving the grandmother paradox. We suggested that a time traveler who manages to eliminate her grandmother before her mother is born might vanish from the universe altogether. It is also possible that the time traveler creates an alternate universe, like those predicted by Everett and Dewitt's parallel universe theory, in which she exists, but her grandmother and her mother do not. Another possibility is that the time traveler arrives in the past in some sort of incorporeal form. In this disembodied form she can see and hear the events as they occur but can never interact with them.

Perhaps the most convincing solution of the grandmother paradox is one offered by J. Craig Wheeler in a recent article in *Astronomy* entitled "Of Wormholes, Time Machines, and Paradoxes." Wheeler labels this solution the **Novikov consistency conjecture**, after its originator Igor Novikov of the University of Copenhagen. In its simplest form the Novikov consistency conjecture states that the past cannot be altered no matter what a time traveler knows or how hard she tries to make such things happen. It is even possible, perhaps likely, that the time traveler's trip into the past is the event that causes her grandmother and grandfather to meet each other in the first place. According to this perspective, the time traveler's trip is a historical event that cannot be altered. If the time traveler visits the past, her visit is a part of those past events, even though the visit itself originates in the future. The past is like a movie that can be played over and over without being altered. No matter how many times the time traveler replays the movie of the past, the events will remain unchanged.[89]

None of this seems overly troubling when applied to the past. However, we must recall that, according to the principles of the space-time cube and the conclusions of relativity, one individual's past is a second individual's present and a third individual's future. Moreover, according to Gödel and Rucker, the whole of space-time exists as a complete unit. This means that all lifelines are complete within the space-time cube. However, if all lifelines exist, then there can be no possibility of altering the future of a particular lifeline. *The future is just as immutable, just as permanent, as the past.* Moreover, to preserve the PFP, we must reach the same conclusion. After all, if a past event in my lifeline is a future event in your lifeline, then in order not to violate the PFP that "past" event cannot change simply because it is in your future. The principle of the fixity of the past must, therefore, become the **principle of the fixity of time** (PFT).

Volition and the Space-Time Cube

Many people, including Rucker, have argued convincingly that the principle of the fixity of time means that we must abandon any belief in volition. Because all time is immutable, nothing we can do will change any event on our lifeline. The sensation of volition is just as illusory as the sensation of the passage of time. Consider the predetermined fixity with which actors move in a taped drama. The actors appear to have volition. They seem to be making decisions, but they are actually playing out a predetermined set of events encoded on the tape. A viewer need only hit the rewind button on the remote control to watch the actors perform the same exact ritual over and over again. Each time the actors appear to be making choices, yet each time they make the same decisions. Each time they act exactly as they did the last time.

At first blush, the principle of the fixity of time appears rather dark and pessimistic. In fact, it seems very similar to the belief in predestination promoted by John Calvin and his successor, Theodore Beza, in the sixteenth century. In its most basic form, the doctrine of **predestination** eliminates the idea that individuals have the ability to exercise volition. Predestination, then, is the theological equivalent of determinism. According to both perspectives, free will is an illusion. According to Calvin and Beza, God in His justice has already determined who will be saved and who will be damned, and it does little good for anyone to suppose that he or she can substitute his or her will for that of God.[90]

Calvin and Beza were not the only ones to express a belief in the fixity of time. Several centuries before Calvin and Beza, St. Thomas Aquinas had also attempted to deal with the paradox caused by the clash between humanity's volition and God's omniscience. **Omniscience** is the ability to know everything. How can God know everything without robbing individuals of their freedom to act? Aquinas and Boethius, another medieval Christian thinker, resolved the problem by noting that God exists beyond time. This perspective allows Aquinas and Boethius to explain that all time is actually present for God to view, much like a painting in a museum is really present for an art lover to admire. Yet, since God exists beyond time, His knowledge of all time does not coexist with any time, past, present or future. In this way, God's knowledge does not interfere with human volition.[91]

The modern reader may be somewhat confused by the distinction between the idea of God's knowledge of all time as presented by Aquinas and the doctrine of predestination as preached by Calvin. In fact, there seems to be little difference between the two ideas. Perhaps even more troubling is the fact that the picture painted by Aquinas and Boethius of time laid out before God as a vast panorama of simultaneous events is reminiscent of both the principle of the fixity of time and the space-time cube. In fact, it is all too easy to imagine God existing as a white-bearded old man observing the totality of the space-time cube from the comfort of His heavenly throne. However, before we abandon the notion of volition entirely, we must place the space-time cube and the principle of the fixity of time within the context of modern science. The theory of relativity is, after all, not the only great breakthrough of the modern era. The other great theory, that of quantum physics, may yet rescue our belief in human volition.

The Theory of Relativity vs. Quantum Theory

As noted previously, the space-time cube and the principle of the fixity of time suggest that we act in much the same way as the actors in a taped drama. Like the actors on the tape, we seem to be making decisions, but we are actually playing out a predetermined set of events embedded within our lifeline. This is not a new problem, of course. The advocates of determinism have been battling the proponents of volition for centuries. There is no reason to believe that the modern era would be any different. In fact, the timing is right for a renewed contest between determinism and volition. This is because the two most influential scientific theories of the modern era, the theory of relativity and quantum theory, appear to clash on this very point. As we saw in the last unit, a primary ingredient of quantum theory is the uncertainty principle, which tells us that we cannot simultaneously measure both the position and the momentum of subatomic particles. We also saw that the Copenhagen Interpretation, the most widely accepted interpretation of quantum theory, holds that this uncertainty is inherent within the nature of the subatomic universe. How is it possible for the inherently uncertain subatomic universe to be frozen within the fixity of the space-time cube?

The Copenhagen Interpretation answers this question by noting that subatomic particles exist in all possible states until they are forced to assume a given state by the intervention of an observer. This is the observer-created universe of Niels Bohr that we discussed at length in Chapter 5. Note, however, that the observer-created universe gives a central role to consciousness in general and to human consciousness in particular. If this is the case, the mind must have the ability to choose freely because, according to quantum theory, it is that very faculty that actually creates the universe. In fact, denying that the mind has the ability to choose would seem to be, at the very least, counterproductive and, at the most, completely pointless. Is it possible to tie the theory of an observer-created universe to the creation of duration within the space-time cube?

Duration as a Product of Human Consciousness. The conclusions that Rucker and Gödel come to clearly reflect the quantum theory of an observer-created universe. The relationship between the mind and the space-time dimension as expressed by Rucker and Gödel is very similar to the ideas of Niels Bohr and David Bohm that we discussed in our examination of quantum theory in Chapter 5. In Bohm's theory, the entire reality of the universe is embedded within each flow of the holomovement. The implicate order then is the enfoldment of the entire universe in upon itself; the enfolded universe is then unfolded or projected into an infinite variety of forms, objects, and other entities. The true reality is the enfolded universe. The forms, objects, and other entities that we experience in our everyday lives are the unfolded manifestations or projections of that implicate order.[92]

If we replace Bohm's term "implicate order" with Rucker's term "space-time" and if we replace Bohm's process of unfolding with Rucker's and Gödel's process of perception, we can see a remarkable similarity between the two theoretical positions. In the theory proposed by Rucker, the entire reality of the universe is embedded within the fixity of the space-time cube. The essence of the universe then is the movement of the entire universe in upon itself, which occasionally comes to rest whenever conscious beings perceive that reality and impose duration upon it. However, the true reality is, according to Rucker, the eternal dimension of the space-time cube. The sensations that we experience in our everyday lives are, according to Rucker, created by our instinctive need to impose duration on those perceptions. According to Bohm, those sensations are simply the unfolded projections of the implicate order.

It is at this precise point that Rucker's, and therefore Gödel's, philosophical conclusions appear to suffer from an internal contradiction. How is it possible for Gödel and Rucker to argue that the mind has the power to shape time and space and at the same time limit the power of the mind by denying the existence of volition? The idea that humanity interacts with the universe to create the illusion of the passage of time is clearly an element of the philosophy proposed by Gödel and Rucker. Yet, if we carry their idea about the simultaneous existence of time to its logical conclusion then we are forced to admit that the principle of the fixity of time leads to a denial of volition. Can we argue that simultaneous time eliminates volition, while with the same breath we say that the mind creates space and time? As noted, this seems to be a contradiction because it both gives and takes away the creative power of the mind. Supporters of Reichenbach might point to this paradox and argue that such a critical internal contradiction demonstrates that Gödel's interpretation of relativity is wrong and Reichenbach's is, therefore, correct. But this is a false conclusion.

Gödel's supporters might simply point out that a single internal contradiction does not necessarily invalidate an entire theory. Perhaps more importantly, however, Gödel's advocates could make a distinction between the conscious decision-making powers of the mind and the intuitive action of the a priori forms of space and time in creating the universe. In fact, the final conclusion might be that the type of activity within the mind that creates the reality of space-time is proof that the mind really has no choice. It reacts instinctively, creating a recognizable space-time dimension with no volition at all. Perhaps then, Reichenbach and Gödel are both saying the same thing after all. Reichenbach says that relativity demonstrates that space-time has a real existence in the physical universe, whereas Gödel is saying that space-time exists only within the mind. Yet, the mind is a part of the physical universe and, if in creating space-time the mind operates intuitively rather than volitionally, the physical existence of space-time is just as inevitable within Gödel's universe as it is within Reichenbach's. The difference, then, may be simply one of degree rather than substance.

CONCLUSION

All of this brings us back to where we started. At the outset of this chapter, we decided that relativity had a definite impact upon philosophy. Moreover, we concluded that, in the case of relativity, Einstein's influence came from his unalterable faith in positivism. This positivistic approach led Reichenbach to the conclusion that physics must precede philosophy. According to this point of view, the task of the philosopher became the interpretation of science in general and relativity in particular. Reichenbach believes that, as a result of this interaction between physics and philosophy, objective truth will be reached. Yet, as we have seen here, this is not always the case. Here we have two eminent scholars who have taken the same theory, relativity, and reached two opposing conclusions. Reichenbach has concluded that relativity has eliminated the idea that space and time are mental constructs, whereas Gödel has concluded that relativity proves that space and time exist only as a priori mental concepts.

The solution to this dilemma may have been offered by Henry Aiken of Harvard University, who, in an essay entitled "The Fate of Philosophy in the Twentieth Century," pinpoints what he believes to be the ultimate purpose of modern philosophy.[93] As was true of Reichenbach before him, Aiken examines the fate of modern philosophy within the context of modern science. In doing so, he comes to the same conclusion that Reichenbach reaches concerning the relationship between science and philosophy, namely that philosophers must follow the path laid down by the scientists.[94] In reaching this conclusion Aiken is rather emphatic, ordering the reader to "make no mistake; if there are principles of science, no philosopher can pretend that there are distinctive principles of philosophy."[95]

The philosopher must, therefore, accept the objective validity of any theory that has been proven correct to the satisfaction of the scientific community. On this much Aiken and Reichenbach can agree. At this point, however, they part company. Reichenbach moves from this conclusion to the implicit belief that philosophers should reach identical, or at least similar, interpretations of any scientific discovery. Yet, as we have seen here in the contrast between Reichenbach and Gödel, such a conclusion is, at best, suspect. Consequently, there must be some other solution to this dilemma. That solution is provided by Aiken, who concludes that the role of the philosopher has become highly individualized. Philosophy no longer creates objective truth. That job has been delegated to science. Instead, philosophy has become a highly individualized activity by which each person reaches his or her own individual truth.[96]

This does not mean that philosophers cannot interpret science or make pronouncements about the meaning of life. It does mean, however, that such interpretations and pronouncements must be seen for what they are, that is, individualized assessments of one person's understanding of his or her own life and destiny.[97] Aiken says it this way:

> Philosophy remains, rather, a primordial activity whereby self-respecting men seek to come to terms, not only with the ends of science or of art or of government, but with their own destinies. . . . This is not to deny the right of the philosopher to do science or logic or history or anthropology. . . . It is only to say that as a philosophical critic he can properly speak only as an individual and private person determined to find the wisdom necessary to his own salvation. In a word, the aim of philosophy once again becomes

merely the Socratic aspiration toward self-knowledge, self-discipline, and self-realization.[98]

To Aiken, the fact that Reichenbach and Gödel reach different conclusions about relativity is not surprising. In fact, such different conclusions are inevitable. Each of them has looked at relativity and has come away with an understanding that best suits his personal vision of reality.

Review

8-1. Einstein's influence on philosophy is felt because of his unwavering faith in positivism. Einstein's positivism implies that philosophy must evolve out of science rather than the other way around. Philosophy can be looked at as a product of physics. The physicist reveals a new way of understanding how the physical universe operates. At that point the philosopher of physics takes over and interprets the implications of this new way to view the universe. The physicist, in the meantime, continues to explore the physical laws that govern the operation of the universe. Philosophy can no longer claim to be a separate discipline that seeks an understanding via a route that is different from that followed by science. The direction that all philosophic thought must take is carefully laid down first by the scientist. This does not mean that the philosopher no longer has a function. Nor does it subordinate the function of the philosopher to that of the scientist. Instead, it alters the role of the philosopher from that of originator to that of interpreter. In essence, philosophers must now accept the objective validity of certain theories that have been developed by scientists. However, the interpretation of the impact of that theory on human knowledge, or on the meaning of life, is still open to a wide variety of philosophical perspectives.

8-2. According to Kant, the physical universe is not something that the mind must struggle to perceive either correctly or incorrectly. Instead, the physical universe must correspond to the activity of the mind. The mind shapes physical reality. Lodged within the mind are certain a priori forms that construct reality according to the pure activity of the mind. The mind places all sensations within time and space to give them coherence. Through the activity of the mind, the unstructured sensations become organized perceptions. According to Kant, there are also other categories within the mind that help shape experience. These categories take the perceptions that have been shaped by the intuitive forms of space and time and transform them into conceptions that follow certain predictable patterns. Once perceptions have become conceptions, they can be seen as universally valid and, therefore, objective. According to Reichenbach, Einstein's work with relativity has shown that Kant was wrong. Relativity has taught us that space and time do not exist within the mind, as Kant believed, but have a real existence that depends upon matter and energy. However, in saying this we must be careful not to go too far. Even Einstein believed that space and time are human constructs. According to Einstein, space and time are created by the mind, but they still correspond to existing physical relationships.

8-3. According to idealism, all reality is the product of the mind. Kant can be considered an idealist, at least in the sense that he believed that the physical universe was organized by the a priori forms of space and time, which exist only in the mind. Gödel argues that relativity requires an idealistic view of reality. Gödel begins his argument by stating that change results from the passage of successive temporal intervals. Moreover, the passage of successive temporal intervals requires the movement of successive present moments. Time, then, is nothing more than an infinite sequence of nows. The theory of relativity, however, has demonstrated that simultaneity is relative. This means that the concept of nowness is not universal. On the contrary, there is no universal present moment, that is, no universal now. Since the experience of now can differ from observer to observer, an infinite sequence of nows is not possible. Consequently, there can be no objective lapse of time and no change anywhere within the universe. Gödel clearly does not believe in the passage of absolute time for the entire universe. He sees, instead, a universe in which all time exists simultaneously. Rudy Rucker presents a new way to visualize Gödel's explanation of simultaneous time. Rucker suggests that, to remain consistent with the theory of relativity, space and time must be seen as wrapped up together in a single, unified reality called the space-time cube, or the block universe.

8-4. If Gödel is correct in his conclusion regarding the simultaneous existence of all time, then the assumption that human beings have free will must be discarded. Free will or volition is the ability of human beings to make decisions for themselves without being controlled by outside influences. A philosophy that stands in opposition to free will is determinism, which holds that all human actions are subject to forces that predetermine the decisions that are made by individuals. Choosing whether volition or determinism characterizes human behavior is not without serious consequences. In fact, if determinism is accepted and volition abandoned, then concepts such as democracy, law, morality, and creativity must follow. Many people have argued convincingly that the principle of the fixity of time means that we must abandon any belief in free will. Since all time is immutable, nothing we can do will change any event on our lifeline. The sensation of free will or volition is just as illusory as the sensation of the passage of time. However, the two most influential scientific theories of the twentieth century, the theory of relativity and quantum theory, appear to clash on this very point. The Copenhagen Interpretation states that subatomic particles exist in all possible states until they are forced to assume a given state by the intervention of an observer. The mind must, therefore, have the ability to choose freely since, according to quantum theory, it is that very faculty that actually creates the universe. The conclusions that Rucker and Gödel come to clearly reflect the quantum theory of an observer-created universe. It is at this precise point that Rucker's, and therefore Gödel's, philosophical conclusions appear to suffer from a confusing, internal contradiction.

Understanding Key Terms

determinism	Novikov consistency conjecture	quality
elsewhen (elsewhere)		quantity
foreknowledge	omniscience	relations
future light cone	past light cone	sensations
idealism	perceptions	solipsism
lifeline (lifeworm, worldline)	predestination	space-time cube (block universe)
Minkowski diagram	principle of the fixity of the past	spatialization
modality	principle of the fixity of time	unidirectional
		volition (free will)

Review Questions

1. How does Einstein's theory of relativity affect philosophy?
2. What is the relationship between philosophy and relativity?
3. What is Kant's theory of space and time?
4. How does relativity verify positivism?
5. What is idealism?
6. What is the relationship between idealism and relativity?
7. How does the simultaneous existence of time lead to the theory of the space-time cube?
8. What is the principle of the fixity of time?
9. How does the simultaneous existence of all time create problems with volition?
10. What is the role of philosophy in the modern world?

Discussion Questions

1. Explain what Reichenbach means when he says, "(t)here is no separate entrance to truth for philosophers." Do you agree or disagree with Reichenbach's conclusion about the relationship between science and philosophy? Why or why not?
2. One of the unexpected conclusions of Einstein's theory of relativity is that time is not necessarily unidirectional. According to Reichenbach, such a conclusion was undreamt of by philosophers. Reichenbach says that this shows that philosophers did not know how to deal with time. Moreover, according to Reichenbach, this fact also proves the validity of positivism. Do you agree or disagree with Reichenbach on each of these points? Explain.

3. Idealism holds that all reality is simply the product of the mind. Gödel argues that relativity proves the validity of idealism. Moreover, Gödel also uses relativity to show that time is not unidirectional and that time does not really "flow" in any direction. According to Gödel, then, all time exists simultaneously so that the past, the present, and the future are all one. Do you agree with each of Gödel's propositions? Why or why not? According to Rucker, the simultaneous existence of time leads to the principle of the fixity of time (PFT). Moreover, the PFT leads to the conclusion that free will does not exist. Do you agree or disagree with this proposition? Explain.
4. Reichenbach says that relativity demonstrates that space-time has a real existence in the physical universe, whereas Gödel says that space-time exists only within the mind. Yet, the mind is clearly a part of the physical universe, and if in creating space-time, the mind is operating intuitively rather than volitionally, then the physical existence of space-time is just as inevitable within Gödel's universe as it is within Reichenbach's. The difference, then, may be simply one of degree rather than substance. Do you agree with this conclusion? Explain why or why not.
5. Aiken concludes that the role of the philosopher has become highly individualized. Philosophy no longer creates objective truth. Instead, philosophy has become an individualized activity by which each person reaches his or her individual truth. Do you agree or disagree with Aiken's assessment? Explain.

ANALYZING *STAR TREK*

Background

The following episode from *Star Trek: The Next Generation* reflects some of the issues that are presented in this chapter. The episode has been carefully chosen to represent several of the most interesting aspects of the chapter. When answering the questions at the end of the episode, you should express your opinions as clearly and openly as possible. You may also want to discuss your answers with others and compare and contrast those answers. Above all, you should be less concerned with the "right" answer and more with explaining your position as thoroughly as possible.

Viewing Assignment—
***Star Trek: The Next Generation*, "Cause and Effect"**

In this episode the Starship Enterprise is caught in a "temporal causality" loop created by a powerful space-time explosion that occurs when the Enterprise collides with another starship that emerges from a temporal rift in space. As a result of this temporal disturbance the crew is compelled to repeat the same events over and over, each time leading to the ship's destruction near the space-time rift. Paradoxically, the space-time explosion that destroys the Enterprise also throws the ship back through the loop. During one of these cycles through the loop, Dr. Crusher begins to experience a sensation of déjà vu. She then begins to hear voices in the night. During one of the successive loops she records the voices and Data determines that they are the voices of the crew. He also manages to pick up the voice of the captain ordering the crew to abandon ship. Eventually the crew speculates that they may be caught in a temporal causality loop. However, each time they come out of the loop and begin again, they forget everything that happened before. They contrive a plan by which they will send themselves a message in the next loop. Data becomes the carrier of that message. He tells himself to remember the number "three." Finally, on yet another journey through the loop, Data realizes that he should defer to the orders of Riker concerning the evasive maneuvers that the ship takes to avoid a second ship, which emerges from the space-time distortion, creating the explosion. The number "three" refers to the number of pips on Riker's collar reminding Data that Riker outranks him and is, therefore, presumed to be more knowledgeable about the operation of the ship. Riker's tactic succeeds. The explosion is avoided, and the loop is severed.

Thoughts, Notions, and Speculations

1. As we saw in this chapter, Gödel's belief in the simultaneous existence of all time leads to the principle of the fixity of time (PFT). The PFT, in turn, leads to the abandonment of the idea that humans can exercise free will. Do the events in the *Star Trek* episode entitled "Cause and Effect" support or contradict the idea that all time exists simultaneously? Do these events support or contradict the notion of human volition? Explain.
2. Based upon what you know about Einstein's ideas on the nature of space-time, as explained by the theory of relativity, would you say that the type of temporal causality loop depicted in "Cause and Effect" is possible? Why or why not?
3. What time-related paradox is created by the fact that the massive explosion that triggers the temporal causality loop also causes the destruction of the Enterprise? Explain.
4. After the final loop, the Enterprise encounters the USS Bozeman. The Bozeman was also caught in an explosion that created the temporal causality loop. Captain Picard immediately invites the captain of the Bozeman on board the Enterprise to discuss what has been happening. Given that the Bozeman is a ship that has traveled eighty years into the future, is it advisable for Picard to explain everything to its captain? Discuss the time-related problems of having such a conference. What other course of action might Picard take? Explain.

5. If the Enterprise is actually repeating the same period of time over and over again, then time in the rest of the universe should not flow forward in relation to those events. Yet, when the Enterprise finally emerges from the loop, the crew discovers that the ship's chronometers are off by seventeen days. Is this an accurate observation on the part of the writers, or should the chronometers be unaffected? Explain.

NOTES

1. Abraham Pais, *Subtle Is the Lord*, 308.
2. Clifford M. Will, *Was Einstein Right?* 75–76.
3. Pais, 308.
4. Ibid.
5. Ibid.
6. Andrew Paul Ushenko, "Einstein's Influence upon Philosophy," 609–14. For additional accounts that explain how the theory of relativity influenced philosophy see Bertrand Russell, "Philosophical Consequences," 210–20, and L. Susan Stebbing, "Interpretations," 196–212.
7. Ibid., 612.
8. Ibid., 612–13.
9. T. Z. Lavine, *From Socrates to Sartre*, 100.
10. Ibid.
11. Ushenko, 613. Despite the convincing tone of Ushenko's argument, there are contrary points of view. For example, in his book, *The ABC of Relativity*, Bertrand Russell points out that, although the tendency to identify the concept of "observer" with a "human observer" is understandable, it is not absolutely necessary. In fact, an observer, Russell points out, could just as well be a clock or a photographic plate. Whether this fact defeats Ushenko's argument is problematic. Nevertheless, Russell's proposition is interesting to ponder. Bertrand Russell, "Philosophical Consequences," 210–11.
12. For example, see the following essays: "Good and Evil," "The State and the Individual Conscience," "The Religious Spirit of Science," "Science and Religion," "The Need for Ethical Culture," "The Question of Disarmament," "Active Pacifism," and "Religion and Science: Irreconcilable?," reprinted in Albert Einstein, *Ideas and Opinions*; see also: "Moral Decay," "Morals and Emotions," "Towards a World Government," "Science and Civilization," and "Why Socialism?," reprinted in Albert Einstein, *Out of My Later Years*; for a complete list of Einstein's nonscientific writings see Paul Arthur Schlipp, ed., *Albert Einstein*, 730–46.
13. Hans Reichenbach, "The Philosophical Significance of the Theory of Relativity," 290.
14. Ibid., 291.
15. Ibid.
16. Stephen Cole, *Making Science*, 3–7.
17. Reichenbach, "The Philosophical Significance of the Theory of Relativity," 291.
18. Ibid.
19. Ibid., 291–92.
20. Ibid., 292.
21. Ibid.
22. Ibid.
23. Ibid., 292–93.
24. Ibid., 310.
25. Ibid.
26. James Trefil, *Reading the Mind of God*, 38.
27. Reichenbach, "The Philosophical Significance of the Theory of Relativity," 302.
28. Kurt Gödel, "A Remark about the Relationship Between Relativity Theory and Idealistic Philosophy," 557.
29. Reichenbach, "The Philosophical Significance of the Theory of Relativity," 310.
30. E. L. Allen, *From Plato to Nietzsche*, 119; Robert Paul Wolff, ed., *Ten Great Works of Philosophy*, 296.
31. Allen, 101; Wolff, 296.
32. Allen, 101.
33. Ibid., 119.
34. Immanuel Kant, "Prolegomena to Any Future Metaphysics," 319–21; Steven M. Cahn, ed., *Classics of Western Philosophy*, 847–48; Will Durant, *The Story of Philosophy*, 294.
35. Kant, 334–36; Durant, 295.
36. Kant, 336; Durant, 295; Stephen Korner, "Immanuel Kant," 160; Lavine, 194.

37. Robert Paul Wolff, "Immanuel Kant," 296–97.
38. Cahn, 847.
39. Durant, 293.
40. Ibid., 291–94.
41. Ibid., 294.
42. Kant, 332, 334–36.
43. Ibid.
44. Ibid., 332; Cahn, 848; Lavine, 196.
45. Reichenbach, 302.
46. Ibid.
47. Ibid., 302–3.
48. Pais, 303–5.
49. Reichenbach, 291–92.
50. Ibid., 306.
51. Ibid.
52. Ibid., 310.
53. Lavine, 143–44; A. C. Ewing, "Idealism," 146.
54. Ewing, 146.
55. Lavine, 144.
56. Ewing, 146–47.
57. J. D. Mabbott, "Thomas Hill Green," 121.
58. J. O. Urmson, "John Ellis McTaggart," 189.
59. Gödel, 557–62.
60. Ibid., 557–58.
61. Ibid., 558.
62. Ibid.
63. Henri Bergson, *Time and Free Will: An Essay on the Immediate Data of Consciousness*, 121.
64. Gödel, 559–61.
65. Ibid.
66. Bergson, 108–10, 120–21.
67. Bergson, 120–21; Gödel, 559–61.
68. Gödel, 559–60; David Millar, ed., et al. *The Cambridge Dictionary of Scientists*, 171–72.
69. Gödel, 559–60.
70. Ibid., 560.
71. See Chapter 7, "Past and Future Time Travel"; see also: Barry Chapman, *Reverse Time Travel: The Exciting Revelation that Traveling Backwards through Time Is Possible* and Paul Parsons, "A Warped View of Time Travel," 202–3.
72. Gödel, 560–61.
73. Ibid., 561–62.
74. Rudy Rucker, *The Fourth Dimension*, 135.
75. Paul Davies, *About Time*, 72.
76. Rucker, 136.
77. Ibid.
78. Ibid., 135–36.
79. Ibid., 145.
80. Ibid., 149.
81. Gödel, 561.
82. Stephen Hawking, *A Brief History of Time*, 23–30; Barry Parker, *Cosmic Time Travel*, 35–39; Rucker, 156–57.
83. Rucker, 149–51.
84. Nigel Calder, *Einstein's Universe*, 190–91; John Gribbin, *Time-Warps*, 88–89; Hawking, 23–30; Parker, 35–39; Rucker, 151.
85. Hawking, 23–30; Parker, 35–39; Rucker, 156–57.
86. Davies, 75.
87. Calder, 191–93; Davies, 75–76.
88. David Widerker, "Providence, Eternity, and Human Freedom," 242. For a series of articles that discuss this and similar concepts within a theological context see Brian Leftow, "Eternity and Simultaneity," 148–79; James Patrick Downey, "On Omniscience," 230–34; David P. Hunt, "Divine Providence and Simple Foreknowledge," 394–414; Tomis Kapitan, "Providence, Foreknowledge, and Decision Procedure," 415–20; David Basinger, "Simple Foreknowledge and Providential Control: A Response to Hunt," 421–27; David P. Hunt, "Prescience and Providence," 428–38.
89. J. Craig Wheeler, "Of Wormholes, Time Machines, and Paradoxes," 56.
90. Karen Armstrong, *A History of God*, 282–83; Archie J. Bahm, *The World's Living Religions*, 296–97.
91. Brian Leftow, "Eternity and Simultaneity," *Faith and Philosophy*, 148–49. Leftow is careful to point out that neither Aquinas nor Boethius explain how God can exist beyond time, yet exist simultaneously with time. He does assume, however, that God's coexistence with all time and his existence beyond time represent two different types of simultaneity. Leftow, 149–50.
92. David Bohm, *Unfolding Meaning*, 12–13.
93. Henry Aiken, "The Fate of Philosophy in the Twentieth Century," 233–52.
94. Ibid., 250.
95. Ibid.
96. Ibid., 250–52.
97. Ibid., 250.
98. Ibid., 250–51.

Unit VI Life and Evolution

Chapter 9
The Origin of Life

COMMENTARY: EXTRATERRESTRIAL LIFE IN THE FAST LANE

On several occasions in recent memory, the scientific community and the world at large have held their collective breath as discoveries have indicated that extraterrestrial life may exist somewhere in the universe. The most recent occasion occurred on August 8, 1996 when NASA scientists announced that they had discovered evidence that life may have existed on the planet Mars about 3.6 billion years ago. The evidence was gathered by scientists using state-of-the-art microscopes to examine a meteorite of Martian origin. A **meteorite** is an extraterrestrial object made of rock or metal that has passed through the atmosphere and hit the Earth. The meteorite in question, which had been discovered in Antarctica in 1984, contains small globules of carbonate, a mineral that reacts with water to form crystals. This indicates that liquid water may have been present on Mars 3.6 billion years ago. Water is, of course, an essential element for life. Other evidence indicates that these globules contain minerals that could be the remains of biological processes. In addition, experiments performed on the meteorite have revealed the possible presence of organic molecules that are often found in fossil fuels. Finally, the most controversial piece of evidence involves the discovery of tiny tubular objects that some scientists speculate may be tiny fossilized microbes. None of this is direct evidence of life, and it would be wildly speculative to conclude that life once existed on Mars based simply on this one discovery.[1] Nevertheless, the hope is that this evidence will lead to the conclusion that life in the universe is not limited to the planet Earth. This is not the first time that scientists have looked to the planet Mars in anticipation of finding life. Another spellbinding occasion occurred in the 1970s when the Viking spacecraft conducted experiments designed to detect the existence of life on the surface of Mars. The Viking lander collected soil samples and discovered that oxygen was being emitted from those samples, indicating the existence of a biological process within the soil. Once again, the scientific community waited in anxious anticipation as verification experiments were conducted. In this series of experiments, equipment on board the lander subjected the soil sample to enough ultraviolet radiation to sterilize the compartment and kill any living material. When the sample continued to emit oxygen, observers knew that the process occurring in the chamber was a simple chemical reaction rather than a biological process. Despite this "near miss," the search for extraterrestrial life goes on. Perhaps before we pursue this line of thought, it would be best to define life and to uncover the origin of life on the planet Earth. What mysterious quality separates living from nonliving matter? What are the characteristics of life? Will humanity ever be able to create artificial life? These are just a few of the questions that form the focus of this chapter.

CHAPTER OUTCOMES

After reading this chapter, the reader should be able to accomplish the following:

1. Identify the characteristics of life.
2. Explain vitalism and reductionism.
3. Discuss some of the problems with these theories.
4. Define the complexification process.
5. Outline the conditions and the forces present on the prelife Earth.
6. Explain the interaction of these forces as a prelude to the first appearance of life.
7. Identify and explain the significance of the Miller-Urey experiment.
8. Describe the creation of synthetic protolife and the manufacture of the handmade cell.
9. Outline the essential features of the theory of the extraterrestrial origin of life.
10. Contrast the creation of a star with the creation of life.

Figure 9.1 The Planet Mars.
The planet Mars has been the source of much controversy in recent years concerning the origin of life. One controversial claim has focused on the discovery that a meteorite, which had found its way to Earth and which had originated on Mars, harbored evidence that life had existed on Mars about 3.6 billion years ago.
Photo Credit: The National Aeronautics and Space Administration

9-1 LIFE VS. NONLIFE

The question that lies at the heart of this chapter is "What is life?" At first glance, the question may appear to be a simple one to answer. Living systems are easy to distinguish from nonliving systems. Most of us can list some of the characteristics that best describe living things. For instance, living things need food. They grow and develop. They interact with the environment according to a set of complicated operations. They have the capacity to reproduce themselves, and eventually they cease to exist. These characteristics are quite adequate for explaining what to look for if someone is on a scavenger hunt for living things. Yet, these characteristics and others like them avoid the real question. That question is "What makes one entity living and another nonliving?" For instance, what gives life to bacteria but not to a computer? Both entities are capable of complicated operations, yet one is clearly alive and the other is not. Both an amoeba and a tornado can duplicate themselves. Yet, we say that the amoeba is alive, but the tornado is not.

The differences in each of these examples may appear to be self-evident. However, when we actually attempt to formulate a definition of life, problems arise. Unfortunately, when most nonscientists discuss the legal and the ethical issues that affect life, they operate on the erroneous assumption that all people share the same unspoken definition of life. This assumption can be misleading and may actually cause many controversies. Countless legal and ethical questions might be easier to solve if people could agree on a common definition of life. The abortion debate, the concern over living wills, the controversy involving euthanasia, and the issue of genetic engineering all require answers to the questions "What is life?" and "How did life originate?" For this reason, we devote the next four chapters to the study of life. In this chapter, we search out and attempt to establish a definition of life. We also look at the most common alternative viewpoints on the nature of life. From that point, we examine the origin of life on the planet Earth and look at the quest for the creation of artificial life. This groundwork allows us to discuss several legal and ethical questions that presuppose an understanding of life.

The Characteristics of Life

To arrive at a universal definition of life, it is helpful to explore some of the essential characteristics of living things. Paul Davies addresses this question in *God and the New Physics* when he says, "To the physicist the two distinguishing features of living systems are *complexity* and *organization*. Even a simple, single-celled organism, primitive as it is, displays an intricacy and fidelity unmatched by any product of human ingenuity."[2]

Unfortunately, both of these characteristics, whether taken by themselves or considered together, are far too limiting. For instance, the characteristic of organization also describes cities, nuclear submarines, and computers, all of which are organized but none of which could be said to be alive. The same could be said for the characteristic of complexity. Certainly anyone who has attempted to work on a modern automobile, a VCR, or a cellular phone will testify to their complexity. Yet, nobody would claim that these technological marvels are alive. Clearly, then the characteristics of organization and complexity cannot stand in isolation. More detail must be added to make the picture complete. Davies responds to this challenge by providing the reader with an intriguing example of a simple living system:

> Consider, for example, a lowly bacterium. Close inspection reveals a complex network of function and form. The bacterium may interact with its environment in a variety of ways, propelling itself, attacking enemies, moving towards or away from external stimuli, exchanging material in a controlled fashion. Its internal workings resemble a vast city in organization. Much of the control rests with the cell nucleus, wherein is also contained the genetic 'code', the chemical blueprint that enables the bacterium to replicate.[3]

Implicit in Davies's bacterium example are two additional characteristics that separate living from nonliving organisms: (1) the ability to analyze and respond to outside stimuli, and (2) the ability to replicate.[4] Herein lies the critical distinguishing characteristics that separate life from nonlife. Living things do not exist in isolation. Rather, they interact with and use the environment with three crucial objectives in mind: "to survive, to compete, to reproduce."[5] So it is not these characteristics in isolation, but in dynamic interactive unity, that give us a living thing. Life, then, is not a condition, but a process.[6] To summarize, a living thing is characterized by complexity, organization, the ability to analyze and react to outside stimuli, and the ability to reproduce.[7]

In the *Star Trek* episode "Home Soil," which is summarized at the end of this chapter, Dr. Crusher defines life by listing the following characteristics: "It must be able to assimilate, respire, reproduce, grow and develop, move, secrete, and excrete." In many ways, this definition is simply a long version of Davies's definition. Although it emphasizes the individual processes that are associated with life, the definition does imply that a living organism must have a certain level of complexity and organization to perform the functions of assimilation, reproduction, growth, development, secretion, and excretion, all of which involve the ability to respond to the environment. Neither Davies's nor Crusher's definition, however, pinpoints the means by which these functions are carried out. It is to this issue, the question of means, that we now turn our attention.

Life at the Subatomic Level

Before addressing the question of how living things carry out their life functions, it is necessary to point out a key ingredient that appears on neither Davies's nor Crusher's list of life-giving characteristics. As Davies points out:

> It is important to appreciate that a biological organism is made from perfectly ordinary atoms. Indeed, part of its metabolic function is to acquire new substances from the environment and to discard degenerated or unwanted

substances. An atom of carbon, hydrogen, oxygen, or phosphorus inside a living cell is no different from a similar atom outside, and there is a steady stream of such atoms passing into and out of all biological organisms.[8]

In other words, when we reach the atomic and subatomic levels, there is no difference between animate and inanimate objects. In fact, in some cases even at the molecular level, there is no difference between a living and a nonliving entity. As Raymond Daudel, president of the European Academy of Science, Letters, and Art, points out in his book *The Realm of Molecules*, "Every living being—every animal, plant, bacterium, and virus—is made of molecules, some of which are identical to those in non-living matter. There is no difference, for example, between the water molecules in our blood and those in a mountain stream."[9]

To dramatize and reemphasize this point, consider, for a moment, the neutrino. Recall that a neutrino is "a zero-mass uncharged particle emitted in the process of beta decay."[10] A neutrino can "pass through a block of lead several light-years thick without disturbing a single atom."[11] Suppose that you (or someone else, if you don't want to take any chances) were sitting on top of this vast block of lead. Now picture millions of neutrinos flowing through your body and into the lead block. From the neutrino's point of view, there would be no difference between you and the block of lead. The neutrino would not be able to tell where your body ended and the lead block began. Therefore, at the subatomic level there is no difference between you and the lead block, because there is no difference between your atoms and the atoms of the lead block. Another way of saying this is to note that there are no such things as living atoms. As Davies says, "though we may not doubt that a cat or a geranium is living, we would search in vain for any sign that an individual cat-atom or geranium-atom is living."[12] Yet, there is clearly a difference between you and the imaginary lead block upon which you are perched. If the difference between life and nonlife does not exist at either the atomic or the subatomic level, at what level does it exist?

9-2   ALTERNATIVE THEORIES ON THE NATURE OF LIFE

Several explanations have been put forth to differentiate between life and nonlife. Two of these explanations, vitalism and reductionism, are the most talked about and, therefore, the most familiar. A third idea, emergentism, is much more recent and may, therefore, seem to be a bit exotic.

Vitalism and Reductionism

Perhaps the most familiar explanation for the nature of life is the idea of vitalism. **Vitalism** states that life results from a mysterious life force that is infused into matter, thus transforming it from inanimate to animate matter.[13] As Dominique

Lacourt notes in the introduction to Martin Olumucki's book *The Chemistry of Life*, "vitalism, in its metaphysical and religious form, consists in attributing the specificity of living phenomena to some particular force that escapes the laws of physics and manifests what is alleged to be a design of the Creator."[14]

In his book, *Seven Clues to the Origin of Life*, A. G. Cairns-Smith defines vitalism as a "deeply mysterious unifying power, a principle of life, an essential magic about living things that divides them from everything else."[15] Vitalism is a time-honored theory that has been promoted by such philosophers as Plato, Thomas Aquinas, René Descartes, and Henri Bergson. Perhaps even more interesting is that vitalism has also been endorsed by scientists like the Swedish chemist Jöns Jakob Berzelius, the French chemist Charles Gerhardt, and the German organic chemist Justus von Liebig.[16] It is certainly in agreement with the understanding of life found in many Western religions. For instance, the idea of vitalism is plainly demonstrated in the words found in Genesis 2:7 of the New Revised Standard Version of The Bible. "Then the Lord God formed man from the dust of the ground and breathed into his nostrils the breath of life; and the man became a living being." Clearly, the "breath of life" in this passage is meant to portray the spiritual entity by which God animated the first human being. Vitalism is also in line with the philosophical concept of the soul, and it fits in nicely with theological notions of immortality. All of this makes vitalism a familiar explanation for the nature of life. As a result, many nonscientists are quite comfortable with vitalism and accept it without question.

The fact that vitalism is internally consistent and enjoys widespread acceptance does not, however, legitimize either the original concept of a "life force" or the ideas that flow logically from that original concept. To verify these ideas, we must submit them to the close scrutiny of the scientific method. Recall that the scientific method is an objective process by which scientists use observation and experimentation to verify tentative explanations about the operations of nature. Every time an empirical law is suggested, tested, and verified, it joins that body of knowledge known as science. However, its membership is guaranteed only as long as it has not been falsified by a new and equally objective test.[17] There is an unstated **premise** here that before a proposition can even be considered for membership in the body of scientific knowledge, it must be verifiable. Or to state it negatively, the premise must be capable of falsification. Seemingly if a statement can be neither verified nor falsified then that statement has no place in science.[18] This, then, is the problem with vitalism. The statement that a supernatural life force energizes inanimate matter to give it life cannot be verified.

However, the problem of verification is not quite this simple. As the Austrian mathematician and philosopher Kurt Gödel demonstrated with his Incompleteness Theorem, there are some statements, mathematical and otherwise, which are true, yet unverifiable. The paradox created by such a statement can be explained however, once we step outside the system that generated it in the first place. Thus, in M. C. Escher's famous lithograph, *Drawing Hands*, it is impossible to explain how the right hand can draw the left hand while the left hand draws the right hand, while remaining within the system depicted on the print. Once outside the system, however, it is easy to see that the entangled image was created by an outside agent, in this case Escher. Such may be the case with many of the puzzles associated with the nature of life and consciousness.[19] However, for the time being we will adhere to the strict verification process required by the scientific method.

Remember also that a successful empirical law or theory must be both clear and open to public scrutiny. Laws and theories that require some sort of hidden knowledge, special talent, or mysterious magic are immediately suspect. This is another problem with vitalism. It requires a special, unknowable, unreachable, supernatural power. Rupert Sheldrake, former research fellow of the Royal Society and scholar of Clare College, Cambridge, makes the nature of this problem clear in his text *A New Science of Life*, when he notes that no verification for vitalism exists despite repeated attempts over the last one hundred years to state predictions that can be tested by workable experiments.[20] In his text, *The Limitations of Science*, the noted mathematician and philosopher, J. W. N. Sullivan notes that vitalism introduces an element of supernatural creativity into the study of life that runs counter to the scientific need for verifiable results.[21]

The clear lack of scientific proof for vitalism seriously reduces its credibility. However, as Sullivan correctly points out, journeying to the opposite end of the spectrum does not solve the problem either. At that mythical "other end of the spectrum" is a doctrine called reductionism. According to **reductionism**, "(t)he whole is composed of nothing but its parts. One does not have the parts plus some mysterious new substance. . . . An organism is nothing but the molecules of which it is made. There is not some lifeforce or suchlike thing which is needed to make a living being. In this sense . . . humans are like machines."[22] In *A New Science of Life*, Sheldrake, who uses the term mechanism rather than reductionism, points out that, from the mechanistic point of view, life can be reduced to the processes found in physics and chemistry because all living things are simply physical and chemical machines.[23]

Reductionism, or more specifically **ontological reductionism**, finds its historical roots in methodological reductionism. **Methodological reductionism** is an approach to scientific investigation that insists on explaining any phenomenon by examining its parts.[24] In fact, despite its philosophical shortcomings, methodological reductionism remains an accepted time-honored approach to the scientific method. Many highly respected scientists, including the British biologist Francis Crick, champion the reductionist approach to science.[25]

Still, there are many others who point out the difficulties inherent in both ontological and methodological reductionism. As Sullivan points out in *The*

Limitations of Science, the fact that vitalism has been eliminated from serious scientific consideration does not automatically mean that reductionism will take its place. In fact, reductionism has its own set of difficulties, not the least of which is that it offers no explanation for the peculiarities of certain biological functions. Another major difficulty is that reductionism leaves little with which to work. Life, according to the reductionist, is the same as nonlife. There really is no point in asking any further questions. This conclusion may by itself cause no particular problem. However, if we follow the reasoning of reductionism to its logical end, we are also forced to conclude that there is no point in distinguishing human life from all other forms of life. As a result, reductionism can be as unsatisfying in the search for life as vitalism.[26] This does not eliminate all our options, however. A third alternative theory is available for explaining the existence of life. That theory, known as emergentism, may hold the key to our understanding of the nature of life.

Emergentism and Complexification

According to **emergentism**, life should be viewed as an emergent quality of the complex organization of matter. Life arises from the complex organization of matter, from the molecular stage on up.[27] The emergent theory of life finds its roots in the holistic approach to scientific investigation. **Holism** seeks to comprehend the complex interaction of the interconnected parts of the universe. In a sense what the reductionist takes apart, the holist puts back together.[28]

A key element in the understanding of emergentism is to focus on the fact that in the emergent view of life, an entire living organism is not simply the sum of its individual parts. Instead, life arises from the complexity without being equal to it.[29] If this holistic aspect is not grasped sufficiently, then the difference between emergentism and reductionism will be obscured. To the reductionist, life is nothing more than physical and chemical processes; but to the emergentist, life arises from chemical and physical processes. A representation of the two theories in mathematical terms would look something like this:

The reductionist view: life = chemical and physical processes
The emergentist view: life > chemical and physical processes

To the reductionist, life equals chemical and physical processes, whereas to the emergentist, life is greater than the chemical and physical processes. The chemical and the physical processes make life possible, but they must not be equated with life. To the emergentist, life exists on another dimensional level arising from these chemical and physical processes.

A Literary Analogy. Another way to understand the difference between the two theories is to look at the reductionist and the emergentist as two very different types of literary critics. The reductionist literary critic would look at a novel and see a list of words, nothing more. In fact, if we were to take our analogy to its logical extremes, the very existence of a reductionist literary critic would be nonsense. He would not be able to look beyond the words on the page, and the analysis of the novel would abruptly halt. Similarly, the reductionist looks at the chemical and physical processes, and his analysis of life suddenly stops.

In contrast, the emergentist literary critic would not stop the analysis of the novel at the level of the words. Instead, he or she would see the plot that arises from the words, the characters that arise from the plot, the actions that arise from the characters, and so on. In short, the words make the novel possible, but the novel must not be equated with the words. The plot emerges from the words and takes on a real existence that is separate from the words. The critic can review the plot of a novel and can comment on the actions of the characters, without the actual novel in hand. Similarly, the plot can be made into a movie or a miniseries. It can be read out loud and recorded as a book-on-tape. It can even be translated into dozens of different languages. In the novel *Fahrenheit 451*, Ray Bradbury's characters must memorize literary works word-for-word to preserve them from the destructive campaign of the firemen of the future. Because plots emerge from words, all those literary works continued to exist, despite the physical destruction of the books themselves.

An Emergentist Definition of Life. One of the clearest definitions of life according to the emergent theory has been provided by Stuart Kauffman of the Sante Fe Institute in New Mexico. Kauffman has said, "life is a natural property of complex chemical systems . . . when the number of different kinds of molecules in a chemical soup passes a certain threshold, a self-sustaining network of reactions—an autocatalytic metabolism—will suddenly appear. Life emerged . . . not simple, but complex and whole, and has remained complex and whole ever since."[30] In line with this approach, William K. Hartmann and Ron Miller in *The History of the Earth* note that, "(t)he quest for the origin of life thus becomes an attempt to understand how organic chemical processes built ever more complex molecules, and how these led to biochemical processes."[31] The famous paleontologist Teilhard de Chardin, has called the process of the organization of matter **complexification**. According to Teilhard, complexification makes the appearance of life almost inevitable, given certain physical conditions.[32] More importantly, however, is the fact that the complexification process can provide us with a solution to the puzzle of when the threshold between living and nonliving systems has been crossed. Recall that when we reach the atomic and subatomic levels, there is no difference between animate and inanimate objects. In fact, in some cases, even at the molecular level, there is no difference between a living and a nonliving entity. If the difference between life and nonlife does not exist at either the atomic or the subatomic level, at what level does it exist?

According to the emergent tradition, the threshold is crossed when the molecular structures become complex enough to create certain key structures and perform certain key functions. This point is described rather succinctly by Robert Pollack, an award winning biologist who worked for a time with James Watson, the codiscoverer of DNA's structure. Pollack explains in his recent study, *Signs of Life*, that living things have four distinct traits in common:

> (T)hey can produce offspring, they have a history of common descent from shared ancestors, they are made of invisible soft building blocks called cells, and each cell carries within itself a singular chemical called deoxyribose nucleic acid, or DNA, a large molecule assembled from the atoms of just five elements—carbon, phosphorus, nitrogen, hydrogen, and oxygen.[33]

Notice, however, that the first three traits actually result from the fourth. In other words, the power to replicate, the shared history, and the common cellular structure of all living systems can all be traced to the complex DNA molecule. We can, therefore, conclude that the passage from nonlife to life occurred when the complex molecular structure of DNA allowed the other three key characteristics of life, replication, shared history, and cellular structure, to appear.[34]

In his study of evolution, *Interpreting Evolution: Darwin and Teilhard de Chardin*, H. James Birx, professor of anthropology at Canisius College in Buffalo, has pointed out that the complexification process need not be limited to the passage from nonlife to life but can also be seen as encompassing the entire evolutionary process. According to Birx, Teilhard saw the process as involving three major stages:

> According to Teilhard the process philosopher, there are three discernible stages of planetary evolution thus far: prelife, life, and thought. Within each stage there is a great spectrum of diversity and degrees of consciousness. Yet, while ascending the atoms of Mendeleev's periodic table and following through organic and psychosocial emergence, one finds that each successive manifestation of evolution is always more complex in its internal configuration of elements, and at the same time each displays increasing degrees of radial energy.[35]

Following Teilhard's lead, this chapter and the next two adopt his progressive approach to the study of life from prelife to life to thought. Viewed in this way the question "How did life on Earth begin?" can be rephrased to state "How did the original complexification process take place?" Implied within this question are three even more fundamental inquiries: (1) What were the conditions during the prelife era? (2) What forces might have been present during the prelife era to influence those conditions? (3) How did the forces and the conditions work together to initiate the complexification of prelife into life?

9-3   PRELIFE AND LIFE

As noted, the question "How did life on Earth begin?" will be the focus of the rest of the chapter. One reason it has been difficult to answer this question is that there is no certainty about either (1) the conditions prevalent during the prelife era or (2) the forces that might have been present to influence those prelife conditions. Nevertheless, over the years, one theory has gained prominent acceptance within the scientific community.

Conditions and Forces during the Prelife Era

The most popular theory concerning the prehistoric conditions and the primitive forces at work on the prelife Earth is the theory of the **prelife soup**. This theory, which is sometimes referred to as the **primeval soup**, describes a turbulent prelife period. The prelife Earth had no oxygen, was bombarded by radiation from space, and churned in the enormous amounts of energy produced by electrical storms and volcanic activity.[36] This view of prelife Earth is described by H. James Birx in *Interpreting Evolution*:

> (T)he atmosphere of our primitive planet contained almost no free oxygen and had no upper ozone layer to act as a shield against ultraviolet radiation. At that time in earth history, the surface of our early planet contained both seas and pools of so-called probiotic soup. They were rich in chemical elements that reacted to the intense radiation from outer space as well as terrestrial sources of energy. As a result of prolonged photochemical activity, these inorganic mixtures gave rise to organic compounds (including amino acids). Through time and chemical selection, these eobionts or organic systems increased in complexity and stability, becoming the immediate precursors of living things.[37]

With this picture in mind of the conditions and the forces present on the prelife Earth, the next step in the exploration of the origin of life would be to answer the following question: How did these forces and these conditions work together during the prelife era to initiate complexification and give rise to living things?

From Prelife to Life

The prelife soup theory then goes on to explain the move from prelife to life as a series of progressively more complex leaps in organization. During the first stage in the development of life from prelife material, beginning between 3.5 and 4.2 billion years ago, the primeval soup came under the energetic influence of the **ultraviolet (UV) radiation**, volcanic heat, and/or lightning. The interaction of theses forces created more and more complicated molecules, leading inevitably to amino acids and other complex molecules. A key feature of these complex molecules was their capacity to replicate themselves. Embedded within this replication process was the newly acquired ability of the complex molecules to pass their characteristics from one offspring to the next. This pattern of replication provided the underlying mechanism for heredity, mutation, and natural selection, all of which furnish the driving force of the evolution from prelife to life.[38] In the next chapter we explore just how these factors interact to allow for the evolution of complex life-forms on the planet. For now, however, we trace the earliest stages in this evolutionary process.

Precellular Organisms. During the next leap forward, beginning at about 3.5 billion years ago, **precellular organisms** appeared. These precellular organisms, or **protocells**, provided a bubble-like environment for the protection and replication of the complex molecular structures that were now present in the primeval environment. The protocells also developed a transport system allowing the molecular structures within to take advantage of the protective nature of the membrane while interacting with and feeding upon the environment.[39] Support for this scenario has been provided by the work of Sydney Fox at the Institute of Molecular Evolution of the University of Miami. Fox showed that the heating of amino acids compels them to combine with one another according to certain well-defined patterns. Moreover, the addition of water to these complex molecules causes the appearance of objects known as **proteinoid microspheres**.[40] These microspheres are virtually identical to the protocells previously described. In fact, comparisons between some of the oldest microfossils on record and sets of artificially fossilized microspheres reveal identical characteristics. The proteinoid microspheres created by Fox are not, however, actually alive. Fox himself describes them as a form of protolife, which places them in a twilight region between living and nonliving things.[41]

Cellular and Multicellular Organisms. The earliest protocells lacked a self-contained nucleus and are, therefore, referred to as **prokaryotes**.[42] The next dramatic leap forward occurred around 3 billion years ago when some of the protocells developed the ability to transform sunlight into energy via photosynthesis. This gave rise to a primitive form of life called **cyanobacteria**. The appearance of cyanobacteria precipitated a traumatic transformation in the atmosphere of the planet Earth. Specifically, the photosynthetic process led to the production of large amounts of oxygen.[43] At first, this new blanket of oxygen was poisonous to most of the protocells, even those that actually produced the oxygen. Initially, much of this oxygen was absorbed into mineral pockets within the Earth, slowing the transformations of the atmosphere just enough to allow some of these precellular organisms to adapt to the new environment. By the time these mineral pockets were filled and oxygen began to saturate the atmosphere, many protocellular organisms had successfully adapted to the presence of that oxygen. Those that did not adapt either died out or took refuge in the nonoxygenated areas of the planet.[44]

The next leap may have been triggered by this high concentration of oxygen in the atmosphere. At some point along the way, but certainly by 1.5 billion years ago, some of these precellular organisms had developed an independent, self-contained nucleus that housed the essential mechanism of the cell's operation. This was the birth of the cell as we now know it today. These newly developed cells, with clearly separated nuclei, are called **eukaryotes**. Eventually these cellular organisms began to colonize, creating the first multicellular organisms.[45] The abundance of oxygen in the atmosphere also led to the creation of the **ozone layer**. The newly developed photosynthetic cells stripped away the hydrogen atoms from water molecules, releasing O_2 oxygen molecules into the atmosphere. Ultraviolet light from the Sun interacted with some of these O_2 molecules, splitting them into free-floating oxygen atoms that later joined with other O_2 molecules to create O_3, giving Earth its protective blanket of ozone. This protective shield led to another leap forward. Beginning about 700 million years ago, more and more complex life-forms began to abound within the environment, leading to the mantle of diverse life-forms that blankets the Earth today.[46]

Experimental Verification

It is not at all certain that the process just described is really what happened on the early Earth. However, experimental evidence indicates that the development of life could have happened this way. For instance, in the early 1950s at the University of Chicago, two chemists, Stanley Miller and Harold Urey, performed an experiment that clearly supports the primeval soup theory. At that time Miller and Urey attempted to reproduce the conditions of the early Earth. They recreated the atmosphere of ammonia and methane and subjected that mixture to a series of electrical discharges designed to duplicate the intense energy discharges of the early Earth and to stimulate the photochemical processes described. The results of the now famous **Miller-Urey experiment** are legendary. After several days, in their pool of chemical soup, Miller and Urey observed the appearance of a number of organic compounds. Included in these compounds were at least twenty-five amino acids and urea. **Amino acids** are, of course, among those chemical compounds that are essential to life as we know it.[47]

The Miller-Urey experiment was followed by a series of similar ones that further supported the theory that life originated as the result of the photochemical processes described. These experiments include those done by two Russian scientists, T. E. Pavlovskais and A. G. Pasynskii, who formulated amino acids

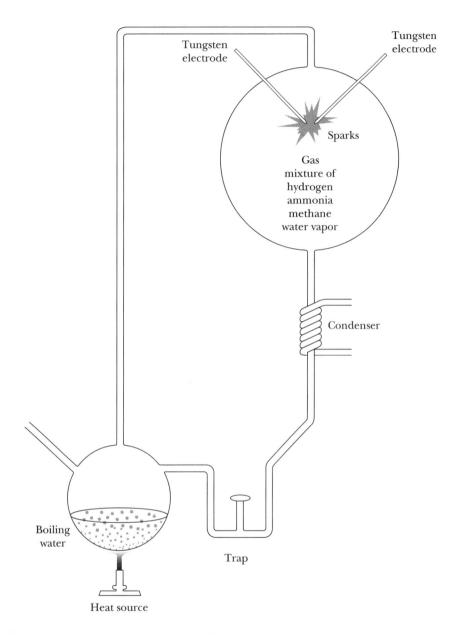

Figure 9.2 The Laboratory Apparatus Used in the Miller-Urey Experiment.

In the famous Miller-Urey experiment at the University of Chicago in 1953, two chemists, Stanley Miller and Harold Urey, reproduced the conditions of the early Earth by recreating the primitive atmosphere of methane and ammonia and subjecting that gaseous mixture to electrical discharges to duplicate the intense energy charges of the early Earth and to stimulate the photochemical process thought to have occurred there at that time. After several days, in their pool of chemical soup Miller and Urey observed the appearance of a number of organic compounds, including at least twenty-five amino acids and urea. Amino acids are among those chemical compounds that are essential to life as we know it to exist on Earth.

using a technique similar to that used by Miller and Urey. Another corroborating experiment was performed in the 1950s by Sidney Fox. The Fox experiment went a step further and produced intricate organic molecules resembling **proteins**. Another series of experiments, performed by J. Oro in the 1960s, went even further and managed to fabricate purine, ribose, deoxyribose, and pyrimidine, the constituents of **deoxyribonucleic acid (DNA)** and **ribonucleic acid (RNA)**. During the 1970s, Cyril Ponnamperuma performed a series of experiments that generated adenosine triphosphate, a key element in developing the energy necessary for life.[48]

To maintain intellectual honesty here, it is necessary to point out that the original experiments performed by Miller and Urey may have been somewhat flawed. New evidence indicates that their estimate of the composition of the Earth's early atmosphere may have been in error. Recent discoveries have demonstrated that the primeval atmosphere may have contained more carbon dioxide and molecular nitrogen than was suspected at first. In addition, they may have overestimated the amount of molecular hydrogen in the atmosphere. Still, this is not the final word on the subject. Research in this area persists and, at some time in the near future, current estimates on the composition of the early atmosphere may change once again.[49]

Alternate Theories

Despite this experimental evidence, we are not absolutely positive that the process described is the only line of evolutionary development on the Earth. Recent research has uncovered the existence of life in the dark depths of the ocean.[50] This discovery has opened several lines of inquiry, one of which involves the question of how these creatures could have survived in the absence of sunlight. The solution to this puzzle may lie in the fact that many of these life-forms have been found close to **deep sea vents**, which emit a volcanic gas called hydrogen sulfide. The proximity of the deep sea bacteria to this source of hydrogen sulfide seems to indicate that the bacteria on the ocean floor uses a system known as chemosynthesis rather than photosynthesis to survive. **Chemosynthesis** involves a chemical process by which the life forms on the ocean floor break down the hydrogen sulfide into useful energy. Consequently, the life-forms in the dark depths of the ocean use geothermal energy rather than sunlight as a source of food. It is entirely possible that these creatures represent an alternate line of evolutionary development on the Earth. The existence of chemosynthetic bacteria in the early days of the Earth may explain, for example, how life could develop and survive before the ozone layer protected life forms from the harmful effects of ultraviolet radiation.[51]

The process described does not include any detailed mention of the role of DNA or RNA, the complex molecules that guide the operation of the cells. This is because there is also some disagreement in the scientific community about the role of DNA, RNA, and protein in the appearance of life on Earth. One theoretical position holds that proteins developed first in the creation of proteinoid microspheres.[52] This is essentially the process that was just described. However, another school of thought has proposed the idea that RNA developed before protein and that the early Earth was ruled by this dominant form of RNA.[53] Since the jury is still out in this regard, we will simply sidestep the issue, and move on to the next great question: Can artificial life be created in the laboratory?

9-4 MAKING LIFE FROM NONLIFE

It is important to note, however, that none of the experiments noted, nor others like them, have actually produced life. Rather, they have conjured up the basic building blocks of life and have, thus, helped support the prelife soup theory of life's origin. Research in this area proceeds today. Two of the key researchers in the area are Steen Rasmussen of the Los Alamos National Laboratory in New Mexico and Jack Szostak of the research division of Massachusetts General Hospital.

Synthetic Protolife

Rasmussen is attempting to produce something that he calls **synthetic protolife** or, more simply, the Metabolism. Synthetic protolife is "a cooperative jumble of chemical activity: each of the many reactions lends a hand to one of the others . . . the reactions should reinforce each other until an ongoing, self-maintaining symphony of interactions has pulled itself up by its bootstraps. This cooperative chemical network will then sustain itself."[54] If Rasmussen is correct, this synthetic protolife is a **metabolism** in the sense that it processes matter from its surrounding environment and turns that matter into energy to maintain itself independently and to help itself grow and develop.[55] Whether synthetic protolife will actually be alive is still problematic. However, even if it simply represents a stage between living and nonliving matter, it exemplifies a monumental leap forward in the search for the origin of life.

The Handmade Cell

Microbiologist Jack Szostak is working on the development of what he calls a **handmade cell**. His research team is working with RNA, rather than DNA. Today RNA functions as a sort of messenger molecule that is responsible for transferring the instructions of DNA for the making of proteins. Szostak theorizes that, under conditions that were present in the early Earth, RNA may have been capable of functioning independently. In today's living things, RNA and enzymes are interdependent. **Enzymes** are molecular structures that are needed

• 300 • Chapter 9 The Origin of Life

Figure 9.3 Comet Halley: Did Life Have an Extraterrestrial Origin?
  The theory of panspermia states that the first molecules on the Earth may have actually had an extraterrestrial origin. These organic molecules may have been delivered to the Earth by the impact of a comet such as Comet Halley pictured above. Recent studies of Comet Halley have revealed that about thirty-three percent of that astronomical body is composed of organic molecules.
Photo Credit: Copyright, Association of Universities for Research in Astronomy, Inc. (AURA), all rights reserved. AURA operates The National Optical Astronomy Observatories for The National Science Foundation.

for the production of RNA. However, RNA is necessary for the creation of the enzymes, and so we have an unending circular process. RNA creates the enzymes that create more RNA that, in turn, creates more enzymes and so on.[56] What Szostak hopes to do is develop a self-replicating strand of RNA that can act as both enzyme and RNA. If he succeeds, he will have taken a giant step toward the production of artificial life and toward demonstrating how organic molecules arose from the inorganic prelife soup.[57]

The Extraterrestrial Origin of Life

The prelife soup theory is not the only theory put forth for the origin of life. An increasingly popular theory, often referred to as **panspermia**, states that the first organic molecules on the Earth actually had an extraterrestrial origin. These organic molecules may have been delivered to the early Earth in a number of different ways. One method might have been by the impact of a comet with the Earth. Researchers estimate that many comets are made up of large clusters of organic molecules. Recent studies of **Comet Halley** have revealed that about thirty-three percent of that astronomical body is composed of organic molecules

(Fig. 9.3). **Asteroids**, which may have collided with the early Earth, also appear to contain large numbers of organic molecules. In less destructive fashion, tons of interplanetary dust particles, many of which contained organic material, blanketed the Earth during its formative period and continue to do so today.[58] Supporters of this theory include the famous astronomers Fred Hoyle and Chandra Wickramasinghe and the world-renowned biologists Francis Crick and Leslie Orgel.[59] Hoyle and Wickramasinghe even theorize that this bombardment of organic molecules from space continues today. Such bombardments, they argue, would account for the sudden, inexplicable appearance of new viral strains that lead to worldwide epidemics like those associated with the Asian flu and similar illnesses.[60]

Naturally, the simple depositing of the organic molecules on the Earth would not have, by itself, guaranteed the development of life as we know it on our planet. However, these interplanetary organic molecules may have provided some of the essential raw materials that could have been acted upon by other forces in the primeval environment to supplement the organic compounds that were already developing within the primitive atmosphere.[61] Still, some researchers are eager to go even further. Clifford Matthews of the University of Illinois, for example, is prepared to say that some of the most basic ingredients of protein can develop even more readily and rapidly on asteroids and comets than on Earth. The accumulation of this organic material on Earth via asteroid and comet impacts may have provided a chemical shortcut to the creation of life on this planet.[62]

9-5   LIFE: AN ACCIDENT OR AN INEVITABILITY

Before leaving this chapter, it might be a good idea to explore a question that has persisted through the ages: Is life an accidental occurrence that resulted from chance or is it a natural process that is inevitable whenever and wherever conditions are right? There are certainly arguments on both sides of this issue. However, several facts should remain foremost whenever the question of the origin of life is considered. First, life has existed on Earth for approximately 3.45 billion years, though the solar system is only 4.6 billion years old.[63] Consequently, by the time the early turbulent events in the history of the solar system ended, prelife crossed the threshold to life rather suddenly. Evidence indicates that extremely well-defined cells were abundant during that early time period.[64] Life not only arose very quickly but also most thoroughly. Naturally, there were long periods of time when little or no development took place. However, although the pace may have varied, the overall complexification process did not. Once life began on Earth, there was no stopping it.[65] The rapid appearance of life on Earth could mean that, once a planet has developed the right conditions, life will emerge as an inevitable result of those conditions. The process of the development of life on a properly conditioned planet might be compared to the birth of a star in a properly conditioned nebula.

The Birth of a Star

Most astronomers agree on the process by which a star is born. The raw material for a star, or a star cluster for that matter, is found in the large interstellar clouds of dust and gas, usually called nebulae, that drift about the universe. However, not all nebulae become stars. To have any prospect of actually developing into a star or a star cluster, a nebula must have a density of about 10,000 atoms per cubic centimeter.[66] Anything smaller does not have sufficient mass for the necessary gravitational collapse that always precedes the origin of a star. Not all nebulae with sufficient mass become stars or star clusters either. Shockwaves from some sort of outside event, such as a nearby supernova, must trigger the collapse of the dust cloud.[67] Again, however, not all collapsing nebulae become stars. Those that are to become stars must have a mass of at least 8 percent of the Sun.[68]

A nebula with less than 8 percent solar mass (or 80 times the mass of Jupiter) will experience neither the necessary heat nor the needed pressure to become a star. Instead, it will become a phenomenon known as a **brown dwarf**. A brown dwarf is a substellar astronomical body that does not undergo nuclear fusion and cannot, therefore, be classified as a star.[69] At the other end of the spectrum, nebulae that have too much mass (usually those that exceed 100 solar masses) will have so much energy that they will ignite rapidly and explode suddenly. If such nebulae spend any time at all as stars, that time is very brief.[70] In between these two extremes are those nebulae that have enough mass but not too much and thus become stars. The point here is this: If certain conditions exist regarding a particular nebula, a star or a star cluster will form; if those conditions are not satisfied, neither a star nor a star cluster will form.[71] There is no magic or mystery about it. Natural law, in this regard, is perfectly predictable.

The Origin of Life

In fact, natural law in many respects is perfectly predictable. Many other examples could be added to the one given. For example, the composition of a planet in our solar system is directly related to its position within the original solar nebula that gave birth to our system. The same could be said of the comets at the outer edges of our solar system and the asteroids and meteoroids that inhabit the inner system. On Earth, the formation of mountains and volcanoes can be anticipated by studying plate tectonics and sea floor spreading. All of these are perfectly predictable, perfectly natural events. Could not the same be said for life? It is possible that, as long as certain conditions exist within a star system, life will inevitably arise on a planet properly situated within that star system.

In fact, some astronomers estimate that Epsilon Eridani, a star less than 11 light years from the Earth, may be such a star system. Evidence indicates that

Epsilon Eridani resembles the Sun as it may have existed 4 billion years in the past. At 1 billion years of age, Epsilon Eridani is a relatively young star and, as such, may now be in the same developmental stages that the Sun went through as life began to assert itself on Earth.[72] Although Epsilon Eridani is somewhat smaller, colder, and less massive than the Sun, it has other characteristics that draw some astronomers to the conclusion that it may be the hub of a life-bearing star system. Chief among these characteristics is the fact that it contains sufficient amounts of carbon, nitrogen, oxygen, and iron, all of which are part of the recipe for the formation of planets. Because Epsilon Eridani is smaller and colder than the Sun, a life-harboring planet would have to orbit somewhat nearer than the Earth orbits the Sun.[73]

Nevertheless, a planet located about 108 million kilometers from Epsilon Eridani would be in the right position to provide for the existence of liquid water, another key ingredient for life. Currently, no direct evidence exists to show that Epsilon Eridani has a planetary system. However, observations supplied by the Infrared Astronomical Satellite, indicate that excess infrared radiation is being emitted from Epsilon Eridani, providing circumstantial evidence of some sort of orbiting phenomenon around that star.[74] Whatever the case, the data that has been accumulated regarding Epsilon Eridani indicates that one of the first missions of some real-but-future Starship Enterprise might be to set sail for Epsilon Eridani to explore the possibility that a newly developing planet exists within that star system which could serve as a window to overlook life as it formed on Earth eons ago. In addition, recent findings by Guillermo Gonzales of the University of Washington have indicated that within the galactic plane of the Milky Way there may exist a limited habitable life zone. This galactic habitable zone represents that area of the Milky Way which exhibits the most favorable conditions for the development of life. Interestingly enough, the solar system's galactic orbit follows a path through the central regions of the habitable zone, which appears to be located midway between the galactic center and the edge of the Milky Way. Consequently, when some future Enterprise ventures out of the solar system on its first life seeking journey, it may not have that far to go after all.[75]

CONCLUSION

The question of how life originated on the planet Earth is not an easy one to answer. Still, by taking a careful step-by-step approach, the secret may one day be revealed to a patient research scientist. In the meantime, there are many questions that remain unanswered. One of these questions involves the origin of human life. It would seem that, so far at least, humanity sits at the apex of the evolutionary process. Prelife and life, then, are just the first two stages in a long process, the goal of which is the emergence of a sentient species that can reflect upon, explore, and unlock the secrets of the universe. Human evolution, then, serves as the subject of the next chapter in our quest for the ultimate secrets of the universe.

Review

9-1. To explore the origin of life, it is necessary to begin with a definition of life. To do that it will be helpful first to explore some of the essential characteristics of life. Two distinguishing characteristics of life are complexity and organization. Two additional characteristics are the ability to analyze and respond to outside stimuli and the ability to replicate. Almost as intriguing as the characteristics that make the list is one that does not appear on it—the existence of some sort of "living" atom. At the atomic and subatomic levels, there is no difference between animate and inanimate objects.

9-2. Several theories have been put forth to explain this difference between life and nonlife. Two of these theories, vitalism and mechanism, are the most talked about and, therefore, the most familiar. A third theory, the theory of emergentism, a holistic approach, is much more recent and may, therefore, seem to be a bit revolutionary.

9-3. The most popular theory concerning the early conditions and the primitive forces at work on the prelife Earth describes a turbulent planet, lacking oxygen, bombarded by radiation from space, and churning in the energy produced by electrical storms and volcanic activity. The prelife soup theory goes on to explain the process of the origin of life in the following way: The chemical compounds making up the prelife atmosphere (or perhaps the early oceans) interacted with the active energy sources bombarding the early Earth (any combination of lightning, volcanic activity, and ultraviolet radiation from the Sun) to drive the molecules into ever more complex structures until they reached a stage of complexification that allowed them to feed on their environment and reproduce themselves. These primitive molecular structures then gave rise to more and more complex structures that grew, developed, and evolved as they interacted with their environment, eventually becoming the wide variety of life forms present on the planet today. In the early 1950s at the University of Chicago, two scientists, Stanley Miller and Harold Urey, reproduced the conditions of the early Earth in an experiment and created amino acids, thus offering support for the prelife soup theory. The now-famous Miller-Urey experiment was followed by a series of similar experiments that further supported the theory that life originated as the result of the photochemical processes described above.

9-4. Research into the origin of life continues today. Of course, the prelife soup theory is not the only one put forth for the origin of life. Another popular theory states that the first organic molecules on Earth actually had an extraterrestrial origin.

9-5. Is life an accidental occurrence that resulted from blind chance, or is it a natural process that is inevitable whenever and wherever conditions are right?

There are certainly arguments on both sides of this issue. If certain conditions exist regarding a particular nebula, a star or a star cluster will form; if those conditions are not satisfied, neither a star nor a star cluster will form. There is no magic or mystery about it. Natural law in this regard is perfectly predictable. The arguments for the natural occurrence of life suggest that the same may be said for life. If certain conditions exist regarding a particular planet, life will eventually arise on that planet, and if those conditions are not satisfied, life will not arise.

Understanding Key Terms

amino acids
asteroid
brown dwarf
chemosynthesis
Comet Halley
complexification
cyanobacteria
deep sea vents
deoxyribonucleic acid (DNA)
emergentism
enzyme
eukaryotes
handmade cell
holism
mechanism
metabolism
meteorite
methodological reductionism
Miller–Urey experiment
nebulae
neutrino
ozone layer
panspermia
photosynthesis
precellular organisms
prelife (primeval) soup theory
prokaryotes
protein
protocells
reductionism
ribonucleic acid (RNA)
synthetic protolife (the Metabolism)
ultraviolet (UV) radiation
vitalism

Review Questions

1. What are the characteristics of life?
2. What is the difference between vitalism and reductionism, two of the major theories on the nature and origin of life?
3. What are some of the problems with vitalism and reductionism?
4. What is the complexification process?
5. What conditions and what forces were present on prelife Earth that may have led to the creation of life?
6. How did the conditions and the forces on prelife Earth act together as a prelude to the first appearance of life?
7. What was the significance of the Miller–Urey experiment?
8. What is synthetic protolife? How will the handmade cell be created?
9. What are the essential features of the theory of the extraterrestrial origin of life?
10. What is the point of comparing the creation of a star to the origin of life?

Discussion Questions

1. The chapter points out that four characteristics are necessary for life. They are complexity, organization, replication, and the ability to respond to the environment. Do you agree that these are the necessary characteristics of life? Explain. Which of the three theories for the nature of life do you prefer: vitalism, reductionism, or emergentism? Explain your position.
2. Do you agree with the idea that countless legal and ethical questions might be easier to solve if people could agree on a common definition of life? Do you agree that the abortion debate, the concern over living wills, the controversy involving euthanasia, and the issue of genetic engineering all require answers to the questions "What is life?" and "How did life originate?" Explain why you agree or disagree with this position.
3. What do you think about the feasibility of the primeval soup theory? Are you inclined to accept or reject the theory. Which of the three theories—vitalism, reductionism, or emergentism—do you see as most compatible with the primeval soup theory? Explain.
4. Steen Rasmussen of the Los Alamos National Laboratory in New Mexico and Jack Szostak of the research division of Massachusetts General Hospital are both working on the problem of creating artificial life. After reviewing the accounts of both experiments, determine which of the two you feel is more likely to produce life. Explain how the work of Rasmussen and Szostak differs from the popular, albeit inaccurate, view of the creation of life in a laboratory, as depicted by Mary Shelley in her famous novel, *Frankenstein*. Given that it is possible that either Rasmussen, Szostak, or some other biologist may succeed in creating artificial life, speculate on whether such a project would be an ethically correct undertaking.
5. Speculate on the possibility that life on Earth had an extraterrestrial origin. What do you think about the possibility that life exists within our solar system on a planet other than Earth? What about the possibility that life exists on a planet now thought to be in the Epsilon Eridani system? Speculate on the possibility that life may exist somewhere else in the universe. Should the government spend money on research into the possible existence of life on other planets? Explain. How would the discovery of extraterrestrial life affect religion? Explain.

ANALYZING *STAR TREK*

Background

The following episode from *Star Trek: The Next Generation* reflects some of the issues that are presented in this chapter. The episode has been carefully chosen to represent several of the most interesting aspects of the chapter. When answering the questions at the end of the episode, you should express your opinions as clearly and openly as possible. You may also want to discuss your answers with others and compare and contrast those answers. Above all, you should be less concerned with the "right" answer and more with explaining your position as thoroughly as possible.

Viewing Assignment— *Star Trek: The Next Generation,* "Home Soil"

The Enterprise has been dispatched to Velara III to check on the progress of a group of specialists who are involved in **terraforming** the planet to Earthlike conditions. The away team arrives on Velara III just in time to witness a new phase of the project. The team has begun to pump underground water from beneath the crust of the planet into the artificial lakes on the surface of the planet. Not long after this phase of the project has begun, the terraformer who began the pumping operation is killed when the laser drill malfunctions. The laser also attacks Data when he attempts to remedy the problem. The terraformers are beamed aboard the Enterprise for their own safety. Data and LaForge return to the surface of the planet to investigate the situation further. Their investigation leads them back to the laser drill where they discover a small crystal emitting light. Suspecting that the crystal may have something to do with the malfunction, Data and LaForge transport it to the Enterprise for further research. Once it is on board the Enterprise, the crystal begins to replicate. Captain Picard is certain that the crystal is living, perhaps even intelligent. At that point, the crystal entity suddenly replicates, adding further evidence to Picard's belief. The ship's universal translator inexplicably activates with a message from the intelligent crystal beings. As it turns out, the crystal life form depends upon the underground water as its source of nourishment. Consequently, when the terraformer drained the underground water, he, in effect, attacked the crystal society. The crystal life form has decided to retaliate. It continues to replicate itself, growing in size and in power as it does so. Eventually, the crystal society controls the ship's computer. Data determines that its primary energy source is light. Riker is sent to the medical laboratory where the crystal has been isolated. When he turns down the lighting, the crystal loses energy and almost dies. The crystal makes peace with the crew of the Enterprise, and Picard has it returned to its home environment.

Thoughts, Notions, and Speculations

1. In their discussion concerning the possibility that the crystal may be alive, Dr. Crusher offers the following definition of life: "It must be able to assimilate, respire, reproduce, grow and develop, move, secrete, and excrete." Would you agree or disagree with Dr. Crusher's definition? Explain. In this chapter we identify two additional characteristics of life. These include complexity and organization. Do these

two characteristics coincide with or contradict Dr. Crusher's definition? Explain. Can you think of any nonliving systems that exhibit one or more of these characteristics? Why are these systems not considered alive? Explain.
2. If the crystal's primary source of energy is light, how does the crystal society manage to exist underground in a relatively dark environment? How did the crystal beings manage to survive during the periods of night that occurred on this planet? Do you know of any living things on Earth that can survive in an environment that is devoid of light? What source of energy could such a living thing use for food? Explain.
3. A portion of this chapter is devoted to an exploration of how life arose on Earth. After reading that explanation, speculate on whether the same or a similar process could have taken place on Velara III. How would a Velaran process of evolution be similar to Earth-bound evolution? How would the Velaran process differ from the Earth-bound process? Explain. Some scientists also speculate that life on Earth could have had an extraterrestrial origin. Could this also be true of Velara III? Explain.
4. Another theory proposed in this chapter is the idea that life may not be the result of random chance but may be inevitable given the right set of circumstances. After rereading that section of this chapter, speculate on whether the existence of the crystal life form on Velara III would support this theory.
5. Three theories of life—vitalism, reductionism, and emergentism—are proposed in this chapter as an explanation for the existence of life. After rereading an explanation of each theory, speculate on which one of the three could apply equally as well to life on Earth as to life on Velara III.

NOTES

1. Robert Naeye, "Was There Life on Mars?" 46–49.
2. Paul Davies, *God and the New Physics*, 59.
3. Ibid.
4. Ibid.
5. A.G. Cairns-Smith, *Seven Clues to the Origin of Life*, 2.
6. William K. Hartmann and Ron Miller, *The History of Earth*, 74–75.
7. Davies, 59; Hartmann and Miller, 74–75.
8. Davies, 59.
9. Raymond Daudel, *The Realm of Molecules*, 29.
10. James Trefil, *From Atoms to Quarks*, 237.
11. Ibid., 188.
12. Davies, 60.
13. Hartmann and Miller, 74.
14. Dominique Lacourt, Introduction to *The Chemistry of Life*, by Martin Olomucki, 25–27.
15. Cairns-Smith, 2.
16. Martin Olomucki, *The Chemistry of Life*, 16, 23.
17. John L. Casti, *Searching for Certainty: What Scientists Know about the Future*, 25–27.
18. Steven Goldberg, *Culture Clash*, 8.
19. Davies, 93; Hofstaelter, *Gödel, Escher, and Bach: An Eternal Colder Braid*, 688–92.
20. Rupert Sheldrake, *A New Science of Life*, 12.
21. J. W. N. Sullivan, *The Limitations of Science*, 94.
22. Michael Ruse, *Philosophy of Biology Today*, 24.
23. Sheldrake, 11.

24. Ruse, 25.
25. Francis Crick, *The Astonishing Hypothesis*, 7.
26. Sullivan, 92.
27. Davies, 62; Stuart Kauffman, *At Home in the Universe*, 47–48. See also Stuart Kauffman, "'What is Life?': Was Schrödinger Right?" 83–114.
28. M. Mitchell Waldrop, *Complexity*, 60–61.
29. Kauffman, 47–48.
30. Ibid., 47.
31. Hartmann and Miller, 75.
32. Julian Huxley, Introduction to *The Phenomenon of Man*, by Pierre Teilhard de Chardin, 15.
33. Robert Pollack, *Signs of Life*, 17.
34. Ibid.
35. H. James Birx, *Interpreting Evolution*, 200.
36. Ibid., 30.
37. Ibid., 30. (Note: Amino acids are among those chemical compounds that are essential to life as we know it.)
38. Tim M. Berra, *Evolution and the Myth of Creationism*, 73–74; Christian de Duve, *Vital Dust*, 18–21; Hartmann and Miller, 75–80.
39. Berra, 74–75; William K. Hartmann, *The Cosmic Voyage*, 438.
40. Berra, 74–75; Hartmann, 438.
41. Berra, 75.
42. Berra, 79; Hartmann and Miller, 101.
43. Berra, 77–78; Hartmann and Miller, 104.
44. de Duve, 133–36; Hartmann and Miller, 108–9.
45. Berra, 79; Hartmann and Miller, 103.
46. Hartmann, 439.
47. Birx, 31; Davies, 69; John Gribbin, *In the Beginning*, 57; Hartmann and Miller, 80.
48. Birx, 31–32.
49. de Duve, 19.
50. Hartmann and Miller, 93.
51. Ibid.
52. Berra, 74–77.
53. de Duve, 21–22; Hartmann and Miller, 93–95.
54. David M. Freedman, "Molding the Metabolism," 36, 38.
55. Ibid., 38.
56. David M. Freedman, "The Handmade Cell," 47–52.
57. Ibid.
58. de Duve, 6–7; Julie Paque, "What Makes a Planet a Friend for Life?" 49.
59. de Duve, 6–7.
60. Hartmann and Miller, 89.
61. Yvonne Pendleton and Dale P. Cruikshank, "Life from the Stars," 42.
62. Ibid.
63. Davies, 68; Hartmann, 273, 439; Kauffman, 10.
64. Davies, 68; Hartmann, 439.
65. Kauffman, 10–12.
66. Hartmann, 274–75.
67. Hartmann and Miller, 27.
68. Hartmann, 277–78. (The Sun is used as a yardstick for measuring solar mass. The Sun is said to have one solar mass. A star that has the same mass as the Sun is called a one solar mass star. The astronomical notation would look like this: 1 $M_\odot$. See Hartmann, 259, 262, 476.)
69. Hartmann, 278; Colin A. Ronan, *The Natural History of the Universe*, 78–79.
70. Hartmann, 279.
71. Hartmann, 273–78; Hartmann and Miller, 27–28.
72. Ken Croswell, "Epsilon Erandi," 46.
73. Ibid., 46–48.
74. Croswell, 46–49; Hartmann, 122.
75. George Musser, "Here Come the Suns: Stars with Planets Seem to Harbor 'Heavy' Metals," 20.

Chapter 10
Evolution and the Emergence of Humanity

COMMENTARY: THE AGE OF ROCKS OR THE ROCK OF AGES?

In 1960, the legendary film star Spencer Tracy appeared in the now-famous movie version of the play *Inherit the Wind*. The play is a fictional account of the Scopes monkey trial of 1925. Tracy plays Henry Drummond, a defense attorney whose job it is to defend a high school teacher who has been charged with teaching Darwinian evolution to his science class. In the film the case is prosecuted by a former presidential candidate and Biblical scholar named Matthew Brady, played by Fredric March. Brady, of course is modeled after the attorney, presidential candidate, and Biblical scholar, William Jennings Bryan, who helped prosecute the real Scopes case, while Drummond is clearly patterned after the renowned attorney, Clarence Darrow, who defended the young teacher. At one dramatic point in the film, Drummond asks Brady how old the Earth is. Brady replies that, according to the famous Biblical scholar Archbishop James Ussher, the Earth was created on Sunday, October 23, 4004 B.C. at 9 A.M. Drummond tries to disprove Brady's claim by showing him a fossil in a rock that an Oberlin College professor has calculated to be well over a million years old. Brady cleverly replies, "I am more interested in the Rock of Ages, than in the age of rocks." Drummond presses Brady for a reconciliation of the contradictions inherent in his testimony. "Is that 9 A.M. Eastern Standard Time or 9 A.M. Rocky Mountain Time?" he asks. Brady has no answer. How does he know that the Earth was created on the date he specifies? Again Brady points to the work of Archbishop Ussher, who carefully calculated the date by counting up the ages of the prophets as they appear in Biblical accounts. Brady is satisfied to take this calculation on faith, as are most of the spectators in the courtroom. But the fossil that Drummond has produced is evidence that this conclusion is false.

Both ideas cannot be correct. Which idea are you most comfortable with? Has life existed in its present form unchanged for four thousand years, or has it evolved over millions, perhaps even billions of years? Who thought up the theory of evolution in the first place? What happens if different scientists disagree on the way evolution works? What are some of the competing ideas explaining evolution? These questions and others like them are addressed in this chapter.

CHAPTER OUTCOMES

After reading this chapter, the reader should be able to accomplish the following:

1. Explain the roles of vitalism, reductionism, and emergentism in the definition of humanness.
2. Identify the evidence that led Darwin to the idea of descent and modification.
3. Contrast the role of reproduction and mutation in the natural selection process.
4. Relate the mechanism by which new species appear.
5. Trace the appearance and the development of humanity.
6. Specify the primary trait that pinpoints the threshold of acquired humanness.
7. Define the encephalization quotient, and explain its relationship to the appearance of humanity.
8. Speculate on the relationship between modern humans and the Neanderthals.
9. Correlate decimation with contingency, and explain their relationship to evolution.
10. Clarify the theory of mass extinctions in relation to evolution.

10-1   WHAT IS HUMAN LIFE?

In the last chapter we explored several points of view regarding the nature of life. These theories—vitalism, reductionism, and emergentism—however, represent only the threshold question. Although they offer some basic ideas on the difference between living and nonliving systems, they do not address the most crucial issue of all—the need to determine guidelines that define the nature of human life. In this chapter and the next we explore how scientific research into the origin and nature of human life is shaping our understanding of two controversial problems facing humanity today: procreative rights and genetic engineering.

The search for an answer to the mystery of human life is not yet complete. In fact, the search for an understanding of human life, like the search for an understanding of the origin of the universe, is in a period of revolutionary science. Consequently, we cannot say with a great deal of confidence which solution to the problem will ultimately be the correct one. Whatever the final scientific solution turns out to be, however, scientists and nonscientists alike will be unable to separate those scientific solutions from the impact that they have on the ethical, philosophical, theological, and social aspects of procreative rights and genetic engineering. This book is designed to introduce the nonscientist to the scientific aspects of these issues. Therefore, we focus on meeting these discoveries head on and consider how they might help us unravel the complexities of procreative rights and genetic engineering.

The Concept of Acquired Humanness

The initial question that must be addressed in this chapter is "What makes human life different from all other forms of life?" In our approach to this question, we take a cue from the work of Harold J. Morowitz and James S. Trefil, who in their work, *The Facts of Life: Science and the Abortion Controversy*, address the same issue while focusing primarily on the problems associated with procreative rights. In their treatment of procreative rights, Morowitz and Trefil suggest that the central question that must be addressed is, "When does the fetus acquire those properties that make humans uniquely different from other living things?"[1]

In their quest for a solution to the problems associated with procreative rights, Morowitz and Trefil have settled on what they call the search for "the acquisition of humanness."[2] They believe that this approach is appropriate for two reasons. First, the acquisition of humanness is a relatively impartial phrase that helps to emphasize the objective nature of their study.[3] Second, the question of when the fetus reaches the point of acquired humanness is one that can be resolved completely inside the boundaries of the scientific method.[4] Given this basic premise, Morowitz and Trefil explain that they intend to pursue their study in the following way: "We first look at human beginnings and determine what property or properties distinguish them from other living things. Then, we examine the development of a human being from a single fertilized egg to birth, and determine when those properties are acquired."[5] In this chapter and the next we also attempt to make these determinations; however, we broaden the focus of the study. In addition to asking the question of acquired humanness in relation to the fetus, we ask when, or perhaps, whether, a genetically created psuedohuman hybrid species acquires the essential qualities of humanness.

Vitalism and Reductionism

It will be helpful to determine how each of the theories discussed in the last chapter—vitalism, reductionism, and emergentism—would approach the question of the acquisition of humanness. As we have seen, vitalism attempts to answer the question about what defines life by postulating the existence of a supernatural life force. Unfortunately, this approach gives us very little to work with when attempting to pinpoint the acquisition of humanness. The life force is, by definition, a product of the supernatural world and, therefore, untestable by the scientific method. Consequently, the application of the life force to issues such as procreative rights or genetic engineering is useless. For instance, suppose we try to resolve the procreative-rights debate by using vitalism. To do this, we would have to figure out when the life force enters the fetus, thus giving that fetus its acquired humanness. However, the life force is, by definition, intangible. Therefore, it cannot be detected, and there is absolutely no way to determine when it enters the body.

In another case suppose we attempt to determine when the life force enters a genetically engineered form of life. Again, we cannot solve this problem by using the life force. Since the life force is supernatural, we have no way to detect it and, therefore, no way to use it to measure when life begins. In fact, advances in medical science have made the existence of a life force even more problematic. Consider, for example, a comatose patient kept alive by a life-support system. How can the vitalist explain this process? Has the life-support system somehow trapped the life force so that it cannot leave the body? If so, the vitalist must admit to some sort of interaction between the physical world of the life-support system and the supernatural world of the life force. Yet such an admission would contradict the basic premise of vitalism—the life force is not physical. If the life-support system has not trapped the life force, then presumably the life force can leave at any time. If that is the case, why does it choose to remain only as long as the life-support system forces the lungs to breath and the heart to pump?

The vitalist could argue that the life force has already left the body even though the life-support system continues to keep the body alive. But then, how do we explain the patient who recovers from the comatose state? Did the life force leave and then somehow return? Even if this is the case, vitalism has provided us with no viable answers to the original questions. We are, in essence, back where we started. The life force, which is completely spiritual, cannot be detected; thus, the entire concept is useless in making ethical judgments in relation to euthanasia. In fact, asking someone to determine the beginning or the end of human life by gauging when the life force enters or leaves the body is a little like asking a blind tailor to measure an invisible customer for an intangible suit with a nonexistent tape measure.

At first blush, reductionism might appear to be more promising. After all, reductionism views human life as a set of chemical and physical processes. Therefore, we should be able to determine when those chemical and physical processes begin and when they end. Yet, this approach also proves unsatisfactory. The error of reductionism may be that it attempts to explain life solely in terms of those chemical processes and nothing more. What the reductionist misses is the fact that it is possible for something to be "real" while being neither physical nor spiritual. For example, no one would deny that truth, justice, and honor really exist. However, these qualities are neither physical nor spiritual. They emerge from the action and interaction of individuals within a set of circumstances.[6] In Chapter 9, we observed how the plot of a novel can arise from words without being equal to those words. We suggested in that context that life can emerge from the chemical and physical processes of the body without being equal to those processes. The same can be said for the quality of acquired humanness. The quality is not purely physical and chemical, but neither is it spiritual. Rather, the quality of acquired humanness emerges from the chemical and physical processes.

Emergentism and Acquired Humanness

By focusing on the quality of acquired humanness, the theory of emergentism provides a logical approach in the question, "What makes humanity special?" At some point along the evolutionary path from the original single-celled organism to the present state of human evolutionary development, a line was crossed, at which point the qualities of humanness emerged from a previous life form that was not yet quite human. As we have seen, this is the point that Morowitz and Trefil labeled the acquisition of humanness.[7] They are not the only researchers, however, to come up with this concept of acquired human traits. The noted paleontologist and Catholic priest Pierre Teilhard de Chardin also discusses the concept of acquired humanness in *The Phenomenon of Man*. Teilhard, however, at times calls it the process of **hominisation**. At other times, he refers to it as a "leap of intelligence" or a "threshold of thought."[8] Still, like Morowitz and Trefil, Teilhard sees this stage in evolution as the most profound stage experienced by humanity.[9]

The issues of procreative freedom and genetic engineering both affect the lives of scientists and nonscientists. Before reaching these issues, however, we must consider two preliminary questions. First, what is the mechanism behind the process of evolution? Second, how did human life evolve from these primitive life forms? The remainder of this chapter will be devoted to these questions.

10-2 THE THEORY OF EVOLUTION

Very few words in science cause more controversy than the term *evolution*. Yet, the theory itself does not have an unusual origin. In fact, the development of the theory of evolution and the search for a mechanism behind evolution can be seen as classic examples of the scientific method at work. Accordingly, we first look at Charles Darwin's search for a method behind the evolutionary process. Then we apply that process to our search for acquired humanness. This study includes an examination of the ancestral roots of modern humanity as well as a look at the differences among and the ultimate fates of those ancestors. Finally, we examine two additional views of the mechanism behind evolution. Those two views are the decimation/contingency explanation of evolution and the theory of periodic mass extinctions.

The Concept of Evolution

Nonscientists are familiar with the name Charles Darwin, the naturalist generally credited with (or blamed for) the theory of evolution. Nevertheless, they are surprised to learn that Charles Darwin was not the first individual to suggest that life as it exists on the Earth today is actually the result of a long history of development. Nor did Darwin originate the phrase "survival of the fittest," an expression generally attributed to him. Darwin was not some sort of antireligious heretic who took it upon himself to battle the forces of fundamentalist Christianity. In fact, when Darwin set off on his historic voyage on the HMS Beagle, he was a devout Christian whose stated intention was to demonstrate the splendor of God's creative work in the world.[10] What Darwin can be credited with, however, is determining the mechanism behind evolution, a mechanism that Darwin called natural selection.[11]

The Initial Evidence. Darwin's association with the HMS Beagle began because Captain FitzRoy needed a gentleman friend for companionship, dinner company, and conversation during the long voyage. British naval regulations at that period in history forbade the captain of any vessel from direct contact with the crew, except to discuss the official business of the ship. To maintain his sanity on the five-year mission, FitzRoy knew that he would need someone outside the chain of command with whom he could converse on a regular basis. However, he could not simply invite someone along without that individual filling some sort of official capacity on board ship. As a result, even though the HMS Beagle had an official naturalist on board, the captain advertised for a gentleman naturalist. Darwin applied for the job, and since he and the captain had an instant rapport, he joined the expedition.[12]

Three pieces of evidence faced Darwin as he gathered specimens during the Beagle's five-year voyage. First, he noted that there was an enormous amount of diversity among the living things that he cataloged on the various islands and on the mainlands that he visited during the voyage. There seemed to be no end to the various forms of life. Moreover, the living things that existed on the islands and on the different mainlands seemed perfectly suited to thrive in those environments.[13] Second, Darwin also noted that living things had an amazing number of similar characteristics. Darwin began to suspect that the differences that he cataloged were, while highly diverse, also superficial. Consequently, the real key to understanding life might lay in the similarities rather than the differences.[14] Finally, Darwin realized that the fossil record of the Earth might very well provide a valuable piece of evidence regarding the relationships among the animals that had existed at various points of time in the past.[15]

This last piece of evidence is crucial, because it extends beyond Darwin's own research. Fossil records had originally been discovered in the seventeenth century, but at that time no one knew what they were. It was not until the eighteenth century that fossils were acknowledged to be the remains of living things.[16] Later in the eighteenth century, a geologist named William Smith used fossil records to match up certain types of strata in diverse locations. Later research demonstrated that these fossils represented animals that had disappeared eons ago. However, even more crucial was the fact that an expanding body of data seemed to establish that, as the strata moved forward in time, the fossils found in them advanced in complexity from primitive forms to more elaborate structures.[17] As Del Ratzsch explains in his text *The Battle of Beginnings*, "Once fossils were accepted as organic remains and once they were ordered geologically, it was insistently apparent that the plants and animals that had inhabited this planet had been different in different periods. . . . There seemed to be some sort of progression."[18] As a result of all of this evidence, Darwin became convinced that some sort of progressive movement lay behind the development of life on the Earth. With this understanding in mind, he focused on the mechanism behind that progression.

The Reasoning behind Descent with Modification. The voyage of the Beagle had lasted from 1831 to 1836, and by May of 1837, Darwin was making rough notes on a theory that would explain evolution, which was then called *transmutation*.[19] Despite Darwin's growing belief in the transmutation of species, up to that point in time, the mechanism behind the transmutation process had escaped his understanding. Then in October of 1838, he chanced to read the well-known monograph entitled *An Essay on the Principle of Population* written in 1798 by Thomas Robert Malthus.[20] Instantly, Darwin saw the connection that was to make his ideas on transmutation crystal clear. In his autobiography, he describes the experience in this way:

> I happened to read for amusement Malthus on *Population*, and being well prepared to appreciate the struggle for existence which everywhere goes on from long continued observation of the habits of animals and plants, it at once struck me that under these circumstances favorable variations would tend to be preserved and unfavorable ones to be destroyed. The result of this would be the formation of new species.[21]

The circumstances that Darwin refers to formed the basis of the reasoning pattern that led to his belief that **natural selection** was the mechanism behind evolution. The first circumstance that Darwin saw as critical was the fact that plants and animals produce significantly more offspring than can ever hope to live full lives. Because so many offspring enter into the local environment, they will be compelled to compete for limited local resources.[22] The second circumstance that Darwin saw as critical is that there are distinct differences among all living things, even among the individuals within a given species. These differences mean that some of the individuals in a group will have advantages over other individuals as they compete for the scarce resources.[23]

Darwin concluded that a struggle for existence results when those individuals with advantages in the local environment survive long enough to pass their characteristics on to their offspring. In contrast, those individuals without advantages are unable to compete and, therefore, do not survive long enough to have any offspring. Thus, their characteristics eventually disappear. This is the process that Darwin labeled **natural selection**.[24] Darwin, however, states it much more succinctly when he writes:

> If such do occur, can we doubt (remembering that many more individuals are born than can possibly survive) that individuals having any advantage, however slight, over others, would have the best chance of surviving and of procreating their kind? On the other hand, we may feel sure that any variation in the least degree injurious would be rigidly destroyed. This preservation of favourable variations and the rejection of injurious variations, I call Natural Selection. Variations neither useful nor injurious would not be affected by natural selection, and would be left a fluctuating element.[25]

Darwin used the term natural selection in a metaphorical sense to distinguish it from the type of **artificial selection** that occurs when breeders work to develop a particular trait in a group of plants or animals. The term *natural selection* was meant to imply that the actions and the interactions of the various individuals within a local environment operated to promote the survival of some individuals and not others.[26] However, in contrast to the complex natural selection process, Darwin saw the efforts of human breeders as barely second-rate. Human breeders are severely limited in what they can accomplish, if only because of the short time allotted to them.[27]

Genetic Variants, Reproduction, and Mutations

What Darwin did not know, but what we do understand today, is how variations arise. The variations arise as a result of random changes in the genetic code. A particular genetic variant may be better suited to survive within the environment than another variant. Natural selection occurs when the individuals exhibiting the favorable genetic variant grow to maturity and produce offspring that pass on that genetic variant to their offspring and so on. This process is also known as **differential reproduction**.[28] Some species pass these changes in the genetic code from one generation to the next asexually; others do so sexually. The sexual technique is more advantageous for creating variations because with sexual reproduction there can be a wide variety of mixing and matching among individuals within a particular population.[29]

The genetic variants are made possible by the process of sexual reproduction, in which the genetic information of the mother and father are mixed into a new combination giving rise to variations. This means that the offspring that are produced will be similar to but not identical with their parents.[30] Sexual reproduction, however, is not the only way that these changes can come about. Another phenomenon is mutation. A **mutation** is "a change in the structure or amount of DNA in the chromosome induced by an environmental factor such as cosmic rays, heat, or chemicals."[31] Mutations impact on evolution only when they affect the genetic designs sent to the sperm or the eggs. The problem with mutations, or the advantage depending on your point of view, is that they are totally unpredictable.[32] A mutation may be neutral or harmful to an embryo. However, it could also be helpful. The mutation may actually allow the offspring that carries it to deal with the environment more successfully than other members of the species that have not experienced the mutation. In this way the mutation is passed on from one generation to the next, affecting the evolutionary development of the species.[33]

The Prelife Soup Revisited. It is time to return briefly to the prelife soup theory that we discussed at length in Chapter 9. Recall that the prelife or primeval soup theory explained the move from prelife to life as a series of progressively more complex leaps in organization. During the first stage in the development of life from prelife material, the primeval soup came under the energetic influence of the ultraviolet radiation, volcanic heat, and/or lightning. The interaction of these forces created more and more complicated molecules, leading inevitably to amino acids, and other complex molecules. An important characteristic of these complex molecules was the newly acquired capacity to

10-2 The Theory of Evolution • 319 •

Figure 10.1 Charles Darwin, at about age 46.
 The English naturalist Charles Darwin is generally credited with developing a theoretical explanation for the mechanism behind evolution. That theory, which Darwin first labeled transmutation and later natural selection, was explained in his landmark work *On the Origin of the Species by Means of Natural Selection* of 1859. In Figure 10.1 we see Darwin as he appeared around the time that *On the Origin of the Species* was first published. In contrast, Figure 10.2 depicts Darwin in later years.
Photo Credit: Corbis-Bettmann

replicate themselves. Embedded within this replication process was the newly acquired ability of the complex molecules to pass their characteristics from one offspring to the next. It is this ability to pass characteristics from one generation to the next that provides the underlying mechanism for heredity, mutation, and natural selection, all of which furnish the driving force behind the process of evolution.[34]

• 320 • Chapter 10 Evolution and the Emergence of Humanity

Figure 10.2 Charles Darwin in later life.
Photo Credit: Corbis-Bettmann

The Appearance of New Species

How do the changes that occur through sexual reproduction and mutation add up to the development of an entirely new **species**? The formation of a new species, which is known as **speciation**, requires both genetic variation and reproductive isolation.[35] To form a new species, a **population**, that is a group of interbreeding individuals of the same species, must be split apart, at least temporarily, to such an extent that the individuals in each subgroup of the original population can no longer interact with each other.[36] Each group will continue to pass on genetic variations to their offspring in isolation from one another. If the total of the genetic variations experienced by each group is such that they prevent interbreeding, then two new species will have formed. On the other hand, if the members of the two subpopulations can still interbreed, then they are still a part of the same species.[37]

For example, imagine a population of lizards that has been split into two subpopulations over a long period of time by the formation of an enormous canyon. The two populations can no longer interact with one another. Now imagine that one side of the canyon is covered by a lush forest, while the other is mostly barren rock. Each subpopulation must deal with different environmental conditions. The traits that allow the lizards in the forest to survive may involve changes in coloration, improvements in eyesight, the ability to climb trees, a dietary preference for a certain type of insect, and so on. In contrast, an entirely different set of traits may help the lizards on the barren, rocky side of the canyon. Each subpopulation would pass on a different set of the favorable traits to successive generations of offspring. If, when the two subpopulations meet each other, they can no longer interbreed, two new species will have developed.[38]

The Appearance of Humanity

Now, we reach the most crucial question of all. At what point and under what circumstances did humanity cross that threshold to become a separate species? It is difficult to pinpoint the exact circumstances under which the human branch of evolution separated from the other primates because the data compiled about this period of time is scarce and hard to come by. However, about twenty million years ago, a wide variety of apelike creatures roamed the Earth. Any of these apelike creatures might have been the ancestor of both the modern apes and humanity. One candidate is an apelike creature, known as **Proconsul**, whose fossilized remains were found in Africa. The creature was relatively small by today's standards, perhaps as tall as a modern baboon; but it had a skull and teeth that resembled those found in apes.[39] Moreover, *Proconsul* displays an interesting combination of ancient and advanced traits. For instance, the *Proconsul* hand and arm were very similar to those of the monkey. In contrast, the skull and teeth remained apelike.[40] Because of this mixture of advanced and primitive traits, many anthropologists believe that *Proconsul* is a likely ancestor of apes and humans.[41]

Before moving on, we should identify a few terms that will help us understand the relationship between apes and humans. The term **hominoids** is used by zoologists and anthropologists as a broad classification, including apes (gibbons, orangutans, gorillas, and chimpanzees); a group of apelike creatures called **Australopithecines**, **ape-men**, or **missing links**; and humans. The apes are considered to be in the family Pongidae and are generally referred to as **pongids**. The ape-men and humans are considered to be in the family Hominidae and are grouped together in the classification known as the **hominids**. The ape-men are referred to as *Australopithecus*, and the humans are given the designation **Homo**.[42]

The Transitional Phase. The Australopithecines are the transitional apelike creatures that provide a bridge between humans, nonhuman members of the genus *Homo*, and apes. This transitional group of ape-men appeared about four million years ago. Predating even the genus *Australopithecus* is *Ardipithecus ramidus*, a species that lived in Ethiopia about 4.4 million years ago. However, anthropologists are not certain that *A. ramidus* is an early representative of the evolutionary line that led to humans. Although it is evident that *A. ramidus* lived in the dense wooded areas of Africa, it is unclear whether they were bipedal. Research continues on this question today.[43]

The oldest member of Australopithecines appears to be a species called *Australopithecus anamensis*, which lived in Kenya about 4.2 million years ago. Members of *A. anamensis* were apparently bipedal and made their home in the grasslands and wooded areas of Africa.[44] Next in line is an important pivotal species referred to as *Australopithecus afarensis*. Perhaps the most well-known member of this group is a female skeleton known as "Lucy," which was discovered in Ethiopia over twenty years ago.[45] *Australopithecus afarensis* lived in Africa 3.2 million years ago. Members of this species were bipedal and made their homes in forests and bushlands. Despite the fact that their evolutionary line may have led to humans, *A. afarensis* still exhibited a number of apelike characteristics and is not considered within the same genus as humans.[46]

A later representative of the Australopithecines, known as *Australopithecus africanus*, lived from 2.8 to 1.9 million years ago. From *A. africanus* came another species known as *Australopithecus robustus*, which lived from 2.0 to 1.5 million years ago, and a third species *Australopithecus boisei*, which lived from 2.0 to 1.2 million years ago. Some evidence suggests that *A. africanus*, *A. robustus*, and *A. boisei* are not ancestral to humans but are part of two different branches that share, along with *Homo*, *A. afarensis* as a common ancestor.[47]

In fact, it may be that *A. afarensis* is a common ancestor of three evolutionary lines of development, only one of which eventually led to humans. One of the other two lines led to *A. africanus* and *A. robustus* and the third led to *A. boisei*. Evidence for the third line leading to *A. boisei* came in the form of a discovery known in popular literature as the Black Skull, but referred to in technical journals as KNM-WT 17000 (Kenya National Museum-West Turkana 17000). This skull may represent a different evolutionary line. This species has been named *Australopithecus aethiopicus*. Some anthropologists believe that it is ancestral to *A. boisei*.[48]

Whatever the case, the appearance of the Australopithecines heralded a significant divergence in the history of human evolution. The branch of evolutionary development that began with the Australopithecines eventually led to human beings. In fact, humans are the only representatives of that branch alive today.[49] At least that is the way it appears now. New fossil discoveries are being made at a rapid rate, but the record is still incomplete and not all anthropologists and zoologists agree with this picture. In fact, some very reputable anthropologists, including the noted paleoanthropologist Richard Leakey, are not convinced that *Australopithecus* is a direct ancestor of humanity.[50] Instead, Leakey believes that a common ancestor predating the Australopithecines has yet to be discovered. He places this ancestor 6 million years in the past.[51]

Still, the view that *A. afarensis* led to **Homo sapiens** persists, and it is the view that we will adopt as the basis for the remainder of our discussion. It is also the view that Morowitz and Trefil have settled upon as the basis of their search for the qualities present in the acquisition of humanness.[52] What, then, are the characteristics that have led some anthropologists and zoologists to the conclusion that *A. afarensis* was the transitional bridge to humanity? First, there are some obvious physical characteristics that would demonstrate that this group was different from earlier primates. One conspicuous difference was that the Australopithecines had the ability to walk erect on two feet. This is an important evolutionary development because it left the hands unencumbered and allowed them to be used for other tasks. Another noticeable physical difference was in the shape of the face. The Australopithecines had much flatter faces, which helped to distinguish them from their cousins on the nonhuman branch of the evolutionary tree. Another less obvious, but nonetheless significant, change involved the shape of the sinuses. So, as we trace the development of humanity, we note that the farther we advance along that evolutionary tree the fewer characteristics these early humans share with their neighbors.[53]

The Road to Acquired Humanness. Approximately two million years ago, *Homo habilis* appeared in Africa. *Homo habilis* is considered to be the first genuine species of the genus *Homo*. Members of the species *H. habilis* are also referred to as the toolmakers and the handymen because they used crudely devised stone tools. Members of *H. habilis* were relatively short, standing approximately 3.5 to 5 feet tall. Of greater significance than their height, however, was the size of their skulls. In 1972, Richard Leakey discovered a skull belonging to a member of *H. habilis* with a cranial capacity of 775c, a size that clearly places this species within the genus Homo rather than Australopithecus.[54] Not all anthropologists are satisfied with this classification, however. Some experts would still prefer to see *H. habilis* classified as *Australopithecus habilis*. Such disagreements however, are to be expected in a field of science that must deal with discoveries that frequently change our understanding of the fossil record.[55]

Following *H. habilis*, which disappeared about 1.5 million years ago, is the species that immediately preceded *H. sapiens*. This species, which has been designated *Homo erectus*, was the first of the human line to migrate from Africa into Asia and Europe. As the name implies, *H. erectus* could walk upright and had a brain capacity that was about 70 percent that of modern humans. *Homo erectus* may have had the power of speech and was the first of our kind to use fire. *Homo erectus* led to *H. sapiens*—exactly when is problematic. However, estimates range from 500,000 to 250,000 years ago.[56]

As noted, the distinguishing characteristic that set *Homo* apart from the other primates was their much enlarged brain capacity. Although many of the other physical characteristics are critical indicators of a new branch point in evolutionary progress, it is the development of the large brain, specifically the enlarged **cerebral cortex** (the outer layer of the brain, sometimes called the neocortex), that was the most crucial event.[57] The development of an enlarged cerebral cortex is found only in *Homo*. This enlarged brain capacity gave them a distinct advantage over their cousins. Eventually, it helped them develop tools and weapons, as well as the ability to exploit the environment in ways that best suited their survival.[58] This enlarged brain capacity probably also gave rise to the human sense of self as a unique and distinct entity.[59] As soon as the larger cerebral cortex had developed, other talents including the ability to analyze and the capacity to think abstractly were just around the evolutionary corner.[60]

The Encephalization Quotient. The use of brain size alone as a distinguishing characteristic in our search for acquired humanness is not, however, without its difficulties. Chief among these difficulties is the fact that some animals, notably whales and elephants, have brains that are somewhat larger than the human brain. At first glance, this may appear to argue against the idea that the size of the brain can make a difference in drawing the line between human and nonhuman characteristics. This is not the case, however, when we consider the development of the brain in relation to the overall size of the body.[61]

To determine the nature of the relationship, we must use an equation designed to compare the actual size of the brain with its expected size. The relationship between the actual size of the brain and its expected size is called the **encephalization quotient** (EQ). The higher the EQ the larger the brain in relation to what is expected, given the overall size of the animal. The baseline measurement is an EQ of 1. An animal with an EQ of 1 would have a brain size that is exactly equal to the size predicted by the equation. An elephant, for example, has an EQ of 1.03. This means that the brain of the elephant is almost exactly what would be expected given its relative size. In contrast, humans have an EQ of 7.0. This means that, given the relative overall size of humans, they have a brain that is 7 times the size that would be expected.[62]

In a recent study entitled *Humans before Humanity*, Robert Foley, the noted British biological anthropologist, explains the importance of the EQ measurement, when he writes, "EQs do demonstrate that humans have brains that are in size as unique as the behaviours which they generate. Primates appear to be generally large-brained and confirm an intuitive understanding of them as intelligent animals."[63]

A critical distinction comes into play here, a distinction that is directly tied to the emergent view of life and human nature. The importance of the development of the cerebral cortex is not in the fact that humans possess it. All mammals share this feature. The critical distinguishing characteristic is in the size of the human cortex. It appears that the size has allowed the cerebral cortex to become more and more elaborate and thus capable of functions that are impossible with smaller versions of the same structure.[64] As Morowitz and Trefil explain, it is within the cerebral cortex, the outer layer of the cerebrum, that, "the properties that distinguish human beings from the other animals are ultimately found to reside. In the human brain, specific areas of the cerebral cortex are involved in speech, conscious movement, the processing of visual information, taste, and other sensory information from different parts of the body."[65]

These functions are possible, however, only when the brain reaches a certain degree of complexity, and this is possible only when the cerebral cortex reaches a certain size. When the brain does, in fact, reach that size, the higher intellectual functions can emerge from the complex activity within the brain.[66] "Something qualitatively different happens when the cortex gets to a certain size—something that only human beings have so far attained. In the language of biologists and engineers, the distinctly human functions of the cortex are *an emergent feature* of brain development" (emphasis added).[67]

It is also enlightening to compare the EQ of modern humans to that of *Homo erectus*. Naturally, calculating the EQ of *Homo erectus* is complicated by the fact that we must depend on estimates of the relative sizes of the body and the brain. Such estimates rely on a determination of average body mass from the calibration of individual bones. Although this is not the most reliable of methods, it is the best we can do given the present state of technology. Even taking this limitation into consideration, the results are interesting. *Homo erectus* appears to have had an EQ of from 3.5 in the early stages of development to 4.0 in the later stages. This indicates a rapid development in the complexity of the brain especially given the fact that at their highest state of development the Australopithecines had an EQ of only about 2.9.[68]

The Mystery of the Neanderthals. Another interesting aspect of human development involves some unanswered questions about the place of the **Neanderthals** in the long history of human evolution. The Neanderthals, who lived between 35,000 and 150,000 years ago, had a remarkably advanced culture that involved toolmaking, cloth making, herbal medicine, and elaborate burial ceremonies that indicated some concern for the spiritual world.[69] Interestingly enough, there is also some evidence that the Neanderthals also had a brain that was some 10 percent larger than that found in modern humans.[70] Unfortunately, the Neanderthals have been plagued with an undesirable image as a stooped-over hulking brute, with a massive skull, beetle brows, and a malevolent disposition. This inaccurate image almost certainly resulted from the fact that some of the first Neanderthal remains that had been unearthed belonged to individuals who suffered from severe arthritis. The resulting reconstructions of these arthritic individuals contributed to this incorrect image.[71]

The question of whether the Neanderthals were our direct ancestors or were actually part of a different species has yet to be answered. There is persuasive evidence on both sides of the issue. Richard Leakey, for instance, cites the discovery of human fossils 40,000 years older than the majority of Neanderthal fossils as solid evidence for the conclusion that "Neanderthals cannot be ancestors of modern humans."[72] Stephen Jay Gould agrees with Leakey on this point noting that the Neanderthals probably did not donate anything to the genetic development of modern humans, although they may have been our distant European relatives. Gould is convinced that human ancestry, on the other hand, leads back to Africa.[73] In contrast, in *The History of Earth*, William K. Hartmann and Ron Miller refer to the Neanderthals as a "variant" of *Homo sapiens*.[74] Similarly, Christopher Wills in his recent study, *The Runaway Brain: The Evolution of Human Uniqueness*, states that the Neanderthals are probably members of our species. Accordingly, he prefers the designation **Homo sapiens neanderthalensis**.[75]

Tim Berra, a zoologist at Ohio State University, on the other hand, presents a more balanced view, suggesting that the jury is still out in the question of the relationship between Neanderthals and modern humans.[76] Accordingly, Berra suggests several possibilities. For example, the Neanderthals could have actually evolved into modern humans. Or the Neanderthals might have been a separate species that was replaced by modern humans. The third possibility is that modern humans interbred with the Neanderthals and, by virtue of greater numbers, simply absorbed them genetically.[77] Robert Foley presents a similarly well-balanced approach to the question of the relationship between modern humanity and the Neanderthals, noting that the Neanderthals, "may be a separate line in the development compared with anatomically modern humans."[78] Additional evidence supporting the idea that the Neanderthals represent a separate species came in 1997 when researchers in Germany revealed the results of a comparison made between human and Neanderthal DNA. The study involved a 379 base-pair sequence taken from a Neanderthal skeleton. When compared to human DNA, the Neanderthal DNA turned out to be unlike that found in modern humans. Although this finding is not the final word on the subject, it represents a very persuasive piece of evidence.[79]

The possibility that the Neanderthals represented a separate intelligent species that coexisted with our human ancestors can lead to some interesting philosophical and ethical speculation. The immediate question that comes to mind is why did the Neanderthals become extinct? Some individuals have suggested that the extinction of the Neanderthals is one of the earliest examples of genocide. This image of the modern humans invading the territory of the Neanderthals and using their superior weapons and their advanced intellect to destroy that evolutionary dead end is both compelling and terrifying.[80] It is also probably false. A more likely scenario has been suggested by Richard Leakey who indicates that the Neanderthals were probably victimized by overspecialization. According to Leakey, over many thousands of years the Neanderthals had become biologically, behaviorally, and culturally adapted to dealing with the harsh life on the glaciers of the last ice age. That adaptation was so ingrained in them that when a warmer climate asserted itself, the Neanderthals simply could not cope with the changes and gradually died out.[81]

Another intriguing question involving the Neanderthals is based on the suggestion that they did, in fact, represent an entirely different species. If this were true and the Neanderthals had not died out, the historical development of human civilization would have been drastically different. What would be the philosophical and ethical relationship between two different intelligent species inhabiting the same planet? Would one species assume the dominant position, or would they cooperate? Would the Neanderthals be granted "human" rights, or would they be categorized with the apes and other lesser species? What direction might religion have taken had humanity shared the globe with another intelligent species? Many religions are based on the assumption that humans are created in the image of God. What happens to this assumption when humans are compelled to share the Earth with another intelligent species? Or, to put it bluntly, would Neanderthals have souls? Would the killing of a latter-day Neanderthal be considered murder?

Such questions invade an uncomfortable, ambiguous area of human ethics. Right now we do not have to deal with such issues because the Neanderthals did, in fact, become extinct. However, rapid developments in areas of medicine and biology concerned with creating artificial life and the development of cloning technology may lead us to a future world inhabited by an intelligent, nonhuman species. That is why the question of acquired humanness is so crucial to an examination of such moral issues. This study will be the objective of the next chapter. Before moving on to that chapter, there is one final issue that we must consider in our study of evolution. That issue is the relationship between the theory of progressive evolution and the theory of mass extinctions.

10-3 DECIMATION, CONTINGENCY, AND MASS EXTINCTIONS

Can the development of Darwin's theory of evolution by natural selection be seen as the creation of a new paradigm within science? Clearly Darwin initiated some revolutionary ideas on the development of life on the Earth when he established his theory of evolution based on natural selection. Moreover, if we review the state of biology and geology in the eighteenth century, we will be compelled to admit that Darwin's theory did much to resolve some of the troubling anomalies that had plagued scientists in both biology and geology. We need only review how Darwin's work helped unravel the puzzle of the fossil records to recall that this is true. It is also indisputable that the theory of evolution has undergone close scrutiny over the last century. Given the extensive testing that evolution has been subjected to and the fact that it has been repeatedly verified, it is relatively safe to conclude that we have entered a period of normal science during which anthropologists, zoologists, and paleontologists are working out the details of the theory of evolution.

The Decimation/ Contingency View of Evolution

This is not to say, however, that the theory of evolution has sailed along smoothly without facing any challenges. On the contrary, no modern scientific theory, with the possible exceptions of the big bang theory and quantum physics, has faced more challenges. However, none of these challenges has attacked the very nature of the evolutionary process. Instead, each challenge focuses on the mechanism behind evolution or the rate at which the evolutionary process occurs.[82] As is usually the case, each of these challenges arises from an anomaly that appears to contradict or at least to question seriously the accuracy of the traditional theory of evolution. One such anomaly involves a bed of strange fossils that exists in British Columbia. This group of fossils, known as the Burgess Shale fossils, is the focus of an exhaustive study written by the noted paleontologist Stephen Jay Gould.[83]

In this study, which is entitled *Wonderful Life: The Burgess Shale and the Nature of History*, Gould challenges the traditional view of evolution as a smooth progression of ever increasing complexification. According to Gould, while complexification does take place, the process is by no means smooth, upwardly mobile, and uninterrupted. On the contrary, the evidence presented by the Burgess Shale fossils indicates that the development of life began with a group of animals that featured a vast variety of complex anatomical designs, most of which have long since vanished. Thus, instead of looking at evolution as a process of diversification and growth, Gould sees it as the elimination of most early anatomical designs, followed by a vast diversification within the designs that survived the elimination process. For reasons that will become clear as we move along, he calls this process **decimation**.[84]

The **Burgess Shale fossils** are not a new discovery. Instead, they represent a reexamination of a previously cataloged group of fossils that were located and first cataloged in 1909 by Charles Doolittle Walcott, an American paleontologist and the director of the Smithsonian Institution.[85] Walcott's classification process was based on a model that saw evolution as the progressive development of more and more intricately designed anatomical structures. He believed, as did most of his contemporaries and as do most biologists today, that evolution is a gradual process of increasing diversity and complexity. Gould explains that, because of this preconceived notion, Walcott forced his classification of the Burgess Shale fossils to fit into a view of evolution based upon a model called the **cone of increasing diversity**. According to the cone of increasing diversity model, at its origin life is limited and undeveloped, but as evolution advanced, living things became more and more complex and, therefore, improved as time went by.[86] In the cone of increasing diversity model, the cone is frequently depicted upside down, that is, with the apex, or point of the cone facing downward. At the inverted apex of the cone we find a single, common ancestor from which all the complex organisms on the branches have evolved.[87] This starting point, that is, the inverted apex, divides, but only into a relatively limited number of branches. Each of these branches then divides over time, leading to many more categories that are actually subdivisions of the initial groupings.[88]

With this model in mind Walcott cataloged the Burgess Shale fossils by forcing them into accepted categories. All of the fossils had to be classified close to the inverted apex. This meant that the fossils had to be seen as primitive, simple ancestors of modern animals. Gould's extensive study reviews Walcott's cataloging efforts and reevaluates the fossil record in light of a recent reclassification of those fossils by the noted paleontologist Harry Whittington of Cambridge University and his graduate students Derek Briggs and Simon Conway Morris. Instead of showing that the Burgess Shale fossils are simple ancestral forms of existing animals, Gould suggests that Whittington's work demonstrates that the fossils represent a wide range of unique structures that blossomed forth suddenly during a period of original diversification. From that point of early diversification, the evolution process eliminated many of the forms.[89]

Thus, the driving force behind evolution is not expansion through diversification, but diversification followed by eradication. According to this view, when multicellular life forms first appeared on the planet, they burst into life in a widely diverse number of different anatomical forms so that at this early stage, life arrived at a high point of anatomical diversity. From that stage forward, many of these anatomical designs were removed allowing the few remaining ones evolve into those that exist today.[90] This decimation process, according to Gould, immediately followed the point of which life had reached its highest point of complexification. After this first stage, eradication, rather than elaboration, ruled evolution.[91]

The next question that might be asked is what process was involved in the elimination of the anatomical designs that disappeared. According to Gould, the driving force behind the elimination process is **contingency**. Gould believes that any small, unpredictable event along the path of evolution might have deflected the selection process in a different direction. Thus, he sees nothing inevitable about the particular evolutionary history that the Earth has experienced. In fact, he believes that chance events rule the process of evolution. It is this dependency upon chance in the evolutionary process that Gould chooses to call contingency. In other words, the pattern of life that exists on the Earth today results from an unpredictable series of contingent events. These contingent events eliminated some of the early structural designs and allowed others to survive, thus reducing the available anatomical structures to the few that do in fact give rise to modern anatomical designs.[92]

Gould further suggests that, if we erase, rewind, and replay the so-called tape of life's history, that another set of contingent events would eliminate a different set of anatomical designs, and life would have evolved in an entirely different direction. As a result, the living world would look quite different today. Erase, rewind, and replay the tape a second time, and you get a different result. Each time the process is repeated, a different set of contingent events occurs, giving rise to a new and entirely different history.[93] Does this view mean that all evolutionary steps are completely reduced to blind chance? Gould is quite emphatic in denying such a fatalistic view of evolution. He prefers to think that some events in the development of life are inevitable. He firmly believes, for instance, that the precise set of physical circumstances that existed on the primitive Earth set the stage for the eventual rise of life.[94] Gould even goes so far as to say that the rules of good design and the dictates of effective construction will limit the types of body forms that will develop in multicellular organisms despite contingent events.[95] Still, Gould firmly believes that there are limits to what is inevitable. For example, the rules may say that whatever evolves will be **bilaterally symmetrical**, but those same rules would not specify any anatomical details that adhere to or flow from that need for bilateral symmetry.[96]

Naturally, the idea of contingency would also apply to human evolution. Gould suggests that, had an unpredictable contingency event wiped out *H. sapiens*, a replay of the tape of life would fail to give rise to any intelligent species. Gould does not support the idea, promoted by several other paleontologists, that either the Neanderthals or some other representative of humanity would have filled the niche left by the disappearance of *H. sapiens*. On the contrary,

Gould believes that only *H. sapiens* showed the necessary tendency toward the development of the higher brain functions necessary to the development of acquired humanness. According to Gould, neither *Homo erectus* nor the Neanderthals exhibited the necessary genetic disposition toward mentality at the modern human level. Instead, *H. sapiens* appeared as a separate genetic oddity, alone possessing the predisposition that could lead to the level of intelligence found in modern humans. Gould labels this approach to human evolution the **entity theory**.[97]

Gould does admit, however, that the entity theory is not the only theory that purports to explain the evolutionary rise of humanity. Another theory, which Gould labels the **tendency theory** holds that the tendency toward intelligence was present in all forms of primitive humanity. In this view then, the evolutionary rise of *H. sapiens*, or some similar cousin with the same intellectual predisposition, was inevitable once a certain level of development was reached.[98] This level of development was the emergence of the ability to walk upright, which led to better hand–eye coordination, which then led to the enlarged brain capacity and to the appearance of a larger cerebral cortex. As soon as the larger cerebral cortex had developed, other talents, including the ability to analyze and think abstractly, were just around the evolutionary corner. Whether those abilities would ultimately be developed by *H. sapiens* or some closely related cousin would not matter to those who support the tendency theory. The point is not who would develop those abilities but rather, once the tendency had appeared, that it would continue to develop somehow.[99]

An even stronger version of the tendency theory has been proposed by Simon Conway Morris, a paleontologist at Cambridge University, who has directly challenged Gould's decimation/contingency theory. Morris has pinpointed numerous instances in which paleontologists, himself included, have misclassified fossils that should have been identified as falling within the line of evolutionary convergence. From these discoveries, and others like them, Morris has concluded that Gould's decimation/contingency theory may be seriously flawed, at least to the extent that it implies that evolutionary trends result from random chance. Instead, Morris proposes that certain evolutionary trends are inevitable, regardless of contingent events. In fact Morris goes so far as to suggest that intelligent life would have eventually evolved on Earth, even if every human ancestor had been destroyed by one of Gould's contingent decimations.[100] Accordingly, this view could be labeled the **strong tendency theory**.

The Mass Extinction View of Evolution

The Burgess Shale fossil record is not the only anomaly to challenge the mechanism of evolution. Another recent anomaly is the mystery of **mass extinctions**. We have already noted that, while the development of life on Earth does involve a progression of ever increasing complexity, that complexification process has at times been interrupted. The Burgess Shale fossils may, for instance, represent the destruction of a large number of anatomical designs by some sort of large-scale catastrophe. It now appears that such mass extinctions may have occurred more frequently than was previously suspected. These mass extinctions represent sharp breaks in the continuity of the fossil record upon which so much of Darwin's original work was based. As if this mystery were not puzzling enough, there is also some evidence that these mass extinctions take place with regularity. A few studies show that the Earth undergoes a mass extinction roughly every twenty-six million years.[101]

The most widely accepted theory, as of this time, on the cause of these mass extinctions suggests that the Earth is periodically bombarded by asteroids or comets that result in enormous destruction that leads to the mass extinctions. This extraterrestrial theory, as it is sometimes called, was first presented by Luis Alvarez in 1979, when he suggested that the impact of an asteroid with the Earth had caused the extinction of the dinosaurs. According to Alvarez, 65 million years ago an average-size asteroid (approximately 5 miles in diameter) traveling at approximately 20 miles per second collided with the Earth, releasing a burst of energy equivalent to an explosion of a 100 million megaton nuclear bomb.[102]

The immediate result was a crater 60 miles wide and 20 miles deep, from which flew an enormous amount of burning material that later rained down on all parts of the Earth like a battery of intercontinental ballistic missiles. Hot molten rock also flowed from the crater and shock waves from the initial impact caused earthquakes and tidal waves on all parts of the Earth. New volcanoes would also have been created at unstable sites on the Earth's surface. The fireball that resulted from the initial impact would have cooled as it reached the upper atmosphere, creating a dust cloud that would have blocked out the sunlight. Once the sunlight disappeared, photosynthesis would cease, and most plant life would die out causing the plant eaters to starve. The carnivores would soon follow when there were no plant eaters left within the food chain.[103]

Although these conditions were obviously quite horrible and ultimately deadly to the dinosaurs, they obviously did not wipe out all life on the planet. A wide variety of plants survived the period of cold and darkness and grew from spores, seeds, and roots that had not been destroyed. Most animal life was not that lucky. Evidence suggests that only those land-bound creatures that weighed less than 50 pounds survived the cold, dark "nuclear winter." Birds did not perish probably because they retained the ability to fly, which allowed them to find food far from their original habitat. Many small mammals persevered because they simply went into hibernation as they might have during a normal winter. Once the dust began to clear and the sunlight began to filter in once again, these animals discovered that they now lived in a paradise of vegetation that had been swept clean of their most serious enemies, the dinosaurs. As a result, the survivors flourished, repopulating the Earth with their descendants.[104]

This view of the evolutionary process is reminiscent of Gould's decimation theory. The difference is, of course, that Alvarez has focused on the mechanism of the mass extinction, rather than the evolutionary process of moving from diversification to elimination, that is, Gould's process of decimation.[105] Nevertheless, the idea that evolution is based on adaptation to catastrophic events is supported by both Alvarez and Gould.[106] Of course, what Gould would immediately point out is that there is no way to know in advance what characteristics are advantageous for surviving a particular type of catastrophe. Therefore, the traits that help the survivors make it through the catastrophe may have nothing to do with why those traits were developed in the first place. According to Gould, this would, of course, simply be another example of contingency at work. In the case of the asteroid impact, for example, the mammals survived because they were small and could hibernate, whereas the birds survived because they were mobile. Paradoxically, the traits that allowed the dinosaurs to dominate the Earth for millions of years may have caused their extinction.[107]

One anomaly that threatened the Alvarez asteroid theory involved data compiled by two well-known paleontologists from the University of Chicago, David Raup and J. John Sepkoski. Raup and Sepkoski compiled data indicating that mass extinctions do not happen at random. Rather, the evidence showed quite clearly that mass extinctions occur every 26 million years.[108] This was the anomaly that Luis Alvarez handed over to Richard Muller, a professor of physics at the University of California at Berkeley, one morning in 1983.[109] The anomaly clearly threatened the asteroid impact theory of mass extinctions that had been put forth by Alvarez. After all, random asteroid impacts might cause a single mass extinction or even a series of mass extinctions. The problem was not in the multiple nature of the mass extinctions, but in the regularity of those extinctions. Asteroid impacts might explain several mass extinctions, but a regular pattern of mass extinctions could not be attributed to the random nature of asteroid activity.[110]

As we have seen earlier in the text, when scientists are faced with an anomaly, there can be several outcomes. For instance, the anomaly may be explained by readjusting the paradigm in some small way. Such adjustments preserve the paradigm and the anomaly is accounted for. The anomaly then becomes a part of the predictive pattern for the paradigm. These types of minor anomalies result in emerging discoveries that readjust the shape of the paradigm. Some parts of the paradigm may be destroyed, but other aspects are constructed in their place. The overall paradigm benefits because it is the emerging discovery that has shaped it in such a way that it is now a much more accurate representation of nature.[111]

This describes almost precisely what happened in the case of the Raup/Sepkoski anomaly. Muller attempted to readjust the Alvarez hypothesis by adding another factor to the equation. Muller suggested that the Sun may have a small companion star that has some sort of strange gravitational effect on the Earth or the asteroid belt every 26 million years. This would demonstrate why the asteroid impacts occurred in such a regular pattern, thus preserving the Alvarez mass extinction theory.[112] The companion-star solution had the added benefit of explaining the fact that the mass extinction of the dinosaurs was accompanied by a rhenium deposit that indicated that the impacting body had originated within the solar system. Several types of rhenium exist. One type is the unstable rhenium-187. Rhenium-187 has a half-life of 40 billion years, which means that about 8 percent of the rhenium-187 must have decayed since the solar system was created about 4.5 billion years ago. This 8 percent factor was found in the rhenium deposit associated with the asteroid impact that was responsible for the extinction of the dinosaurs. It is not likely that an extraterrestrial body would have this same 8 percent decay factor unless it had been created during the same era that saw the creation of the solar system. It, therefore, made sense to assume that the hypothetical companion star operated solely within our solar system.[113]

The companion-star hypothesis went through several incarnations until Muller and Alvarez settled on the idea that the star's orbit had a gravitational effect on comets rather than asteroids. **Comets** consist mostly of frozen water, ammonia, and methane. However, mixed in with that frozen material is an enormous amount of rock and dust. In fact, the composition of most comets is at least one-third rock, dust, and minerals. Consequently, a sufficiently massive comet colliding with the Earth would produce the same destructive force as an asteroid. Additionally, there is an enormous storehouse of comets located approximately 0.5 to 1 light year from the Sun. This storehouse of comets, which is known as the **Oort Comet Cloud**, contains from 100 billion to 10 trillion comets. Muller and Alvarez suggest that the Sun's companion star, now designated **Nemesis**, might pass through the comet cloud every 26 million years. The gravitational shock caused by the passage of Nemesis would disturb thousands, probably millions, perhaps even billions of those comets kicking them into the inner solar system. During the passage of Nemesis through the comet cloud, a process that would last about one million years, the inner solar system would be subjected to a shower of comets, drastically increasing the odds that one or more of these comets would hit the Earth.[114]

Another explanation for the periodic comet showers that bombard the inner solar system has been proposed by Michael R. Rampino and Richard Stothers of NASA's Goddard Institute for Space Studies. Rampino and Stothers believe that the gravitational disruptions which disturb the comet cloud may correspond to the cyclical movement of the solar system up and down through the galactic plane of the Milky Way. This theory has been referred to as the **carousel theory** because the movement of the solar system as it rotates about the galactic center resembles the activity of a wooden horse on a circular carousel. According to the carousel theory, every 30 million years, a figure which now appears to measure more accurately the period between extinction level events on earth, the solar system moves through the densest part of the galactic plane. Gravitational disruptions caused by this movement may be severe enough to affect the Oort Comet Cloud in such a way that billions, perhaps even trillions, of comets are kicked toward the inner region of the solar system increasing the probability of extinction level collisions with the Earth.[115] The advantage of this explanation over the Nemesis theory is that it depends upon a known cyclical event rather than the existence of a yet undiscovered companion star for the Sun.

Speculation about the cause of these periodic mass extinctions did not cease with the advent of either the Nemesis theory or the carousel theory. Another recent theoretical position has suggested that one of the causes of such extinctions might be related to gamma-ray bursts that have been detected by NASA's orbiting Compton Gamma Ray Observatory. According to some experts, these gamma-ray bursts may be caused by the merging of two neutron stars. One variation of gamma-ray burst theory, proposed by Nir Shaviv and Arnon Dar of the Israeli Institute of Technology, suggests that under certain conditions the gamma-ray burst caused by a merger between two neutron stars may be followed by an energetic burst of cosmic rays. If a merger of this type were to occur within 3000 light years of the solar system, the resulting outburst of gamma and cosmic rays might be sufficient to destroy most life on the Earth. Figures compiled by Shaviv and Dar indicate that, within the 3000 light year danger zone, such collisions may occur about every 100 million years.[116] Although this figure does not match the 26 to 30 million year cycle of extinctions, since the final word on the cause of the mass extinctions has yet to be uttered, it pays to keep an open mind about all possibilities.

CONCLUSION

The search for a mechanism behind evolution is a classic example of the scientific method at work. Charles Darwin's search for a method behind the evolutionary process began with the accumulation of evidence on his five-year mission on board the Beagle and ended when he managed to piece the puzzle together with a little help from the work of Thomas Malthus. As we have just witnessed, however, Darwin's work represents just the start of the search for the origin of life in general and human life in particular. The search for the origin of life through evolution has given us a means for distinguishing human life from other forms of life on the planet. This dividing line, that is the acquisition of humanness, will be the focus of the next chapter, as we explore procreative rights.

Review

10-1. The initial question that was addressed in this chapter was "What makes human life different from all other forms of life?" In our approach to this question, we took a cue from the work of Harold J. Morowitz and James S. Trefil, who in their recent work *The Facts of Life: Science and the Abortion Controversy* address the same issue while focusing primarily on a search for the acquisition of humanness. Vitalism attempts to answer this question by postulating the existence of a supernatural life force. Unfortunately, this approach gives us very little to work with when attempting to pinpoint the acquisition of humanness. Reductionism might appear to be more promising than vitalism in the search for acquired humanness. However, reductionism often fails to take into consideration that life, while dependent upon chemical and physical processes, need not be reduced to those processes. In contrast, by focusing on the quality of acquired humanness, the theory of emergentism provides a logical approach to the question "What makes humanity special?" At some point along the evolutionary path, from the original single-cell organism to the present state of human evolutionary development, a line was crossed. At this point, the qualities of humanness emerged from a previous nonhuman life form.

10-2. Three pieces of evidence faced Darwin as he gathered specimens during the Beagle's five-year mission. First, he noted that there was an enormous amount of diversity among the living things that he cataloged on the various islands and on the mainlands that he visited during the voyage. Second, Darwin noted that living things had an amazing number of similar characteristics. Third, Darwin began to suspect that the fossil record of the Earth might very well provide a valuable piece of evidence regarding the relationships among the animals that had existed at various points of time in the past. While reading a work by Thomas Malthus, Darwin noticed several circumstances that, combined with his observations on the voyage of the Beagle, led him to his theory of natural selection. The first circumstance that Darwin noticed was the fact that plants and animals produce significantly more offspring than can ever hope to survive. Because so many offspring enter into the local environment, they will be compelled to compete for limited local resources. Darwin also saw that there are distinct differences among all living things, even among the individuals within a given species. These differences mean that some of the individuals in a group will have advantages over other individuals as they compete for the scarce resources. Darwin concluded that a struggle for existence results when those individuals with advantages in the local environment survive long enough to pass their characteristics on to their offspring. This is the process that Darwin labeled natural selection.

10-3. According to Stephen Jay Gould in his study *Wonderful Life: The Burgess Shale and the Nature of History*, although complexification does take place during evolution, the process is by no means smooth, upwardly mobile, and uninterrupted. On the contrary, the evidence presented by the Burgess Shale fossils indicates that the development of life began with a group of animals that featured a vast variety of complex anatomical designs, most of which have long since vanished. Thus, instead of looking at evolution as a process of diversification and growth, Gould sees it as the elimination of most early anatomical designs, followed by a vast diversification within the designs that survived the elimination process. Gould calls this process decimation. According to Gould, the driving force behind the decimation process is contingency. Gould believes that any small, unpredictable event along the path of evolution might have deflected the selection process in a different direction. Another recent anomaly is the mystery of mass extinctions. These mass extinctions represent sharp breaks in the continuity of the fossil record. There is also some evidence that these mass extinctions take place approximately every 26 million years. The three most popular mass extinction theories state that these extinctions are extraterrestrial in nature. These theories are the nemesis theory, the carousel theory, and the gamma-ray burst theory.

Understanding Key Terms

artificial selection
Australopithecines
Australopethicus
bilateral symmetry
Burgess Shale fossils
carousel theory
cerebral cortex
comet
cone of increasing diversity
contingency
decimation

encephalization quotient (EQ)
entity theory
gamma-ray burst theory
genetic variant
hominids
hominoids
Homo
Homo sapiens
Homo sapiens neanderthalensis
mass extinction
mutation

natural selection
Neanderthals
Nemesis
neocortex
Oort comet cloud
pongids
population
Proconsul
speciation
species
strong tendency theory
tendency theory

Review Questions

1. How would vitalism, reductionism, and emergentism deal with defining humanness?
2. What evidence led Darwin to the idea of descent and modification?
3. How do reproduction and mutation fit into the natural selection process?
4. How do new species appear?
5. What are the various steps in human evolution?
6. Which primary trait pinpoints the threshold of acquired humanness?
7. What is the encephalization quotient, and how does it relate to the appearance of humanity?
8. What is the relationship between modern humans and the Neanderthals?

9. What are decimation and contingency, and how do they relate to evolution?
10. What is the theory of mass extinctions?

Discussion Questions

1. What is your reaction to the use of the Morowitz-Trefil characteristic of *acquired humanness* as a guideline for determining the difference between human life and all other life? Does this seem like a practical, straightforward approach, or do you see some difficulties with it? Explain. Do you believe that the guideline can be successful in settling the legal and ethical battles that surround abortion, euthanasia, cloning, and genetic engineering? Explain.
2. Can you build a convincing case using either vitalism or reductionism as the basis for dealing with the abortion debate or the challenge of euthanasia? Explain.
3. What do you think about the theory of evolution as proposed by Darwin? Were you surprised to learn of Darwin's early religious inclinations? Do you see the evidence he used to lead him to natural selection as valid? Explain. What about the logic of his reasoning in reaching his conclusion that natural selection provides the mechanism behind evolution? Do you see his thought process as rational and logical or is he grasping at straws? Explain.
4. Are you comfortable with the idea that the Neanderthals are human ancestors or would you be more comfortable seeing them as a separate species? Explain. Assume for a moment that the Neanderthals were a separate species. Now try to answer some of the following questions about the Neanderthals. What would be the philosophical and ethical relationship between two different intelligent species inhabiting the same planet? Would one species assume the dominant position or would they cooperate? Would the Neanderthals be granted "human" rights or would they be categorized with the apes and other lesser species? What direction might religion have taken had humanity shared the globe with another intelligent species? Many religions are based on the assumption that humans are created in the image of God. What happens to this assumption when humans are compelled to share the Earth with another intelligent species? Would Neanderthals have souls? Would the killing of a latter-day Neanderthal be considered murder? Explain each of your answers.
5. Do you see any validity in Gould's theory of evolution by decimation? Gould also plays up the role of contingency in the evolutionary process. What effect would contingency have on the development of *Homo sapiens*? Explain. Are you more comfortable with the tendency or the entity theory of human development? Explain. What about the theory of mass extinctions? Are you comfortable with the Muller-Alvarez theory of Nemesis as the cause of the periodic mass extinctions charted by Raup and Sepkoski? Explain.

ANALYZING *STAR TREK*

Background

The following episode from *Star Trek: The Next Generation* reflects some of the issues that are presented in this chapter. The episode has been carefully chosen to represent several of the most interesting aspects of the chapter. When answering the questions at the end of the episode, you should express your opinions as clearly and openly as possible. You may also want to discuss your answers with others and compare and contrast those answers. Above all, you should be less concerned with the "right" answer and more with explaining your position as thoroughly as possible.

Viewing Assignment—*Star Trek: The Next Generation*, "The Measure of a Man"

The USS Enterprise arrives at Starbase 173 on routine administration and logistic business. While orbiting the Starbase, the crew receives a visit from Commander Bruce Maddox. Maddox is an expert in cybernetics who has a plan to dismantle Lt. Commander Data to study his positronic brain. Maddox hopes that by doing so he can learn how to reproduce the work of Data's creator, Dr. Noonian Soong. Ultimately, Maddox hopes to create an entire army of androids with the capabilities of Data. Each starship in the fleet will then be equipped with an android who will do the unhealthy, unpleasant, and dangerous jobs that fall to that Starship. This will prevent the human crew members from taking hazardous risks and facing unnecessary dangers. Additionally, it will eliminate the need to consider the feelings and the welfare of the android crew members who are sent out to face hazardous situations. Unfortunately, Maddox is not certain that he can duplicate Data's positronic brain. If he is not successful, he will not be able to reassemble Data who will, in effect, die on the operating table. To avoid this dangerous operation, Data resigns from Starfleet. In an attempt to prevent Data's resignation, Maddox takes his case to the Judge Advocate General (JAG) at the Starbase. Maddox succeeds in convincing the JAG officer to grant his request to prevent Data's resignation. Maddox argues that Data is merely a machine and, therefore, cannot resign from Starfleet any more than the ship's computer can resign. Captain Picard objects to this decision. The JAG officer, therefore, is compelled to hold a hearing to determine Data's status. Since Starbase 173 is in a relatively remote section of the galaxy, the JAG office is understaffed. According to regulations, Captain Picard must act as Data's advocate and Commander Riker must represent Maddox's point of view. Riker does an extremely good job representing Maddox's position. In a relatively disheartened state, Picard talks to Guinan, who points out to Picard that what Maddox plans to do is produce a race of disposable people. Picard immediately sees that what Maddox is proposing amounts to slavery, and makes this argument at the hearing. Basically, Picard convinces the court that Data possesses an emergent humanness that cannot be taken away from him. This emergent humanness gives him the same rights as any other sentient being in the Federation. Accordingly, he cannot be reduced to a piece of property, but must, instead, be treated with the same respect and the same dignity with which all sentient life forms are treated. Moreover, he must be granted the same rights as any other sentient being in the Galaxy. The final decision of the court is to allow Data to choose his own fate. Accordingly, Data refuses to undergo the procedure suggested by Commander Maddox.

Thoughts, Notions, and Speculations

1. In this episode, Commander Riker argues that Data is simply a physical representation of a human being created by Dr. Noonian Soong. Riker argues further that Data's purpose is to serve human interests. Moreover, Riker states that Data is simply a collection of neural nets and heuristic algorithms, whose hardware was developed by Dr. Soong and whose responses are dictated by a software program also developed by Dr. Soong. Riker's arguments are very effective in convincing the court that Data is simply a piece of property owned by Starfleet. As a piece of property he has no rights and no freedom to choose his own fate. How convinced are you by this argument? Explain your answer.
2. Guinan suggests to Picard that Riker's treatment of Data would allow Starfleet to create an entire generation of disposable people who would do the dirty jobs and the dangerous work that no one else wants to do. She suggests that these disposable people could be used without any concern about their welfare or their feelings. Picard concludes that such treatment would amount to slavery. Do you agree with Picard's assessment? Explain.
3. The term "sentient being" is used in this episode by both Picard and Maddox. Presumably, what Picard and Maddox mean by sentience is the point at which a living machine, like Data, crosses a line and begins to exhibit human characteristics. In this chapter we refer to the threshold of sentience as the point of acquired humanness. Lt. Commander Maddox sets down the following criteria for determining when this threshold has been passed. He notes that to be sentient a being must exhibit intelligence, self-awareness, and consciousness. First define each of these characteristics. Then consider whether you agree or disagree with these criteria as set down by Maddox. Can you come up with a set of characteristics besides intelligence, self-awareness, and consciousness?
4. Consider whether these criteria should be applied in other controversial areas. Consider, for example, whether these criteria could be used to determine when a fetus becomes human. Also consider whether this criteria could be used in cases of euthanasia. Suppose scientists determine how to engineer human beings genetically. At what point would these genetically engineered individuals be

considered human? Would the criteria proposed by Maddox work in such a situation? Consider the possibility that in the future scientists may be able to create clones. Could the Maddox criteria be applied to clones?

5. The JAG officer suggests another criteria. She states that the question is whether Data has a soul. What is meant by the concept of the soul? Can an android like Data have a soul? What about genetically manufactured individuals? Would they have souls? What about clones? Would clones have souls? Consider what you have learned about the scientific method. Is the use of the scientific method compatible with the concept of the soul? Explain.

NOTES

1. Harold J. Morowitz and James S. Trefil, *The Facts of Life*, 9.
2. Ibid.
3. Ibid.
4. Ibid.
5. Ibid.
6. Paul Davies, *God and the New Physics*, 82–83.
7. Morowitz and Trefil, 9.
8. Pierre Teilhard de Chardin, *The Phenomenon of Man*, 174.
9. Ibid.
10. Boyce Rensberger, *Instant Biology*, 179.
11. Charles Darwin, *The Origin of Species*, 67–68.
12. Stephen Jay Gould, *Ever Since Darwin*, 28–30; Rensberger, 179.
13. Del Ratzsch, *Battle of Beginnings*, 24; Rensberger, 184.
14. Ratzsch, 24; Rensberger, 184.
15. Ratzsch, 17; Rensberger, 186.
16. Ratzsch, 17.
17. Ibid., 17–18.
18. Ibid., 18.
19. H. James Birx, *Interpreting Evolution*, 133–34.
20. Ibid., 134.
21. Charles Darwin, Autobiography in *The Life and Letters of Charles Darwin*; Gould, *Ever Since Darwin*, 21.
22. Darwin, *Origin*, 53–54.
23. Ibid., 67–68.
24. Ibid., 68.
25. Ibid., 67–68.
26. Darwin, *Origin*, 68–70; Rensberger, 189.
27. Darwin, *Origin*, 69–70.
28. Tim M. Berra, *Evolution and the Myth of Creationism*, 8.
29. Rensberger, 190.
30. Andre E. Hellegers, "Fetal Development," 194–95; Rensberger, 190–91.
31. Berra, 172.
32. Rensberger, 191–92.
33. Ibid., 192–94.
34. Berra, 73–74; Christian de Duve, *Vital Dust*, 18–21; William K. Hartmann and Ron Miller, *The History of the Earth*, 79–80.
35. Berra, 12–13; Rensberger, 195.
36. Berra, 12–13.
37. Berra, 12–13; Rensberger, 195–97.
38. Berra, 12–15; Rensberger, 195–97.
39. Berra, 91, 99; Richard E. Leakey and Roger Lewin, *Origins*, 47; Carl Sagan, *The Dragons of Eden*, 101–3.
40. Berra, 91.
41. Berra, 91, 99; Leakey, *Origins*, 47; Sagan, 101–3.
42. Berra, 91–93, 100.
43. James Shreeve, "Sunset on the Savanna," 123.
44. Ibid., 121–22.
45. Berra, 100–3; Richard Leakey, *The Origin of Humankind*, 130.
46. Berra 100–106; Morowitz and Trefil, 70.
47. Berra, 104–7.
48. Ibid., 106–7.
49. Berra 100–6; Morowitz and Trefil, 70.
50. Berra 106–7; Leakey and Lewin, 77; Morowitz and Trefil, 70.
51. Berra, 106–7; Leakey, 30–33.
52. Morowitz and Trefil, 70.
53. Ibid., 69–72.
54. Berra, 110; Leakey and Lewin, 76–77.
55. Berra, 110.
56. Ibid., 112–13.

57. Morowitz and Trefil, 69–70; Sagan, 88–89.
58. Morowitz and Trefil, 69–70; Sagan, 88–109.
59. Sagan, 103.
60. Morowitz and Trefil, 70; Teilhard de Chardin, 169–74.
61. Robert Foley, *Humans before Humanity*, 161–63.
62. Foley, 163–64.
63. Ibid., 165.
64. Morowitz and Trefil, 98–100.
65. Ibid., 98–99.
66. Ibid., 100.
67. Ibid.
68. Foley, 164–65.
69. Berra, 115; Morowitz and Trefil, 71; Leakey and Lewin, 117–19.
70. Hartmann and Miller, 193.
71. Berra, 113–15; Leakey and Lewin, 117–18.
72. Leakey, 86. (Note: See also Leakey's extensive discussion of the Neanderthals in *Origins: What New Discoveries Reveal about the Emergence of Our Species and Its Possible Future*, 117–19.)
73. Stephen J. Gould, *Wonderful Life*, 319.
74. Hartmann and Miller, 193.
75. Christopher Wills, *The Runaway Brain*, 55–57. (Note: See also Wills's extensive discussion of the relationship between the Neanderthals and *Homo sapiens*, 150–61.)
76. Berra, 117.
77. Ibid., 115–17.
78. Foley, 72.
79. P. Kahn and A. Gibson, "DNA from Extinct Human," 5323; News and Editorial Staffs, "Breakthrough of the Year," 2041.
80. Leakey, 118; Sagan, 107.
81. Leakey, 118. (Note: For an interesting novel about the intriguing possibility that Neanderthals still exist in remote places on the Earth see John Darlton, *Neanderthal* (New York: Random House, 1996).
82. Berra, 48.
83. Gould, 23–25.
84. Ibid., 25–49.
85. Ibid., 24.
86. Ibid., 38.
87. Ibid., 38–43.
88. Ibid.
89. Ibid., 14, 45–47.
90. Ibid., 45–47.
91. Ibid., 46–47.
92. Ibid., 51.
93. Ibid., 48–51.
94. Ibid., 289.
95. Ibid.
96. Ibid., 289–90.
97. Ibid., 289, 319–20.
98. Ibid., 319.
99. Morowitz and Trefil, 70.
100. Simon Conway Morris, "Showdown on the Burgess Shale: The Challenge." 49–51.
101. Berra, 16–17; Rensberger, 210–11; Richard Muller, *Nemesis*, 3–5.
102. Muller, 3, 10.
103. Ibid., 10–13.
104. Ibid., 10–14.
105. Gould, 45–47.
106. Muller, 14.
107. Gould, 306–7.
108. David M. Raup, *The Nemesis Affair*, 115–26.
109. Muller, 3–5; 89.
110. Ibid., 3–4
111. Thomas S. Kuhn, *The Structure of Scientific Revolutions*, 52–53.
112. Muller, 7–8.
113. Ibid., 8.
114. Ibid., 104–8.
115. Michael R. Rampino, "The Shiva Hypothesis: Impacts, Mass Extinctions, and the Galaxy," 10–11.
116. Peter J. T. Leonard and Jerry T. Bonnell, "Gamma-Ray Bursts of Doom," 29–34.

Unit VII Procreation, Biotechnology, and the Law

Chapter 11
Juriscience and Human Life

COMMENTARY

The world of the future described by Axel Madsen in his novel, *Unisave*, is rather dismal and depressing.[1] By the end of the twenty-second century, the planet Earth is burdened by a population that exceeds twenty-four billion people. The resources of the planet are stretched to their limit. The United Nations has established Unisave, an agency given the authority to impose strict controls, including an extensive contraceptive program, on Earth's entire population. Some of these controls include strict quotas on the number of children allowed each couple, compulsory sterilization under certain circumstances, the mandatory use of sex suppressants by young people, extensive electronic surveillance to discourage violations of antisex laws, and eventually, a program of coercive gericide, which is defined in the book as the systematic killing of older people. While Madson's account is fictional, it is clearly in line with predictions made by other writers concerned with the future of humanity. For example, the economist Robert Heilbroner has expressed a remarkably similar point of view in his book *An Inquiry into The Human Prospect*. As Heilbroner explains, the challenges facing humanity now and in the future are products of undisciplined scientific and technological progress.[2] According to Heilbroner, science and technology have developed the means to prolong life, to cure infertility, to reduce infant mortality, and to eradicate many deadly diseases, but have not developed any way to feed, clothe, or provide resources for the people who survive as a result of these advances. Consequently, the population grows, the environment deteriorates, and the Earth's natural resources are devoured haphazardly, while science and technology remain helpless. But is it really science and technology that are to blame? Might the responsibility more properly be placed on the shoulders of those who make the policies and enforce the laws that regulate science and technology? Does the law have any responsibility to step into the scientific arena to control such developments? To what extent should the government be allowed to control scientific research and technological progress? In *Unisave*, the government must regulate the reproductive activities of every person on the planet. To what extent should the government be permitted to interfere in such private matters? These questions and others like them form the basis for our examination of how science and the law intersect.

CHAPTER OUTCOMES

After reading this chapter, the reader should be able to accomplish the following:

1. Explain the need to study science in relation to issues such as abortion and genetic engineering.
2. Define the law.
3. Explain the ethical roots of the law.
4. Relate the contemporary sources of the law.
5. Explain the similarities between science and the law.
6. Identify the current state of the law in relation to abortion.
7. Explain the interplay between the law and science in relation to abortion.
8. Specify the relationship between acquired humanness and the law of abortion.
9. Explain how the right to privacy and responsibilities relate to procreative freedom.
10. Identify the legal issues surrounding the use of contragestive agents.

11-1   SCIENCE, ETHICS, AND LAW

In the last two chapters, we explored the origin and evolution of life on the planet Earth. We also examined the evolution of humanity and located the point at which some experts believe our primitive ancestors acquired humanness. After exploring these topics the nonscientist may once again face the scientist and, in all sincerity, ask how such knowledge can be put to immediate practical use. Several explanations might satisfy this need for immediate results. In this chapter and the next, however, we concentrate on the ethical and legal consequences of research into the origin of life. Specifically, our focus will be on the impact of such research on procreative matters and genetic engineering. Both of these issues have their origin in the life sciences. Without advances in neonatal medicine and maternal health care, for example, the issue of procreative choice would not have initiated such turmoil. It is the ability, or at least the perceived ability, of medical science to push back the limits of viability that has given rise to much of this controversy. Similarly, if microbiologists had not developed genetic engineering, the debate surrounding recombinant DNA, genetic engineering, and gene therapy might not have materialized with the intensity that we witness today.

Neither scientists nor nonscientists can escape these issues. They will impact on each of us now and in the future. The peculiar difficulty associated with these particular problems, however, is their multidimensional nature. They require not only a knowledge of science but also a basic understanding of ethics and law. For that reason, before turning to a detailed look at the issues themselves, we pause to examine the ethical foundation and contemporary sources of the law as well as the interaction between science and the law. We then apply what we have learned about the scientific nature of life to the ethical and legal aspects of procreative rights and genetic engineering. The objective of this study is to examine how our present understanding of the science of life can address the legal and ethical intricacies of these two controversial issues, procreative freedom and genetic engineering.

In this chapter, we will focus primarily on **procreative freedom**, which includes the right to use contraception, the right to give up a child for adoption, the right to adopt a child, the right to seek treatment for infertility, the right to assisted reproduction, the right to have children, and the right to have an abortion. We focus on the abortion debate because this is the most controversial of procreative matters and, therefore, the area that receives the most attention in the courts and legislatures. However, it is generally unwise to separate the abortion question from other aspects of procreative freedom because all reproductive matters are bound together by the same set of legal principles. After all, if the law has the power to outlaw abortion, it may also have the power to coerce abortion, order sterilization, or seize a person's children. In short, the law may have the power to create the type of totalitarian nightmare envisioned by Axel Madsen in *Unisave*.

11-2 WHAT IS THE LAW?

If any area of human knowledge suffers more from misinformation than science, it is law. Like science, the law has been misrepresented by the news media and the entertainment industry. As a result, when science and the law interact, many people are justifiably confused. All of this would be nothing more than a bit of bad news were it not for the fact that some of the most controversial scientific issues of the modern age are entangled with the law. For the sake of simplicity we use the term **juriscience** to refer to the treatment of issues that stand at the intersection of science and the law. However, before addressing these juriscientific issues directly, we need to reach a basic understanding of the law.

The Ethical Roots of the Law

The **law** is a set of rules and regulations established by the government of a nation or the subdivision of a nation in order to maintain order and justice. **Order** is established when there is harmony and stability within a nation, whereas **justice** is reached when the law is applied with impartiality and equity. Ideally, the law-making process should be guided by a set of objective ethical concepts that can be used to distinguish right from wrong. To be of any real use in establishing a legal system, a set of ethical concepts must be objective. Ethical concepts based upon the belief that right and wrong can change from person to person or from situation to situation, a concept frequently referred to as **ethical relativism** or **subjective ethics**, is of limited use in any attempt to establish an unbiased legal system.[3]

Consequently, some other standard must be adopted as the ethical basis for law. Two ethical concepts that might be applied in this regard are utilitarianism and reciprocity. **Utilitarianism** or consequentialism states that right and wrong can be determined by measuring an action in terms of the good or evil that it produces. If the action produces more good than evil in relation to the people affected, then the action is right. On the other hand, if the action produces more evil than good, it is wrong. The principle of reciprocity can be used to judge the moral substance of a rule. **Reciprocity** is based on the belief that sentient beings recognize their own self-worth and, therefore, strive to be treated with dignity and respect. Ideally, sentient beings would also recognize that all other sentient beings prize their own self-worth and, therefore, also wish to be treated well. Consequently, the primary guiding principle for sentient beings would be to treat others as they would like to be treated themselves. This principle is frequently called the **golden rule**.[4]

Contemporary Sources of American Law

The five contemporary sources of American law are common law, judicial decisions, statutory law, the Constitution, and administrative regulations. The American legal system finds its roots in the **common law** of England. The principles of English common law were developed over the centuries by the king's magistrates, each of whom was assigned an annual circuit throughout a district of the English countryside. The only guideline the king established for these magistrates was to be as consistent as possible in making decisions. The goal, according to the king, was to establish a body of common law for all England. Periodically, the magistrates would share their decisions. In this way, whenever a magistrate faced a novel situation, that magistrate could rely on the previous decision of another magistrate. This process of relying on prior case rulings created an enormous body of common law throughout all England and later in the English colonies in America. Many of these principles are still part of the legal system today.[5]

The process of relying on past authority as a guideline for present cases is called ***stare decisis***. Literally translated, *stare decisis* means let the decision stand. Although the idea of creating law by allowing previous judicial decisions to stand originated with common law, the process is now applied in a number of different ways. For example, in addition to establishing common law, judges can also create law through **judicial decisions** that interpret statutory law or that determine the constitutionality of statutes, regulations, and executive actions. The body of law that consists of these judicial decisions is known as **case law**. This is true whether the court is applying common law, interpreting a statute, determining issues of constitutionality, or any combination of the three. A previous case that is used as authority in a present one is called a **precedent**. A precedent that must be followed by a court is a **binding precedent**, whereas one that can be either followed or ignored is a **persuasive precedent**. Generally, a precedent is binding on a court if that court is under the **jurisdiction**, or authority, of the court that made the original decision. Thus, all courts in this country are bound by decisions made by the U.S. Supreme Court.[6] As we shall see, the role of precedent is extremely crucial in many juriscientific issues. However, it is especially critical in the procreative rights debate, which has focused on a 1973 U. S. Supreme Court case known as *Roe v. Wade*.[7] The reasoning in this case guides most arguments on procreative matters today but especially in relation to the abortion debate.

Statutory law is the law made by a legislature, such as the U. S. Congress or a state general assembly. Generally, statutory law will command, prohibit, regulate, or declare something.[8] A statute, for instance, that outlaws abortion completely would be a statute which prohibits an activity, whereas a statute that requires a physician to inform a woman of all options to abortion would be a statute that commands something. If a legislature delegates some of its authority to a regulatory agency, such as the Food and Drug Administration (FDA), then the rules made by that agency are said to be administrative regulations.[9]

A legislature has the power to alter or even eliminate a line of precedent. The legislature can do this by passing a new statute that contradicts, alters, or eliminates the legal principles established by that line of precedent. However, for that statute to survive a legal challenge, it must also be in line with constitutional principles. A **constitution** is the fundamental law of a nation (or state). Accordingly, a constitution will establish the government of a nation, outline the rights and liberties of the people of that nation, and set down some basic principles that will guide the operation of that nation.[10]

This is precisely what the U. S. Constitution does. The government of the nation is established by the articles, whereas the people's rights and liberties are guaranteed by the Bill of Rights.[11] The Constitution also sets down the basic principles of the nation, one of which is that the Constitution is the supreme law of the land. Any statute, regulation, or precedent that conflicts with the Constitution is unconstitutional and therefore void. The power to determine constitutionality resides in the courts, with ultimate authority resting with the U.S. Supreme Court.[12]

Herein lies the critical importance of nationally known cases like *Roe v. Wade*. In *Roe v. Wade* the United States Supreme Court struck down a state statute that had outlawed abortion. This means that, by principles of *stare decisis*, *Roe v. Wade* became binding precedent. All courts must, therefore, treat *Roe v. Wade* with the utmost respect when interpreting any statute related to procreative rights. Moreover, any legislature that enacts such a statute must also make certain that it is in line with constitutional principles as laid down by the Supreme Court in *Roe v. Wade*. However, the issue of procreative rights, like many legal problems associated with science, cannot be decided solely within the courtroom. The very nature of the court's reasoning in *Roe v. Wade* creates a potential conflict between the culture of science and the culture of law.

11-3 THE INTERSECTION OF SCIENCE AND THE LAW

The central holding of *Roe v. Wade* is that a state cannot interfere with a woman's right to seek an abortion before the fetus becomes *viable*. Before moving on to the reasoning of the court in this case, several key terms must be defined. These include zygote, blastocyst, preembryo, embryo, fetus, and viability. A **zygote** is the single cell that exists after an egg has been fertilized. A **blastocyst** is a specialized, multicell structure that develops approximately two days after an egg is fertilized. In the blastocyst cells are sufficiently differentiated to be able to tell which cells will develop into the embryo and which will attach to the uterine lining and develop into the placenta. The **placenta** connects the embryo to the pregnant woman so that the embryo can receive nourishment from and eliminate waste into the pregnant woman's blood stream. During the time that the placenta is formed up until about day 15, the term **preembryo** is used to describe the entity. After the 15th day the entity is referred to as an **embryo**, and at some point between day 56 and day 60 it can begin to be referred to as the **fetus**.[13]

A fetus has reached **viability** when it can survive outside the womb, even if it is only with artificial assistance. At the time of *Roe v. Wade*, the court ruled that the line of viability could be drawn at the end of the second trimester (three-months period), or six months of pregnancy. The court's reasoning went like this: A fetus can survive outside the womb, albeit with artificial assistance, after it has entered the third trimester of prenatal life. At this point in fetal development, the point of viability, the government has a **compelling interest** in preserving that life. The government's compelling interest at the point of viability outweighs the right that the woman may have to undergo an abortion.[14]

The compelling-state-interest standard does not apply only to cases involving procreative issues. Rather, it applies to any situation in which a statute infringes upon a fundamental right guaranteed by the U.S. Constitution. As such, the compelling-state-interest standard represents a step beyond the traditional **rational basis standard** that previously governed statutes that somehow affect the Equal Protection Clause of the Constitution. Under traditional equal protection principles, in defending a statute that appeared on its face to discriminate against a certain class of people, the government had to show only that the statute in question bore a rational basis to a legitimate governmental interest.[15] Thus, a state statute that requires physicians to have reached a certain educational level in order to be licensed might discriminate against certain people who are unable, for whatever reason, to obtain that educational level. However, the educational and licensing requirements for physicians bear a rational, that is a reasonable, relationship to a legitimate governmental objective. In this case, that objective would be to protect the health of the people.

As a method for limiting governmental power, the compelling interest principle provides a standard for determining the constitutionality of governmental action that is stricter than the rational basis standard. As noted, the compelling interest standard is generally applied when a statute limits a fundamental right guaranteed by the U.S. Constitution. The standard was firmly established by the U.S. Supreme Court in the case of *Shapiro v. Thompson*, which involved state laws in Pennsylvania and Connecticut that required a one-year residency before an individual could receive welfare payments. The U.S. Supreme Court decided that, since the statutes limited the fundamental right of interstate travel, they would be declared unconstitutional, unless each state could prove that their statute was necessary to promote a compelling state interest. Both states failed in this regard and the statutes were declared an unconstitutional violation of the Equal Protection Clause.[16]

In procreation cases the fundamental right at stake is the right of privacy, which the Supreme Court has declared arises under the Ninth Amendment to the U.S. Constitution. As such, no statute touching on procreative rights can be allowed to stand unless it is necessary to meet a compelling state interest. Moreover, that compelling state interest, which the Court identified as the interest in preserving the potential life of the fetus, does not even arise until the third trimester of pregnancy. Consequently, in *Roe v. Wade* the Court decided that a woman's constitutional right to privacy in procreative matters could not be infringed by the government during the first trimester. In other words, at that early stage in a pregnancy there is no compelling interest for the government to promote. Even in the second trimester the government can regulate abortion only to preserve the life or the health of the pregnant woman. It is not until the third trimester that the state has a compelling interest that can be measured against the woman's constitutional right to privacy in procreative matters. At that point the state may enact legislation necessary to promote the state's compelling interest in preserving the potential life of the fetus. Thus, during the third trimester a statute could limit and even, under some circumstances, prohibit abortions, provided that the provisions of such a statute are necessary to promote the state's compelling interest. If certain provisions unnecessarily infringe on the woman's fundamental right to privacy, then those provisions might still be declared unconstitutional, even if they apply only during the third trimester.[17]

Far from settling the abortion conflict, however, *Roe v. Wade* simply fanned the flames of the controversy. For close to three decades both sides in the debate have waged a war in the courts, in state legislatures, and in the media. Part of the cause of the continuing debate is that science and law approach the issue from different perspectives. Before looking at the differences, however, it will be somewhat enlightening to look at the similarities between science and law.

Similarities between Science and Law

As noted in previous chapters, the goal of science is to uncover the truth about the universe, by applying the objective standards of the scientific method.[18] We value science not only because of the results but also for the process used to obtain those results.[19] A crucial part of this process is that any successful theory will meet three criteria: predictability, clarity, and continuity. First, an effective theory will make *predictions* about the world that can be verified. Verification occurs when other scientists test a theory experimentally.[20] Second, a successful theory must be *clear* and open to public scrutiny. To be of value, a theory must be concise enough to be measured according to objective criteria. Theories that require some sort of hidden knowledge, special talent, or mysterious magic are immediately suspect. Finally, a genuine theory will be based on a line of *continuity*. In other words, the new theory must evolve logically from what is already known about that area of science. In this way the theory extends our understanding of the world and adds to our storehouse of scientific knowledge.[21]

Interestingly enough, the law also requires the same three qualities of continuity, clarity, and predictability. First, an effective law will make *predictions* about the future. A precedent will, for instance, predict, at least in a general sort of way, how courts will address similar situations in the future. Moreover, a statute that prohibits or commands an action also predicts how people will act in the future, even, as the famous jurist Richard Posner points out, if those people must be coerced by law enforcement agencies.[22] Second, like scientific theories, the law must be *clear* and open to public scrutiny. A law that is hidden, obscure, or ambiguous is worthless as a guide for human conduct.

Finally, *continuity* requires that a law be directly connected with the past. This principle is seen most clearly in the process of *stare decisis*. A judge's decision in a present case must be in line with all binding precedents that deal with that area of law. Most courts have made it abundantly clear that a line of precedent can be overturned only when serious conditions warrant such an action. In a 1993 case, the U.S. Supreme Court stressed the vital nature of precedent within the legal system. Underlining the significance of precedent, the court summarized a series of questions that judges should address as they consider whether or not to overrule an earlier case. These questions include: (1) Has the previous precedent become impractical? (2) Have so few individuals relied on the previous precedent that overturning it would not cause an unusual amount of difficulty or injustice? (3) Has the previous precedent become merely a shadow of an outdated legal rule? (4) Finally, has the previous precedent become extinct because of changes in society? Only when a judge can affirmatively answer one or more of these questions should that judge even consider overruling a precedent.[23]

In addition to predictability, clarity, and continuity, science and law share other substantive characteristics as well. Both science and law help shape the world in which we live. Scientists, for instance, add to our storehouse of knowledge as they discover how the universe operates at all levels of existence, from the depths of outer space to the subatomic world of quantum physics. Many of these discoveries can be channeled into practical technological advances that shape our culture and our social institutions. One need only recall how our understanding of biology has led to developments in **genetic engineering** to see that this is true. Whether they are working in a medical lab to formulate a cure for cancer, digging for dinosaur bones in a dusty desert to explain mass extinctions, or launching an orbiting space telescope to measure the age of the universe, scientists are on the cutting edge of knowledge.

Attorneys, judges, and legislators, however, are also, at times, on this cutting edge. In this way the law also helps shape our culture and social institutions. Legislators, for instance, can introduce legislation aimed at solving problems within the socioeconomic structure. Historical examples of this process can be seen in the antitrust legislation of the 1890s, the New Deal programs of the 1930s, and civil rights legislation of the 1960s. Legislators also create administrative agencies, such as the Environmental Protection Agency (EPA) and the Nuclear Regulatory Commission (NRC), which have the power to regulate some portion of the socioeconomic system. When a legislature delegates some of its authority to a regulatory agency, the rules made by that agency are said to be administrative regulations. Nevertheless, these regulations have the force of law.[24]

The executive branch also has an active role in the law-making process. The primary role of the chief executive is to enforce the statutes that the legislature has passed. This does not mean, of course, that the president or the governor must carry out each law personally. On the contrary, the chief executive of the nation or a state heads an enormous bureaucracy charged with the responsibility of carrying out the acts of the legislature. Beyond this, however, the chief executive can propose a legislative package at the beginning of each legislative session, as the president does every year in his annual state-of-the-union address. Moreover, the president has the power to negotiate treaties with foreign nations and to enter executive agreements. Also, because he is identified as the ceremonial head of the federal government, the president can lead the country on national campaigns to reshape our social institutions. This is what President Kennedy did, for example, when in the early 1960s he proposed that the United States would place an American on the Moon by 1970.

Even judges, as interpreters of the law, participate in the law-making process. Although most judges face a case only if an individual or a group of individuals has filed a lawsuit in that judge's court, they still write opinions that have a lasting impact on all subsequent cases that involve the same legal principles. Moreover, in some states judges are permitted to issue advisory opinions to legislators. In these states a senator or representative can take a bill that is pending before the state general assembly to a judge, usually a judge or justice who sits on the highest appellate court of that state, and ask that judge to issue an official advisory opinion on the constitutionality of that bill. Judges also frequently have the opportunity to evaluate the constitutionality of a law, if that law ends up at the center of a controversial lawsuit that reaches the court.[25]

Despite these similarities, science and the law operate in different venues and in many situations the two disciplines should remain in those separate venues. For example, in Chapters 2 and 3, we discussed several competing theories that seek to explain how the universe began. The law has absolutely no authority to endorse or overrule any of these theories. The origin of the universe simply falls outside the jurisdiction of the law. Similarly, science cannot be used to undermine the authority of the law. For example, while a scientific expert may be permitted, perhaps even encouraged, to give testimony based on a revolutionary new theory, on whether to admit that testimony must be left to the judge.

The Interaction between Science and Law

Although science and law frequently operate within different spheres of influence, neither discipline operates within a cultural vacuum. Consequently, they interact more often than we might, at first, imagine. In fact, they frequently operate together to promote a jointly held set of common objectives. For example, the government, which is, after all, an extension of the law, provides financial support for scientific research and development. The government awards such funding through the National Science Foundation, the Department of Defense, the National Institute of Health, and the Department of Energy.[26] Second, a scientific advance or a technological development might need legal approval before it can be marketed to the public. This is what happens when an agency, such as the Food and Drug Administration (FDA), is called upon to evaluate and approve new pharmaceutical developments. Other agencies with similar regulatory responsibilities include the Environmental Protection Agency (EPA) and the Nuclear Regulatory Commission (NRC).[27] Third, the law allows individuals to preserve their rights to any new inventions that they develop. These rights can be protected by applying for a **patent**. As long as that new technological development meets certain criteria, the patent will be granted and the inventor's rights will be preserved.[28]

Fourth, a scientific advance or a technological development might provide a revolutionary new means for gathering or evaluating evidence at trial. This is what is involved in the controversy surrounding the reliability of procedures such as voiceprint identification and DNA typing.[29] The standard for making such an evaluation was first established in 1923 when a federal court in *Frye v. United States* elected to forbid the admission of evidence provided by a polygraph (lie-detector) test.[30] The standard established in the *Frye* case prohibits the admission of any scientifically gathered evidence unless that evidence is based upon principles that are generally accepted by the scientific community.[31] Although the general-acceptance standard has since been superseded by the Federal Rules of Evidence, the approach taken by the court in *Frye* demonstrates the deliberative process of the law in relation to the acceptance of new scientific principles, at least in regard to the admissibility of evidence. Unless a scientific principle has been so firmly established within the system that it is generally accepted by the scientific community, that principle probably will not be used to admit evidence at trial.[32] On the other hand, in a recent case the U.S. Supreme Court emphasized that federal trial judges have wide discretion in decisions regarding the admissibility of scientific evidence. Undoubtedly, this approach will lead to a case-by-case analysis of scientific evidence, especially when that evidence involves innovative theories that oppose the orthodox scientific paradigm.[33]

A fifth way that science and the law can legitimately interact occurs when an invention or a new process causes harm to people or property. In a lawsuit based upon such an injury the court will be called upon to establish a standard to measure liability. Establishing a liability standard is crucial because such a standard will determine who will bear the financial burden of any loss caused by an injury to an innocent party. Often when a new technology is first introduced to the courts, the inclination is to limit its use or at least levy a stiff penalty against those who are involved in the operation of that new technology. An interesting case in point involves the development of the court's attitude toward aviation. In the early stages of aviation the courts considered the practice of flying aircraft not only dangerous but also foolhardy, and they were, therefore, quite ready to impose absolute liability on any aviator who happened to harm someone or something on the ground.[34] This meant that if an aircraft caused harm, the pilot and the owner of that aircraft were held responsible, regardless of how careful the pilot might have been.[35] By the middle of the twentieth century, however, some courts were more readily persuaded that flying had become somewhat safer and, as a result, those courts relaxed the absolute standard. Nevertheless, even as late as the 1950s, some jurisdictions still applied the absolute liability standard in aircraft cases, clearly demonstrating the deliberative nature of the law in regard to hazardous activities initiated by new technological developments.[36]

The same deliberative process of the law can be witnessed in a variety of situations involving the introduction of other new technological developments. In recent years, the courts have had to deal frequently with new technological developments. Some of these new technological developments have been declared inherently dangerous. These would include crop dusting, refining oil in a heavily populated area, storing flammable liquids, and using explosives. Other technological developments have been found not to be inherently dangerous. Some of these developments include transmitting electricity in power lines, using radiological (x-ray) equipment, and distributing gas in gas lines.[37]

The final way that science and the law can legitimately interact is when a scientific theory or a technological development either creates a new legal problem or adds a new twist to an already existing legal issue. It is in this way that science and the law intersect in regard to procreative freedom. In the case of procreative freedom, advances in our understanding of the neocortex could lead us to a more accurate identification of the moment at which a fetus reaches the point of acquired humanness. Some commentators have suggested that the moment of acquired humanness should replace viability as the dividing line for the government's power to place restrictions on abortion. This would constitute a new twist on an already existing issue. For the remainder of this chapter we explore the complex interaction of science and law in procreative matters.

11-4   JURISCIENCE AND PROCREATIVE FREEDOM

The interaction between science and law can be seen rather dramatically in one of the latest in a long line of procreative freedom cases heard by the U.S. Supreme Court over the last twenty-five years. The case, *Planned Parenthood of Pennsylvania v. Casey*, involves a Pennsylvania state statute that limited a woman's right to have an abortion by requiring that she give her informed consent to the procedure.[38] Another provision required a married woman to notify her spouse of her plan to seek an abortion. The law also compelled a juvenile to obtain the informed consent of at least one parent before she could undergo an abortion. Finally, the statute allowed these provisions to be bypassed if a medical emergency made the abortion necessary.[39]

The Law, Science, and Procreative Freedom

In *Planned Parenthood of Pennsylvania v. Casey*, several clinics and one physician representing an entire class of physicians filed a federal lawsuit asking the court to stop the implementation of the statute. The clinics and the physician wanted the court to declare the statute unconstitutional. The trial court did, in fact, declare most of the act unconstitutional and as a result issued an injunction forbidding the enforcement of the law. The appeals court reversed the findings of the trial court on all provisions except the spousal notification clause, which it held to be unconstitutional.[40]

The U.S. Supreme Court agreed to hear the case because it represented an opportunity for the Court to clarify some of the doubts and uncertainties surrounding *Roe v. Wade*. To understand the relationship between *Roe v. Wade* (*Roe*) and *Planned Parenthood of Pennsylvania v. Casey* (*Casey*), it is important to examine first the premise and the corollaries upon which *Roe* rests. The Court's reasoning in *Roe* is based upon the very clear premise that a fetus is not a person and, therefore, need not be accorded the rights guaranteed by the U.S. Constitution.[41] From this premise two corollaries arise. The first is that a woman has a right to control her own body even to the extent that she can have an abortion, unless a compelling governmental interest counteracts that right.[42] The second is that the government's interest in protecting the life of the fetus does not automatically become a compelling interest at all times during pregnancy. Instead, that compelling interest arises at a point when certain criteria have been met. In *Roe*, the U.S. Supreme Court decided that the point at which the government's interest is compelling enough to allow it to interfere in a woman's right to control her own body in pregnancy is the moment at which the fetus becomes viable.[43]

After considering the fundamental constitutional questions in *Roe*, the Supreme Court in *Casey* concluded that the essential holding of *Roe* should be reaffirmed. The Court went on to explain that before the fetus reaches viability, a woman has the right to seek an abortion free from the undue interference of the state. However, this does not mean that the government has no interest in protecting fetal life. On the contrary, the Court states quite explicitly that the government's legitimate interest in protecting the life of the fetus and the health of the woman extends to the beginning of the pregnancy. Thus, the government can regulate a woman's decision to seek an abortion even before viability. However, any

Figure 11.1 The U. S. Supreme Court at the Time of *Planned Parenthood of Pennsylvania v. Casey*.
In a landmark case known as *Planned Parenthood of Pennsylvania v. Casey*, the U.S. Supreme Court concluded that the essential holding of *Roe v. Wade* should be upheld. At the time the Supreme Court consisted of (back row from left) Associate Justices David Souter, Antonin Scalia, Anthony Kennedy, Clarence Thomas, and (front row from left) Associate Justices John Paul Stevens, Byron White, Chief Justice William Rehnquist, Associate Justices Harry Blackmun, and Sandra Day O'Connor. In an unusual move, the majority opinion in *Planned Parenthood of Pennsylvania v. Casey* was written by three of the associate justices, Anthony Kennedy, Sandra Day O'Connor, and David Souter.
Photo Credit: UPI/Corbis-Bettmann

governmental regulation that impacts on a woman's right to choose to have an abortion may not place an undue burden on that right.[44] According to the Court's final analysis, "An undue burden exists, and therefore a provision of law is invalid, if its purpose or effect is to place a substantial obstacle in the path of a woman seeking an abortion before the fetus obtains viability."[45]

Some legal experts see this shift to the undue burden standard as the fatal flaw in *Planned Parenthood v. Casey*. They argue that the undue burden standard represents a legal step backward, undoing much of the progress that had been made during the last three decades in supporting a woman's right to make decisions freely in procreative matters. There may be much merit in this position. However, the shift to the undue burden standard does not disturb the central holding of *Roe v. Wade* in regard to viability. The Court in *Planned Parenthood*

v. *Casey* emphasized that the point at which the fetus becomes viable still represents the dividing line before which the government cannot take away a woman's right to make the final choice to end her pregnancy.[47] The Court said it this way: "Our adoption of the undue burden standard does not disturb the central holding of *Roe v. Wade* and we affirm that holding. Regardless of whether exceptions are made for the particular circumstances, a State may not prohibit any woman from making the ultimate decision to terminate her pregnancy before viability."[48] As we have seen, viability is defined as the moment at which the fetus can live, albeit with artificial assistance, outside the womb. Consequently, the viability standard does not depend upon any innate qualities or characteristics of the fetus itself. Instead, the viability standard targets the ability of physicians to keep the infant alive.

The Acquisition-of-Humanness Standard

This distinction has led some commentators, notably biologist and philosopher, Harold J. Morowitz, and physicist James S. Trefil, in their work *The Facts of Life: Science and the Abortion Controversy* to conclude that viability is an improper standard for measuring the point before which the government cannot prevent a woman from making the ultimate choice in abortion related decisions. Morowitz and Trefil suggest that the viability standard should be replaced with a guideline that is more directly related to some biological quality inherent within the fetus itself.[49] Other commentators have also suggested that some criteria more closely akin to the quality of humanness or the characteristic of personhood should be used to measure the point of the government's compelling interest in the life of the fetus.[50] Accordingly, some of these commentators suggest that the court should not ask when the fetus can survive outside of the mother's womb but what makes human life different from all other forms of life. Or as Morowitz and Trefil, suggest, "When does the fetus acquire those properties that make humans uniquely different from other living things?"[51]

In their quest for a solution to the abortion debate, Morowitz and Trefil focus on the acquisition of humanness.[52] In the last chapter we addressed the issue of the acquisition of humanness in relation to the evolutionary development of the human race. At that point in the discussion we noted that the most critical distinguishing characteristic that set our first human ancestors apart from the other primates was their enlarged brain capacity. Although many of the other physical characteristics are critical indicators of a new branch point in evolutionary progress, it is the development of the large brain, specifically the enlarged cerebral cortex, that was the most crucial event.[53] This enlarged brain capacity gave our ancestors a distinct advantage over their cousins. Eventually, it led to the development of tools and weapons, as well as the ability to exploit the environment in ways that best suited their survival.[54]

With the development of the cerebral cortex as their guideline for the acquisition of humanness, Morowitz and Trefil have what they consider to be a biologically realistic way to determine when a fetus makes the leap to humanness. They contend that the development of the cerebral cortex, not viability, should be the standard by which the government's compelling interest should be measured because the development of the cortex is in tune with scientific reality. Viability, the argument goes, is based less on the development of the fetus and more on the capability of medical technology to keep the fetus alive outside of the mother's womb.

The unanswered question is whether the law should embrace this new scientific guideline or should it resist changing the established viability rule. Our examination of this question will involve three steps. First, we explain the development of the cerebral cortex. Then we see why Morowitz and Trefil, among others, believe that a more objective, scientific criteria should replace the viability standard in abortion cases. Finally, we look at some of the problems that arise when the law considers moving to a new, scientifically created standard.

The Development of the Cerebral Cortex. Like the construction of most complex structures, the development of the brain takes place in several major stages. First, the raw materials must be assembled. Second, those raw materials must be made into the basic building blocks of the structure. Third, the building blocks must be transported to the site and assembled. In this case the raw materials are present once the egg and sperm unite. Consequently, we focus our attention on the second stage, the creation of the building blocks, in this case the nerve cells, and the third stage, the creation of the connections among those nerve cells giving rise to the cerebral cortex. It is at this third stage that the acquisition of humanness occurs.[55]

The Creation of Neurons. The second stage in the development of the cerebral cortex involves the creation of the basic building blocks of the cortex. In this case those basic building blocks are the brain cells or **neurons**. Each neuron consists of a main cell body from which extend two types of filaments. One filament is the **axon**, which can be considered the transmitter of the neuron. The other filament is called the **dendrite**, which is roughly analogous to the receiving antenna.[56] Each axon starts as one long filament flowing from the central cell body of the neuron. The axon begins as a single filament and ends by spreading out like the branches of a tree. At the end of each branch is a synaptic terminal. The **synaptic terminal** houses tiny bubbles filled with **neurotransmitter molecules**. When stimulated by an electric current flowing down the axon, the bubbles, or **synaptic vesicles** as they are called, travel to the edge of the synaptic terminal, where they attach themselves and release the neurotransmitter molecules into the **synaptic cleft**, a space between the axon and the dendrite of another neuron. The axon and the dendrite are not in direct contact with one another. Instead, the connections are made by the neurotransmitter molecules, which are sent into the synaptic cleft.[57]

The neurons and all of their constituent parts make up the building blocks of the brain. The neurons begin to develop in an embryo that is seven weeks old. However, the cells are quite simple at that stage, lacking both axons and dendrites. The immature cells then move to that part of the body that will develop into the cerebral cortex. When the neurons are in their proper positions, they begin to grow axons and dendrites as they mature.[58] Curiously, not all neurons mature in the same way. The development of a neuron depends on its location within the maturing cortex. Six different types of cells develop at six different levels within the cortex. The neurons are supported by a network of cells, the **glial cells**, which provide nutrition for the neurons.[59]

The Origin of the Cortex. At this point in the growth of the cortex, the neurons are in place. The final stage involves connecting the neurons to fire up the cortex so that the cortex operates as an active brain. The six complex layers of neurons together are not yet a cortex, any more than a plot outline is a novel. The key element in firing up the neurons is the formation of the synapses. Once the synapses have developed, a fully operational brain emerges.[60] The development of the synaptic connections among the neurons is so important that any abnormal interference in their proper development can produce behavioral disorders later in life.[61] Once the synaptic connections have appeared, we have crossed the threshold and reached the level of sentience so crucial for the determination of emergent humanness.[62]

Several important studies conducted on the brain tissue of stillborn infants along with a number of similar studies performed on the fetal brain tissue of monkeys and other primates have led scientists to two crucial conclusions regarding this all-important stage in the development of the cortex. First, the synapses do not form haphazardly or at different rates. Rather, once the synapses are ready to form, they do so throughout the entire brain. Second, there is a definite point in the development of the fetus at which the process of synapse building takes a quantitative and qualitative leap forward.[63] This point, which has been designated as the origin of the cortex, occurs between *twenty-five and thirty-two weeks after conception*. Morowitz and Trefil conclude that this period of time can, therefore, be considered the crucial threshold point for *the acquisition of humanness*.[64]

Other Standards of Humanness

The critical conclusion that Morowitz and Trefil come to here is that scientists can now identify a point in fetal development that corresponds to the point in evolution at which humanness is achieved. According to Morowitz and Trefil, this point of acquired humanity does not depend upon the viability of the fetus but on an innate biological quality within the fetus. Recall that viability depends upon the capabilities of medical science to keep a fetus alive outside the womb. In sharp contrast, the origin of the cortex relies, instead, upon a biologically identifiable moment within the life of the fetus itself. Therefore, Morowitz and Trefil conclude that it makes much more sense, from a scientific point of view, to focus on the development of the cerebral cortex and, therefore, the acquisition of humanness as the governing principle in abortion decisions.[65]

Although Morowitz and Trefil appear to be among the first to suggest the origin of the cortex as a proper replacement for the viability standard, they are not the first to suggest the idea of acquired humanness as the correct guideline in abortion cases. One commentator, Baruch Brody, in his essay "On the Humanity of the Fetus," has expressed the point of view that humanness cannot be defined by any outside, arbitrary criteria. Rather, humanness can only be discovered by an evaluation of what it means to be human. According to Brody, when we discover the essential characteristic that makes people human, we can solve the problem of determining when the fetus becomes human. The equation works like this. Once the fetus is in possession of that essential characteristic, it has become human and is entitled to all human rights. Brody identifies the essential characteristic of humanness as the possession of a functioning brain. He, therefore, concludes that the beginning of encephalographic activity represents the threshold of humanness.[66]

All this sounds very much like Morowitz and Trefil and, in fact, Brody's argument does follow the same line of logic as that of the other two scholars. There is, however, one significant difference. That difference lies in the timing of the brain's development. As we have seen, Morowitz and Trefil focus on the birth of the cortex, which occurs at about twenty-five weeks. In contrast, Brody pinpoints the beginning of encephalographic activity, which occurs at six weeks.[67] This represents a difference of eighteen weeks. Moreover, even Brody's line of demarcation, that is, the sixth week, cannot be taken as an absolute starting point for the acquisition of humanness, because encephalographic activity can begin as late as eight weeks.[68]

At best then, Brody's criteria represent the beginning of a period of ambiguity. Before six weeks, Brody would say that the fetus has not acquired the essential characteristic of humanness. After six weeks, the fetus has reached a point at which it has a greater possibility of crossing that line and becoming human. However, that line is different for each fetus and can only be determined when encephalographic activity actually begins. Therefore, Brody's criteria has limited application as a generalized principle. Brody's logic in choosing the onset of encephalographic activity as the starting point for the acquisition of humanness is that the same criteria are used to measure the point of death. Since the end of life is measured by the cessation of brain activity, then the beginning of life should be determined by the onset of brain activity. At the very least, the balanced use of the same criteria for life and death gives Brody's answer to the problem of determining the onset of humanness a logical symmetry that is lacking in most other formulas.[69]

Two other scholars, Michael Tooley and Mary Anne Warren, also focus on the activity of the brain as the point for determining the onset of humanness. Taking a remarkably existential point of view, Tooley prefers to accentuate the onset of self-awareness as the threshold of acquired humanness.[70] Included in Tooley's concept of self-awareness is the ability to envision the future, as well as the capacity to have a personal stake in that future. Tooley also places emphasis on the ability to conceive of the self as a continuous entity.[71] Warren's criteria, though a bit longer and more involved than Tooley's, nevertheless depend upon the development of certain capacities within the human mind. Warren actually lists five abilities of the human mind which, once achieved, mark the onset of humanness. These abilities are, in order of critical importance, consciousness, reasoning, self-motivated activity, the capacity to communicate, and the presence of self-concepts.[72] Warren seems to believe that consciousness and reasoning are the most essential of the five criteria, and she appears to hold that they alone might be sufficient to establish the existence of humanness. She also at times designates the ability to carry out self-motivated activity as a necessary characteristic.[73]

Advantages of the Acquisition-of-Humanness Standard. Before moving on to an evaluation of these criteria, let us review what we have established thus far. First, Morowitz and Trefil have proposed that the viability standard of *Roe* and *Casey* be replaced with a more scientifically accurate criterion that relies not on the ability of technology to keep a fetus alive outside of the womb, but on some biological criterion that is inherent within the fetus itself. Morowitz and Trefil go on to suggest that this biologically based criterion should be the birth of the cortex, which occurs at about the twenty-fourth week of pregnancy. Others who support the notion that viability should be eliminated in favor of a scientifically accurate criterion include Brody, Tooley, and Warren. Brody suggests the onset of encephalized activity, Tooley suggests self-awareness, and Warren suggests five criteria including consciousness, reasoning, self-motivated activity, the capacity to communicate, and the presence of self-concepts.

There are certain advantages to the course of action suggested by these scholars. Chief among these advantages is the fact that the acquisition-of-humanness standard focuses *not* on something external to the fetus, as viability does, but instead on a quality that is inherent within the fetus itself. As such, the acquisition of humanness represents a more objectively measurable standard than viability. Viability can change, depending on advances in medical technology. Conceivably, the barrier of viability can be pushed back in time further and further, thus making the point at which the government's compelling interest in the life of the fetus arises changeable and uncertain. In fact, according to the Supreme Court at least, the last word may never be spoken on this issue because science may be able to push the door of viability back further and further to earlier and earlier stages of pregnancy. In discussing the future of the viability standard of *Casey*, the Court states that viability may continue to move to earlier points in pregnancy "if fetal respiratory capacity can somehow be enhanced in the future."[74]

According to Morowitz and Trefil, no such problem exists with the acquired humanness standard they have proposed. Since the standard of acquired humanness, specifically the standard based upon the birth of the cortex, depends upon a biological quality inherent within the fetus, the standard becomes much more concrete and less subject to change than the criterion of viability. To their credit Morowitz and Trefil do not claim that all problems connected with abortion will be solved by adopting their standard. The exact moment that the cortex of a particular fetus has developed will still remain somewhat uncertain. Accordingly, the line of demarcation, that is the twenty-fourth week, does not represent an absolute starting point for the acquisition of humanness. This problem is similar to the one encountered by Brody in his attempt to pinpoint the line of demarcation as the onset of encephalographic activity. Both criteria seem to represent the beginning of a period of uncertainty during which the fetus is becoming more and more human all the time.[75]

As Morowitz and Trefil say, "[b]efore [the beginning of the third trimester], we can present convincing arguments that the fetus has not acquired humanness. Once the burst of synapse formation starts, however, this denial is no longer possible."[76] What then is gained by the adoption of the acquisition-of-humanness standard? According to Morowitz and Trefil, the primary advantage of their scientific standard is that it is "in tune with biological reality."[77] Rather than being based upon the technological support systems of a particular scientific era, the acquisition-of-humanness standard is based upon what happens within the body of the fetus. This is a quality that is inherent within the biological development of the fetus, not an arbitrary standard based upon what medical support systems may be technically feasible at the time.

Problems with the Acquisition-of-Humanness Standard. The argument that the acquisition-of-humanness standard provides a more scientifically accurate view than the viability standard is a powerful one. After all, as we have seen, one of the primary goals of the law is to establish a predictable set of standards that can guide the behavior of the people. Proposing a guideline that is scientifically in tune with biological reality and that is inherent within the fetus itself would seem to meet this goal admirably. Despite the fact that the standard proposed by Morowitz and Trefil has its own shortcomings, it is still a workable guideline. Indeed, all of the standards suggested thus far are workable. Moreover, the underlying idea of using humanness or personhood, rather than viability, as a guideline in abortion cases also appears to be workable. What then is the problem?

The problem lies not in the underlying theory but in its execution. Even those who agree that the acquisition of humanness should be used as the guideline in abortion cases cannot agree on what constitutes humanness. Brody suggests that personhood is achieved at six weeks at the onset of encephalization activity. Tooley suggests it occurs with the advent of self-awareness. Warren suggests humanness must be tied to consciousness, reasoning, self-motivated activity, the capacity to communicate, and the presence of self-concepts. And Morowitz and Trefil suggest that the fetus acquires humanness at the birth of the cortex. Each of these guidelines appears convincing when considered within the context of its own set of finely tuned, scientific arguments. However, when they are compared to one another, they all seem more or less arbitrary. Each standard claims to be in line with biological reality, and indeed, all the arguments have a firm foundation in the science used to defend them. But why should we prefer one guideline over another?[78]

In fact, there are other guidelines that could be suggested that might push the point of personhood even further back into the opening stages of pregnancy. For instance, it could be argued that humanness is imprinted within our DNA and that the blueprint for a life is completed with the formation of the zygote.[79] Recall that a zygote is the single cell that exists after an egg has been fertilized.[80] All that is needed after that point is enough time and the proper environment for the fetus to fully develop and become a person. Thus, according to this view, humanness would begin at conception.[81] A similar guideline would place the moment of acquired humanness at the point at which the zygote has been successfully implanted in the womb. This approach would place the moment of humanness about six to seven days after conception.[82]

Naturally, each suggested cut-off point for humanness has some basis in biological reality. To their credit, Morowitz and Trefil are especially convincing in arguing that the origin of the cortex should define the moment that the fetus becomes human. Two critical pieces of biological evidence form the cornerstone of their argument. First, they point out that the birth of the cortex occurs at about the twenty-fifth week of pregnancy. This point in time corresponds almost exactly to the end of the second trimester and is thus exactly in line with the criteria established in *Roe* and *Casey*. Therefore, the origin of the cortex criterion seeks not to change the timetable established by the viability standard of *Roe* and *Casey*, but to move the focus of that timetable from the abilities of medical technology to the biological characteristics of the fetus itself. Therefore, without altering the timetable, Morowitz and Trefil argue that their standard can replace technological possibilities with biological reality.[83]

The second reason that Morowitz and Trefil are confident that their standard best fits biological reality is that the birth of the cortex corresponds almost exactly to the moment at which the fetus crosses the line to survivability. This is significant because recent studies have demonstrated that there is an absolute limit to the ability of medical technology to keep the fetus alive outside of the mother's womb. Studies of the survival rates of premature infants indicate that no advances in medical technology will be able to push back the date of viability earlier than the twenty-fourth week of gestation.[84] Medical advances may increase the survival rate of infants born after the twenty-fourth week; but, at the present time, it is not possible to extend the limit any earlier.[85]

This twenty-four week barrier, before which the baby cannot survive outside the womb, is caused by the fact that, before that time, the fetal organs, especially the kidneys, heart, and lungs, are not developed to the point that they can sustain the infant without support. However, such support cannot be provided to the infant because any attempt to interfere with the baby's primitive organs at this point makes the infant's condition worse. At this time, the problem does not seem to be a technological shortcoming. Rather, it appears that nature itself has set up a boundary line of survivability before which even the most sophisticated technological support is useless.[86]

Are Morowitz and Trefil correct? Does the twenty-four week limit spell an absolute barrier beyond which science will be unable to penetrate in its quest to extend fetal viability? Many people are skeptical whenever an absolute limit is placed upon the ability of science to solve a problem. It is always safer, and usually more accurate, to state that there is no ultimate barrier to the ability of science to unravel a mystery. Specifically, in this case, the arguments presented by Morowitz and Trefil suggest that fetal development requires a natural womb. This is certainly true today. However, it may be somewhat overconfident, perhaps even foolhardy, to say that the creation of an artificial womb will never be realized. If an artificial womb can be constructed, then the twenty-four week barrier is broken and the question of survivability is no longer at issue.[87]

The deliberative nature of the law makes it virtually certain that none of the suggested criteria for the measurement of the moment of acquired humanness will be adopted by the courts in the near future. The tendency of the law to weigh all points of view equally also makes it unlikely that any legislature will adopt a human life statute or that the Congress will introduce a fetal rights amendment to the Constitution. But this is as it should be. After all, one of the law's highest functions is to weigh and balance new ideas as carefully as possible. The law seeks to assimilate new ideas, but in a deliberate fashion so as to preserve an orderly and stable state. The law's approach to the abortion debate is an excellent example of this attempt to keep things stabilized. If judges and legislators were swayed by each new scientific argument that came along, the viability standard would probably change on a weekly, perhaps even a daily, basis. Instead, the law takes a steady and deliberative approach and, as a result, will adhere to the predictable and workable standard of viability, until there is irrefutable proof that such a standard is no longer applicable.

11-5 JURISCIENCE AND PROCREATIVE RIGHTS

Another reason that the law has rejected any attempt to replace viability with the acquired-humanness standard, whatever the scientific basis for that standard, is that such a standard ignores a fundamental side of the abortion debate—the woman's right to be free from unnecessary and harmful governmental interference in the control of her own body. It is important to remember that the law has a dual function. In addition to establishing order and stability, the law must strive to treat people with justice. The acquired-humanness standard, with its focus on fetal development, sidesteps the treatment of women and thus ignores the whole question of justice for women in relation to procreative freedom. Science may not be concerned with the concept of justice, but the law cannot afford to ignore the fundamental question of equity and fairness. The question of justice in relation to procreative rights is multidimensional. One dimension involves the woman's privacy rights. Another concerns her responsibilities toward others in relation to procreative freedom.

Procreative Freedom and the Right to Privacy

The right to privacy has several different legal meanings. The right to privacy can mean the right to confidentiality. Generally, such a right arises when one person has access to privileged information, for example the medical records of another person. In such a situation the right to privacy gives rise to a duty not to reveal that restricted information. The right to privacy can also involve locality. This concerns the right that individuals have to use their property without unwarranted interference. Neither of these privacy rights is what is meant in the case of procreative freedom, however. Rather, the right to privacy in procreative matters, as articulated by the Court in both *Roe* and *Casey*, means each person's right to sovereignty over the most intimate decisions in life.[88] These decisions include any choice that impacts on procreative matters. As we have seen, these procreative matters involve not only the right to use contraceptives, the right to have an abortion, and the right to choose adoption over child rearing, but also the right to have children in the first place.

This right was first established by the U.S. Supreme Court in *Griswold v. Connecticut*. In *Griswold* the Court held that the government could not criminalize the use of contraceptives by married couples.[89] In *Eisenstadt v. Baird*, the Court extended the same protection to unmarried couples.[90] The Court in *Eisenstadt* noted that, "[i]f the right of privacy means anything, it is the right of the *individual*, married or single, to be free from unwarranted governmental intrusion into matters so fundamentally affecting a person as the decision whether to bear or beget a child."[91] Finally, in *Carey v. Population Services International* the Court ruled that the sale and the distribution of contraceptives were also protected activities.[92] It is crucial to recall that the procreative liberty that the court spoke of in these cases includes not only the right to avoid reproduction through the use of contraceptives and abortion, but also the freedom to procreate without unjustifiable governmental interference.

The right to procreate is fundamental to our existence as individuals because it goes to the heart of what it means to be a human being. It contributes to our identity as individuals and as a part of the human family. Procreative decisions also shape our relationships with others and involve a number of responsibilities that are both moral and social. Women are affected in a profound way because it is tied directly to their physical and psychological well-being. Courts have ruled that individuals should be permitted to make these personal decisions free from the intrusive interference of the government. Such decisions, the courts have ruled, are best made by those individuals who are directly affected by the consequences of those decisions. None of this has anything to do, except in the most indirect way, with a scientifically determined moment of acquired humanness.[93]

Science may advance, as it did in Axel Madsen's fictional world of *Unisave*, to the point that we will be able to control all aspects of human procreation. Or science may eventually be able to manufacture people artificially as the Alphas and Betas at the Central London Hatchery and Conditioning Center did in Aldous Huxley's *Brave New World*.[94] In *Brave New World*, all procreative choices were made by the government according to a master plan designed to maximize the stability of the social order. The world imagined by Huxley may be just around the corner from a scientific perspective. From a legal point of view, however, such a world is not likely, at least not in the United States. In *Roe* and *Casey*, and in all other cases touching on the nature of procreative rights, the U.S. Supreme Court has explicitly established the right of privacy in procreative matters with such force and such authority that nothing short of a constitutional amendment or an armed insurrection will alter those rights in the foreseeable future.

Procreative Freedom and Responsibilities

The right to privacy in procreative decisions is not based solely upon a woman's personal concerns. Such decisions cannot be made in a vacuum. Rather, they are made in the context of family, community, and professional relationships. As Robin West, professor of law at the University of Maryland Law School, points out, a woman making procreative choices must contend with many factors that limit her freedom. She may be burdened by the absence of her spouse or by his irresponsibility in fiscal and family concerns. She may work for a corporation that is unresponsive to the requirements of working mothers. She may be trapped in a social setting that is apathetic when it comes to the problems of day care. In such cases a procreative decision may be made so that the woman can live up to her responsibilities as a contributing member of society. Such considerations are not a part of the acquired-humanness standard, whatever its scientific foundation.[95] According to West, focusing on the acquired-humanness factor obscures the real nature of procreative decisions. She argues, quite convincingly, that procreative decisions, even the decision to have an abortion, can be a morally responsible choice.[96] As West writes, "[t]he abortion decision typically rests not on a desire to destroy fetal life but on a responsible and moral desire to ensure that a new life will be borne only if it will be nurtured and loved."[97]

A study conducted by Carol Gilligan, associate professor of education at the Harvard University Graduate School of Education, supports West's contention that women tend to make moral judgments in terms of specific responsibilities rather than generalized ethical principles.[98] The study also demonstrates that most women caught in a procreative dilemma do not think in terms of the humanness of the fetus. In fact, in some cases the humanness of the fetus is assumed, and the woman feels that she must balance the responsibility that she has to the unborn child with the responsibility that she owes to herself and to other people who depend upon her for support. In many cases Gilligan found that women characterize the decision to have a child as the "selfish" decision, whereas their decision to give up the child is seen as the responsible choice.[99]

It is, therefore, unlikely that any change in the viability standard would have any effect on the moral reasoning involved in procreative decisions. Thus, the law appears to be correct in its deliberative approach to this controversial and highly volatile subject. Unfortunately, the law's deliberative process is not always beneficial in dealing with juriscientific issues. Sometimes, undue deliberation causes hardship and unnecessary delays in areas that would be better handled in a more direct and efficient manner. One example of the negative impact of the law's deliberative process involves the availability of RU486 and other contragestive agents.

Juriscience and Contragestive Agents

Before examining the issue of contragestive agents, we should step back a moment and remind ourselves of the Court's position in *Planned Parenthood of Pennsylvania v. Casey*. In brief, the Court in *Casey* supported the viability standard as the point before which the government cannot prevent a woman from making the ultimate choice in abortion related decisions. Moreover, any governmental restriction imposed upon this decision making process before viability must not place an undue burden on the woman. Thus, in *Casey*, the

Court said that requiring a woman to inform her husband before she seeks an abortion would be an undue burden, whereas requiring a 24-hour waiting period would not impose such a burden.[100] Commentators have referred to this point of view as the "middle position" because it *permits* but *does not encourage* abortion.[101]

By assuming the middle position, the Court has honored past precedents such as *Roe v. Wade*, *Griswold v. Connecticut*, *Eisenstadt v. Baird*, and *Carey v. Population Services International*, as it should. Moreover, the Court has not jumped on any new scientific bandwagon, as has been suggested by some commentators. Instead, the Court has chosen to maintain a firm and steady dedication to the viability standard as its primary guideline in limiting procreative decisions that involve abortion. For the most part, the Court's willingness to consider all new scientific theories, coupled with its reluctance to endorse a single theory over all others without overwhelming proof has successfully balanced order and justice within the social system. However, this is not always the case.

In one area in particular, the law's deliberative attitude has caused some problems. This area involves the introduction of new birth control devices such as contragestive agents. A **contragestive agent** prevents the implantation of a fertilized egg on the uterine wall. If used at a slightly more advanced point in a pregnancy, the contragestive agent interrupts the implantation of the embryo. Contragestive agents include the morning-after pill, intrauterine devices, and low-dosage birth control pills. However, the contragestive agent that has received the most attention lately is RU486, an agent manufactured by the French company Roussel-Uclaf.[102]

There are several advantages to the use of contragestives. First of all, contragestives are safer than surgical abortions. They are also more efficient and more economical than the more conventional methods of abortion. Also, since contragestive agents operate so early in a pregnancy, their use offends fewer people than other techniques. In fact, if used early enough, a contragestive agent actually operates more like a postcoital contraceptive than an abortion method. Despite these advantages, the manufacturers of some contragestive agents, notably RU486, have been reluctant or unable to market their product in the United States.[103] The manufacturer's reluctance to market RU486 is not based upon any failure on the part of the agent itself. On the contrary, RU486 has proven itself to be highly effective in other markets. The agent has proven to be successful in France, Great Britain, and the Netherlands. In fact, in France, RU486 is responsible for thirty-three percent of all abortions. Moreover, the effectiveness of RU486 has been recently successfully tested in both China and Sweden.[104]

Why then has the manufacturer been hesitant to market the agent in the United States? Part of this reluctance was no doubt due to the volatile nature of the abortion debate here. However, another source of this hesitancy can be traced to the law itself. One aspect of this has been the reluctance of the federal government to allow the importation of RU486. In one instance, the Justice Department under the Bush administration denied permission to a woman who asked to be allowed to import minute quantities of RU486 into the United States. The refusal was issued even though it was a general practice to permit Americans to bring small amounts of unapproved drugs into the United States.[105]

Added to this deliberative process is the rigorous testing procedure required by the Food and Drug Administration (FDA). Since its inception, the FDA has developed a reputation for moving quite cautiously in its approval of new drugs and medical devices.[106] However, in the 1980s, the FDA became even more guarded when it was faced with the challenge of newly developing techniques of biotechnology. At that time, the contrast between the cautious attitude of the FDA and the aggressive stance of the biotechnology companies became readily apparent. Moreover, the FDA's testing procedures began to place an economic strain on these relatively new biotechnology companies.[107]

This approach undoubtedly had a chilling effect on the manufacturer of RU486, which was still reluctant to submit the new contragestive agent to the FDA despite the Clinton administration's invitation to do so and despite the Supreme Court's clear message in *Casey* that *Roe* will not be overruled. Eventually, the manufacturers of RU486 made a remarkable decision. In 1994, Roussel-Uclaf agreed to give a nonprofit organization known as the Population Council exclusive rights to RU486 in the United States.[108] Under terms of the agreement the Population Council was charged with the task of supervising the clinical trials of RU486. Moreover, the Population Council also promised to find an American company which would be responsible for seeking final FDA approval of RU486.[109]

The law generally debates all sides of an issue before adopting new technologies and new scientific theories even when it might be inappropriate to do so. Such has apparently been the case with RU486.[110] Some commentators have been critical of this deliberative process because it prevents many helpful drugs and medical devices from reaching the public in a timely fashion. In addition, a highly structured deliberative process tends to support the established scientific community and thus stifles progress and innovation. Does the RU486 case and the critical comments of some experts mean that legislators, judges, and regulators should adopt a more rapid approach to the approval of scientific theories and technological developments? Should regulators be more willing to permit the introduction of new drugs and new medical devices into the marketplace? Should judges be more liberal in their tolerance of new scientific theories that may change legal rules? Should legislators be more willing to introduce statutes that mandate the teaching of particular scientific theories in our schools and universities?

The answer to each of these questions is—probably not. If legislators, judges, and regulators do throw caution to the wind, they run the risk of endorsing unproven theories, introducing dangerous and untested products, or unwittingly incorporating pseudoscience or junk science into the body of the law. Additionally, there is the added problem of promoting scientific solutions to problems that may actually do more harm than good. These solutions may seem clinical and objective, but they also might inadvertently damage long-held legal rights. Such has been the case with the Norplant system.

The Norplant Controversy. The Norplant contraception method involves the implantation of six tiny silicone tubes under the skin of a woman's upper arm. Each of the tubes is approximately as big as a match stick. The implantation procedure lasts about 10 to 15 minutes. After the tubes have been placed in the woman's arm they begin to discharge an artificial type of progestin, which prevents ovulation and results in temporary infertility. The hormones also restrict the movement of sperm into the uterus, thus adding an additional protection against fertilization. The effects of the original implantation last for five years. Figures on the effectiveness of the Norplant system vary from 1.5 to 5 percent possibility of pregnancy over the 5-year period of its effectiveness.[111]

Almost as soon as Norplant was released on the market, several legislators seized upon its availability as a means of establishing social policy. In Louisiana, a state legislator proposed a bill that would reward mothers on welfare with a payment of $100 if they would agree to have the Norplant system implanted. Similarly, in Kansas, legislation was introduced that would authorize an initial fee of $500 and a $50 maintenance payment each year to welfare mothers who agreed to the Norplant treatment.[112] Another state legislator has suggested requiring fertile women who are also drug abusers to be implanted with the Norplant system as a prerequisite for probation.[113] Other suggestions that have surfaced in relation to Norplant include requiring implantation as a condition of welfare, using the Norplant system as compulsory contraception to prevent congenital disease, or implementing Norplant as a means to reduce pregnancy of adolescents.[114]

These cases provide examples of lawmakers who have moved decisively to use a new scientific development to shape social and legal policy. The question is whether such a move is beneficial to all society in the long run. Certainly, there are arguments on both sides of this issue.[115] However, no matter what point of view a person supports in these situations, there can be no blinking at the fact that both the Louisiana and Kansas legislators have proposed radical solutions that should have been considered in a more cautionary fashion. The area of procreative freedom is much too fundamental and much too precious to surrender for the sake of quick solutions. The deliberative legal process prevailed in all of the Norplant situations. Both the Louisiana and the Kansas state legislatures refused to rush to judgment and enact the Norplant welfare incentive programs. In both instances, the processes of the law acted as they should to curb the tendency to incorporate every new scientific idea and each new technological development into the law. However, this is not the end of the story. In the case of procreative rights, the courts have been dealing with a familiar subject area. After all, procreation has been with us since the dawn of time. But what happens when an entirely new scientific endeavor enters the social arena, bringing with it new ethical and legal problems? This is what has happened with recombinant DNA and genetic engineering. It is to these controversial topics that we will turn our attention.

CONCLUSION

In this chapter we have examined several widely diverse topics. First, we entered the area of the law to understand better how it operates and to see how the law and science interact. Our purpose was to examine the similarities between the goals of law and those of science. We also focused on science and law as partners in shaping social and cultural issues. We then explored the impact that science has on law. This intersection of law and science is known as juriscience. One key issue on the cutting edge of juriscience is the issue of procreative rights, which involve deeply complex legal and ethical issues. Some commentators have suggested that a definition of humanness might be the first step toward the solution of the ethical and legal problems associated with procreative rights. Several different standards have been suggested as a basis for defining humanness. All suggested criteria, however, appear to be somewhat arbitrary. Moreover, none of these standards take into account the woman's rights in procreative matters.

Review

11-1. Nonscientists may face scientists conducting research into the origin and evolution of life and ask "Why are you so concerned about issues that seem not only remote, but also of no immediate practical value?" We can address these concerns in a number of ways. First, there is a practical side to research into the origin and evolution of life. Determining how life began may help save individual lives, prolong life in general, and perhaps even create artificial life. Understanding the far-reaching implications of these issues, however, requires, not only a knowledge of science, but also a basic understanding of ethics and the law.

11-2. Juriscience refers to issues that stand at the intersection of science and the law. The law is a set of rules and regulations established by the government of a nation or the subdivision of a nation in order to maintain harmony and equity. Ideally, the law-making process should be guided by a set of objective ethical concepts that can be used to distinguish right from wrong. Two ethical concepts that might be used in this regard are utilitarianism and reciprocity.

Utilitarianism states that right and wrong can be determined by measuring an action in terms of the good or evil that it produces. The principle of reciprocity can be used to judge the moral substance of a rule. Reciprocity is based on the belief that sentient beings recognize their own self-worth and, therefore, strive to be treated with dignity and respect. The contemporary sources of American law are common law, judicial decisions, statutory law, the Constitution, and administrative regulations.

II-3. The law and science interact in a number of different ways. First, the government, which is after all an extension of the law, may be called upon to provide financial support for scientific research and development. Second, also at the legislative and regulatory level, a scientific advance or a technological development might need legal approval before it can be marketed to the public. Third, the government allows individuals to preserve their rights to any new invention that they develop. These rights can be protected by applying for a patent. Fourth, a scientific advance or a technological development might provide a revolutionary new means for gathering or evaluating evidence at a trial. A fifth way that science and the law can legitimately interact within the courts occurs when an invention or a new process causes harm to people or property. The final way that science and the law interact is when a scientific theory or a technological development either creates a new legal problem or adds a new twist to an already existing legal issue.

II-4. The central holding of *Roe v. Wade* is that a state cannot interfere with a woman's right to seek an abortion before the fetus becomes *viable*. Viability can be defined as the point at which the fetus can live, albeit with artificial assistance, outside the womb. The viability standard, at least the way it has been defined by the Court thus far, does not depend upon any innate qualities or characteristics of the fetus itself. Instead, the viability standard targets the ability of physicians to keep the infant alive. Some commentators have suggested focusing on the inherent qualities of human life rather than on the abilities of a physician to preserve that life. Many different standards have been suggested for determining the point at which the fetus gains human status. All of these standards appear to be somewhat arbitrary, and none of them take into account the woman's rights in procreative matters.

II-5. One reason that the law has rejected any attempt to replace viability with the acquired-humanness standard, whatever the scientific basis for that standard, is that such a standard ignores the woman's right to be free from unnecessary and harmful governmental interference in the control of her own body. The question of justice in relation to the abortion debate is multidimensional. One dimension involves the woman's privacy rights. The right to privacy in abortion cases, as articulated by the Court in both *Roe* and *Casey*, means each person's right to sovereignty over the most intimate decisions in life. These decisions include any choice that impacts on procreative matters. Another concerns her responsibilities toward others in relation to reproductive freedom. Many outside factors impact upon a woman's procreative decisions, which may be made so that the woman can live up to her responsibilities as a contributing member of society. Such considerations are not a part of the acquired-humanness standard, whatever its scientific foundation. It is, therefore, unlikely that any change in the viability standard would have any effect on the moral reasoning involved in procreative decisions.

Understanding Key Terms

axon	golden rule	preembryo
binding precedent	judicial decisions	procreative freedom
blastocyst	juriscience	rational basis
case law	jurisdiction	reciprocity
common law	justice	RU486
compelling state interest	law	*stare decisis*
Constitution	neuron	statutory law
contragestive agent	neurotransmitter molecules	subjective ethics
dendrite		synaptic cleft
due process of law	normality	synaptic terminal
embryo	order	synaptic vesicles
ethical relativism (subjective ethics)	patent	utilitarianism (consequentialism)
fetus	persuasive precedent	
	placenta	viability
fundamental right	precedent	zygote

Review Questions

1. Why is it necessary to study science in relation to issues such as abortion and genetic engineering?
2. What is law?
3. What are the ethical roots of law?
4. What are the contemporary sources of law?
5. How are law and science similar?
6. What is the current state of law in relation to abortion?
7. How do law and science interplay in relation to abortion?
8. What is the relationship between acquired humanness and the law of abortion?
9. How are the right to privacy and the need to meet responsibilities related to procreative freedom?
10. What legal issues surround the use of contragestive agents?

Discussion Questions

1. The text points out that far from settling the abortion conflict, *Roe v. Wade* simply fanned the flames of the controversy. For more than twenty-five years, both sides in the debate have waged battle in the courts, in state legislatures, and in the media. Part of the cause of the continuing debate concerns the court's insistence that viability be used as the guideline for determining when a fetus becomes human. What is your reaction to the use of the viability standard? Would you favor maintaining the viability standard or shifting to the acquired-humanness standard as the guiding principle? Explain.
2. Some commentators have suggested that the time at which the cortex forms coincides with the time at which the fetus can survive outside the womb. This phenomenon appears to support the idea that humanness emerges at a point that corresponds with these two events, that is at the twenty-fourth week of gestation. What do you make of this coincidence? Does the fact that the two events correspond to one another support the birth of the cortex theory, or are the two events mutually exclusive? Do you agree with statements that seem to indicate that there is a limit to the point at which human intervention can help a fetus survive outside the womb? Explain. What do you think it would take to convince the Supreme Court to shift from a viability standard to a birth-of-the-cortex standard? Explain. Would you be willing to conduct a campaign aimed at making the change? Explain.
3. In Aldous Huxley's well-known novel, *Brave New World*, all procreative choices are made by the government according to a master plan designed to maximize the stability of the social order. Some commentators suggest that, from a scientific point of view, the world imagined by Huxley may be just around the corner. Do you agree or disagree with this assessment? Explain. Some experts say that, from a legal perspective, such a world is not likely, at least not in the United States. Do you agree or disagree with this assessment? Explain. Some people have suggested that the future civilization of Huxley's *Brave New World* might actually be preferable to the world as it exists today. These commentators point to the fact that in Huxley's universe most of the problems that face modern civilization have been eradicated. First make a list of all of the problems that have been eliminated by the genetically engineered and behaviorally conditioned world of Huxley's *Brave New World*. Then argue that eliminating these problems is immoral.
4. The law has generally been slow to incorporate new technologies and new scientific theories even when it is appropriate to do so. The textbook suggests that this is what happened in the case of RU486. Do cases like this mean that legislators, judges, and regulators should adopt a more proactive stand in relation to scientific theories and technological developments? Explain. Should regulators be more willing to permit the

introduction of new drugs into the marketplace? Explain. Should judges be more liberal in their tolerance of new scientific theories that may change legal rules? Explain. Should legislators be more willing to introduce statutes that mandate the teaching of particular scientific theories in our schools and universities? Explain.
5. In *Griswold v. Connecticut* the U.S. Supreme Court held that the government could not criminalize the use of contraceptives by married couples. In *Eisenstadt v. Baird* the Court extended the same protection to unmarried couples. In *Carey v. Population Services International* the Court ruled that the sale and the distribution of contraceptives were also protected activities. Do you agree or disagree with the Court's ruling in each of these cases? Explain your position in relation to each case.

ANALYZING *STAR TREK*

Background

The following episode from *Star Trek: The Next Generation* reflects some of the issues that are presented in this chapter. The episode has been carefully chosen to represent several of the most interesting aspects of the chapter. When answering the questions at the end of the episode, you should express your opinions as clearly and openly as possible. You may also want to discuss your answers with others and compare and contrast those answers. Above all, you should be less concerned with the "right" answer and more with explaining your position as thoroughly as possible.

Viewing Assignment—*Star Trek: The Next Generation*, "The Offspring"

Lt. Commander Data attends a cybernetics conference at which he learns about a new submicron matrix transfer technology. A proper application of the new technology may allow him to transfer his programming to another android. Accordingly, on his return from the conference he conducts an extended project which results in the creation of a new android. In effect, the new android, which Data names Lal, is his daughter. Immediately, controversy surrounds the creation of Lal, and Starfleet sends Admiral Haftel to investigate Data's fitness to act as a proper father. Haftel meets with the Enterprise crew and interviews a very confused Lal. Apparently, Starfleet has given Haftel the authority to remove Lal from Data should he conclude that Data is an unfit father. Lal's distress at the possibility of leaving the Enterprise and being separated from Data causes her positronic brain to malfunction. As a result she suffers an emotional breakdown. Data's attempts to save her are futile and her neural system undergoes a total failure. Before she passes away, Data transfers her memory into his positronic brain. Data's final action prevents Lal's complete demise.

Thoughts, Notions, and Speculations

1. Recall that in a previous episode, "The Measure of a Man" (see chapter 10 page 341), Picard convinces the court that Data possesses an emergent humanness that cannot be taken away from him. This emergent humanness gives him the same rights as any other sentient being in the Federation. Accordingly, he cannot be reduced to a piece of property but must, instead, be treated with the same respect and the same dignity with which all sentient life forms are treated. Moreover he must be granted the same rights as any other sentient being. The final decision of the court is to allow Data to choose his own fate. Now, however, Data seems once again to be treated as a piece of property. Apparently, Starfleet has the right to take his child away from him without due process of law. Does this mean that Data does not enjoy the same rights as other members of the Federation? After reading this chapter and learning how the law works, speculate on what may have happened to change the law since Data's hearing in "The Measure of a Man."

2. Many people argue that the government should be involved in regulating abortion rights. Such steps do, of course, set a precedent for regulating other family-oriented issues. This may be what happened in "The Offspring." It is possible that Data has not been singled out for special treatment. For all we know, Starfleet has the power to remove any parent's child after a simple interview, such as the one conducted by Admiral Haftel. Do you believe that the government should have the kind of power that Starfleet appears to have in this case? Explain.

3. In this chapter, we defined the threshold of sentience as the point of "acquired humanness." We also identified this level of acquired humanness with the development of the neocortex. Recall also the characteristics of sentience laid out by Lt. Maddox in "The Measure of a Man." Maddox stated that to be sentient, a being must exhibit intelligence, self-awareness, and consciousness. Do these characteristics appear in line with your understanding of "acquired humanness"? Does Lal qualify as sentient according to these criteria? Explain.

4. Should Data have the right to transfer Lal's memories into his own brain? Explain. Now reconsider your position in light of the following argument. In effect, by removing Lal's memory while she still exists, Data has robbed her of both continuity and identity, that is, the very thing that make her a person. Has Data not, therefore, committed euthanasia? Explain.

5. Should artificial beings, like Data and Lal, be granted equal status with human beings? Explain. Now consider the possibility that in the near future scientists may be able to engineer human life genetically. Should such genetically manufactured individuals be granted equal status with human beings? Explain.

NOTES

1. Axel Madsen, *Unisave*, 111.
2. Robert L. Heilbroner, *An Inquiry into the Human Prospect*, 56–57.
3. Gordon W. Brown and Paul A. Sukys, *Business Law with UCC Applications*, 9th ed., 3–4, 17.
4. Gordon W. Brown and Paul A. Sukys, *Understanding Business and Personal Law*, 10th ed., 7–10.
5. Brown and Sukys, *Business Law*, 23.
6. Ibid., 23–24. (For a good example of a case in which a court creates a common law principle while simultaneously interpreting a statute see *Greeley v. Miami Valley Maintenance*), 49 Ohio St. 3d 229 (1990).
7. *Roe v. Wade*, 410 U.S. 113, 93 Sup. Ct. 705 (1973).
8. Brown and Sukys, *Business Law*, 21.
9. Ibid., 25–26.
10. Ibid., 18.
11. Ibid., 18–19.
12. Brown and Sukys, *Understanding Business and Personal Law*, 16.
13. Harold J Morowitz and James S. Trefil, *The Facts of Life*, 46; Boyce Rensberger, *Instant Biology*, 102–6.
14. *Roe v. Wade*, 410 U.S. 113, 93 Sup. Ct. 705 (1973).
15. Bernard Schwartz, *Constitutional Law*, 290–91.
16. *Shapiro v. Thompson*, 394 U.S. 618, 89 Sup. Ct. 1322, 22 L.Ed. 600 (1969).
17. *Roe v. Wade*, 410 U.S. 113, 93 Sup. Ct. 705 (1973).
18. Leslie Stevenson and Henry Byerly, *The Many Faces of Science*, 2–3.
19. Ibid., 3.
20. Jeremy Bernstein, *Cranks, Quarks, and the Cosmos*, 20–24.
21. Bernstein, 17–20; John L. Casti, *Searching for Certainty*, 29.
22. Richard A. Posner, *The Problems of Jurisprudence*, 67.
23. *Planned Parenthood of Pennsylvania v. Casey* 112 Sup. Ct. 2791, 2808–9.
24. Brown and Sukys, *Business Law*, 25–26.
25. While federal judges cannot issue advisory opinions, some state judges, those in Massachusetts, for instance, are allowed to issue such opinions. See Henry Campbell Black, *Black's Law Dictionary*, 75; Gordon W. Brown, North Shore Community College, Beverly, MA, September, 1996.
26. For a detailed look at this funding process see: Steven Goldberg, *Culture Clash: Law and Science in America*, Chapter 4.
27. For a detailed look at this regulatory process see Goldberg, *Culture Clash*, Chapter 6.
28. Brown and Sukys, *Business Law*, 296; 633–35. The criteria for a patent includes the following: The invention must qualify as a device. This means that mathematical formulas and laws of nature cannot be patented. Second, the invention must involve a novel, nonobvious, useful feature that had not been in use before the invention of this specific device.
29. For a detailed article on voiceprint identification, see James Hennessy and Clarence H. A. Romig, "A Review of the Experiments Involving Voiceprint Identification," 183–98.
30. *Frye v. United States*, 54 U.S. App.D.C. 46, 293 Fed. 1013 (1923).
31. *Frye v. United States*, 54 U.S. App.D.C. 46, 293 Fed. 1013 1014 (1923).
32. Fed. Evid. R. 702–6; *Daubert v. Merrell Dow Pharmaceuticals*, 113 Sup. Ct. 2786 (1993); Goldberg, 20–23.
33. *General Electric Co. v. Joiner*, 96–188; Harvey Berkman, "High Court Defers to Judges on Scientific Evidence," sec. A, p. 10.
34. Simeon E. Baldwin, "Liability for Accidents in Aerial Navigation," 20. In explaining the logic behind imposing absolute liability on pilots regardless of care, Baldwin writes, "An aeronaut, for his own advantage or amusement, flies over a town, and accidentally falls, causing injury to life or property as he reaches the earth. He is engaged in a dangerous pursuit: dangerous to himself and others. If unprotected by any authority of positive law, he ought to be held responsible for whatever misadventures, of a kind not unusual in such a pursuit, may befall him. In other words, it would seem that the aeronaut must absolutely assume, on principles of general jurisprudence, the risk of any such accident causing loss to others." (Baldwin, 20–21.)
35. Baldwin, 20–21; John Augustine Eubank, "Land Damage Liability in Aircraft Cases," 188, 190.
36. Eubanks, 191–93. In his article Eubanks makes it clear that the absolute liability standard in aircraft cases was under serious challenge in the 1950s.
37. Len Young Smith and G. Gale Roberson, *Smith and Roberson's Business Law*, 135.
38. *Planned Parenthood of Pennsylvania v. Casey*, 112 Sup. Ct. 2791, 2803 (1992).

39. Ibid.
40. Ibid.
41. *Roe v. Wade*, 410 U.S. 113, 93 Sup. Ct. 705 (1973); Ronald Dworkin, *Life's Dominion*, 109–110; John A. Robertson, *Children of Choice*, 57.
42. *Roe v. Wade*, 410 U.S. 113, 93 Sup. Ct. 705 (1973); Dworkin, 105; Robertson, 57.
43. *Roe v. Wade*, 410 U.S. 113, 93 Sup. Ct. 705 (1973); Dworkin, 105; Robertson, 57.
44. *Planned Parenthood of Pennsylvania v. Casey*, 112 Sup. Ct. 2791 2820–21 (1992).
45. Ibid., 2821.
46. Marcia Cole, "25 Years of 'Roe v. Wade,'" sec. A, p. 24.
47. *Planned Parenthood of Pennsylvania v. Casey*, 112 Sup. Ct. 2791, 2821 (1992).
48. Ibid.
49. Morowitz and Trefil, 131, 154.
50. Ruth Macklin, "Personhood and the Abortion Debate," 81–102. Although Macklin herself is opposed to this idea, she does an excellent job of presenting the opposite point of view.
51. Morowitz and Trefil, 9.
52. Ibid.
53. Morowitz and Trefil, 69–70; Carl Sagan, *The Dragons of Eden*, 88–89.
54. Morowitz and Trefil, 69–70.
55. Ibid., 105–6.
56. Rush W. Dozier, Jr., *Codes of Evolution*, 123–24; see also Francis Crick, *The Astonishing Hypothesis* and Roger Penrose, *The Emperor's New Mind*.
57. Dozier, 123–25.
58. Morowitz and Trefil, 113; Gordon M. Shepherd, *Neurobiology*, 203–5.
59. Morowitz and Trefil, 113–16.
60. Ibid., 116.
61. Shepherd, 215.
62. Morowitz and Trefil, 116.
63. Morowitz and Trefil, 117; Shepherd, 659.
64. Morowitz and Trefil, 119.
65. Ibid., 119, 154–55.
66. Baruch Brody, "On the Humanity of the Fetus," 230–40.
67. Ibid., 230.
68. Andre E. Hellegers, "Fetal Development," 196–97.
69. Brody, 230–31.
70. Michael Tooley, "A Defense of Abortion and Infanticide," 72. For a brief introduction to existentialism and the notion of the importance of self-consciousness see Frederick R. Karl and Leo Hamalian eds., *The Existential Imagination* and Walter Kaufmann, ed., *Existentialism from Dostoevsky to Sartre*. Specifically in Kaufmann's work see Jean-Paul Sartre, "Self-Deception," 241–70; and Sidney Finkelstein, *Existentialism and Alienation in American Literature*.
71. Tooley, 72.
72. Mary Anne Warren, "On the Moral and Legal Status of Abortion," 224.
73. Ibid.
74. *Planned Parenthood of Pennsylvania v. Casey* 112 Sup. Ct. 2791, 2811 (1992).
75. Morowitz and Trefil, 155.
76. Ibid.
77. Ibid., 154.
78. Jane English, "Abortion and the Concept of a Person," 241–42. English argues that trying to determine the acquisition of humanness is not only difficult but also unnecessary. In fact, she believes that pursuing that line of inquiry just confuses the issue making everything about the abortion debate just that much more difficult.
79. William Despain, North Central State College, Mansfield, OH, July 29, 1996.
80. Morowitz and Trefil, 46.
81. Despain.
82. Dworkin, 32–33; Morris Fishbein, ed., *Medical and Health Encyclopedia*, 808; Hellegers, 196–97; "Reproduction," *Collier's Encyclopedia*, (1990).
83. Morowitz and Trefil, 119, 159–60.
84. Ibid., 133–34.
85. Ibid., 139.
86. Ibid., 139–43.
87. Steve Abedon, Ohio State Univ., Mansfield, July 1996.
88. Dworkin, 53.
89. *Griswold v. Connecticut*, 381 U.S. 479, 85 Sup. Ct. 1678 (1965).
90. *Eisenstadt v. Baird*, 405. U.S. 438 (1972).
91. *Eisenstadt v. Baird*, 405. U.S. 438, 453, (1972).
92. *Carey v. Population Services International*, 431 U.S. 678 (1977).
93. Robertson, 24.
94. Aldous Huxley, *Brave New World*; see also Aldous Huxley, *Brave New World Revisited*.
95. Robin West, "Forward," 82–88.
96. Ibid., 83.
97. Ibid.
98. Carol Gilligan, *In a Different Voice*, 73–84.
99. Ibid.
100. *Planned Parenthood of Pennsylvania v. Casey* 112 S.Ct. 2791 (1992).
101. Robertson, 46, 60–62.
102. Dworkin, 177; Robertson, 63–64.
103. Dworkin, 177; Robertson, 63–64.
104. Dworkin, 177; Robertson, 63–64.

105. Dworkin, 178.
106. Gordon W. Brown, Paul A. Sukys, and Mary Ann Lawlor, *Business Law with UCC Applications*, 8th ed., 278–79; Gordon Brown, Paul Sukys, and Lois Anderson, *Understanding Business and Personal Law*, 9th ed., 237; George D. Cameron, *The Legal and Regulatory Environment of Business*, 488–89; Betty Brown and John Clow, *Introduction to Business*, 342. See also John Nguyet Erni, "AIDS Science: Killing More than Time," 400–430.
107. Robert Teitelman, *Profits of Science: The American Marriage of Business and Technology*, 199.
108. Leonard A. Stevens, *The Case of Roe v. Wade*, 172–73.
109. Ibid.
110. Some commentators are not as convinced that the manufacturer of RU486 (the French company Groupe Roussel Uclaf) has been entirely innocent in comparable situations in the past. There have been some allegations, for instance, that Roussel Uclaf orchestrated a series of events that led to the French government's intervention in the threatened withdrawal by Roussel Uclaf of RU486 from the marketplace several years ago. Edouard Sakiz, the chairman of Groupe Roussel Uclaf, and Claude Evin, the French minister of health, denied these allegations. Hoechst, the German parent company of Groupe Roussel Uclaf, also denied the charge of collusion. For details see R. Alta Charo, "A Political History of RU-486," 43–93, see especially 62–69.
111. Carol Levine, ed., "Should the Courts Be Permitted to Order Women to Use Long-Acting Contraceptives," 2; Robertson, 69.
112. Robertson, 71.
113. Levine, 3; see also Jim Persels, "The Norplant Condition," 4–13; for the opposing view see an article by the American Medical Association also in *Taking Sides: Clashing Views on Controversial Bioethical Issues*, ed. Carol Levine, 14–19.
114. Robertson, 84–89, 91–92.
115. See Kenneth R. Foster and Peter W. Huber, *Judging Science* and Peter W. Huber, *Galileo's Revenge*.

Chapter 12
The Regulatory Environment and Genetic Engineering

COMMENTARY: THE LAW AND THE MEANING OF MANUFACTURED LIFE

A microbiologist working for the General Electric Company developed a new strain of bacteria by adding a DNA plasmid to an already existing strain of bacteria. A **plasmid** is a form of DNA that carries a particular gene. The modification created a type of bacteria capable of disrupting the basic components of crude oil. This capability made the new strain especially helpful in mopping up oil spills. As such, the new bacteria could prove to be a very valuable commodity. The microbiologist attempted to patent both the process of producing the bacteria and the bacteria itself. The patent examiner had no difficulty granting a patent for the process, but balked at granting patent rights to the bacteria itself. The examiner argued that, as a new life form, the bacteria was not a proper subject matter for a patent. The Patent Office Board of Appeals upheld the examiner's rejection. On further appeal, the Court of Customs and Patent Appeals reversed the decision of the Board and the examiner. The Commissioner of Patents and Trademarks asked the U.S. Supreme Court to hear the case. The Supreme Court, in *Diamond v. Chakrabarty*, ruled that the new life-form could be patented.[1] In allowing the patent of this new life-form, the Supreme Court had taken a monumental step forward. Some commentators believed that the Supreme Court not only went beyond its authority but also opened the door to possible abuse in the area of biotechnology. Others argued that the Supreme Court's decision would have little practical effect because the patent-application process is so slow that few biotechnology companies will bother to apply for patents. Still others found it curious that the Supreme Court allowed a life-form to be patented, whereas in *Roe v. Wade* the Court had refused to define life.[2] Whatever point of view the legal and scientific commentators took, they all appeared to agree on one thing: the Supreme Court had certainly crossed into a new frontier. Few people could have predicted the fallout from the invention of the new bacteria, let alone from the attempt to patent the new life-form. Even before that decision, the legal problems associated with biotechnology had multiplied faster than anyone could have imagined. These problems and their solutions form the basis for our study in Chapter 12.

CHAPTER OUTCOMES

After reading this chapter, the reader should be able to accomplish the following:

1. Explain the need to regulate scientific research.
2. Identify the constitutional basis for the regulation of science and technology.
3. Explain the process of peer review in the granting of funds for scientific research.
4. Identify the advantages and the disadvantages of the peer-review process.
5. Identify the arguments for and against a science court.
6. Contrast the law's attitude toward basic research with its attitude toward technology.
7. Explain some of the basic concepts involved in genetics.
8. Outline some of the procedures involved in genetic engineering.
9. Identify some of the benefits and concerns associated with genetic engineering.
10. Name some of the regulatory efforts made in relation to biotechnology.

12-1 HOW THE LAW REGULATES SCIENCE

Before moving on to a more detailed discussion of the specific juriscientific issues involved in genetic engineering, we should pause for a moment and attempt to unravel the complicated puzzle of interacting legal and scientific forces. In our attempt to make the picture somewhat clearer, we first explore the legal theory behind the regulatory powers of the government. We examine the process by which the regulatory apparatus of the government awards money for scientific research and then explore some of the advantages and disadvantages of this process and contrast it to some alternative procedures, specifically the establishment of a science court. We look at the method by which the government regulates technology. Only then will we be equipped to examine some of the legal developments that have occurred in the world of genetic engineering.

The Basic Plan of the Constitution

In the summer of 1787, a group of delegates from each state in the United States met in Philadelphia to revise the Articles of Confederation. The Articles had served as the country's constitution for eight years. However, its shortcomings were becoming more and more unmanageable and most people realized that it was time for a change. When the Founding Fathers met, their assignment was merely to revise the Articles. They could have taken the safe route and simply patched the holes in the fabric of the Articles. Instead, they took a bold step forward and abandoned the entire document in favor of a new constitution.[3]

The new Constitution was a remarkable piece of legal craftsmanship. One key element of this craftsmanship lies in the original blueprint used by the delegates in their plan for the new government. That blueprint was based upon two guiding principles: a separation of powers and a system of checks and balances.[4] The goal of the **separation principle** is to prevent governmental power from falling into the hands of a single person or a small group of people. To prevent this sort of tyranny, the three primary powers of the government, to make, enforce, and interpret the law, are divided among three branches of government. The legislative branch makes the law, the executive branch enforces it, and the judicial branch interprets it.[5]

The goal of the **balancing principle** is to prevent each branch from exercising absolute power within its sphere of influence. To prevent this, each branch shares a small portion of the power set aside for each of the other branches.[6] One example of this is the president's power to veto legislation. The veto allows the president to influence the lawmaking power of the legislature. Legislators, however, can override the veto if they can muster a sufficient number of votes. This is an example of the legislature's check over the power of the executive.[7]

The framers of the Constitution also limited governmental power in another way. Under the original plan of the Constitution, the national government was to be a government of **enumerated powers**. This meant that the powers of all of the branches of the government, but especially the legislative branch, would be limited to those actually granted in the Constitution. This raises an interesting question. Because neither scientific research nor technological developments are specifically listed as areas over which the national government has power, can that government spend tax dollars in support of such projects? A brief historical review of the powers of the national government reveals that the government does indeed have such authority. Several of the enumerated powers in the Constitution have been used to justify the national government's interest in science and technology. Some of these include the power to raise an army and navy, the power to grant patents, the power to take the census, and the power over weights and measures.[8]

However, by far the most powerful clause in the campaign to allow the national government to spend money in support of science is the **general welfare clause**.[9] Using the power to spend money for the general welfare as the basis of its authority, Congress established the Department of Agriculture in 1862. As a part of the establishment of this department, Congress also authorized the hiring of chemists and botanists.[10] The constitutionality of this interpretation of the general welfare was confirmed by the U.S. Supreme Court in 1936 in *United States v. Butler*.[11]

The government does more than simply support science and technology with tax dollars, however. It also regulates both activities. The power of Congress to regulate has also been firmly established throughout the history of the republic. The most instrumental clause behind the regulatory power of Congress is the **commerce clause**, which states that "Congress shall have Power . . . To regulate Commerce with foreign Nations and among the several States."[12] The courts have gradually broadened congressional power to regulate under the commerce clause so that today the national government can regulate just about any activity that somehow impacts upon commerce. In today's complex world of mass transportation, mass communication, and mass marketing, there are very few things that remain outside congressional regulatory power.[13]

Generally, congressional regulatory power is exercised by the creation of a regulatory agency such as the Food and Drug Administration (FDA) or the Environmental Protection Agency (EPA). The process is relatively simple. Congress first surveys an area of society or the economy and decides that regulation is needed. Then Congress passes legislation setting up an agency to police that part of the social structure. The legislation will set down some general guidelines that the agency is to use in exercising its regulatory powers. The rules made by the agency pursuant to those guidelines are called **regulations**. These regulations generally have the force of law. The agency also has the power to enforce those regulations.[14]

Congress feels justified in creating these agencies for two reasons. First, the socioeconomic arena is so vast and so complex today that it would be virtually impossible for any one group to have the time or the energy to delve into all areas successfully. There is simply too much to do and too little time in which to do it. Second, most legislators are generalists. They do not possess the expertise necessary to regulate the highly specialized areas of the socioeconomic arena as it exists today. Consequently, they feel compelled to rely on the expertise of the specialists who run the agencies and who make crucial judgments in relation to establishing and enforcing regulations.[15] No areas of government regulation are touched more directly by experts than the funding of scientific research and the regulation of technology. What may be surprising is that these two areas are treated very differently by the law.

The Regulation of Basic Research Funding

The major federal agencies involved in funding scientific research include the National Science Foundation, the Department of Defense, the Department of Energy, and the National Institutes of Health (NIH).[16] Of course, this brief list of funding agencies presents a rather simple and somewhat incomplete picture of a very complex bureaucratic set up. This complexity can be appreciated if we focus on just one of these agencies, the NIH, which consists of a variety of support divisions and several separate institutes. Each institute is charged with research and development within a particular area. The National Cancer Institute, for instance, is charged with cancer research. The NIH is actually part of the Public Health Service, which is, in turn, a subcomponent of the Department of Health and Human Services.[17]

Funding decisions are delegated to the experts of the agencies because such decisions generally require scientific knowledge that is somewhat beyond the background of the typical member of Congress. Moreover, delegating the decision-making process to the experts also helps prevent Congress from attempting to control the projects, the universities, and the individuals who receive such funding. The idea is to place the decisions on science funding in the hands of those who should know best where the money should be spent.[18]

The Peer-Review Procedure. This decision-making process is known as **peer review**. As a part of the peer-review process, a scientist will submit a grant application to the appropriate federal agency. The agency will then see that the grant application is reviewed and evaluated by prominent scientists in that area. The reviewers are directed to consider the value of the research project as well as its feasibility.[19] In addition, reviewers are directed to take into account the competency of the primary member of the research team applying for the grant.[20]

In considering the competency of the primary research scientist, peer reviewers, or referees, may consider the success of the researcher's previous projects, specifically those projects formerly funded by the agency in question. Included in the evaluation of the previous projects might be a consideration of whether the researcher followed the recommendations of the peer-review group on certain aspects of those earlier projects, such as the need to obtain supporting data.[21] Moreover, it is not unheard of for the reviewers to also consider the status of the researcher's institution.[22] However, studies have indicated that, although peer-review members are somewhat more inclined to treat researchers from prestigious institutions more favorably than those from smaller institutions, that tendency is not significant.[23]

The initial peer-review team either approves or rejects the grant proposals. The review team may also be asked to assign a rating to all approved proposals. The ratings will help the next review level, generally an administrative group of some sort, to weigh the relative merits of various research projects.[24] The upper-level administrative groups of the various institutes can send a grant back to the peer group for reconsideration or approve a grant despite the peer group's rejection. The administrative group will take into consideration not only the merits of the grant itself but also the requirements of the individual institute, as well as those of the entire agency. For example, in making its grant allocations, the National Cancer Advisory Board would take into consideration the needs of the National Cancer Institute as well as those of the National Institutes of Health in making its funding recommendations.[25]

Of course, each agency conducts the process according to its own individual procedures. This means that the details of each agency's peer-review process will differ from that conducted by other agencies. For example, one agency might include many stages in the process, whereas another might award a grant after a single peer-review process. One agency might make use of scientists who are not a part of the agency; another might use only inside reviewers.[26] It is also possible that a special study group may be convened to review a research grant, especially if the grant applicant is refiling a previously rejected grant proposal.[27]

Peer Review: Support and Criticism. There are several good reasons to favor peer review as a part of the scientific-funding procedure. First, peer review is the accepted procedure within scientific circles. All researchers know, for example, that their work must be submitted to a peer-review committee of some sort if they hope to have the results of their research published in a reputable scientific journal. Therefore, they are on familiar ground when applying for governmental grants.[28] A second advantage is that peer review takes the grant-approval process out of the political arena and places it into the domain of science, where many people believe it belongs. This feature seems to be increasingly important because Congress was criticized severely for its attempt in the 1980s to introduce legislation aimed at funding scientific projects for specifically named institutions. Several institutions of higher education have also been criticized for going directly to Congress for financial support.[29]

A third advantage is that peer review ensures that research-grant applications are judged primarily on their merits, at least the first time through the process. The peer groups are directed to make decisions based exclusively on scientific merit and on the reputations of the researchers, rather than on funding considerations. Although the peer groups do rate the various grant approvals based upon the goals and the needs of the individual institutes, they do not make the actual funding allocations. The funding decisions are left to higher-echelon administrators. By the time a grant approval reaches this level, however, its scientific merit has already been rated, making it more difficult for an administrator to eliminate scientifically beneficial projects without substantial reason.[30]

For all its merits, however, peer review has been severely criticized and challenged. One of the chief criticisms is that the process turns out to be somewhat biased against younger researchers and against those from smaller institutions. In essence, the claim is that, because most of the members of peer review groups are from the more prestigious universities, they are more inclined to approve grant applications from those who also hail from such institutions. The tendency is sharpened because in some agencies the peer-review team members are explicitly directed to take into consideration the reputation of at least the primary research scientist attached to the grant.[31] One former congressman went as far as to charge that the peer review system actually discourages innovation and creativity.[32]

Professor Stephen Cole of the State University of New York at Stony Brook has compiled some rather impressive figures that contradict this criticism. Cole's conclusions suggest that there is little favoritism exercised by reviewers despite that the majority of them come from major institutions and that they are specifically directed to consider the past scientific reputation of the grant applicant. Cole's study demonstrates that, although some peer reviewers do treat researchers from well-known institutions a bit more favorably than those from other places, that tendency is not significant.[33] The results also indicate that older applicants are no more likely to receive an approval than younger applicants.[34] Cole explains the impression that the peer-review process appears to favor experienced researchers from larger institutions because they are more likely to apply for such grants and thus make up a much greater proportion of the applicant pool. This is a process that Cole labels self-selection.[35]

Even more significant was Cole's conclusion that the explicit direction to peer reviewers to consider the scientific reputation of the applicant did not significantly affect the chances of the grant application. Cole discovered that most reviewers first evaluate the project on its scientific merit and only later turn to the credentials of the applicant. According to the data compiled by Cole, negative reactions to grant applications were only rarely explained in terms of the reputation of the researcher. Most negative reactions were based firmly on a critique of the scientific aspects of the grant proposal.[36]

The Law's Involvement in Peer Review. Another criticism leveled at the peer-review process is that it falls outside the regular pattern of the legal system. In fact, this criticism, which is leveled at the entire process of funding scientific research in the United States today, finds its roots in the original governmental plan laid down by the Founding Fathers in 1787, which was based on a separation of powers and a system of checks and balances.[37] The goal of separation was to prevent governmental power from falling into the hands of a single person or a small group of people. The goal of the balancing principle was to prevent small groups of people from exercising absolute power even within their small sphere of influence.[38] The argument is that the process of funding scientific research violates both principles by placing funding decisions in the hands of a few people who have absolute authority over those decisions.

This criticism overlooks the fact that as a part of the federal bureaucracy these funding agencies are subject to the **Administrative Procedure Act**, which guarantees that all such agencies will follow certain procedures that are set down by Congress.[39] Moreover, under provisions of the Administrative Procedure Act, many of the decisions made by these funding agencies are subject to an appeal to the federal courts. In fact, Section 702 of the Administrative Procedure Act states quite explicitly that "(a) person suffering a legal wrong because of agency action, or adversely affected or aggrieved by agency action within the meaning of a relevant statute, is entitled to judicial review thereof."[40]

Unfortunately, this judicial appeal process has proven less than satisfactory for some grant applicants who believe that they have been treated unfairly by a funding agency.[41] The source of this dissatisfaction, however, seems to come from a misunderstanding of the role of the courts in the appeal process. The court is not charged with reviewing the scientific value of a grant proposal that has been rejected by an agency review process. Judges cannot be expected to have the qualifications needed to review decisions made by a panel of experts who have focused primarily on the scientific merits of the proposal.[42] As one legal scholar explained the problem, the court's review of a scientific issue is often costly and inefficient. Moreover, such proceedings can also produce unwelcome, even troublesome solutions.[43]

Although reviewing the scientific merits of a grant proposal falls outside the court's job description, that does not mean that the court has no role in the appeal process. Rather, the court's role is to determine whether the agency treated the applicant fairly and whether the agency has followed its own review procedure. As one court explained it: "The court's only legitimate function in a case such as this is the extremely limited one of insuring that the agency's determination was arrived at through procedures which were reasonable and fair."[44] In essence, this means that the scientific decisions of an agency in the peer-review process are left up to the discretion of the agency. Therefore, unless the agency's process violates an applicant's constitutional rights, or runs counter to statutory or procedural rules, that decision will not be successfully challenged in court.[45] There are, of course, sound reasons for this approach, as the court in *Grassetti v. Weinberger* explains, not the least of which is that the evaluation of a scientific research proposal lies outside any court's area of expertise.[46] The court in *Grasseti v. Weinberger* goes on to list several additional reasons why it would be unwise to allow the courts to intervene in such matters.

For instance, if grant applicants could appeal all adverse decisions on a scientific basis, most of the funding agencies would be subject to extensive litigation. Such a heavy litigation schedule would not only occupy much of their time but also further delay the actual dispersal of funds. These delays would ultimately hurt all grant applicants and would clearly damage the actual goals and objectives of the agencies.[47]

The Science Court Proposal

Nevertheless, continued dissatisfaction with peer review has led some commentators to conclude that the entire system should be overhauled. One suggestion has been the establishment of a **science court** that could be charged with, among other things, the responsibility of reviewing the scientific merits of grant proposals. The science court proposal is not a new one. It was first advocated in 1976 in a report issued by the Task Force of the Presidential Advisory Group on Anticipated Advances in Science and Technology.[48] The report, which was entitled *The Science Court Experiment: An Interim Report*, suggests the establishment of a science court that would employ the traditional adversarial procedure to settle disagreements over issues of science. The science court would provide a forum for advocates on all sides of a scientific controversy, who would present their arguments before a panel of neutral judges with scientific backgrounds.[49]

The science court could be convened by Congress, the president, or an administrative agency whenever they faced a pressing issue that could only be resolved by a proper understanding of the science involved in the issue.[50] Other potential petitioners before the science court might include special interest groups such as the Sierra Club and the Planetary Society, institutions like the American Association for the Advancement of Science and the New York Academy of Sciences, state governments that might be affected by a scientific policy matter, or individual scientists who might have a grievance over a funding decision by a federal agency. For example, if the science court was in existence at the time, it could have adjudicated a dispute over the safety of storing radioactive waste in Yucca Mountain in southern Nevada.[51] Certainly, the state of Nevada would have had standing to bring the issue before the science court, if one existed, which would consider scientific evidence on both sides of the safety issue.

The value of the science court proposal has been debated for decades. Proponents see it as a means for solving some of the more technical aspects of scientific issues that plague the nation today. Other supporters believe that a science court would provide a forum for some individuals and institutions that have a stake in such issues but that are virtually without a public voice today. For example, a science court could have provided a forum for opponents of the Antiballistic Missile (ABM) program in the 1960s. As it was, opponents of the ABM system had to be content with the presentation of their views in a published article that escaped the attention of most of the public.[52] Still other supporters of the science court see the court as a possible alternative to the present appeal process that follows the rejection of a grant proposal by a federal funding agency. Rather than focusing on the procedural aspects of the peer-review process, which is what the federal courts are constrained to do under the present system, the science court would actually look at the scientific merit of the proposal.

Opponents of the science court argue that the establishment of such a court would simply duplicate functions that are already carried out by other parts of the government. The president, for example, might find little use for a science court, since he already has a science advisor.[53] Similarly, the review process of most funding agencies already involves several appeal levels within the organization as well as a chance to resubmit rejected applications. Adding a science court to these levels would, therefore, seem redundant. Another difficulty involves the choice of judges. Should the judges be chosen on a case-by-case basis or should they have the kind of permanent stature reserved for other federal judges? Will the judges need to be completely neutral on a position before they can hear a case or will they actually be chosen to represent different sides of the issue? What areas of expertise will the judges be called upon to exhibit? Will zoologists be competent to judge issues in chemistry? Will medical researchers be able to address problems of cosmology?[54] All of these unanswered questions seem to make the establishment of a science court somewhat unlikely in the near future.

Moreover, these questions represent only a series of hypothetical problems facing the science court proposal. Whenever the government has had the opportunity to establish an actual science court, even more unexpected stumbling blocks have surfaced. A case in point is the attempt to establish a science court in the state of Minnesota to settle a dispute between two utility companies, the United Power Association and the Cooperative Power Association, on the one hand, and an organized group of Minnesota farmers, on the other. The dispute arose when the utility companies invoked eminent domain in order to take certain parcels of land from the farmers to construct a series of power lines across the state. The farmers objected to what they perceived as the unjust seizure of their land and actively, at times violently, opposed the construction of the power lines. As a way to settle the dispute, the governor of Minnesota proposed the establishment of a science court to determine whether the power lines represented a health and safety risk to the farmers.[55]

The science court proposal became a source of controversy but ultimately failed to emerge for two reasons. First, as proposed by the governor, the science court was a temporary forum established simply to settle this single dispute. It was, therefore, very difficult to convince the parties to cooperate with the science court when they had already placed the issue in the hands of the existing authorities, including both the regular court system and the appropriate regulatory agencies. This was especially true of the utility companies, which had been successful in both traditional forums. Second, the supporters of the science court proposal had not provided any way to limit the issues before the science court to scientific concerns nor to ensure the equal involvement of all the participants in the dispute. The utility companies had the financial resources needed for arguing the case before the science court; the farmers did not. As a result of these problems, the science court in Minnesota never materialized.[56]

One of the most convincing criticisms of the science court proposal, however, is one that goes to the heart of the matter. As we saw in Chapter 11, the primary role of the law is deliberative in nature. The law by and large exhibits a deliberative attitude when asked to step into controversial issues that are beyond the scope of its authority and expertise. Most of the time, judges and regulators are wise enough to stay out of areas that do not concern them or that they are ill-equipped to handle. We saw this attitude at work in the examination of procreative freedom in Chapter 11. We also saw this deliberative approach in this chapter when we discussed the limitations placed upon the courts in their assessment of funding decisions made by peer groups in governmental agencies. This is not to say that the law does not at times venture into areas that are best left alone. We saw this happen in the preceding chapter when we examined the legislative proposals made in Kansas and Louisiana regarding the Norplant system. Most of the time, however, the law avoids tackling scientific issues, at least at the level of basic scientific theories and pure scientific research. The situation is quite different, however, when the issue is technology. Here, the law is willing, in fact, almost eager, to exercise its regulatory powers.

12-2 HOW THE LAW REGULATES TECHNOLOGY

Thus far, we have focused on the law's involvement in the funding of basic scientific research. Moreover, we have seen that, by and large, the law exercises a deliberative "hands-off" approach toward funding decisions. Now we turn to the other side of the question. What approach does the law assume when basic research turns into technology, as happened when Ananda Chakrabarty developed a new strain of bacteria capable of disrupting the basic components of crude oil? Who decides whether it is safe to use this new bacteria? Who has the right to profit from the development of this new bacteria? How are these decisions made? How are they appealed? Does technology receive the same type of protection granted to basic research?

How Science and Technology Are Different

Before we explore the juriscientific regulation of technology, we need to explain the difference between science and technology. Generally, the term **science** is used to describe an activity that involves basic research aimed at understanding how nature works. As noted earlier, the goal of science is to uncover the truth about the universe. However, that does not mean that scientists compile data and conclusions like a retailer lines the shelves of a store with merchandise. Instead, scientists seek to interpret that data and those conclusions in an attempt to uncover an understanding of how the universe operates.[57] Yet, as we have seen, other disciplines, such as philosophy and theology, also seek an understanding of the universe. What then is the difference? The difference lies in the process by which that truth is sought. The scientific method is an objective process by which scientists use observation and experimentation to verify theoretical explanations about the operation of nature. (See Chapter 1 for a detailed explanation of the scientific method.)

In contrast, **technology** involves a practical application of knowledge usually, but not necessarily, associated with science.[58] Technological developments are not always based on an understanding of scientific principles. Thus, medieval archers did not need to understand the principles of aerodynamics or the law of gravity to develop a more efficient version of the crossbow. On the other hand, an understanding of aerodynamics and gravity certainly helped in the technological development of the aircraft industry in the twentieth century. Some commentators would prefer to apply the term **craft** to any practical development that is not based on an understanding of science. This view is rather limited in its outlook because it eliminates the creation of many tools, artifacts, and processes that were developed prior to the age of science. Thus, the medieval archer who invented a new, improved crossbow would be exercising his craft, not refining a new technological development.[59]

Other commentators, however, would define technology to include any useful application of knowledge. This definition recognizes that, though an understanding of science will generally accelerate technological progress, science does not always lay behind technological developments. Technology may or may not include the creation of some sort of material artifact. Certainly, many technological developments rely upon the invention of or an improvement in some sort of material artifact. Examples include the electric light, the airplane, the automobile, the radio, the computer, the transistor, the silicon chip, and so on. However, advances in technology can also come simply by approaching a problem in a novel and inventive way. Using this broader approach to the term *technology* then, Henry Ford's invention of the assembly line would involve technology as would the creation of a new computer program.[60]

Because of its relationship to practical applications of knowledge, technology usually impacts upon society in ways that are more direct, more evident, and thus more easily regulated than those of science. Since technology also involves a useful application of some sort, there is often a market for that technology. This means not only that new technology has an impact on the economy, but also that someone intends to profit from the sale and distribution of that technology. Thus, again we can see that technology can be more easily regulated than science.

The Regulation of Technology

As was true for the regulation of science, the legal basis for the regulation of technology is found in the United States Constitution. As we saw earlier in this chapter, several of the enumerated powers in the Constitution have been used to justify the national government's interest in science and technology. In particular, the power to grant patents has been very instrumental in the government's power to promote as well as regulate the development of technology. Patents are actually property rights awarded to inventors by the national government. Once an inventor has secured a patent he or she has the exclusive right to manufacture, use, distribute, and/or sell that invention for seventeen years, a period that can be extended under certain circumstances. The advantage of having a patent is that it grants exclusive rights to the inventor for a specified term of years. The downside is that in a patent application the inventor must lay out all of the details of that invention for the patent office. In fact, the patent application must be so detailed that anyone with sufficient background in that discipline can duplicate the invention without substantial delay.[61]

The law sets down the following criteria for a patent. The invention must qualify as a **device**. This stipulation does not, however, necessarily mean that the invention must be an artifact. The courts have, for example, approved patents for processes, including those that involve computer programs. However, the criteria do mean that an inventor cannot patent a law of nature, a mathematical formula, or an abstract theory. Second, the invention must involve something that is *novel and nonobvious*. This feature means that the invention must be unique and must not be so self-evident that any person having ordinary skill in the discipline could come up with the same idea. Finally, to be patentable, the invention must be *useful*. The quality of usefulness means that the invention must demonstrate some pragmatic application of knowledge.[62]

Patents and Genetic Engineering. The immediate question that comes to mind in relation to patents and biotechnology is "Are genetically engineered products patentable?" As noted in the commentary at the opening of this chapter, this question was answered in the case of *Diamond v. Chakrabarty*, in which the U.S. Supreme Court approved a patent for the new strain of bacteria.[63] As revolutionary as this case may appear on the surface, a careful reading of the Supreme Court's rationale reveals that the case, once again, demonstrates the law's deliberative role in the social system as well as the Court's largely cautious attitude, especially toward science.[64] The Court in *Diamond v. Chakrabarty* actually does nothing more than look at the plain language of the statute as written by Congress and determine whether the new strain of bacteria can be included within that language. The statute, the Court says, allows for the patenting of "any new and useful process, machine, manufacture, or composition of matter."[65] Since it is clear that the new strain of bacteria does not qualify as either a "process" or a "machine," the question is whether it qualifies as a "manufacture" or a "composition of matter."[66] Since the new strain of bacteria did not exist naturally and since it now exists because of Chakrabarty's genetic engineering, it qualifies as a "manufacture" and is patentable.[67] Although this ruling may seem dramatic, it is, in fact, simply the result of a straightforward interpretation of a statute.

This restrained and straightforward attitude is especially obvious when the final decision of the Court is contrasted with what some members of the legal and scientific communities had asked the Court to do in the first place. As noted, the Court did little more than interpret the wording of the federal law that applied to this case. However, many experts, including some scientists who had been awarded the Nobel Prize, filed arguments asking the Court to go far beyond this routine task. Instead of being asked simply to interpret patent law, the justices were urged to outlaw the parenting of genetically engineered products altogether, not on statutory authority but because of the dangers posed by genetic research.[68] The Court, in fact, makes a point of outlining the severe opposition that it faced when making this decision:

> The briefs present a gruesome parade of horribles. Scientists, among them Nobel laureates, are quoted suggesting that genetic research may pose a serious threat to the human race, or, at the very least, that the dangers are far too substantial to permit such research to proceed apace at this time. We are told that genetic research and related technological developments may spread pollution and disease, that it may result in the loss of genetic diversity, and that its practice may tend to depreciate the value of human life.[69]

The Court, however, maintained its prudent approach to such matters and refused to adopt the strict regulatory role suggested by these arguments. Two reasons were presented for this ruling: First, the Court wisely pointed out that scientific research is not likely to halt or even slow down no matter how it rules in *Diamond v. Chakrabarty*. The Court noted that such research had been carried out for many years without any input from the Court on whether the results of that research would or would not be subject to patent protection. It therefore concluded that any ruling at the time would be unlikely to stop any research efforts in this area. Second, the Court recognized that social-policy decisions involving something as vast and complicated as genetic engineering are best left to legislative and regulatory agencies. Such agencies have the resources to conduct the type of in-depth and far-reaching investigation that such issues demand.[70] Following the Court's suggestion, it is to this regulatory activity that we now turn our attention.

Agencies and the Regulation of Technology. The regulatory power of the national government comes primarily from the commerce clause. Over the history of the Constitution, the courts have extended the power of Congress to regulate the socioeconomic arena under the commerce clause. The power has been strengthened to such a degree that today, the national government can regulate just about any activity that somehow affects interstate commerce. This, of course, includes the power of Congress to regulate technology.[71] In general, Congress delegates this power to the regulatory agencies. This is true in technology as well as in science. We have already noted that Congress delegates this regulatory power because it lacks the time and the expertise to handle such regulatory activities on a daily basis.

Nevertheless, there is a significant difference between the way that the law handles science and the way that it regulates technology. As we noted earlier, a good part of the regulation of basic research involves funding decisions, and most of these funding decisions are made by the scientists themselves via the peer-review process. Moreover, we also saw that the Courts are willing to give an enormous amount of latitude to those scientific decisions. The same is not true of the regulation of technology, where the law plays a much more pivotal and much more intrusive role.

There are, of course, good reasons for this difference. One reason is the fact that technological developments are bound to have a much more direct and immediate effect on the public than does basic research. If a scientist receives a grant to study whether a particular type of vaccine is effective in preventing AIDS, there will be little initial impact on the public at large. If, however, a pharmaceutical firm proposes to market that vaccine, then many people become targeted almost instantly. Therein lies the qualitative difference between the law's interest in science and its interest in technology.

12-3 GENETICS AND THE FRONTIERS OF JURISCIENCE

An important step in our study of genetics is to establish the nature of genetic engineering and then to determine whether it can be considered science or technology or both. As a part of this discussion, we must reintroduce ourselves to the study of genetics. We begin by looking at the historical development of genetics, followed by an examination of the structure, function, and importance of genetics to living things. We then examine recombinant DNA techniques and the Human Genome Project, as well as some benefits and concerns that flow from these activities. At that point, we should be ready to evaluate how the law has attempted to regulate activities in genetic engineering.

An Introduction to Genetics

To appreciate the historical development of modern genetics, we must begin with the work of the Austrian monk Gregor Mendel. In 1865 at the Brünn Society for the Study of Natural Science, Mendel revealed the results of a methodical study that he had conducted on the breeding of pea plants.[72] Mendel's work was significant because he traced the origin of inherited traits to two hereditary units that were passed on from parent plants to their progeny. One parent plant contributed one of these hereditary units and the second parent, the other. Moreover, the two hereditary units did not stand on equal footing with one another. One trait was dominant and the other recessive. For example, if the hereditary unit for tall trees is dominant and that for short trees is recessive, combining the two units will produce tall trees. The name later given to these hereditary units was the **gene**.[73] Around the same time, the Swiss scientist Friedrich Miescher discovered that the nuclei of living cells were at least partially made of deoxyribonucleic acid (DNA).[74]

The Double Helix. The role of inheritance and mutation in the evolution of life on Earth has been recounted in Chapter 10. Ironically, although Miescher predicted that heredity was controlled by complex molecules, he never realized that the material he had discovered within the nuclei of cells, DNA, was that very molecule. At this point, it is enough to note that genes and DNA are integral to that process. What was not known until the middle of the twentieth century was the mechanism of this hereditary process. At that time, several experts discovered that within the nucleus of all living cells were small oblong structures called **chromosomes** that housed the genes. The genes were always arranged in sets of twos. Each member of a set was contributed by each of a child's parents. The substances necessary to sustain life, enzymes and proteins, were produced by the genes. The genes in turn were constructed of DNA. Therefore, the genes, which are made of DNA, contain the information needed to manufacture proteins and enzymes, which are necessary for life. At first, there were difficulties with this model. One difficulty was that DNA was thought to be too simple for the job it seemed to be doing.[75]

This anomaly was cleared up in part by James Watson and Francis Crick, who discovered that DNA is actually constructed as a **double helix**.[76] A double helix looks something like a spiral staircase. Therefore, it is helpful to simplify things by imagining the spiral staircase flattened out so that it looks like a ladder. The double helix consists of sets of precisely matched bases. These matching bases make up the rungs of the ladder. There are four such bases. They are

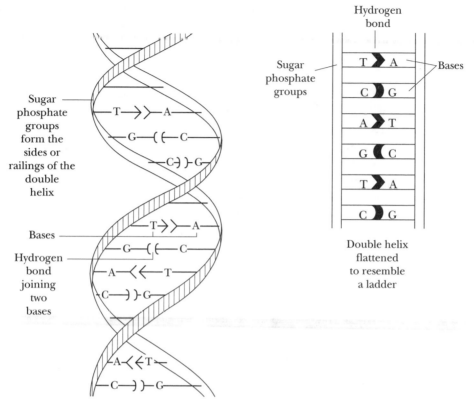

Figure 12.1 The Double Helix.
DNA is structured as a double helix, which looks somewhat like a spiral staircase. It is sometimes helpful to imagine the double helix flattened out so that it resembles a ladder. The double helix consists of sets of precisely matched bases. These matching bases make up the rungs of the ladder. There are four such bases: adenine, guanine, cytosine, and thymine. Each base is supported by a sugar-phosphate group, which correspond to the rungs of the ladder. A base plus its supporting sugar-phosphate group is a nucleotide. The bases will join by a hydrogen bond only in specifically ordered pairs. This is important because, as the double helix splits, each side becomes a mold for the production of a matching strand. An exact duplicate of the strand of DNA is assured because of the restricted pattern in which the bases can join. The double helix will split to perform two crucial functions: replication and the synthesis of protein.

A (adenine), G (guanine), C (cytosine), and T (thymine). Each base is supported by a sugar-phosphate group. The sugar-phosphate groups would correspond to the supporting sides, or railings, of the ladder. A base plus its supporting sugar-phosphate group is a **nucleotide**.[77] The bases will join only in specifically ordered pairs. This pattern is called a **base-paired sequence**. G joins with C, and vice versa, but not with A or T; whereas A joins with T, and vice versa, but not with G or C. This is critical because, as the double helix splits (imagine the ladder being divided down the rungs), each side becomes a mold for the production of a matching strand. Moreover, an exact duplicate of the strand of DNA is assured because of the restricted pattern in which the bases can join.[78] But, the nonscientist may ask, why does the DNA double helix have to come apart in the first place? The double helix must split to perform two functions: (1) replication, and (2) the synthesis of protein.

The Process of Replication. For a cell to duplicate, the DNA sequence must be replicated. This process, called DNA replication, takes place in a broth of free-floating nucleotides. Remember that a nucleotide is a base plus its supporting sugar-phosphate group. As one of the first steps in the replication process, an enzyme uncoils the DNA double helix, while another enzyme splits the two strands. Once the double helix is split, the original right half can serve as a template for the replication of the left half. This is accomplished by enzymes that capture the free-floating nucleotides in the broth surrounding the DNA strand, and then align the exposed base portions of the nucleotides with the exposed bases on the right half the helix according to the proper A–T and G–C complementary sequence. At the same time, following an identical process, a new right half is created using the original left half of the helix as a template. Another enzyme will then check to make certain that no mistakes have been made in the matching of base pairs. This checking process is called **proofreading**. Thanks to the **genetic code**, once cell division is complete, each daughter cell is identical to the original cell.[79]

The Synthesis of Protein. The double helix also splits when the genetic code is read for the synthesis of protein. Proteins are complicated molecules made of smaller groups of molecules called amino acids.[80] Proteins perform a variety of functions for the body. They can act as the catalysts that are responsible for chemical reactions in the cells. They can act as messengers. They can act as building blocks for cells and tissues within the body. In short, proteins are necessary for life.[81] The genetic code is responsible for telling the cells which amino acids are needed to construct which proteins in order to carry out the metabolic tasks of the body. The genetic code is found in the DNA sequence within the cells. All of the cells in a human body contain an exact copy of the individualized DNA sequence that was present in the original fertilized egg. While the double helix is intact, this sequence, that is, the genetic code, is unreadable. In a sense, it is sealed into the heart of the double helix and can be read only when the helix splits. Again, think of the ladder splitting down the middle of the rungs. It is only when the strand has split, that the genes, which are actually set sequences of nucleotides, can be duplicated and carried to that part of the cell where the proteins can be manufactured.[82] Remember that nucleotides are made of a base plus its supporting sugar-phosphate group.[83]

How is this process of protein synthesis carried out? Before a protein can be synthesized, the DNA information (the gene) which is within the nucleus must be transported outside the nucleus into the cytoplasm. The **cytoplasm** is that portion of a cell that exists outside the nucleus but inside the outer membrane of the cell wall.[84] The actual synthesizing of protein takes place within the cytoplasm. An exact copy of the gene, rather than the gene itself, is transported from the nucleus into the cytoplasm. The working copy of the gene is made by a process known as **transcription**. Transcription takes place within the nucleus. A key player in this process of transcription is ribonucleic acid (RNA). An RNA molecule, called an **RNA-polymerase molecule**, splits the appropriate segment of the double helix. The two sides of the strands are now exposed, revealing the sequence of bases (the As, Ts, Cs, and Gs). The RNA-polymerase molecule uses free-floating RNA nucleotides to match the nucleotides on the exposed DNA strand. The only exception here is that instead of using thymine (T), the base uracil (U) is used. When the process is complete, the result will be a working copy of that DNA strand, or to state it in more familiar terms, a working copy of the gene. The working copy of a gene is called **messenger RNA (mRNA)**.[85]

The next step in the protein synthesis process is **translation**, which is the joining of amino acids to make proteins. Since translation takes place outside the nucleus in the cytoplasm, the mRNA must leave the nucleus to enter the cytoplasm. The key players waiting for the mRNA in the cytoplasm are **transfer RNA (tRNA)**, the amino acids, and ribosomes. Strands of tRNA attach themselves to amino acids according to the sequence of three bases on that tRNA. The trick is to match the base sequences found on the mRNA with the complementary base sequences found on the tRNA. Matching these sequences will link the amino acids together in the order needed to create the protein that has been ordered by the metabolic activity of the organism. This trick is carried out by ribosomes. A **ribosome** is an enormous molecule constructed from a large number of proteins and another kind of RNA called **ribosomal RNA (rRNA)**. Thousands of these ribosomes are available in the cell. The ribosome catches and holds fast to the mRNA. The ribosomes perceive the mRNA in sets of three bases called a **codon** or a **triplet**. The ribosome will then capture a strand of tRNA with a three-base sequence, called an **anticodon**, which complements the codon on the mRNA. Remember, along with the tRNA, the ribosome has also captured the amino acid attached to the strand of tRNA.[86]

The ribosome will then read the next codon on the mRNA and capture a strand of tRNA with the complementary anticodon. Again, along with the tRNA, the ribosome has captured the amino acid attached to that strand tRNA. The two amino acids can now bond with one another. This is the beginning of the construction of the needed protein. The ribosome can now discard the unneeded tRNA, move to the next codon on the mRNA, and capture a strand of tRNA with the complementary anticodon. Again, the ribosome has also captured the amino acid attached to that strand of tRNA. The ribosome continues to read codons on the mRNA and to capture appropriate strands of matching tRNA, and along with the tRNA the attached amino acid, until it reads a "stop" message from the mRNA. At that point the appropriate sequence of amino acids has created the protein that the metabolic activity of the organism ordered in the first place. The protein then carries out its assigned function. The translation process may seem long and laborious. In reality, however, translation takes place quite rapidly. Each second, millions of amino acid reactions occur within thousands of ribosomes. This amounts to over 2000 molecules of protein built up every second.[87]

A Primer on Genetic Engineering

No term in science, except perhaps nuclear energy, causes more unnecessary fear and trepidation than *genetic engineering*. Much of this fear and trepidation, however, is caused by ignorance. Now that we have a basic understanding of genetics, we should be able to better evaluate the benefits and the risks associated with genetic engineering. The best way to begin an examination of genetic engineering is to dissect the phrase itself. The first term, "genetics," involves an understanding of genes, which, as we have seen, carry the basic plan for all processes in living things. The second term, "engineering," is a procedure by which an individual modifies nature hopefully, but not necessarily, for the betterment of humanity. In this sense, an engineer who designs a suspension bridge has modified nature by providing travelers with a simpler and more efficient route over an obstacle, such as a canyon or a river. Genetic engineering also involves a modification of nature. In the case of genetic engineering, however, the modification involves a living thing. The most effective way to modify a living organism is to alter its genetic makeup. **Genetic engineering** then is the process by which modifications to living things are made by manipulating the genes.[88]

The ability to assemble **recombinant DNA** is at the heart of genetic engineering. This process is often referred to in the media as *gene splicing*.[89] You should recall from our discussion of protein synthesis and DNA replication that enzymes play a major role in the manipulation of DNA. Therefore, it should not be surprising that the first tool needed in the production of recombinant DNA is an enzyme. This enzyme, called a **restriction enzyme**, is a biochemical substance with the capacity to make a very precise incision in a strand of DNA. As a result of this incision, a number of bases in the DNA strand are laid open. If another strand of DNA is cut by a restriction enzyme and the bases are complementary to the bases in the original strand, then the two portions of DNA will recombine—hence the term "recombinant DNA." The process can be used to transfer a gene from one strand of DNA to a second, thus altering the genetic makeup of that strand.[90]

The Benefits of Genetic Engineering. At this point, the nonscientist is likely to ask, "What is the point of transferring genes and recombining DNA?" The answer arises from our basic definition of genetic engineering. Recall that genetic engineering is the process by which modifications to living things are made by manipulating the genes.[91] Thus, the process of transferring genes can modify living things in the most profound way. Now, before we begin imagining genetically engineered monsters, let us pause and recall that genetic engineering can be used to make small changes in living things that actually reap enormous benefits.

Consider, for example, the development of genetically engineered tomatoes. When tomatoes are picked, they begin to decompose immediately. Apparently, nature's plan is to release the seeds rapidly, thereby ensuring the reproduction of the tomato plant. This rotting process is very helpful to the tomato plant but very bothersome to farmers and shoppers. It would be beneficial if the decomposition process could be slowed down. This is something that genetic engineering has accomplished. The chemical that triggers decomposition is caused by a gene that is activated when the tomato is picked. To slow decomposition, a gene can be implanted that stops this process.[92]

The genetic engineering of plants can be beneficial in a number of other ways. For instance, some high-yield crops can be engineered to be resistant to pests. A gene from a plant highly resistant to a certain pest is inserted into a plant that suffers from low resistance to that pest. In fact, this approach has successfully protected some tomato and tobacco plants.[93] It may also be possible to engineer more nutritious plants. One project currently under way is aimed at making corn a source of lysine, an amino acid needed by humans but not a part of zein, the protein produced by corn. If the gene that produces zein can be isolated, its DNA can be modified by adding the codons that produce lysine. The modified gene can then be reinserted into the corn plants. The hope is that the newly engineered corn plant will be able to produce zein and lysine. Thus far, the attempts to do this have not been successful, owing in part to the fact that overall protein production drops dramatically in the modified corn. Nevertheless, because so much of the world's population relies on corn as a source of protein, the stakes are too high to abandon such a promising project.[94]

There are, of course, other examples of beneficial applications of genetic engineering. One such example was cited in the commentary at the opening of this chapter in the case of *Diamond v. Chakrabarty*.[95] Another frequently cited example is the development of insulin. Prior to the development of genetic engineering, insulin could be obtained only by removing it from the pancreas of cows and pigs. However, as the incidence of diabetes began to rise in the United States, it began to look as if the production of insulin would not meet the demand. This might have led to a crisis situation, if not for genetic engineering. In this case, the process involves removing the gene that creates insulin in humans and implanting it into a bacterium known as Escherichia coli (E. coli). The genetically modified E. coli then produces insulin that is almost identical to human insulin. The vast majority of insulin used today is produced in this way.[96] Other genetically engineered medical products include somatotropin for the treatment of pituitary problems, Hepatitis B vaccine for immunization against the hepatitis B virus, erythropoietin for treatment of anemia, Interleukin-2 for treatment of cancer of the kidney, whooping cough vaccine for immunization against whooping cough, and Factor VIII for the treatment of hemophilia.[97]

The Human Genome Project. There are also many other benefits that may flow from genetic engineering directly to human beings. Perhaps the most ambitious project associated with genetic engineering and human beings is the **Human Genome Project (HGP)**. A **genome** is the sum total of all genetic information for a particular living organism.[98] The immediate objective of the HGP is to obtain a complete set of human genetic information. In other words, the goal of the project is to create a comprehensive, sequential blueprint of human DNA.[99] Although each person's genome is different, the similarities far outweigh the differences. Thus, once a single human genome has been mapped, it will have universal applications.[100] Moreover, the ultimate plan of the HGP is to map many individual human genomes on a regular basis, as well as the genomes of other living things, so that this genetic information can be contrasted, compared, and put to practical use.[101]

To meet these goals, two initial chores must be completed. First, the site of all genes must be determined. This procedure is known as **DNA mapping**. The map must then be increasingly refined until the exact location and sequence of all base pairs is also determined for the entire human genetic code. The attempt to determine the base sequence is called **DNA sequencing**.[102] This is not an easy goal to accomplish because there are over three billion base pairs within the human genome. DNA sequencing is a critical step, however, because research into genetically caused diseases indicates that many of these diseases are caused by mutations in a gene that affect only a few of these base pairs. Understanding this may lead not only to the ability to diagnose the propensity toward such disease at an early time, but also to the use of recombinant DNA techniques to combat them. The procedure, called gene therapy, involves replacing mutated genes with normal ones.[103]

Gene therapy involves regulating and perhaps curing those diseases that may have a basis within the genetic code. At this early point in the HGP we have strong evidence that diseases such as cystic fibrosis, amyotrophic lateral sclerosis (ALS or Lou Gehrig's Disease), and severe combined immunodeficiency (SCID) have a genetic foundation. Unlocking the genetic source of such diseases may allow scientists to create a means for dealing with these illnesses or at least prevent passing the genetic problem from one generation to the next.[104] Already a scientist at the National Institutes of Health, Dr. W. French Anderson, has devised a genetic technique for dealing with patients who suffer from SCID. Anderson works with a patient's blood cells. He extracts them from the patient and, by means of a **retrovirus**, places healthy genes into the cells and then replaces the repaired cells into the patient's body.[105]

A complete understanding of the pattern embedded within the human genetic network might also allow gene therapists to help individuals structure a healthy lifestyle. Individuals who know early in life the physical afflictions to which they are vulnerable can then plan a proper campaign to minimize the risks associated with those potential problems.[106] However, we must avoid becoming overly optimistic about the outlook for gene therapy. Isolating disease-causing genes and determining the mutated base sequence are not easy tasks. Moreover, the isolation of the disease-causing gene and the mutated base sequence is not the end of the story. The base sequence must be translated into the amino acid sequence which will then identify the protein involved. Before the function of the protein can be determined, it must be matched to a similar protein whose function has already been identified. Even this does not necessarily lead to a cure for the disease.[107] Often the circumstances simply are not correct for the development of an effective therapy. However, the isolation of the disease gene is a critical step in the process of finding any cure to a genetically caused disease, and certainly puts us light years ahead of the old trial-and-error methods used in the past.[108]

Concerns about Genetic Engineering. Now that we have had a chance to look at a few of the benefits of genetic engineering, we should balance the picture by considering some of the concerns that surround this activity. Some of these worries address the dangers inherent in the process of genetic engineering itself. One fear is that an experiment might go astray, leading to the creation of a new virus or disease, which could begin a deadly, uncontrollable epidemic of some sort. One early example of this apprehension involved the work of biochemist Paul Berg of Stanford University, who, in the early 1970s, worked on a recombinant DNA experiment that involved combining a cancer virus with a bacterial virus. The misgivings expressed by some experts over this experimental study was that the bacterial virus might invade the bacteria that lives in the human intestinal tract, thereby creating an infectious strain of deadly bacteria.[109] To his credit, in response to this criticism, Berg became instrumental in insisting upon strict procedures in handling potentially dangerous viruses. Moreover, Berg also played a key role in initiating a conference among molecular biologists that resulted in a protocol of safety procedures for experiments involving cancer-causing viruses.[110] Nevertheless, the risk associated with such experimental work still concerns many people.

Perhaps even more troubling than the predictable problems associated with individual experiments and the unanticipated consequences related to genetic engineering. Whenever new technological developments are introduced, unforeseeable results arise. This is true whether we are speaking of something as overwhelmingly beneficial as the computer or something as obviously dangerous as the atomic bomb. Some consequences are predictable. Others are not. For instance, when aerosol hairspray cans were first developed, it was known that certain types of chemical propellants were dangerous because they can explode under intense heat. This danger was predictable. However, it was later discovered that these same propellants can harm the ozone layer. This risk was not predictable.[111]

The development of refrigeration, the invention of the automobile, and the introduction of radio and television provide further examples of unpredictable results from technological advances. These inventions made our lives more efficient, pleasant, and enjoyable. Such results were predictable. However, no one could have predicted that these inventions would lead to an increasingly isolated lifestyle among the American people. How did this happen? How did these clearly beneficial inventions lead to the unpredictable result of isolation. The pattern can be traced in this way. First, the invention of refrigeration meant that people could preserve food for extended periods of time and, therefore, no longer had to shop as often. This cut down on trips to the market in town. The invention of the automobile had a similar effect. Owning an automobile meant that people no longer had to live close to their workplace. This allowed them to move farther and farther away from the central city. Moreover, as automobiles became more and more affordable, people no longer had to depend upon public vehicles for transportation. Radio and television had similar effects by removing the need to leave home for entertainment. Thus, instead of going into town for a concert or a dance, individuals remained indoors to listen to the radio and, later, to watch television. All of these conditions led to greater and greater isolation, a result that was not predicted when these inventions were first introduced to the public at large.[112]

The same type of unpredictability surrounds genetic engineering. No matter how many precautions are taken and no matter how many safety protocols are initiated, the risks are still unpredictable. This reality was made painfully clear in an experiment run under the auspices of the National Institute of Allergy and Infectious Disease of the National Institutes of Health. The experiment involved the use of a genetically engineered "model mouse" in AIDS research. Model animals are sometimes used in research when it is inadvisable or impossible to subject either humans or other animals to the experiment in question.[113] A predictable risk associated with this research was the possibility that the mice that had been implanted with the HIV virus might escape from the laboratory. Consequently, researchers took enormous precautions to prevent this from happening. What the researchers did not predict was that the HIV virus that had been implanted within the model mice might interact with a normally harmless virus carried by these mice. This interaction might have created a new type of AIDS virus. This new virus could have developed the capability of affecting different types of cells and might have been able to replicate at an alarming rate of speed. Most frightening was the possibility that this variation of the virus, unlike the standard AIDS virus, might have developed the ability to travel through the air. Fortunately, none of this happened.[114] Moreover, an article in *Science* about the problem sparked the creation of a workshop sponsored by the National Institute of Allergy and Infectious Disease, the express purpose of which was to reconsider the level of safety employed at institutions involved in HIV research.[115] Nevertheless, this type of unpredictable hazard continues to cause concern for many people.

Other misgivings about genetic engineering address the application of genetic engineering to humans. We have already noted some of the advantages that can result from genetically engineered medical products, such as insulin. We have also looked briefly at the HGP and gene therapy. Along with these benefits, however, come some risks. Chief among these risks is the possibility that the process will be used not only to avoid hereditary illnesses but also to reshape the definition of human nature. At the present time, by tacit agreement, geneticists work only with the genes that are related to hereditary illnesses and do not work with those that make people "normal." In fact, geneticists scrupulously avoid defining normality. This is certainly a wise course of action. However, it is also a course of action that is fraught with danger. Ironically, the very attempt to avoid defining normality eventually ends up defining it. Thus, **normality** becomes the absence of disease-causing genes.[116]

This may be a convenient definition, but it is also one that is ambiguous and, therefore, open to manipulation. What exactly is a disease? Certainly, there is no question about some debilitating illnesses. Afflictions such as cystic fibrosis, multiple sclerosis, amyotrophic lateral sclerosis, severe combined immunodeficiency, and even schizophrenia would clearly be labeled as such. However, what about color blindness, or deafness, or nearsightedness? Would it be appropriate to eliminate these afflictions from the gene pool? What about shortness or left-handedness? How about gray hair? Should these also be eliminated? Equally as troubling as defining the concept of disease is determining how to eliminate such diseases. Does elimination mean preventing the birth of individuals who carry such disease-causing genes? Or does it require placing reproductive restrictions on "genetically defective" individuals?[117] These questions and similar ones must be handled by the regulatory process.

The Regulation of Genetic Engineering

Now we can address the original question that formed the jumping-off point for this portion of the chapter. That question is whether genetic engineering is a science or a technology. Recall that the term *science* involves basic research aimed at compiling knowledge about nature, whereas the word *technology* includes a useful application of that knowledge. Certainly, genetics involves basic research into the nature of life. For instance, the type of work done by Watson and Crick when they discovered the double helix structure of DNA was basic research and, therefore, qualifies as science. The same could be said of the work done by Mendel and Miescher. In fact, whenever microbiologists discover a new virus or a new bacteria, they are adding to our storehouse of knowledge and are, therefore, engaged in science. On the other hand, when Chakrabarty, while working for General Electric, developed his petroleum-eating bacteria, he was involved in technology. The same could be said of the creation of human insulin by genetically engineered bacteria or the development of long-lasting genetically designed tomatoes. It would seem, therefore, that basic research into genetics may be labeled as science, but much of what we associate with genetic engineering, including recombinant DNA and the HGP, are a part of technology. The term usually associated with such genetic engineering activities is **biotechnology**.

Private Regulation of Biotechnology. The initial move toward the private regulation of biotechnology came in 1973 when, as noted earlier, the microbiologist Paul Berg of Stanford became instrumental in insisting upon strict procedures in handling potentially dangerous viruses. Berg also played a central role in initiating a conference at Asilomar, California, where close to 100 biomedical scientists met to discuss safety procedures for experimentation with cancer viruses. The conference resulted in a protocol of safety procedures for experiments involving cancer-causing viruses.[118]

Following this first conference, a second one was held at Asilomar in 1975. This conference, which was sponsored by the National Academy of Sciences (NAS), involved the leading names in molecular biology at that time, both American and international.[119] The conference was suggested by a study committee set up by the Assembly of Life Sciences (ALS) and chaired by Berg. The ALS study committee also recommended an immediate moratorium on certain types of experimental work. It is to the credit of the community of molecular biologists that many of them voluntarily followed the suggested moratorium. The Second Asilomar Conference, succeeded in putting together guidelines for recombinant DNA research.[120]

These guidelines, which were designed to minimize the potential dangers involved in recombinant DNA research, specified the need to include in any experimental design effective containment procedures. The need to match containment procedures with the level of experimental risk was emphasized in the final report. Accordingly, the more extensive the experiment, the more rigorous the containment system in place to ensure the appropriate safety level. It is also interesting to note that, despite the voluntary spirit of the conference and the self-regulatory nature of the guidelines, the report also recommended that the governments of the world use the guidelines to develop codes of practice and procedure for recombinant DNA research within their borders.[121]

Governmental Regulation of Biotechnology. Soon after the end of the Second Asilomar Conference, the federal government jumped into the regulatory arena. One of the first steps toward governmental regulation occurred when the recently formed Recombinant DNA Advisory Committee (RAC) of the National Institutes of Health met to draft their own set of guidelines for recombinant DNA research. Although initially composed only of research scientists, the RAC eventually incorporated members from nonscientific fields, including law, education, and consumer protection. By June 1976, the RAC guidelines were completed and were officially added to the Federal Register. The guidelines applied to all institutions receiving support from the NIH that conduct recombinant DNA research. These parameters also included research done by the NIH itself, as well as research projects conducted abroad under NIH support. The guidelines specified the roles and responsibilities of those involved in recombinant DNA research and outlined the procedures to be followed to ensure the effective containment of any hazardous biological material. The NIH has revised these guidelines many times over the last twenty years.[122] In 1986, a new government organization, the Biotechnology Science Coordinating Committee, was set up "to coordinate the regulation of recombinant DNA biotechnology in the United States."[123]

Congress also entered the regulatory arena soon after the Second Asilomar Conference. In April 1975, the health subcommittee of the Senate Committee on Labor and Public Welfare began a series of hearings entitled the "Examination of the Relationship of a Free Society to Its Scientific Community." The subcommittee took testimony from scientists on both sides of the issue, those in favor of continued self-regulation and those who supported governmentally established guidelines. Although little of consequence came from these hearings, they did signal an end to the tendency of Congress to look the other way in relation to biotechnology. What was of particular concern to several members of Congress was the fact that the NIH guidelines, although effective, were limited in application to institutions receiving federal support from the NIH. This left a large segment of the genetic engineering community, specifically the privately owned and operated biotechnology firms that were also in the midst of conducting potentially hazardous recombinant DNA research, unregulated.[124]

Although most of our discussion in this chapter has centered on the role of the law in the regulation of genetic engineering, we must also remember that government intervention into biotechnology will eventually need to go beyond the orchestration of recombinant DNA research. For example, gene therapy involves diagnosing and perhaps treating those diseases that are based within the genetic code. Gene therapy may be a rarity today, but the more common it becomes, the more likely it is that the law will be compelled to regulate it, despite the law's tendency to avoid such controversial areas.

Consider, for example, the issue of privacy. Some people may not want to know that they face the possibility of suffering from a debilitating disease at some point in their future. What would a genetic therapist's ethical and legal duty be in such a case? Even more serious is the possibility that other people may have access to an individual's personal genetic makeup. Insurance companies may begin to screen applicants based on certain tendencies found in their genetic makeup. Some insurance companies might go so far as to deny coverage to some people because they display a particular genetic disposition. Employers might also use a person's genetic makeup as a determining factor in decisions regarding hiring, promoting, and discharging their employees.[125]

Decisions such as these will have to be made with care and deliberation, in a nondiscriminatory manner. The law can step in to provide protection; unfortunately, as we have seen, the law is often slow to react. Nevertheless, there are precedents for this sort of action at the legislative, executive, and judicial levels. For example, at the federal level, the Americans with Disabilities Act (ADA) protects people with disabilities from discrimination.[126] This statute applies to acts of discrimination in the workplace, public accommodations, and public services.[127] The same legal principles that establish the protections granted by the ADA could be used by Congress to prohibit discrimination against individuals based solely on the content of their personal genetic makeup.

The states may also get involved in passing legislation that protects people from genetic discrimination. Some states have already taken such steps. Maryland, for example, has established a Commission on Hereditary Disorders (CHD). The CHD has the authority to investigate discrimination against an individual based upon the identification of a hereditary disorder.[128] Moreover, the enabling legislature that set up the commission also defined a hereditary disorder as "any disorder resultant from the genetic material DNA (Deoxyribonucleic acid) which is transmitted from a parent or parents to his or her child."[129] This may indicate that such steps by other governmental agencies are not only possible, but also probable in the near future.

Private Industry and the Human Genome Project. In the spring of 1998, a company known as Celera Genomics, formed by J. Craig Venter, president of The Institute for Genomic Research, began a privately funded project aimed at developing a sizable portion of the sequential blueprint of the human genome within three years at a cost of $300 million. To accomplish this feat, Celera Genomics will utilize an innovative automated gene-sequencing machine supplied by the Perkin-Elmer Corporation. If successful, this venture will be considerably less expensive than the $3 billion dollar government-sponsored Human Genome Project. One reason for such an optimistic prediction is that Celera Genomics will move directly to gene sequencing, rather than moving from gene mapping to gene sequencing, which is the approach used by the Human Genome Project. This method, known as whole-genome shotgun sequencing, will significantly reduce both the time and cost involved in the overall project. Celera Genomics will also save time and money by leaving to others the final stage in the process. This final step involves the time-consuming task of filling in the blanks and eliminating any flaws in the initial sequencing process.[130]

Scientists within the government-sponsored Human Genome Project are skeptical about the claims made by Venter and Celera Genomics. One objection to the approach promoted by Celera Genomics is that eliminating the gene mapping step will reduce the quality of the final product and increase the probability of making serious errors. These concerns explain why the government-sponsored project originally rejected the technique now embraced by Celera Genomics. Another fear voiced by government scientists is that access to the human genome will be severely curtailed because of Celera's interest in making a profit. For its part, Celera Genomics has stated that it plans to release most of its results into the public domain. As for the probable success of the private venture, supporters of Celera Genomics argue that The Institute for Genomic Research used this technique successfully in the sequencing of bacteria genomes. Critics have responded by pointing to the relative complexity of the human genome when compared to the bacteria genomes previously sequenced by the Institute.[131]

Not to be outdistanced by Venter and Celera Genomics, the National Human Genome Research Institute approved grants amounting to $81.6 million for several academic research institutions that were leading the way in the government-sponsored Human Genome Project (HGP). These grants were supplemented by a $57 to $77 million pledge by Wellcome Trust to a British research facility. These grants along with support from the U.S. Department of Energy's Joint Genome Institute has allowed the government-sponsored HGP to accelerate the pace of its project without resorting to the whole-genome shotgun sequencing favored by Celera.[132]

CONCLUSION

In this chapter, we have examined the interaction between science and the law in relation to basic research and in relation to technology. We saw that in the context of basic research, the law tends to treat scientists quite well. They are permitted to sit on peer-review groups that determine how billions of tax dollars are spent each year on basic scientific research. This has led some people to the conclusion that Congress has created a self-governing republic of scientists.[133] Unlike the rest of us, the argument goes, scientists are allowed to police themselves. As members of peer-review groups in national institutes, they are privileged to make decisions that affect billions of dollars annually. Yet, the courts refuse to review those decisions, except to detect procedural rather than substantive difficulties. This has led some people to propose the creation of a specialized science court. While the science court certainly remains a possibility, it is more likely that things will continue as they have. Certainly, the cautious and deliberative nature of the law accounts for some of this stability. However, most of it is due to the fact that, once scientists leave their laboratories and apply science to technological developments, the law regulates them just as much as it regulates the rest of us. The interaction between science and the law, though somewhat complex, nevertheless, follows a predictable pattern.

Review

12-1. The U.S. Constitution is based upon two guiding principles: the separation of powers and a system of checks and balances. The most important power that enables the national government to spend money in support of science is provided by the general welfare clause. The government supports science and technology with tax dollars, and also regulates them. The most instrumental clause behind the regulatory power of Congress is the commerce clause. The major federal agencies involved in funding scientific research include the National Science Foundation, the Department of Defense, the Department of Energy, and the National Institutes of Health. Funding decisions are delegated to the experts of the agencies because such decisions generally require scientific knowledge that is somewhat beyond the background of the typical member of Congress. This decision-making process is known as peer review. There are several good reasons to favor peer review as a part of the scientific funding procedure. However, peer review has been severely criticized and challenged. Continued dissatisfaction with peer review has led some commentators to conclude that the entire system should be revised. One suggestion has recommended the establishment of a science court that could be charged with, among other things, the responsibility of reviewing the scientific merits of grant proposals. There is not much likelihood, however, that such a court will be established in the near future.

12-2. The term *science* is used to describe an activity that involves basic research aimed at understanding how nature works. In contrast, technology involves a practical application of knowledge usually, but not necessarily, associated with science. Because of its relationship to practical applications of knowledge, technology often impacts upon society in ways that are more direct, more evident, and thus more easily regulated than those of science. Since technology also involves a useful application of some sort there is consequently often a market for that technology. This means not only that new technology has an impact on the economy but also that someone intends to profit from the sale and distribution of that technology. Thus, again we can see that technology can be more easily regulated than science. As was true for the regulation of science, the legal basis for the regulation of technology is found in the U.S. Constitution. The power to grant patents has been instrumental in the government's power to promote as well as regulate the development of technology. Patents are actually property rights awarded to inventors by the national government. The question that comes to mind in relation to patents and biotechnology is whether genetically engineering products are patentable. This question was answered affirmatively by the U.S. Supreme Court in the case of *Diamond v. Chakrabarty*.

12-3. Genes, which are made of DNA, contain the information needed to manufacture proteins and enzymes, which are necessary for life. James Watson and Francis Crick discovered that DNA is actually constructed as a double helix. The double helix must split when the cell replicates (when it duplicates itself in the process of cell division) and when the genetic code is read for the production of protein. Genetic engineering is the process by which modifications to living things are made by manipulating their genes. At the heart of genetic engineering is the ability to produce recombinant DNA. This process is often referred to as *gene splicing*. Genetic engineering can be beneficial in a number of ways. Perhaps the most famous project associated with genetic engineering and human beings is the Human Genome Project (HGP). The immediate objective of the HGP is to create a comprehensive blueprint of human DNA. Some concerns also surround genetic engineering. There are the dangers inherent in the process of genetic engineering itself, both anticipated and unanticipated. There are also misgivings about genetic engineering being applied to humans. The initial move toward the private regulation of biotechnology came in two conferences held at Asilomar in the 1970s, which resulted in a set of self-regulatory guidelines for genetic engineering.

Understanding Key Terms

Administrative Procedure Act	double helix	protein
	enumerated powers	recombinant DNA
anticodon	gene	regulations
balancing principle	general welfare clause	replication
bases	genetic code	restriction enzyme
base-paired sequence	genetic engineering	retrovirus
biotechnology	gene therapy	ribonucleic acid (RNA)
chromosomes	genome	ribosome
codon (triplet)	Human Genome Project (HGP)	ribosomal RNA (rRNA)
commerce clause		RNA polymerase molecule
craft	messenger RNA (mRNA)	science court
cytoplasm		
deoxyribonucleic acid (DNA)	normality	separation principle
	nucleotide	technology
device	peer review	transcription
DNA mapping	plasmid	transfer RNA (tRNA)
DNA sequencing	proofreading	translation

Review Questions

1. Why is it necessary to regulate scientific research?
2. What is the constitutional basis for the regulation of science and technology?
3. What is the role of peer review in the granting of funds for scientific research?
4. What are the advantages and disadvantages of the peer-review process?
5. What are the arguments for and against a science court?
6. What is the difference between the law's attitude toward basic research with its attitude toward technology?
7. What are some of the basic concepts involved in genetics?
8. What are some of the procedures involved in genetic engineering?
9. What are some of the benefits and concerns associated with genetic engineering?
10. What regulatory efforts have been made in relation to biotechnology?

Discussion Questions

1. Much of the decision-making process for granting funds for basic scientific research is carried out by peer review. What is your reaction to this? Do you believe that such important funding decisions should be left up to scientists, or should nonscientists have input? Support your opinion. If you would like to see input from nonscientists, explain how the peer-review procedure should be changed.
2. The proposed science court would provide a forum for advocates on both sides of a scientific controversy who would present their arguments before a panel of neutral judges with scientific backgrounds. The science court could be invoked by interested parties whenever they faced a pressing issue that could only be resolved by a proper understanding of the science involved. Assume that you are responsible for putting together a legislative package supporting the science court. What arguments would you make to support its establishment? Now switch sides and propose arguments against the science court.
3. Memories of the Nazi eugenics programs of the Second World War and images promoted by literary works, such as *Brave New World*, *The Island of Dr. Moreau*, *Frankenstein*, *Coma*, and *Unisave*, as well as films like *Jurassic Park* and *The Lost World* have given many people a negative image of genetic engineering. How would you argue that such images are inaccurate and dangerous distortions of genetic engineering? Now switch sides and promote the idea that genetic engineering in general and the Human Genome Project in particular are dangerous undertakings that should be stopped immediately. What arguments

could be made that a Nazi-like eugenics program could never arise in the United States?

4. Are you comfortable with the idea of gene therapy? How would you suggest that legislators, administrators, and judges handle issues of privacy in relation to genetic information? What about employers? Should employers be allowed to screen employees based on their genetic blueprint? Should insurance companies be allowed access to such information for potential clients? Explain.

5. The evolutionary process saved us from the ethical and legal nightmare of having to balance the rights and responsibilities that would inevitably arise when two sentient species share a single planet. Genetic engineering could raise the same problem in different ways. The primary ethical questions would involve the creation of a quasihuman hybrid species. Would you support or oppose the creation of quasihuman hybrids? Explain. Can the concept of acquired humanness that was explored in Chapters 10 and 11 be used to determine the degree to which a quasihuman species can be said to have human characteristics? Can the birth of the cortex criteria be used to determine when a quasihuman hybrid crosses the threshold from nonhuman to human? Why or why not? Some anthropologists have suggested that the awareness of the existence and potential nonexistence of self is the crucial sign that a particular human ancestor had obtained a level of sentience that would mark it as human. The key piece of evidence which indicates that a group of individuals has obtained self-awareness is the ritual burial of the dead. Should this be used as the definitive guideline? Explain. Suppose that in the future, a race of pseudohuman hybrids demands the burial of their fallen brothers and sisters. What would you do in such a case? Explain.

ANALYZING *STAR TREK*

Background

The following episode from *Star Trek: The Next Generation* reflects some of the issues that are presented in this chapter. The episode has been carefully chosen to represent several of the most interesting aspects of the chapter. When answering the questions at the end of the episode, you should express your opinions as clearly and openly as possible. You may also want to discuss your answers with others and compare and contrast those answers. Above all, you should be less concerned with the "right" answer and more with explaining your position as thoroughly as possible.

Viewing Assignment— *Star Trek: The Next Generation,* "Unnatural Selection"

In deep space, the USS Enterprise discovers the USS Lantree, the entire crew of which has died from a mysterious malady that somehow causes rapid aging. In an attempt to solve the mystery, the Enterprise retraces the steps in the Lantree's last mission. As a result, the Enterprise crew find themselves at the Darwin Genetic Research Station, which is engaged in a project aimed at creating genetically perfect humans. At this point in the research project the scientists have managed to develop several genetically perfect children. The crew of the Enterprise soon discovers that the inhabitants of the station are suffering from the same old-age malady that destroyed the crew of the Lantree. The scientists on the research station ask the Enterprise to take the children on board the starship to save the children from the illness. Captain Picard is justifiably wary of such a request and withholds his approval until Dr. Pulaski agrees to examine one of the children. The child is kept in suspended animation in order to safeguard the crew. Dr. Pulaski determines that he is not a carrier. Picard remains skeptical. Dr. Pulaski suggests that she be allowed to examine the child at a safe distance from the ship in a shuttle craft. Once they are a safe distance from the Enterprise she will revive him and conduct the tests again. Picard agrees. Soon after reviving the child, however, Dr. Pulaski contracts the disease. Apparently, Dr. Pulaski's original diagnosis was incorrect. The children on the station actually are carriers of the old-age disease. Dr. Pulaski and the boy are returned to the research station, which is now under quarantine. Once on the station, Data determines that the children, these perfect genetic creations, are not just carriers of the disease, but are actually causing it. Somehow their active immune systems developed a virus that attacks and alters human DNA. Data beams back to the ship and obtains an unaltered sample of the doctor's DNA. Using the transporter, Data reconfigures the transporter pattern of Dr. Pulaski by eliminating the damaged DNA and replacing it with the untainted DNA. The same is done for the rest of the members of the research facility. Consequently, the children are no longer a threat, and the problem has been solved.

Thoughts, Notions, and Speculations

1. The stated objective of the project under way on the Darwin Genetic Research Station is the creation of genetically perfect human beings. Imagine that you are the person who has been assigned the task of defending this project before the Federation Science Council. How would you defend the need for the project. Now switch sides and imagine that you are opposed to the project. How would you argue against it?
2. Based upon your reading in this chapter, what are some of the problems inherent in any attempt to define a genetically "perfect" human being? If you were assigned the task of describing a perfect

human being, what elements of perfection would you include in your definition? Again switch sides and argue that it is difficult, if not impossible, to devise such a definition. Now go a step further: argue that it is not only difficult but also ethically wrong to create such a standard of perfection.
3. The authors of "Unnatural Selection" lay the blame for the aging disease on the genetically engineered children. Are the authors expressing a moral judgment by taking this approach? If so, what is that judgment? Describe an alternative ending that would not blame the children. Explain.
4. Would the genetically perfect children created on the Darwin Genetic Research Station qualify as a quasihuman hybrid species? Speculate on what would be necessary for the creation of such a quasihuman hybrid species. Is the creation of such a species a desirable objective? Does it make any difference that the research is conducted in deep space? Explain.
5. Do you believe the type of genetic work conducted on the Darwin Research Station will one day become a reality? Why or why not?

NOTES

1. *Diamond v. Chakrabarty*, 447 U.S. 303, 100 Sup. Ct. 2204 (1980). The Court's rationale for this decision is explained at length later in the chapter.
2. *Roe v. Wade*, 410 U.S. 113, 93 Sup. Ct. 705 (1973).
3. Gordon W. Brown, Edward Byers, and Mary Ann Lawlor, *Business Law with UCC Applications,* 7th ed., 3.
4. This is, of course, something of an oversimplification. In framing the new Constitution, the delegates to the convention had to deal with many other aspects of the government. Some of these included maintaining the autonomous nature of the states, dividing powers between the national government and the state governments, determining how the states would relate to each other, figuring out the actual apparatus of the national government, and laying down the rights of the people. Most of these activities were accomplished rather well, with the exception of the people's rights. These had to be added later in the first ten amendments, also known as the Bill of Rights. For a more detailed look at the writing and the structure of the U.S. Constitution see Bernard Schwartz, *Constitutional Law*.
5. Len Young Smith, Richard Mann, and Barry Roberts, *Essentials of Business Law*, 44.
6. Ibid.
7. Bernard Schwartz, *Constitutional Law*, 137.
8. For the legislature's census-taking power see the U.S. Constitution, art. 1, sec. 2, cl. 2; for the power over weights and measures see art. 1, sec. 8, cl. 5; for the power to grant patents see art. 1, sec. 8, cl. 8; for the power to raise an army see art. 1, sec. 8, cl. 12; for the power to maintain a navy see art. 1, sec. 8, cl. 13; for the power to regulate both the army and the navy see art. 1, sec. 8, cl. 14.
9. U.S. Constitution, art. 1, sec. 8, cl. 1; Steven Goldberg, *Culture Clash*, 36.
10. Goldberg, 36–37.
11. *United States v. Butler*, 297 U.S. 1, 56 Sup. Ct. 312 (1936); Goldberg, 37; Schwartz, 74.
12. U.S. Constitution, art. 1, sec. 8, cl. 8.
13. Gordon Brown, Paul Sukys, and Mary Ann Lawlor, *Business Law with UCC Applications*, 8th ed., 580.
14. Ibid., 15–16.
15. Ibid.
16. Goldberg, 44.
17. *Grassetti v. Weinberger* 408 F. Supp. 142 (1976).
18. Goldberg, 45–46.

19. Ibid., 56–57.
20. Stephen Cole, *Making Science*, 84.
21. *Grassetti v. Weinberger*, 408 F. Sup. 142 (1976).
22. Goldberg, 56–57.
23. Cole, 145.
24. *Grassetti v. Weinberger*, 408 F. Sup. 142 (1976).
25. Ibid.
26. Goldberg, 56–57.
27. *Grassetti v. Weinberger*, 408 F. Sup. 142 (1976).
28. Goldberg, 57. For a detailed examination of the peer-review process see Kenneth R. Foster and Peter W. Huber, *Judging Science*, 161–205.
29. Ibid., 45.
30. *Grassetti v. Weinberger*, 408 F. Sup. 142 (1976).
31. Cole, 139–41.
32. Ibid., 140.
33. Ibid., 145.
34. Ibid.
35. Ibid., 153–55.
36. Ibid., 152–53.
37. Smith, Mann, and Roberts, 44.
38. Ibid.
39. Brown and Sukys, 16; Smith, Mann, and Roberts, 57–58.
40. 5 U.S.C. Section 702; *Grassetti v. Weinberger*.
41. In addition to *Grassetti v. Weinberger*, see the following cases: *Apter v. Richardson*, 510 F. 2d 351 (7th Cir. 1975); *Kletschka v. Driver*, 411 F. 2d 436 (2d Cir. 1969); and *Cappadora v. Celebreze*, 356 F. 2d 1 (2d Cir. 1966).
42. Harvey Saferstein, "Nonreviewability," 367, 382–83.
43. Ibid., 382.
44. *Grassetti v. Weinberger*, 408 F. Sup. 142 (1976).
45. Ibid.
46. Ibid.
47. Ibid.
48. James A. Martin, "The Proposed 'Science Court,'" 1058, 1072.
49. Ibid., 1072–75.
50. Ibid., 1069.
51. For a discussion of the Yucca Mountain nuclear waste controversy see Chris G. Whipple, "Can Nuclear Waste Be Stored Safely in Yucca Mountain?" 72–79.
52. Barry M. Casper, "Technology Policy and Democracy," 29–30.
53. Martin, 1069.
54. Ibid., 1076–78.
55. Barry Casper and Paul Wellstone, "Science Court on Trial in Minnesota," 282–89. For an in-depth analysis of the science court proposal in Minnesota see Barry Casper and Paul Wellstone, *Powerline*. For a clever and informative fictional account of the difficulties involved in the establishment and operation of a national science court see Ben Bova, *Brothers*.
56. Casper and Wellstone, "Science Court on Trial in Minnesota," 282–89.
57. Leslie Stevenson and Henry Byerly, *The Many Faces of Science*, 2–3.
58. Frederick Ferre, *Philosophy of Technology*, 26.
59. Ibid., 14–17.
60. Ibid.
61. Brown and Sukys, 289–91, 614–15.
62. *U.S. Code*, vol. 35, sec. 101; Brown and Sukys, 289–91, 614–15.
63. *Diamond v. Chakrabarty*, 447 U.S. 303, 100 Sup. Ct. 2204 (1980).
64. Ibid.
65. *U.S. Code*, vol. 35, sec. 101; *Diamond v. Chakrabarty*, 447 U.S. 303, 100 Sup. Ct. 2204 (1980).
66. *Diamond v. Chakrabarty*, 447 U.S. 303, 100 Sup. Ct. 2204 (1980).
67. Ibid.
68. Ibid.
69. Ibid.
70. Ibid.
71. Brown and Sukys, 580.
72. Robert Pollack, *Signs of Life*, 44–45; Robert Shapiro, *The Human Blueprint*, 13.
73. Enzo Russo and David Cove, *Genetic Engineering*, 134–41; Shapiro, 13–22.
74. Russo and Cove, 141, 150; Shapiro, 24–25.
75. Rush W. Dozier Jr., *Codes of Evolution*, 75; Pollack, 25; Shapiro, 46–74; See also James D. Watson, *The Double Helix*.
76. Dozier, 75; Pollack, 25; Shapiro, 46–74.
77. Dozier, 76–77.
78. Boyce Rensberger, *Instant Biology*, 39–44; Pollack, 20–26.
79. Necia Grant Cooper, ed., *The Human Genome Project*, 39–44; Dozier, 76–78; Rensberger, 39–44, 94–96; Russo and Cove, 30–32; James Trefil and Robert Hazen, *The Sciences*, 542.
80. Sybil P. Parker, ed., *McGraw-Hill Concise Encyclopedia of Science and Technology*, 1520.
81. Rensberger, 30, 62.
82. Ibid., 42.
83. Dozier, 77.
84. Parker, 531.
85. Cooper, 45–46; Dozier, 79–80; Rensberger, 44–46; Russo and Cove, 35.
86. Cooper, 46–47; Dozier, 80–81; Rensberger, 49–53; Trefil and Hazen, 544–47.
87. Cooper, 46–47; Dozier, 80–81; Rensberger, 49–53; Trefil and Hazen, 544–47.
88. Russo and Cove, 65.

89. Robert Hazen and James Trefil, *Science Matters*, 239.
90. Cooper, 52–53; Hazen and Trefil, 239; Trefil and Hazen, 551–52.
91. Russo and Cove, 65.
92. Trefil and Hazen, 552; Russo and Cove, 74, 83.
93. Russo and Cove, 82.
94. Ibid., 83.
95. *Diamond v. Chakrabarty*.
96. Hazen and Trefil, 239; Russo and Cove, 94; Trefil and Hazen, 552.
97. Russo and Cove, 95.
98. Russo and Cove, 10, 219; Trefil and Hazen, 550.
99. Cooper, ix, 70; Hazen and Trefil, 550.
100. Russo and Cove, 43–44.
101. Cooper, 70.
102. Cooper, ix, 70; Hazen and Trefil, 240; Trefil and Hazen, 550–51.
103. Cooper, ix, 72.
104. Russo and Cove, 114–117; Goldberg, 124–25.
105. W. French Anderson, "Human Gene Therapy," 810–11; Goldberg, 125; Russo and Cove, 117–18. In addition to his own SCID gene therapy protocol, Anderson also outlines several additional gene therapy protocols, many of which have already been initiated. (Anderson, 808–13.)
106. Goldberg, 126.
107. Cooper, 77.
108. Cooper, 77; Russo and Cove, 118–19.
109. Donald S. Fredrickson, "Asilomar and Recombinant DNA," 267–69.
110. Fredrickson, 268–69.
111. Bernard E. Rollin, *The Frankenstein Syndrome*, 78.
112. The World Future Society, "The Art of Forecasting."
113. Rollin, 115–17.
114. Paolo Lusso, et al., "Expanded HIV-1 Cellular Tropism by Phenotypic Mixing with Murine Endogenous Retroviruses," 848–52; Joseph M. McCune, et al., "Pseudotypes in HIV-Infected Mice," 1152–53; Rollin, 115–17.
115. Gregory Milman, "HIV Research in the SCID Mouse," 1152.
116. Evelyn Fox Keller, "Nature, Nurture, and the Human Genome Project," 362–63.
117. Goldberg, 124–25; Keller, 360–61.
118. Fredrickson, 268–69; Stevenson and Byerly, 192–93.
119. Fredrickson, 274; Fredrickson's article in *Biomedical Politics* gives a detailed history of the events that led to the Second Asilomar Conference, as well as summaries of each day's activities and the conclusions and recommendations that came from the conference. The article also contains a list of all participants. The final report that came out of the Second Asilomar Conference can be found in Paul Berg, et al., "Summary Statement of the Asilomar Conference on Recombinant DNA Molecules."
120. Fredrickson, 275–84.
121. Paul Berg, et. al., "Summary Statement of the Asilomar Conference on Recombinant DNA Molecules."
122. Judith Areen, ed., et al., *Law, Science and Medicine*, 50–63.
123. Fredrickson, 284.
124. Areen, 66, 79.
125. Goldberg, 126–27.
126. Brown and Sukys, 520; Daniel Kevles and Leroy Hood, "The Code of Codes," 368.
127. Kevles and Hood, 368.
128. Commission on Hereditary Disorders, 43 Md. Code Ann. Section 817, in *Law, Science and Medicine*, ed. Judith Areen, 1346.
129. Ibid.
130. Lisa Belkin, "Splice Einstein and Sammy Glick," 28–31, 56; Eliot Marshall and Elizabeth Pennisi, "Hubris and the Human Genome," 994–95.
131. Belkin, 28–31, 56; Marshall and Pennisi, 994–95.
132. Elizabeth Pennisi, "Academic Sequencers Challenge Celera in a Sprint to the Finish," 1822–23.
133. David Bazelon, "Coping with Technology through the Legal Process," 96; Goldberg, 66–68.

Unit VIII Science: The Final Frontier

Chapter 13
The Implications of Science for the Future

COMMENTARY: HISTORICAL STUDY OF NEW YORK CITY, 2022

From a report entitled *The Soylent Corporation: 2017–2037,* **issued on 20 May 2148**

By the year 2022, the population of Greater New York City had surpassed 40 million people. The city itself had spread from Philadelphia to Boston. The added space was, however, no consolation for the people crammed into that urban cesspool. Most of the population was unemployed. Their only source of income was a small welfare entitlement handed out by the government at uneven intervals. Consequently, most of these people were homeless, living in churches, in abandoned cars, and on the stairways of dilapidated buildings. Those who were employed lived in broken-down apartment buildings, most of which lacked even the basic necessities of heat and running water.

In addition to this, the air was permanently stained by a foul layer of green smog that strangled the entire Northern Hemisphere. The only trees in the city were housed in a tiny preserve that had once been Gramercy Park. Air conditioning had been outlawed for all but the very rich as a wasteful use of scarce energy resources.

Worst of all was the food crisis. Years earlier, global warming had reduced the growing season to a few months during which only the hardiest crops could be produced. As a result, in all urban areas, but especially in Greater New York, food supplies dwindled. Meat and vegetables were completely unavailable on the open market. Dairy products were rare. Bread could be obtained only a few times each month, and it took $150 in hard currency to buy a jar of fruit jam.

People stood in long lines for hours to receive their meager ration of a foodstuff known as soylent. Soylent was produced by the Soylent Corporation from plankton that was harvested on underwater farms in the Atlantic Ocean. Unfortunately, a study conducted by the Soylent Corporation between 2017 and 2019 revealed that the oceans were dying and that the available supply of plankton for the world marketplace would be exhausted by the year 2020.

It was at this critical point in our history that the Soylent Corporation, in cooperation with the federal government, took a daring step forward. In a forward-looking move, Soylent executives took advantage of the one resource that the planet had in abundance—people. The Soylent Corporation developed a process by which the bodies of dead people could be transformed into nutritious, flavorful wafer-like cookies, known as Soylent Green. In this way, people became the basic foodstuff of the planet. Looking back to that bold and courageous step, we now see that we can thank the Soylent Corporation for the survival of the human race.—Signed S/5293/433/E310

This story may sound like pure fantasy, and, in fact, it did serve as the plot for the 1973 film *Soylent Green*. But could it be more than this? Could these events actually come to pass? Is this vision of the future simply a fantasy spun from the creative mind of a clever science fiction writer, or is it an accurate prediction of the future? Will science solve the problems of population growth and environmental decay before we face such a crisis, or will we wait too long, precipitating the situation depicted in *Soylent Green*? These and other questions are addressed in this chapter.

CHAPTER OUTCOMES

Upon finishing this chapter, the reader should be able to accomplish the following:

1. Describe the three major external challenges facing the modern world.
2. Identify those parts of the world that have the greatest growth in population.
3. Identify the outcomes of the population explosion that would have the most devastating effects on the various cultures and environments of the world.

4. Explain how Malthusian checks slow population growth.
5. Detail how the threat of nuclear war has changed since the end of the Cold War.
6. Define the allies, the orphans, and the rogues.
7. Relate the options available for dealing with a multipolar world situation.
8. Explain how the challenge of the environment is different from the other two major external challenges facing the modern world.
9. Outline the evidence that exists for global warming.
10. Explore some of the consequences of global warming.

13-1   THE FUTURE ACCORDING TO HEILBRONER

The economist Robert L. Heilbroner published a book in 1973 entitled *An Inquiry into the Human Prospect*. In this work he examined several challenges facing humanity as it entered the final third of the twentieth century. The three major challenges that Heilbroner identified were uncontrolled population growth, the threat of nuclear war, and the destruction of the environment. In addition to identifying these three challenges, Heilbroner also focused on what he saw as the underlying causes of those challenges. Perhaps it will not be a surprise that Heilbroner lays the blame for these crises on the shoulders of science and technology. For example, he identifies the technological advances in medical science and health care as contributing causes of the population explosion. Ironically, Heilbroner notes that it is the lowering of the death rate, rather than the rise in the overall fertility rate, that has led to an imbalance in the rate of population growth. In other areas he focuses on the responsibility of science and technology for nuclear weapons and for an environmental crisis caused largely by industrial growth and development.[1]

To be fair, Heilbroner does not accuse science and technology alone for these problems. Instead, he points to the uneven way that society has allowed science and technology to develop. According to Heilbroner, our civilization has permitted and even encouraged harmful and destructive technological advances while ignoring the opportunity to control those advances or to offset the negative effects of new technologies.[2]

Whether science and technology will be able to solve the problems created by this lopsided approach to their development is, at best, problematic. Heilbroner holds out little hope, although he does discuss, at length, the ability of society to respond to these challenges.[3] Moreover, it will do no good to argue that Heilbroner's conclusions are out of date simply because his book was written in the 1970s. Heilbroner has been diligent enough to rewrite *An Inquiry into the Human Prospect* twice over the last thirty years, once in the 1980s and again in the 1990s. In each revision he has not abandoned his basic premise, despite the events of the intervening decades, but has, instead, added more evidence to support his original claim. Nevertheless, it would be unwise to take Heilbroner's position at face value. It would be much more responsible to search for sources that either corroborate or disprove his premise. That then is the goal of this chapter. We look at evidence supporting the fact that civilization is facing a crisis situation caused by the population explosion, the threat of nuclear war, and the destruction of the environment. We then examine the question of whether a well-developed science and technology policy can help solve these problems.

13-2   THE CHALLENGE OF POPULATION

In *Inquiry*, Heilbroner noted that the population of Earth in 1973 was 3.6 billion people and that the fastest growing areas in terms of population were to be found in the undeveloped world. In these areas, including Latin America, Southeast Asia, and Africa, their population would double every quarter century. At that speed, according to Heilbroner's figures, the population of Earth would reach 40 billion by the year 2073.[4] Three questions about this rate of population growth become immediately apparent. First, are these population figures still accurate at the end of the twentieth century? Second, if the rate of population growth has not slowed, or has slowed very little, why is this so? Finally, what are the short- and long-term effects of an unchecked growth in the population of the planet Earth?

The Growing Population

The news regarding the growth of the world's population has not improved since Heilbroner made his first predictions in 1973. Since that year, the world has added over 2.3 billion people to the overall population, giving the planet Earth close to 6 billion human inhabitants.[5] Figures compiled in 1995 by the United Nations Department for Economic and Social Information Analysis show that the more developed nations of the world account for only 1.2 billion people.[6] The United Nations identifies the following areas as the **more developed world**: North America, Europe, Oceania, and the former Soviet Union. Regions considered a part of the **less developed world** are Africa, Latin America, and that part of Asia which is not within the former Soviet Union.[7] Heilbroner agrees with this division, although he includes Japan among the nations of the more developed world. The addition of Japan to the developed side of the ledger is probably well advised.[8] Clearly then, it is the less developed world that carries the bulk of the world's population.

The real problem, however, lies not in the current figures but in the rate of population growth. According to estimates made by the United Nations, the growth rate among the less developed nations is 1.9 percent per year, whereas the growth rate in the more developed world is 0.4 percent. Assuming that the rate of growth remains about the same, the population of the planet could reach between 7.9 and 11.9 billion people by the year 2050.[9] The actual figure for the population of the world in 2050 predicted by the United Nations Department for Economic and Social Information Analysis is 9.83 billion. This figure is arrived at by assuming that the growth rate remains about the same. If the growth rate increases significantly, the population could reach 11.9 billion. On the other hand, if it drops off, the low-end estimate is still 7.9 billion.[10]

The sheer size of these figures makes it difficult to appreciate the magnitude of the problem. Therefore, it might be helpful to look at the numbers in a different way. One way is to look backward and judge how long it took the human race to amass the population that inhabits the Earth today. The human population reached 1 billion for the first time in 1825.[11] This date means that it took approximately 4 million years for the human race to reach a population of 1 billion.[12] Remarkably, it took only one hundred years to duplicate that feat, so that by 1925, the world's population had reached 2 billion. Between 1925 and 1976, the population doubled again, reaching approximately 4 billion in the year of America's bicentennial.[13] At the current rate of growth, approximately 88 million people are added to the global population each year.[14]

Another way to dramatize these figures is to speak of them in terms of annual numbers. Between 1950 and 1955, approximately 47 million people were added to the population each year.[15] This is roughly equivalent to adding six cities the size of New York to the world annually.[16] In contrast, as noted above, between 1985 and 1990, the world population rose about 88 million each year. This would be the same as adding close to twelve cities the size of New York to the planet. Should the trend continue, the annual population growth rate could soon top 112 million. At this rate, almost fifteen "New Yorks" would be added to the population each year.[17]

The growth rate is clearly the fastest in the less developed world. Since 1950, the overall rate of population growth in the less developed world has been 161 percent, compared to 43 percent in the more developed world.[18] Between 1994 and 2050, the population of the less developed world is expected to grow by 93 percent, while the more developed world will grow by only 4 percent.[19] In sheer numbers this means that out of the 9.83 billion people that will inhabit the Earth by 2050, approximately 8.7 billion will live in the less developed world.[20]

Before we accuse the less developed world of doing the Earth a disservice, however, we must remember that numbers alone are only part of the story. The rest of the story involves overconsumption and underdistribution. Although the more developed world constitutes about 20 percent of the world's population, 80 percent of the world's goods and services are produced and consumed by them.[21] If we phrase it another way, of the 88 million people added to the population annually between 1985 and 1995, approximately 4.5 million were added to the more developed world. Those 4.5 million new souls, however, have as much of an environmental impact on the Earth as the 83.5 million new individuals that were added each year to the less developed world.[22] Heilbroner paints an even more graphic picture of the situation when he vividly points out that the present world situation looks very much like a train in which most of the passengers are jammed into boxcars while a privileged few ride in first-class luxury at the front of the train.[23]

Causes of the Population Explosion

Beyond the simple fact that the more people that exist, the more people will be born, what other circumstances have contributed to the dramatic growth in population, especially in the less developed world? The phenomenal growth rate of the less developed world has been helped by two conditions: the declining nature of infant mortality and an increase in the life expectancy of most people on Earth. **Infant mortality rates** measure the death rate of children who do not reach their first birthday. Recent figures clearly reveal that infant mortality rates have declined dramatically in both the more developed and the less developed worlds. This decline is due largely to scientific advances in medicine and health care. Programs of immunization have eliminated many childhood diseases, and the use of antibiotics has reduced the dangers of other life-threatening illnesses that plague children.[24] In explaining the effects of such modern health practices, Paul Kennedy, professor of history at Yale University, points to Tunisia as an example in his work *Preparing for the Twenty-First Century*. According to Kennedy, between 1965 and 1990 the death rate per thousand infants in Tunisia fell from 138 to 59. Similarly, during the identical span of time, the death rate among children between one and five years of age fell from 210 per thousand to 99 per thousand.[25]

The drop in infant mortality rates, however, is only part of the story. Another part focuses on the dramatic increase in life expectancies in all parts of the globe, but especially in the less developed world. Taking a worldwide perspective, the **average life expectancy** has reached 64.4 years. This represents a rise in the average life expectancy of 6.5 years since 1975. The highest figures are found in Japan, where life expectancy is 79.5 years. Hong Kong follows Japan with 78.6 year; Iceland and Sweden are next, each having 78.2 years. The figure falls only slightly in North America, where most people can expect to live 76.1 years. The lowest life expectancies are found in Africa and some parts of Asia, notably Afghanistan (43 years).[26]

Although these two elements, declining infant mortality rates and rising life expectancies, are key factors in rapid population growth, there are other circumstances that contribute to this trend. For example, most of the less developed countries have agricultural economies that depend upon family-run farms for food production. Moreover, historically, the less developed world has had a high rate of infant mortality. Consequently, to offset the effects of a high infant mortality rate, families in these countries tended to be very large. When infant mortality rates drop and life expectancy increases while family growth rate does not change, a jump in the size of the population is inevitable.[27] Coupled with this is the failure to introduce adequate measures for birth control, especially in the countries that are growing the fastest. Some of the latest figures available indicate that the level of contraceptive use is about 56 percent in the less developed regions of the world compared to 72 percent in the more developed regions of the world. Areas where the use of contraceptives is the lowest include Africa (18 percent), western Asia (43 percent), and Latin America (58 percent).[28]

The Consequences of an Unchecked Population

In 1973, Heilbroner pinpointed several consequences that could result if we fail to check the rapid growth of the world's population. These consequences include urban disintegration, disease, and famine.[29] Looking at the situation today, there is little reason to change these predictions. In fact, they are the same predictions that were made by Paul Kennedy in the 1990s in *Facing the Twenty-First Century*.[30] Both Heilbroner and Kennedy identify urban disintegration as the major problem.[31] Kennedy's figures highlight the danger rather dramatically. He notes that in the less developed world the growth of urban areas is particularly dramatic. Looking first to the past he notes that in the 1980s only 32 percent of the less developed world lived in the cities compared to the current rate of almost 40 percent. If this trend continues, by 2025 almost 57 percent of the population of the less developed world will live in urban areas.[32]

Urban Population Growth. What is even more eye-opening is the fact that the growth rate in the urban areas of the less developed world is accelerating rapidly. According to the United Nations, on a global basis, the population of cities is growing at a rate of 2.2 percent each year. However, the growth rate in the urban areas of the least developed countries is 4.6 percent per year, which is almost eight times the growth rate of cities in the more developed world (0.6 percent). In fact, looking back over the last twenty years, figures indicate that the urban population in the less developed world has doubled in size. By 2025, the urban areas of the less developed world are expected to have increased by 135 percent.[33]

If we express this trend in figures rather than percentages, the picture becomes all the more grim. The rise in urban population centers means that by 2025, 4.1 billion people in the less developed world will be living in cities, compared to the 1.4 billion who live there today.[34] What these numbers mean is that the vast majority of the people in the less developed world will be crammed into megacities like Sao Paulo (16.1 million), Mexico City (15.5 million), and

Shanghai (14.7 million).[35] Projecting only 15 years into the future, Shanghai will add 8.7 million people to its ranks, while Sao Paulo will add 4.7 million, and Mexico City 3.3 million. In contrast, the megacities in the more developed world will not have such a dramatic increase in population growth. New York, for instance, will see a population increase of only 1.3 million. Even Tokyo, the largest urban area in the world, will see an increase of only 2.2 million; it may lose its number-one position to Bombay, which will have a population of 27.4 million, 12.9 million more than it has today.[36]

The problem, however, is not so much the raw numbers as the density with which people are packed into the urban areas. Today, New York City must deal with a population concentration of 11,400 people per square mile. At first glance such a figure seems staggering. However, the number pales into insignificance when compared to the the population concentration in such cities as Lagos, Nigeria, which must currently deal with 143,000 people per square mile, and Djakarta, Indonesia, with 130,000 per square mile.[37] According to both Heilbroner and Kennedy, the population density that will plague the world's largest cities will result in an atmosphere of urban decay in which disease and famine will run rampant.[38] As people pour into these cities, they will need food, clothing, and shelter, all of which will be in short supply. Moreover, they will find a job market stretched to the ultimate limits.[39]

Population Growth, Unemployment, and Disease. The pressure on the job market in the cities of the less developed world will be virtually out of control by the year 2025. In that year, the less developed world will have a labor force that exceeds 3 billion. This figure represents an increase of 1.34 billion workers over the present figure of 1.76 billion. To insure that most of these people have work, between now and 2025, the less developed world will have to add 38 to 40 million new jobs to the economy annually.[40] If the past teaches us anything about such trends, it is that in the future, the lives of city dwellers will become almost unbearable. Kennedy makes this point rather vividly when he compares the situation facing the cities of the less developed world to the plight of Paris in the 1780s. At that time an enormous crowd of 100,000 homeless people wreaked havoc on the streets of the French capital. This number is minuscule compared with the number threatening the cities of the less developed world today.[41] Unemployment figures in these cities has surpassed 25 percent and those figures threaten to rise during the next thirty years.[42]

Moreover, people who cannot find jobs will be unable to afford whatever food and shelter is available. The result will be increased problems with basic sanitation and waste disposal. As waste builds up and sanitation becomes more and more difficult, the cities will become breeding grounds for infectious disease. Already today we can see the beginnings of this problem. On a global basis, over 16.5 million people were killed by infectious disease in 1993. This figure means that over 32 percent of the people who died that year fell to some sort of infectious disease. Among the most serious of these diseases are pneumonia, tuberculosis, malaria, measles, and hepatitis. With the exception of measles, most of these infectious diseases have strains that resist current antibiotics.[43] This means a higher death rate in the future as the cities become more crowded and infection rates rise. Just as distressing are the vast numbers of people that become ill with infectious diseases but do not die. Currently between 300 and 500 million people are stricken with malaria each year, whereas over 8 million people develop tuberculosis.[44]

By far the most pervasive of all infectious diseases, however, is acquired immunodeficiency syndrome (AIDS). What makes the AIDS pandemic so alarming is the fact that it generally takes about eight or nine years before symptoms of the infection begin to show up.[45] Consequently, present figures tracking people with AIDS represent only the beginning of the problem. Kennedy goes so far as to compare the figures on AIDS today as merely the tip of the iceberg. The vast majority of people with AIDS, he argues, are hidden below the imaginary water line. Taking present figures as a baseline, Kennedy extrapolates into the future and predicts that in some areas of the less developed world, notably Africa, the number of people with AIDS could jump from 100 thousand to 2 million by the year 2000. Of course, these figures mask the seriousness of the situation since there is a difference between people with AIDS, and those who have tested positive for the human immunodeficiency virus (HIV) and who may eventually develop AIDS. When we shift the focus from people who have AIDS to those who are HIV positive, the picture becomes even grimmer. On a global basis, the World Health Organization has predicted that the worldwide population of people who are HIV positive will soon top 40 million. Another study, this one conducted by a group of Harvard epidemiologists, predicts an even more dire outcome, declaring that 100 million people will soon be HIV positive. Both studies agree, however, that most of these people will be members of the less developed world.[46]

In 1798, Thomas Robert Malthus wrote a work entitled *An Essay on the Principle of Population*, in which he argued that, while the available food supply in a given region will increase only **arithmetically**, the animal population in that same region will grow **geometrically**. Eventually, the animal population, including humans, will have surpassed the available food supply. This will precipitate a battle for the available resources, a battle that will be won by those individuals best suited to compete within the environment of that region.[47]

As a result, any event that impacts upon the ability of a group to compete for the limited resources of a particular area is termed a **Malthusian check**. Heilbroner argues that the prevalence of infectious disease may act as one of the Malthusian checks on the growth of the human population in the twentieth-first century. However, Heilbroner makes another more startling prediction about these Malthusian checks. That prediction involves the rise of "iron" governments in the less developed world. Such dictatorial "iron" governments may, seeing the unjust distribution of the world's resources, begin a campaign designed to force the more developed world to share those resources with the less developed world. In the event of such a move, the planet will be one step closer to the next challenge, that is, the threat of global war.[48]

13-3 THE CHALLENGE OF WAR

Many people believe that the disappearance of the Soviet Union as the free world's major opponent in a nuclear confrontation means that the threat of war itself has disappeared. Unfortunately, far from decreasing the threat of war, the disappearance of the Soviet Union has actually increased not only the probability of regional wars but also the possibility of a global conflict. Moreover, the politics of the post–Cold War era are more complicated and less predictable than they ever were when the West confronted only a single major opponent in the international arena. Today, the world faces not only a major confrontation between the principal global powers, but also the threat of regional conflicts that will go unchecked by the competing interests of the international superpowers.

The New World Order

During the Gulf War, the phrase "new world order" became a popular rallying cry for the allied nations who faced Saddam Hussein over the invasion of Kuwait in 1990. The first modern use of the phrase can probably be traced to a comment made by Soviet president Mikhail Gorbachev, who, in a speech before the United Nations in 1988, stated that progress in the global community could be realized only if the nations of the Earth moved toward "a new world order."[49] The term was adopted by the Bush administration as the world moved toward a confrontation with Iraq over the invasion of Kuwait in 1990. Secretary of State James Baker used the phrase on national television in September of that year.[50] Moreover, the phrase was employed by President Bush many times during the crisis, notably at a meeting with Gorbachev in September of 1990 and in his State of the Union address on January 29, 1991.[51] Despite the threat of war, the phrase was clearly used to denote a positive outlook for the future. However, in the time since the Gulf War, the phrase has taken on a new meaning that is not quite as optimistic as it may have seemed during that conflict.

Still, the optimistic interpretation of the phrase has been maintained by several political analysts, among them Colonel Harry G. Summers, a member of the Council on Foreign Relations and the International Institute for Strategic Studies, who, in his work *A Critical Analysis of the Gulf War*, notes that, as the single surviving superpower, the United States will be in a position to reshape the geopolitical structure of the globe. Summers sees the **bipolar world** of the Cold War era giving way to a **unipolar world** dominated by American military might.[52] Looking to the past first, Summers notes that during the Cold War the United States had to maintain a containment strategy, the objective of which was to prevent the Soviet Union from expanding its influence in the less developed world. Militarily this meant that the United States rarely engaged in any full-scale offensive operation because to do so would have risked a nuclear confrontation with the Soviet Union. Instead of engaging in offensive operations, the strategy was to use conventional weapons to build a defensive wall against Soviet expansion. This led to a variety of regional conflicts in areas such as Korea, Cuba, and Vietnam.[53] According to Summers, the loss of the Soviet Union as an ally in these regional conflicts means that belligerent nations will have to face the United States without the backing of the Soviet Union. Using the Gulf War as an example, Summers concludes that such confrontations will inevitably end in American victories.[54]

The Dangers of a Multipolar World. If confronted by Heilbroner's prediction that future wars of redistribution will be waged by "iron" governments of the less developed world against the affluent nations of the more developed world, Summers would likely respond that such wars would be fought only at great risk to the less developed nations. Moreover, these wars are likely to be short-lived and futile, and will ultimately backfire, leaving the less developed nation worse off than previously. But what if Summers is incorrect in his analysis? What if the end of the Cold War leads not to a unipolar world dominated by American interests, but to a **multipolar world** beyond the control of any single national will? This is the view supported by Graham E. Fuller, a senior political analyst at the RAND corporation, who in his work *The Democracy Trap: Perils of the Post-Cold War World* foresees a world ripe for regional conflicts, any one of which could erupt into global war.[55]

Fuller's thesis states that while the United States and the Soviet Union habitually found themselves on opposite sides in many regional conflicts throughout the Cold War, the two superpowers, nevertheless, provided a measure of restraint that prevented the escalation of such conflicts into global war.[56] President John F. Kennedy grasped this essential truth as early as 1961 when he said that future wars would involve regional conflicts fought by guerrillas and insurgents.[57] Technology, President Kennedy reasoned, had made global war unthinkable. Consequently, the major powers would find themselves backing either the established regime or the local revolutionaries, depending on the political orientations of each.[58] History has shown that President Kennedy's predictions have so far been correct. During the last thirty years we have witnessed a series of regional conflicts.

It is important to see, however, that these regional conflicts were restrained by the presence of the superpowers. Although the leaders of the Soviet Union might, for example, provide political, economic, and military support for Ho Chi Minh in Vietnam, Muammar Qadhafi in Libya, Daniel Ortega in Nicaragua, Fidel Castro in Cuba, and Kim Il-Sung in Korea, they would also provide the necessary restraint that would prevent any regional conflict from escalating beyond the immediate area.[59] The historical period between 1946 and 1991 is filled with examples of regional conflicts that might have escalated into global war had it not been for the restraint of the superpowers. One need think only of the Korean War, the Cuban missile crisis, and the war in Vietnam to be reminded of that fact.

Fuller argues that the end of the Cold War may also signal the end of the restraining influence of the United States and the Soviet Union. According to Fuller, the superpowers may no longer be interested in local conflicts in which they might have intervened during the Cold War as a matter of course.[60] The breakdown of social order in the former Yugoslavia is a good example of a dispute that, during the Cold War almost certainly would have immediately involved the superpowers as adversaries. In that situation the United States might have been motivated to intervene in order to promote the establishment of democratic states in Slovenia, Croatia, and Bosnia and to ensure a secure European theater of operation.[61] In contrast, the Soviet Union might have supported the Bosnian Serbs who, from a nationalistic point of view at least, would have been their closest cousins.[62] Under post–Cold War conditions, for years neither the Americans nor the Russians were motivated to do much more than offer token support, generally designed only to protect the peacekeeping forces in Bosnia.[63]

Will the future see a series of unchecked regional conflicts as predicted by Fuller, or will the Americans be motivated to intervene in such conflicts as predicted by Summers? The truth probably lies somewhere between the two points of view. While Fuller is correct that the world has become multipolar, that multipolar condition does not necessarily mean that the United States, as the last remaining superpower, will always be content to remain on the sidelines. Such was not the case when in 1999 NATO, which is undeniably dominated by American interests, launched a series of prolonged attacks against Serbia in an attempt to stop the repression of ethnic Albanians by the regime of Yugoslav President Slobodan Milosevic. The atrocities committed by the Milosevic regime in Kosovo, coupled with Milosevic's unwillingness to agree to a political solution, may have left the United States and its allies with little choice but to intervene militarily in this situation.[64] Still the involvement of NATO in this case demonstrates the instability of a multipolar world in which regional conflicts will sometimes go unchecked or which may, depending on the circumstances, escalate in unpredictably dangerous ways by the intervention of the United States.

The Dangers of Nuclear Weapons. No one has yet resorted to the use of nuclear weapons in any of these regional conflicts. However, will this moratorium on the use of nuclear weapons continue in the future? Any answer to this question would be pure speculation, of course. History is of little help here because nuclear weapons have been used only once in wartime. Certainly, we can hope that reason and good will would prevail preventing the deployment of nuclear weapons. On the other hand, those nations who actually possess nuclear weapons have generally considered the use of those weapons as a viable option in planning their military strategies. For example, immediately following the use of the atomic bomb on Hiroshima and Nagasaki in 1945, the U.S. Joint Chiefs of Staff produced a planning document for the American military that endorsed the use of atomic weapons in a first strike against the Soviet Union or any other similarly situated enemy nation. Given the political climate of the post–war era, the plan could very easily have been adopted as the official strategy of the United States, had it not been so diametrically opposed to national principles and to the traditional defensive policies of the American military establishment.[65] Nor were the Joint Chiefs alone in the suggestion that the United States use the atomic bomb as an offensive weapon. In 1946, a first strike using atomic weapons was endorsed by a special panel of military and civilian specialists who had been charged with the responsibility of analyzing the atomic tests held at the Bikini atoll that year. Under the leadership of Karl Compton, the president of MIT, the panel recommended that the United States be ready to use the atomic bomb as an offensive weapon. The panel concluded that the destructiveness of atomic war made the offensive use of such weapons imperative. According to the panel's report, the only defensive strategy in an atomic war would be to destroy an enemy nation before it had the opportunity to destroy the United States.[66] Moreover, the first strike use of atomic weapons was clearly endorsed on many occasions by the commander of the Strategic Air Command, General Curtis LeMay. In fact, military policy makers in the 1940s were so convinced of the need to use atomic weapons in a first strike that they had a special name for the strategy. They called it "killing a nation."[67]

Nor did the idea that atomic weapons could be a viable strategic option vanish in the 1940s. The possible use of atomic weapons continued to be an option for several decades. For example, the United States military forces in Europe in the 1970s included the deployment of tactical nuclear weapons as an option within their military policy if, in a conventional confrontation with enemy forces, the use of such weapons was seen as necessary to stop an attacking enemy army and to compel that enemy to negotiate an end to the hostilities.[68] As a defensive measure, the use of tactical nuclear weapons may appear at least understandable, if not entirely acceptable. In fact, it may even seem conservative, because the policy advocates the use only of "tactical" nuclear weapons.

However, we should not be deceived by the use of the apparently benign term "tactical" nuclear weapon. The type of weapon that was considered tactical by these policymakers would have been more powerful than the bombs that were dropped on Hiroshima and Nagasaki in 1945. If a single tactical weapon of this type had been used in Europe under the circumstances described, the bomb's initial blast would have killed or injured between 1.5 and 1.75 million people. An additional 5 million would have later experienced severe illness and injury from the resulting radiation.[69]

Consequently, it would be not only unwise but also irresponsible for us to ignore the dangers associated with the proliferation of nuclear weapons based solely upon a belief that no one would ever really use such weapons. For this reason we will examine the possibility that Fuller is correct in his prediction that a multipolar world will result from the end of the Cold War. We will ask the following question: How will the multipolar new world order influence diplomatic, economic, and military decision making? To address this matter, we must first examine how the nations have realigned in the wake of the Cold War, and then we will determine what this realignment means in terms of diplomacy, economics, and military activity.

The Global Realignment of Nations

In an article entitled "Lessons of the Next Nuclear War," Michael Mandelbaum, professor of American foreign policy at Johns Hopkins University, presents a scenario that predicts a realignment of nations after the demise of the Soviet Union. The premise upon which Mandelbaum's article is based is similar to Fuller's premise in *The Democracy Trap*, that the geopolitical arena has been realigned from a bipolar arrangement to a multipolar one.[70] This multipolar new world order is described by Mandelbaum as consisting of three groups of nations. Those he calls "the allies," those labeled "the orphans," and those that are termed "the rogues."[71]

The Allies: Germany and Japan. The **allies** include those nations, principally Germany and Japan, that have the technological ability to develop nuclear weapons but have refrained from doing so because they were under the protective mantle of the United States. The United States had treaties with these nations guaranteeing the use of American forces for their protection. Moreover, both Germany and Japan witnessed the constant and reassuring presence of American troops on their home soil. Such an American presence gave the Germans and the Japanese the type of confidence that they needed to abstain from building their own nuclear arsenals despite legitimate concern over Soviet activities. The demise of the Cold War has altered this situation considerably. Both Germany and Japan may be justifiably uncertain of the American commitment to their security, now that the threat of Soviet domination has evaporated.[72]

What would happen if either Germany or Japan or both decided to create their own nuclear arsenals despite their signatures on the latest nonproliferation treaty? Certainly there would be some advantage gained by the United States if such a scenario played itself out. For instance, the nuclear armament of either Germany or Japan would take some of the financial pressure off the United States, which would no longer be required to spend money, commit troops, or share technology in their defense. Despite this advantage, the move toward a nuclear Germany and Japan would not be welcomed, primarily because of the destabilizing effects. As Mandelbaum points out, the nuclear arming of Germany and Japan would create a multipolar world in which no country could be certain of its allies or its enemies. Moreover, there is no guarantee that the trend toward nuclear armament would stop with these two countries. In fact, it is more likely that other nations would follow the German and Japanese example, leading to an even more destabilized global situation.[73]

Fortunately, such a move by either country is unlikely. Recently, both Germany and Japan have revealed little motivation to strike out on their own. For instance, although they contributed monetary support to the Gulf War, neither country committed any military troops to the the conflict. This fact is especially revealing since both Germany and Japan rely on oil from the Middle East.[74] If they were not willing to take military action when their national interests were directly threatened, it is unlikely that they would do so when the threat remains remote and uncertain. Germany's similar inaction in the Bosnian crisis would seem to underscore this lack of commitment.[75] Nevertheless, the possibility remains a real one that should not be overlooked.

The Orphans: Israel and Pakistan. In contrast, a different type of threat may be posed by the orphan states. According to Mandelbaum, the **orphans** include those nations that lacked the type of firm military commitment given to the allies, but which, nevertheless, looked to the United States for protection during the Cold War.[76] Such nations have the added disadvantage of being threatened by neighboring states that do not recognize their legitimate status as independent sovereign nations. The two primary candidates for orphan status are Israel and Pakistan. These states are likely to feel more threatened by the end of the Cold War precisely because they never had the luxury of a defense treaty with the United States nor the comforting presence of American troops on their home soil.[77]

The recent nuclear tests conducted by Pakistan in 1998 in response to similar tests conducted by India bear witness to the insecurity of the orphan states. The orphans feel compelled to protect themselves against the perceived aggressiveness of their neighbors. Pakistan, for example, has long felt threatened by the growing military strength of India. Adding to this tension is the ongoing quarrel between the two nations over the contested territory of Kashmir. Consequently, once India tested five nuclear devices in May of 1998, it was only a matter of time before Pakistan, in its own self-defense, responded in kind. Further complicating this situation is the fact that both nations now candidly boast of their ability to arm ballistic missiles with nuclear weapons. Clearly, the arming of these two powers has destabilized the entire region.[78] Similarly, Israeli membership in the family of nuclear powers threatens to destabilize the Middle East and makes it much more difficult to convince the Arab states to refrain from developing their own nuclear arsenals.[79]

The Rogues: Iraq and North Korea. The rogues pose the greatest threat to world stability precisely because they are poised to wage the **wars of redistribution** that Heilbroner predicts in *An Inquiry into the Human Prospect*.[80] The **rogues** include an aggressive group of nations with expansive goals that run counter to the interests of most other nations and that threaten stability of the global community. According to Mandelbaum, the three primary rogue nations are Iraq, Iran, and North Korea, although at times, Syria, Libya, and Algeria are also capable of acting the role. With the end of the Cold War the rogues have lost the financial support and military backing of their primary friend and benefactor, the Soviet Union. At first glance, this may seem to be a positive turn of events. After all, eliminating the influence of the Soviet Union over a rogue nation like Iraq would diminish the leverage that Iraq might have over its neighbors. However, as noted, the loss of Soviet protection and support also means the loss of Soviet restraint. During the Cold War, the rogues could count on Soviet support, but, as a trade-off, the West could rely on Soviet reasonableness to restrain the rogues. This is no longer the case.[81] The threat posed by North Korea is especially troubling because in 1998 intelligence agencies in the United States discovered that the North Koreans were building a secret nuclear reactor and plutonium reprocessing installation capable of fabricating approximately six nuclear bombs sometime within the next five years. The seriousness of this development was exacerbated by the fact that the establishment of the underground facility violated an earlier agreement made between North Korea and the United States in which the North Koreans agreed to abandon their desire to develop nuclear weapons in exchange for an international aid package valued at $6 billion.[82] Moreover, although we have focused our discussion on the perils of nuclear weapons, we must not forget the potential danger represented by other weapons of mass destruction (WMDs). Such WMDs include biological and chemical weapons, as well as their ballistic missile delivery systems.[83]

Of course, not everyone agrees that the rogue nations represent the type of immediate threat envisioned by Mandelbaum. For instance, John Mueller, professor of political science at the University of Rochester and Karl Mueller, assistant professor of comparative military studies at the School of Advanced Airpower Studies at Maxwell Air Force Base in Alabama, point out that the existence of rogue nations is nothing new to a world which has seen more than its fair share of such states over the last fifty-five years. They point to Castro's Cuba, Nasser's Egypt, and Sukarno's Indonesia as examples of rogue states that were at the height of their power during the Cold War. Mueller and Mueller also contend that the extent of the threat posed by WMDs in the hands of the rogues is often exaggerated. Biological weapons, they argue, are of little use in open warfare because such weapons can easily backfire, claiming as victims, not only the intended targets, but also the rogues themselves. Moreover, Mueller and Mueller point out that chemical weapons are so inefficient that they really should not be categorized as WMDs in the first place, at least not unless we want to label bullets and machetes as WMDs as well.[84] While Mueller and Mueller may be correct in their conclusion that the danger represented by the rogues is remote, the stakes are so high that we must try to anticipate any action they might take, and plan the most suitable response to that action.

Wars of Redistribution. A critical factor here is that, as a part of the less developed world, the rogues feel it is permissible to invade their neighbors or to nationalize Western businesses in order to redistribute the wealth. Moreover, the rogues easily justify such aggression because they believe they have been treated unfairly in the past. A good example of this is Iran, which had its oil taken by the British for over forty years in an arrangement which, at the very best, could be described as inequitable. Consequently, when the Iranians nationalized British oil interests in May 1951, they did so with an air of moral righteousness.[85] Perhaps more to the point, some of the problems that exist between rogue nations and their neighbors have been created by the intervention of the Western powers and Russia, sometimes when these nations have arbitrarily divided a country into separate regimes. In such a case the rightful reunification of an unjustly divided country becomes the stated objective of the rogue nation as it launches a war of redistribution. Certainly, this was the case with the partition of Korea at the end of World War II. This division, which resulted from an impromptu agreement between the Russians and the Allies and which was later exacerbated by the Cold War, led to a shooting war that began when North Korea invaded South Korea in an attempt to reunite the country.[86]

Another case in point is Iraq's relationship with Kuwait. Unlike much of the rest of the world, the Iraqi government did not perceive its military action against Kuwait in August 1990 as an invasion of a sovereign country. Instead, the Iraqis viewed the aggression as a last resort, aimed at obtaining for Iraq a degree of economic and political security that could not be acquired in any way other than by the use of force. Iraq justified its incursion into Kuwait by arguing, with some historical accuracy, that the establishment of the border between the two nations had resulted from a forced agreement that Iraq no longer felt obligated to follow. Iraq also implied that Western policy in Kuwait was just one dimension of an overall Western strategy to fragment the Arab states and thus diminish their global influence.[87] Saddam Hussein, Iraq's leader, has repeatedly made these points clear. He has frequently condemned foreign aggression aimed at disrupting what he calls the Arabism of the Gulf. Moreover, he has often referred to the border conflict between Kuwait and Iraq as a dispute between brothers.[88]

Available Strategic Options

The point here is that the rogues believe that they have perfectly justifiable reasons for using force against their neighbors, and that this aggression might one day involve the use of nuclear weapons. What strategy can the West use to counteract this threat? Unfortunately, none of the available strategies are very attractive. The West could, for instance, offer neighboring states defensive measures that might protect them from a nuclear attack. However, the development of such a defense is by no means without technical problems.[89] Nor would it be satisfactory for the West to offer the rogues themselves economic or diplomatic rewards to entice them to surrender their aggressive aims. Such a move by the West would not only be politically unpopular; it would also be diplomatically dangerous since it could pave the way for future blackmail attempts by the rogues or other nations that may be tempted to gain their own economic or diplomatic advantage by emulating the behavior of the rogues.[90]

Interestingly enough, in their article "Sanctions of Mass Destruction," Mueller and Mueller suggest that the more developed world in general, and the United States in particular, already have a strong weapon that can be used against rogue nations. That weapon is the imposition of economic sanctions. Economic sanctions (Mueller and Mueller actually prefer the label "economic warfare") have been used successfully in the past against nations like Haiti and Serbia, effectively destroying the economic infrastructures of those two nations. Nevertheless, despite the success of such sanctions, Mueller and Mueller are openly critical of this approach because, more often than not, the ultimate victims of economic warfare are the people of the nation rather than its ruling party. They point to Iraq as an example of this problem, laying out a convincing list of detrimental consequences affecting the people of that nation, including a decrease in the availability of food and medicine, severe damage to the country's sewer and sanitation network, and, perhaps worst of all, a terrifyingly rapid rise in infant mortality. Meanwhile, Saddam Hussein remains in power and continues to restore his military machine, presumably including his stockpile of chemical and biological weapons, the destruction of which was one of the avowed goals of the economic sanctions in the first place.[91]

Another option open to the West is the execution of preventative wars against the rogues. A **preventative war** results when one nation attacks another to prevent the second nation from taking an action considered detrimental to the best interests of the first. Such wars are not without precedent. In the wake of the development of the atomic bomb, for example, General Leslie Groves, the military chief of the Manhattan Project, suggested initiating a preventative war to stop any foreign nation that was on the brink of building nuclear weapons from actually producing such weapons.[92] Later in that same decade, General Curtis LeMay, the commander of the Strategic Air Command (SAC), was prepared to risk a preventative war if such a war would nullify the Soviet Union's ability to destroy SAC bombers before they were airborne.[93] In 1954 a special study group convened by the Joint Chiefs of Staff recommended to President Dwight D. Eisenhower that the United States launch a preventative war against the Soviet Union. The objective of the war would have been to destroy the Soviet capability of launching a first nuclear strike against the U.S. Fortunately, not all the Joint Chiefs supported the idea. Army Chief of Staff General Matthew Ridgway, for instance, staunchly opposed a preventative strike against the Soviets. Apparently, President Eisenhower agreed with General Ridgway, because, later that same year, the president expressly rejected the idea of preventative war.[94]

Recently, however, the traditional American resistance to preventative war has seriously diminished. In the summer of 1998, American forces in the Arabian and Red seas launched missiles at a terrorist training installation in Afghanistan and a chemical weapons complex in the Sudan, following the bombing of American embassies in Kenya and Tanzania. Although this was not the first time that the United States launched missile strikes on foreign targets following terrorist incidents, the nature of this attack was quite different from similar episodes in the past. One primary difference was that the attacks against the base in Afghanistan and the factory in the Sudan were characterized by the American administration, not only as retaliatory raids, but also as preemptive strikes carried out to protect Americans from future terrorist aggression. In this way, the American attacks in the Middle East appeared to represent a change of policy based upon the acceptability of preventative war.[95]

Moreover, the United States is not alone in accepting the inevitability of preventative war. The Israeli attack on the Osiraq nuclear reactor in 1981 is a prime example of a preventative strike. The objective of that attack was to cripple Iraq's plan to develop a nuclear arsenal. Although the attack was successful, it was not well-received by the international community. Consequently, it would be difficult to earn the type of international backing that would be necessary to support any future preventative war.[96]

There is a final option open to the West. It is an option which many people think unrealistic and impractical and which others see as inefficient and expensive. Nevertheless, it is the single option that might some day eliminate the threat of war from the face of the globe. That option is to develop a policy of global scientific and technological development aimed at removing, once and for all, the motivation for aggression from the rogues, the allies, and the orphans. If these nations no longer feel threatened, or cheated, or exploited, but were, instead, convinced they have been given their fair share of the world's resources, there would be no need to nationalize industries, fight wars of redistribution, or engage in diplomatic blackmail. This, however, is the subject of the final chapter of the book. Before the problems of population and war can be addressed by science and technology, we must look at the third external challenge—environmental deterioration.

13-4   THE CHALLENGE OF THE ENVIRONMENT

In his book *An Inquiry into the Human Prospect*, Heilbroner accurately points out that the challenge posed by the environment is qualitatively different from the challenge of population and the challenge of war. He explains this difference by noting that, whereas population growth can be slowed and wars can be prevented, the capacity of the environment to absorb pollutants has an unequivocal upper boundary. Moreover, that capacity may be reached in a remarkably short period of time.[97] Generally, whenever environmental difficulties are discussed, three problems receive close attention. They are (1) the prevalence of diseases that can be directly related to the growing rate of pollution, (2) the destruction of delicately balanced ecosystems, and (3) the possibility of global famine due to a rapidly diminishing food supply. It now appears that the third factor is the most immediate, and perhaps the most devastating of the three environmental challenges facing the world today. The principal cause of this drop in food production is the greenhouse effect that has led to a pattern of global warming.[98]

Evidence of Global Warming The primary cause of **global warming** is the increased use of fossil fuels resulting in an escalation in the level of carbon dioxide (CO_2) in the atmosphere. Moreover, there are other gases, notably nitrous oxide, methane, and chloroflurocarbons, that contribute to this problem.[99] All these gases, including CO_2, permit the passage of sunlight during the day but prevent the escape of infrared heat radiation during the night. The net effect is a gradual warming of the upper atmosphere, which, in turn, results in an overall warming of the entire planet. This warming is generally referred to as the **greenhouse effect**. Interestingly, scientists have known of the greenhouse effect since about 1980 but have been unable to convince national governments and the public at large of the seriousness of the threat. Unfortunately, the United States has been one of the nations reluctant to admit the extent of the problem. It was not until 1991 that American officials acknowledged that global warming warranted serious attention. Even with this admission, however, the United States refused to go along with an international agreement aimed at limiting the emission of CO_2 into the atmosphere.[100]

Many other nations, however, did agree to institute controls over the emission of CO_2 into the atmosphere. Despite these efforts, the level of CO_2 in the atmosphere has continued to rise during the 1990s. On a global scale the level of CO_2 has increased 5 percent since 1990 in many industrial nations of the more developed world. However, in many countries in the less developed world, where pollution controls are weak at best, the levels have risen between 10 and 40 percent since 1990.[101] Moreover, there is an enormous amount of evidence that these increases are leading to disastrous results. For example, there is evidence that the average global temperature has risen 0.6 degrees Celsius since 1900. Moreover, the fastest rise has been in the later years of the twentieth century. Figures indicate that the ten hottest years of the twentieth century took place after 1980.[102] Predictions also indicate that in the next 100 years, if nothing is done to reverse these trends in global warming the average global temperature could rise as much as 3.5 degrees Celsius.[103]

This drastic rise in temperature may soon be accelerated by the destruction of forest land. Scientists have discovered that carbon emissions currently equal 6 billion tons per year. Of that 6 billion tons, 1.5 billion tons do not reach the atmosphere but are, instead, absorbed by forests located in the Northern Hemisphere.[104] Unfortunately, the ability of the forests to absorb this much carbon may drop drastically in the near future. Ironically, this drop will be caused by global warming trends already in progress. The forests will be unable to adapt to this unexpected climactic change, leaving the trees weak and, therefore, susceptible to disease and insect infestation. In this weakened condition the forests would be prone to fire.[105] Moreover, fire is not the only danger threatening these forests. Thousands of trees are being cut down each year for the production of lumber, paper, and firewood. Areas affected by this destruction include Southeast Asia, Africa and South America.[106] Once these trees are destroyed, a good portion of that 1.5 billion tons of carbon, once absorbed by the forests, will be released into the atmosphere, thereby pushing the average global temperature even higher.[107]

Rising temperatures alone, however, are only a part of the story. Other pieces of evidence indicate that global warming is no longer the stuff of science fiction. For instance, in 1995, a piece of Antarctica roughly the size of Rhode Island slid into the Atlantic Ocean, evidence of a warming trend in that area of the globe. At the opposite end of the planet, the temperature has increased to such a degree that Siberia is warmer today than it has been for almost 1000 years. In the Alps glaciers are steadily retreating from land that has been under ice for several thousand years. Farther north in Europe, the winters are gradually getting warmer each year. In southern Asia, dangerously intense heat waves in India have increased with regularity over the last few summers.[108] From a global perspective, a yearly rise in sea level of 3 millimeters has been recorded since 1993. Such a change is ample evidence that the warming trend is not localized to any one part of the globe. A similar piece of evidence supporting the global seriousness of the warming trend is a dramatic shift in the timing of the seasons. Drastic climactic shifts from winter to summer and summer to winter with virtually no spring or fall indicate this change in timing. This trend has destabilized a climactic pattern that had ruled the planet for over three hundred years.[109]

Consequences of Global Warming

Contrary to popular opinion, global warming and the greenhouse effect do not simply mean a hotter planet. Perhaps the most serious consequence will be the loss of the Earth's relatively stable climate. In its place we will experience what the meteorologists call a "climate of extremes." Such a climate will be prone to floods, droughts, heat waves, tornadoes, thunderstorms, hailstorms, and hurricanes of immense proportions. These disturbances are caused by the agitation of the atmospheric and oceanic patterns that govern the Earth's weather system. This agitation results from the overall rise in global temperature.[110]

Of these destructive forces the most dangerous and potentially the most deadly will be an increase in the number of droughts suffered in the food-producing countries of the world. Water is, of course, absolutely essential in the production of food. Without water, life on Earth would end rather abruptly. Among the many uses for water, the production of food ranks as number one. More than 60 percent of the water taken from rivers, lakes, and other sources on a global scale is devoted to raising crops. Not even the industrial, urban, and household consumption of water combined compares to the level of use devoted to agriculture.[111]

Each year, on average, every person on the planet consumes 300 kilograms of grain. Since we add over 88 million people to the population of the globe annually, we will need over 27 billion cubic meters of new water each year to grow the crops necessary to feed these additional people. If this rate of consumption remains constant, by the year 2025 we will need an additional 780 billion cubic meters of water for crop production.[112] Otherwise, millions of people will starve. The situation would be precarious even if we could count on the amount of usable water remaining constant. Unfortunately, this is not the case. Rather, at a time when the population is growing so rapidly, the amount of usable water is decreasing due to droughts caused by global warming.[113] This means that the available water supply, which must increase to feed these people, is actually dropping at an alarming rate.

Even at the present moment, the availability of water is dropping dangerously. Across the globe, eighty nations, comprising 40 percent of the population of the globe, are experiencing a formidable loss of water.[114] Regions currently suffering heavy losses include India, northern China, northern Africa, and the Middle East. The American Southwest is also experiencing a disturbing drop in the level of underground water tables. Moreover, several of the most critical rivers on Earth are drying up long before they can empty into the sea. These include the Huang He River in China, the Amu Dar'ya River in Asia, and the Colorado River in the southwestern United States.[115] One of most alarming situations involves the rivers of India, most of which are completely dry at certain times of the year. This unfortunate turn of events includes the Ganges River, which serves as the single most important water supply for much of southern Asia, a region which is growing more rapidly than most in population.[116]

The future looks no brighter. At present rates of consumption, the Ogallala aquifer, located under the Great Plains in the United States, will be depleted soon. Figures record an 11 percent drop since 1986. The present rate of loss amounts to about 12 billion cubic meters per year. Farmers in Texas, Oklahoma, Kansas, and Colorado have already felt the effects of this decline.[117] Because this source of water supplies nearly 20 percent of the irrigated land in the United States, its loss will be significant. Things are not much better in the southwestern United States or in California. In the Southwest, the region located east of Phoenix has already experienced a decline in underground water levels of over 120 meters, and California currently suffers a reduction of 1.6 billion cubic meters of water annually. Other areas near the Southwest that will be affected by this type of loss include Mexico City, which withdraws water from underground sources almost 80 percent faster than it can be replaced.[118]

There are, of course, other negative consequences associated with environmental deterioration. These include the deforestation of much of the world, notably Southeast Asia and South America, the extensive damage being done to the ozone layer, the depletion of the world's supply of oil and minerals, the erosion of soil suitable for the growing of food, the failure to check the disposal of hazardous chemicals into the oceans and the atmosphere, the exhaustion of most oceanic fisheries, and the appearance of environmentally related diseases.[119] Exploring each of these is beyond the scope of this book. For now it is sufficient to point to the statistics and try to determine whether there is any way to avoid this disastrous slide into environmental chaos.

CONCLUSION

Whether science and technology will be able to correct the problems posed by the population explosion, nuclear war, and environmental deterioration will depend less upon our understanding of the laws of nature and more upon the ability of humanity to execute an effective international science and technology policy. The challenges can be met and the problems can be solved if we can overcome our differences and work together. What will it take for the international community to create a workable plan? Some people argue rather convincingly that only some outside threat, such as a planetary plague, will motivate the nations of the world to develop a coherent science and technology policy. Others state that such a goal can be reached through diplomatic channels. Still others believe that the development of any international science policy is beyond our capabilities. What might be helpful is for us to look at examples of successful science policies. Has any nation or state ever developed an effective science and technology policy? The answer to this question is the focus of "The History of the Future," our next and last chapter.

Review

13-1. In 1973, the economist Robert L. Heilbroner published a book entitled *An Inquiry into the Human Prospect*, in which he identified three major challenges facing humanity today: uncontrolled population growth, the threat of nuclear war, and the destruction of the environment. In addition, Heilbroner focused on the underlying causes of those challenges. Heilbroner lays the blame for these crises on the shoulders of science and technology. However, Heilbroner does not accuse science and technology alone for these problems. Instead, he points to the uneven way that society has allowed science and technology to develop. In this chapter we examined evidence supporting the fact that civilization is facing a crisis situation caused by these three major challenges.

13-2. Since 1973, the world has added over 2 billion people to the overall population, giving the planet Earth almost 6 billion human inhabitants. The real problem, however, lies not in these figures but in the rate of population growth. According to estimates made by the United Nations, the growth rate among the less developed nations is 1.9 percent per year, whereas the growth rate in the more developed world is 0.4 percent. Assuming that the rate of growth remains about the same, the population of the planet could reach between 7.9 and 11.9 billion people by the year 2050. This phenomenal rate of population growth can be traced to two principal factors—the declining nature of infant mortality and an increase in the life expectancy of most people on Earth. Several consequences could result if we fail to check the rapid growth of the world's population. These consequences include urban disintegration, disease, and famine.

13-3. The end of the Soviet Union has turned the world's attention away from the possibility of a war between the world's superpowers. The prevalent attitude today is that the disappearance of the Soviet Union as the free world's major opponent means that the threat of war itself has disappeared. However, far from decreasing the threat of war, the disappearance of the Soviet Union has actually increased not only the probability of regional wars but also the possibility of a global conflict. Moreover, the politics of the post–Cold War era are more complicated and less predictable than they were when the West confronted only a single major opponent.

13-4. The challenge posed by the environment is qualitatively different from the challenge of population and the challenge of war because, although population growth can be slowed and wars can be prevented, the capacity of the environment to absorb pollutants has an upper limit. That capacity may be reached in a remarkably short period of time. The possibility of global famine due to a rapidly diminishing food supply now appears to be the most immediate, and perhaps the most devastating environmental challenge facing the world today. The principal cause of this drop in food production is the greenhouse effect that has led to a pattern of global warming. The primary cause of global warming is the increased use of fossil fuels resulting in an escalation in the level of carbon dioxide, nitrous oxide, methane, and chloroflurocarbons in the atmosphere. These gases permit the passage of sunlight during the day but inhibit the escape of infrared heat radiation during the night. The net effect is a gradual warming of the upper atmosphere which, in turn, results in an overall warming of the entire planet. This warming is generally referred to as the greenhouse effect. Global warming and the greenhouse effect do not simply mean a hotter planet. The most serious consequence of global warming will be the loss of the Earth's relatively stable climate. In its place we will experience a "climate of extremes." Such a climate will be prone to floods, droughts, heat waves, tornadoes, thunderstorms, hailstorms, and hurricanes of immense proportions. Of these destructive forces the most dangerous and potentially the most deadly will be an increase in the number of droughts suffered in the food producing countries of the world.

Understanding Key Terms

allies
arithmetic growth
average life expectancy
bipolar world
geometric growth
global warming

greenhouse effect
infant mortality rate
less developed world
Malthusian check
more developed world
multipolar world

orphans
preventative war
rogues
unipolar world
wars of redistribution

Review Questions

1. What are the three major external challenges facing the modern world?
2. What parts of the world are responsible for the greatest growth in population?
3. What outcomes of the population explosion will have the most devastating effects on the various cultures and environments of the world?
4. How do Malthusian checks slow population growth?
5. How has the threat of nuclear war changed since the end of the Cold War?
6. Who are the allies? the orphans? the rogues?
7. What options are available for dealing with a multipolar world situation?

8. How does the challenge of the environment differ from the other two challenges?
9. What evidence exists for global warming?
10. What are the consequences of global warming?

Discussion Questions

1. The less developed world has a much more rapid rate of population growth than the more developed world. However, the population of the more developed world has a more dramatic impact on the resources of the planet than that of the less developed world. Which population, that of the more developed world or of the less developed world, has a greater moral responsibility for the population problem? Explain your response.
2. Several strategies for dealing with the aggressive tendencies of rogue nations, such as Iraq and North Korea, are identified in this chapter. Which of these strategies do you feel would be the most effective? Which of these strategies can you defend from a political perspective? from a moral perspective? Explain your responses.
3. One of the principal causes of global warming and the greenhouse effect is the production of carbon dioxide resulting from the burning of fossil fuels. A good portion of these fuels is burned in the more developed world. Does the more developed world have a moral responsibility to restrict the use of fossil fuels to protect the environment even though such restrictions would cause hardship and possibly lower the standard of living in the more developed nations? Explain.
4. Can wars of redistribution ever be morally justified? Can preventative wars ever be morally permissible? Explain.
5. Can the global challenges of population, nuclear war, and environmental deterioration be solved by science and technology? From a sociopolitical perspective, can the people of Earth redirect scientific and technological activities toward those solutions? Explain.

ANALYZING *STAR TREK*

Background The following episode from *Star Trek: The Next Generation* reflects some of the issues that are presented in this chapter. The episode has been carefully chosen to represent several of the most interesting aspects of the chapter. When answering the questions at the end of the episode, you should express your opinions as clearly and openly as possible. You may also want to discuss your answers with others and compare and contrast those answers. Above all, you should be less concerned with the "right" answer and more with explaining your position as thoroughly as possible.

Viewing Assignment—*Star Trek: The Next Generation,* "Encounter at Farpoint"

In this episode, which served as the pilot for *Star Trek: The Next Generation,* Picard and the crew of the Enterprise are sent to Deneb IV to negotiate the terms of that planet's admission to the United Federation of Planets. As their primary bargaining chip in these negotiations, the Bandi, the inhabitants of Deneb IV, offer to Star Fleet the use of Farpoint Station. The puzzle is that the Bandi do not appear to possess the technological and scientific know-how necessary to build Farpoint. Yet, the station does, in fact, exist. While on the way to Farpoint, the Enterprise is intercepted by a member of the Q continuum. Q orders the human race to stop its exploration of the galaxy. Picard attempts to escape by ordering the Enterprise to maximum warp. When this effort fails, Q sends Data, Yar, Troi, and Picard to a courtroom in the middle of the twenty-first century in the midst of "the postatomic horror." There Q charges that the human race has committed a pattern of savage and barbarous acts that have spread terror throughout the galaxy. Eventually, Picard acknowledges that the human race has had a barbaric history. However, he also argues that this has changed, and he asks that Q submit the crew of the Enterprise to a test to determine their worthiness as a species. Q agrees. The test apparently involves the Bandi and Farpoint Station. Eventually, Picard figures out that Farpoint Station was not constructed by the Bandi after all. Instead, Farpoint is actually an alien life-form that the Bandi captured and have been exploiting for its energy. Picard uncovers this in the midst of an attack from the life-form's companion, which is destroying Deneb IV in order to rescue the captured alien. Picard orders the crew to direct an energy beam to the surface of the planet. The captured alien absorbs the energy from the beam. This energy revitalizes the alien, which then escapes the influence of the Bandi and rejoins the companion. Picard's insight and the salvation of the alien are enough to redeem humanity in the eyes of the Q continuum, at least temporarily.

Thoughts, Notions, and Speculations

1. Q accuses the human race of being a savage and barbaric race. What evidence in this chapter might lead you to agree with Q? Explain.
2. Data, Yar, Troi, and Picard are transported by Q into a twenty-first century courtroom. During their trial we learn that the Earth has undergone a nuclear war and that in the middle of the twenty-first century, the planet has been plunged into an era of social collapse, labeled the postatomic horror. Given what you learned in this chapter about the three challenges facing humanity, speculate on the likelihood

of the future as predicted in "Encounter at Farpoint." In answering this question consider such concepts as the new world order; the allies, the orphans, and the rogues; and wars of redistribution and preventative wars.

3. Based on what you learned in this chapter, speculate on what event or events might prevent the postatomic horror predicted in "Encounter at Farpoint." Explain.

4. The Bandi capture and exploit an alien creature for their own advantage. Part of this exploitation involves taking whatever they need from the alien to maintain their high standard of living, while giving the alien only what it needs to survive and nothing more. How is this situation analogous to the way that the more developed world of the twentieth century treats the less developed world? Can such treatment, that is, both the Bandi's treatment of the alien and the more developed world's treatment of the less developed world, be defended morally? Explain.

5. The alien's companion attacks the Bandi in order to free its exploited companion. Is such an attack morally justified? Explain. Depending on your answer to the first part of this question, could you defend an exploited country that elects to wage a war of redistribution against its more developed neighbor? How would such a war of redistribution be similar to the alien's attack on the Bandi? How would such a war be different? Explain.

NOTES

1. Robert L. Heilbroner, *An Inquiry into the Human Prospect*, 56.
2. Ibid., 57.
3. Ibid., see especially Chapter 3, "The Socio-Economic Capabilities for Response," 77–121.
4. Ibid., 32–33.
5. Christopher Flavin, "The Legacy of Rio," 16–17. (Note: The figures are estimated from the information charted in Figure 1-3, "World Population, 1900–2050, under the Assumption of Population Growth Rate.")
6. United Nations Department for Economic and Social Information Analysis, *Concise Report on the World Population Situation in 1995*, 1.
7. United Nations Department for Economic and Social Information Analysis, *Long-Range World Population Projections*, 5, 28–29. (Note: Despite the implications of these divisions, the United Nations disavows any intent to make value judgments when using terms like "more developed" and "less developed." Instead, such terms are meant to be taken in a statistical sense only. There are also several inconsistencies in the way in which the United Nations classifies certain areas of the world. For example, Oceania is said to include Melanesia, Micronesia, and Polynesia, all areas that would appear, on the surface at least, to belong within the less developed world. Interestingly enough, in earlier surveys the United Nations did not include these areas in the more developed world, but instead considered them as a part of the less developed world. Although the United Nations Department of International Economic and Social Affairs recognizes this discrepancy, it makes no attempt to explain the difference between the earlier and the later classifications. Similarly, Japan, which is clearly a part of the more developed world, is included within the less developed world in recent surveys, whereas in earlier surveys it was

considered a part of the more developed world. Again, the U.N. acknowledges the inconsistency without trying to explain it. For the sake of convenience and in an effort to defer to the greater wisdom of the United Nations, while pointing out the inconsistencies within these classifications, we will not dispute the classifications themselves. See United Nations, *Long-Range World Population Projections*, 5.)
8. Heilbroner, 32.
9. United Nations, *Concise Report*, 1–2.
10. Ibid.
11. Paul Kennedy, *Preparing for the Twenty-First Century*, 22.
12. William K. Hartmann and Ron Miller, *The History of Earth*, 195.
13. Kennedy, 22.
14. Flavin, 16.
15. Kennedy, 23.
16. John W. Wright, ed. *The New York Times 1998 Almanac*, 232. (These figures are derived by using the 1990 population of New York, which was 7.3 million that year, according to statistics compiled by the U.S. Bureau of Statistics.)
17. Kennedy, 23; Wright, 232.
18. United Nations, *Concise Report*, 2.
19. Ibid.
20. United Nations, *Concise Report*, 1–2; United Nations, *Long-Range World Population Projections*, 28–29.
21. Holmes Rolston, III, "People, Population, and Place," 36.
22. Kennedy, 23–24; Rolston, 36.
23. Heilbroner, 39.
24. Kennedy, 24–25.
25. Ibid., 24.
26. United Nations, *Concise Report*, 8–10; Paradoxically, some demographic studies report a drop in average life expectancy in some areas of the less developed world. *See* Paul Gallagher, "Implosion of Population Growth Rate Continues Through 1998."
27. Kennedy, 24.
28. United Nations, *Concise Report*, 21.
29. Heilbroner, 35–38.
30. Kennedy, 26.
31. Heilbroner, 37; Paul Kennedy 26–27.
32. Kennedy, 26.
33. United Nations, *Concise Report*, 26–28.
34. Kennedy, 26.
35. United Nations, *Concise Report*, 27.
36. Ibid., 28.
37. Kennedy, 26.
38. Heilbroner, 37; Kennedy, 26.
39. Kennedy, 27.
40. Ibid.
41. Ibid.
42. Heilbroner, 37.
43. Anne E. Platt, "Confronting Infectious Diseases," 115–16.
44. Ibid., 115.
45. Kennedy, 27.
46. Ibid., 27–28; Some experts argue that the AIDS pandemic, especially as it affects major parts of the less developed world, is actually driving total global population growth toward zero. While this appears to be a minority point of view, it is worthy of consideration. See Paul Gallagher, "Implosion of Population Growth Rate Continues Through 1998." Gallagher points to a 1998 report issued by the United Nations Population Division of the U.N.'s Department of Economic and Social Affairs as one source of demographic information on this trend. Gallagher honestly admits, however, that the United Nations Fund for Population Affairs (UNFPA) reports a different set of figures. See UNFPA, *The State of the World Population 1998*. Figures in the UNFPA report more closely parallel population projections found earlier in Chapter 13. For instance, the UNFPA predicts a world population of 9.4 billion in 2050 while earlier figures in this chapter reported a global population mark of 9.83 billion in that year.
47. H. James Birx, *Interpreting Evolution*, 59.
48. Heilbroner, 35, 39–40.
49. Mikhail Gorbachev, "U.S.S.R. Arms Reduction," 230.
50. James Baker, *Face the Nation*, September 10, 1990.
51. Ed Hinson, *The New World Order*, 15; Harry G. Summers, *A Critical Analysis of the Gulf War*, 248.
52. Summers, 250–54.
53. Ibid.
54. Ibid.
55. Graham E. Fuller, *The Democracy Trap*, 84.
56. Ibid., 84–85.
57. Robert F. Kennedy, *To Seek a Newer World*, 165.
58. Ibid.
59. Fuller, 84. (Note: In fact, the extent of Soviet control over some of these regimes is only now just coming to light. See, for example, Francois Raitberger, "Behind the Scenes in '62 Missile Crisis," *USA Today* 15 August 1995, sec. A, p. 10.)
60. Ibid., 84–85.

61. Sabrina P. Ramet, "The Breakup of Yugoslavia," 109. Professor Ramet, associate professor of international studies at the University of Washington, does an excellent job of predicting the extent of the civil unrest in central Europe following the collapse of Yugoslavia. At the end of the article, she indicates that American intervention to prevent the war might be in the best interests of the United States in order to promote democracy in Slovenia, Croatia, and Bosnia, to divert America's allies from an unbalanced preoccupation with the problems of central Europe to the detriment of other global issues, and to preserve the security and stability of central Europe.
62. R. W. Apple, "How the World Makes Bosnia Safe for War," *New York Times*, 4 June 1985, sec. 4, p. 6.
63. Ibid., sec. 4, pp. 1, 6. Eventually, in August 1998 NATO forces began an air campaign against the Bosnian Serb Army. Interestingly, the resolution of that conflict follows rather quickly. Bruce Clark, "NATO Survey: Knights in Shining Armour?" 14.
64. Francis X. Clines, "NATO Opens Broad Barrage Against Serbs As Clinton Denounces Yugoslav President," *New York Times*, sec. 4, p. 6.
65. Richard Rhodes, *Dark Sun*, 225–26.
66. Ibid., 346.
67. Ibid., 347.
68. Thomas B. Allen, *War Games*, 79.
69. Ibid., 85, 87–88.
70. Michael Mandelbaum, "Lessons of the Next Nuclear War." Mandelbaum actually credits John Mearsheimer for originating this multipolar view. He points to an article written by Mearsheimer, entitled "Back to the Future: Realism and the Realities of European Security," which appeared in the Winter 1990/1991 issue of *International Security*. Moreover, Mearsheimer apparently bases his theories on the work of Kenneth N. Waltz in his book *Theory of International Relations* (Reading, MA: Addison-Wesley, 1979). The difference, however, is that Mandelbaum and Mearsheimer see the multipolar realignment as creating instability, while in some works, Waltz sees the multipolar realignment as causing political stability. According to Mandelbaum, this is the stand that Waltz takes in *The Spread of Nuclear Weapons: More May Be Better*, Adelphi Paper 171, London: International Institute for Strategic Studies, 1981.
71. Ibid., 23–24.
72. Ibid., 24–27.
73. Ibid., 27–28.
74. Summers, 251–52.
75. Apple, 6.
76. Mandelbaum, 28.
77. Ibid., 28–29.
78. John Kifner, "Pakistan Sets Off 6th Atomic Blast, but Urges 'Peace,'" *New York Times* 31 May 1998, sec. 1, 1, 8. (Note: For a more detailed look at the volatile relationship between Pakistan and India see: Lee Michael Katz, "Most Dangerous of Neighbors: India and Pakistan Have Been Enemies Since Their Creation," *USA Today*, 13 August 1997, sec. A, p. 6; Daniela Deane, "One of the Untouchables Rises to Power: 'Common Man' Becomes India's New President," *USA Today*, 13 August 1997, sec. A, p. 6; and "Timeline: Some Key Events for the Indian Subcontinent," *USA Today*, 13 August 1997, sec. A, p. 6.)
79. Mandelbaum, 28–30.
80. Heilbroner, 39–43.
81. Mandelbaum, 33–36.
82. "North Korea's Nuclear Ambitions," *New York Times*, 19 August 1998, A28.
83. John Mueller and Karl Mueller, "Sanctions of Mass Destruction," 43.
84. Ibid., 44–46.
85. David Halberstam, *The Fifties*, 363–64.
86. Richard L. Greaves, et al., 1072–73.
87. Sari Nusseibeh, "Can Wars Be Just?", 77.
88. Saddam Hussein, *On Current Affairs in Iraq*, 110–11; 121.
89. Mandelbaum, 34.
90. Ibid.
91. Mueller and Mueller, 49–51.
92. Rhodes, 225.
93. Ibid., 454.
94. Ibid., 562–63.
95. James Bennet, "U.S. Cruise Missiles Strike Sudan and Afghanistan Targets Tied to Terrorist Network," *New York Times*, 21 August 1998, A1; Philip Shenon, "Hitting Home: America Embarks on a New Style of Global War," *New York Times*, Sunday, 23 August 1998, sec. 4-1.
96. Ibid., 35–36.
97. Heilbroner, 47.
98. Lester R. Brown, "The Acceleration of History," 6–7.
99. Christopher Flavin, "Facing Up to the Risks of Climate Change," 22.

100. Hartmann and Miller, 223.
101. Flavin, "Facing Up," 21.
102. Ibid., 22.
103. Ibid., 23.
104. Ibid., 24; For an opposing view see: James Dunn, "Can the Greens Destroy Nature."
105. Ibid., 26.
106. Brown, 6.
107. Flavin, "Facing Up," 24–26.
108. Ibid., 23.
109. Ibid.
110. Ibid., 25–26.
111. Sandra Postel, "Forging a Sustainable Water Strategy," 41.
112. Ibid.
113. Ibid.
114. Flavin, "Facing Up," 25.
115. Brown, 5.
116. Postel, 43.
117. Brown, 9; Postel, 42.
118. Postel, 42.
119. Brown, 4–7; Hartmann and Miller, 224–26.

Chapter 14
The History of the Future

COMMENTARY: "BOTH: I WILL HAVE THEM BOTH!"

Consider, for a moment, the following excerpt from the poem "The True-Blue American," by the twentieth-century poet Delmore Schwartz:

> Naturally when on an April Sunday in an ice cream parlor Jeremiah
> Was requested to choose between a chocolate sundae and a banana split
> He answered unhesitatingly, having no need to think about it
> Being a true-blue American, determined to continue as he began:
> Rejecting the either–or of Kierkegaard, and many another European;
> Refusing to accept alternatives, refusing to believe the choice of between;
> Rejecting selection; denying dilemma; electing absolute affirmation:
>   knowing
>   in his breast
>   the infinite and the gold
>   of the endless frontier, the deathless West.
> "Both: I will have them both!" declared the true-blue American.[1]

Who among us has not, on occasion, when faced with the either–or choice that confronts young Jeremiah, wanted to cry out, "Both: I will have them both"? The poet Delmore Schwartz has indicated that Jeremiah's desire to "have them both" marks him as a "true-blue American." For decades the more developed world in general, and America in particular, has been quite comfortable with this pattern of overconsumption. But how long can this go on? Is it possible for the more developed world to bring the rest of the world up to its level of development without sacrificing its own standard of living? Will the more developed world do this cooperatively or will the less developed world compel the more developed world to submit to its demand for an equal share of the world's wealth? What role can science and technology play in redistributing the wealth among nations? Who will be responsible for the development of a responsible global science and technology policy? What characteristics will be required to make such a policy a success? The answers to these and similar questions will be discussed in this chapter.

CHAPTER OUTCOMES

Upon finishing this chapter, the reader will be able to accomplish the following:

1. Describe how recent developments in reproductive technology may help deal with overpopulation.
2. Relate how advancements in communication may alleviate the population problem.
3. Explain how advancements in transportation may solve a part of the population problem.
4. Explain how alternative sources of energy may address the problem of war.
5. Detail how the threat of war may be lessened by advancements in genetic engineering.
6. Outline how nanotechnology may help eliminate inequities in resource allocation.
7. Explain how science and technology may deal with environmental problems.
8. Outline recent international efforts to solve environmental problems.
9. Detail the characteristics of an effective science and technology policy.
10. Explain some practical steps in the establishment of a global science and technology policy.

14-1   THE ROLE OF SCIENCE AND TECHNOLOGY

The objective of this book has been to introduce the nonscientist to the world of science not only by explaining some of the most significant scientific discoveries of the last 100 years, but also by exploring how science interacts with other human endeavors such as theology, philosophy, ethics, and the law. There can be no more significant intersection between science and the nonscientific world than the link between science and the global challenges described in the preceding chapter. In a very real sense this link may hold the key to the survival of the human race on this planet. As noted in the preceding chapter, the three global challenges of overpopulation, nuclear war, and environmental deterioration are directly related to the advances made possible by science and technology. However, as Robert Heilbroner correctly points out in his book *An Inquiry into the Human Prospect*, science and technology by themselves are not responsible for these problems. Rather, it is the uneven way that science and technology have been allowed to develop that lies at the basic core of the crisis.[2] This fact leads to the final two questions addressed in this book. The first asks whether the global challenges of overpopulation, nuclear war, and environmental deterioration can, from a technical point of view, be solved by science and technology. The second explores whether, from a sociopolitical perspective, the people of Earth are capable of redirecting scientific and technological activities toward those solutions.

14-2   SCIENCE, TECHNOLOGY, AND THE POPULATION PROBLEM

First, from a practical point of view, can science and technology do anything about the population crisis? In the preceding chapter, we explored some of the causes and consequences of unchecked population growth. The two principal causes were identified as the declining nature of infant mortality and an increase in the life expectancy of most people on Earth. Both of these causes have been instigated by advances in medical, biological, and health science. It is not likely that such trends will disappear in the near future. For one thing, it would be ethically questionable to advance the idea that scientists should call a halt to research designed to save the lives of children or to improve the quality of human life on the planet. Moreover, calling a moratorium on such research would be impractical from a logistical point of view.

However, neither of these alternatives need be pursued because the population bulge caused by the drop in infant mortality and increase in life expectancy is not produced solely by these factors. Instead, the population has continued to climb because there have been few corresponding programs designed to attack the other side of the problem, that is, family growth rate. Recall that the countries with some of the greatest population problems have agriculturally based economies. Traditionally, families in agricultural societies depend upon large families as a labor pool because such societies also generally have high infant mortality rates. When infant mortality rates drop and life expectancy increases, while family growth rates do not change, a jump in the size of the population is inevitable.[3] This is a problem that can be handled by science and technology in a number of different ways. However, one of the most obvious ways is through the development of safer and more effective means of birth control. Unfortunately, while the introduction of new and safer contraceptive devices is technologically feasible, it is also difficult from a social, cultural, and ethical perspective.

New Reproductive Technologies

In Chapter 10 we explored some recent developments in the area of reproductive technology. One of these developments was the introduction of new birth control devices such as contragestive agents. Contragestive agents are effective because they prevent the implantation of a fertilized egg on the uterine wall. The morning-after pill, intrauterine devices, and low dosage birth control pills are all examples of contragestive agents. Contragestive agents have several advantages that make them an effective means of birth control and, consequently, may make them tenable as reproductive alternatives in the less developed world. Principally, contragestives are more economical than other methods of reproductive regulation. This may convince the economically stressed nations of the less developed world to support a program of birth control based on the use of contragestives. Perhaps as important, the use of contragestives offends fewer people than other reproductive control techniques because contragestives are relatively safe and operate very early in a pregnancy. In fact, if used early enough, a contragestive agent actually operates more like a postcoital contraceptive than as a method of pregnancy termination. This advantage could make contragestive agents more attractive in cultures that are reluctant to introduce radical birth control methods into the population at large. Moreover, contragestive agents have already proven to be highly effective in several regions of the world. For instance, contragestive agents have proven to be successful in France, Great Britain, and the Netherlands. In fact, in France, the contragestive agent known as RU486 is responsible for 33 percent of all deliberate pregnancy terminations. Moreover, RU486 has been tested in both China and Sweden.[4]

Another recent development in birth control technology is the Norplant system of contraception. The Norplant system involves the implantation of six tiny silicone tubes under the skin of the upper arm of the patient. Each of the tubes is approximately as big as a match stick. The implantation routine lasts about 10 to 15 minutes. After the tubes have been placed in the woman's arm they begin to discharge a synthetic version of progestin. The progestin prevents ovulation. This results in temporary infertility. The hormones also restrict the movement of sperm into the uterus, thus adding an additional protection against fertilization. The effects of the original implantation last for five years. Figures on the effectiveness of the Norplant system vary from 1.5 to 5 percent possibility of pregnancy over the five-year period of its effectiveness.[5]

The introduction of these and other innovative reproductive technologies in the less developed nations of the world would not only help reduce population growth but would also substantially reduce the health risk to women in these regions, many of whom, when faced with unwanted pregnancies, are forced to seek unsafe abortions. The number of such women is now staggering. A recent study conducted by the World Health Organization revealed that, on a worldwide basis, more than 20 million women undergo unsafe abortions each year. Significantly, the number of unsafe abortions in the less developed world is substantially higher than in the more developed world. For example, in Africa and Latin America for every 1000 births, 30 unsafe abortions are performed annually, while in North America and northern Europe for every 1000 births only two unsafe abortions are performed each year.[6]

Difficulties Related to New Birth Control Technologies. Despite the alarming nature of these figures, the difficulty of introducing any new birth control technology into the less developed nations of the world persists. One obstacle to success is purely logistic. This involves the ability, or inability, of a government to provide its people with access to effective birth control technology. In addition, such programs must communicate to the people accurate information about the use of that new technology. When such programs are not in place, even the most advanced forms of birth control technology will do little good. Such has been the case in Nigeria, Pakistan, and Ethiopia, where access to birth control technology is inadequate at best. In Ethiopia, for example, fertility rates have increased 4 percent over the last thirty years, whereas in Nigeria the drop in the total fertility rate from 6.8 to 6.5 is negligible.[7] In contrast, in nations like Bangladesh, Colombia, South Korea, and Thailand, where strong programs promoting access to birth control technology have been implemented over the last thirty years, the total fertility rates have dropped significantly. Both Colombia and Thailand, for instance, saw a drop of 60 percent in their total fertility rates.[8]

Logistic problems are not the only barrier to the introduction of advanced procreation technology, however. Another obstacle is the fear that such governmental programs would lead to coercive, even compulsory programs of birth control. Such fears are not unfounded. Even the most advanced, and presumably most enlightened, countries of the world have been tempted to promote involuntary birth control projects. When Norplant was first marketed in the United States, several government officials proposed using it for establishing social policy. One state legislator suggested requiring fertile women who are also drug abusers to be implanted with the Norplant system as a precondition for probation.[9] Other proposals for the establishment of involuntary birth control programs in the United States include using the Norplant system as compulsory contraception to prevent congenital disease, implementing the use of Norplant to reduce teenage pregnancies, and authorizing the implantation of Norplant as a prerequisite for the receipt of welfare payments.[10]

These situations show how a government can quickly become abusive as it attempts to use new reproductive technologies to control population growth. Unfortunately, there are even more extreme examples of such abuse. In China, for instance, the government has gone as far as to compel involuntary abortions and sterilizations in the implementation of a population policy that limits each family to one child. Similarly, in India the government of Indira Ghandi implemented a program of compulsory sterilization which was so severe that it eventually contributed to the downfall of that government.[11] Although such drastic measures are rare, nevertheless, they may prevent many well-meaning people from supporting certain governmental actions in the area of birth control, even when those actions are designed to establish effective and responsible reproductive policies. Unfortunately, the fear of government abuse may mean that many official programs involving birth control remain ineffective. This is true despite the fact that 81 percent of the governments of the world provide some sort of support for reproductive control.[12] The bottom line is that, although science and technology can provide improved technology for reproductive control, they cannot change the attitudes of people toward that control and toward any government that attempts to implement such control. On the other hand, science and technology may be able to handle more easily another problem associated with population growth—urban density.

Decentralization through Communication and Transportation

Recall that a major challenge associated with population growth is not so much the raw numbers as the density with which people are packed into the megacities of the world. The migration of people from the country into the cities was initially caused by the advent of industrialization. This migration was made necessary because manufacturing equipment, which was expensive, complex, and enormous in size, required centralized factories as a base of operation. These conditions drew workers into cities, where the factories were located.[13] This trend toward urbanization was augmented by a commitment to central planning and a belief in the beneficial spinoffs from a centralized economy.[14]

By the end of the twentieth century millions of workers had abandoned agriculture and moved into the cities causing an urban crisis characterized by overcrowding, unsanitary conditions, inadequate waste disposal, and infectious diseases.[15] The initial reaction to this problem was for cities to grow larger to accommodate the burgeoning population. Of course, this growth meant a loss of available farmland as the cities paved over agricultural areas. At first, this worked fairly well because, as the cities grew, the farms simply moved elsewhere. Now, however, as the available land diminishes, urbanization results in a direct loss of agricultural acreage.[16] Today, the disappearance of agricultural land is advancing at a staggering rate. In Asia alone, millions of acres of farmland have been lost to urban sprawl. For instance, in recent years, the small island of Java in Indonesia has lost 20,000 hectares of agricultural land. A **hectare** is equal to 10,000 square meters of land. Had this land been used for planting, enough rice would have been produced to feed over 300,000 Indonesians, many of whom suffered as a result of the loss.[17] Similarly, in China between 1987 and 1992, 3.87 million hectares of farmland were destroyed to make way for the growth of cities and the development of industry. Such a loss meant that, during that time period, 15 million tons of grain were not produced that could have been grown, absent the destruction of the farmland.[18]

Could science and technology help solve the problems caused by urbanization and industrialization? Certainly, the potential for such a solution exists. Recall that originally, urbanization was made necessary by the centralization of industry. Workers were compelled to move into centralized cities where expensive equipment and complex factories were located. This type of work required face-to-face communication and localized industrial activities. To put it simply, the workers had to be located in the factories to do the job. If, however, business and industry could be decentralized, people could move out of the cities.[19] This type of decentralization has been made feasible by improvements in electronic communication and advanced transportation systems.

The Revolution in Telecommunications. Perhaps the most obvious link in the telecommunications network is the development of a widespread computer system that can make long-distance business transactions commonplace. Even today, our dependency on computers to transact business is so complete that it is difficult to imagine the civilized world without a complex computerized network to manage telephone calls, credit card sales, check transactions, and airline reservations. At the present time, billions of such transactions are handled annually by computer. Our dependence on this vast computer network was made even more obvious by the wide range of concerns associated with the Y2K situation.[20] Tomorrow, the development of a global network of supercomputers, each of which will have a power level surpassing the combined capacity of 10 million of today's personal computers, will make the present system seem antiquated by comparison.[21]

However, this supercomputer system is only the beginning of the telecommunications revolution. Another dimension is the development of an advanced fiber-optics telecommunications network that will link the four corners of the world. Today, the annual number of telephone calls per person in the United States is 4000. As the cost of telecommunications diminishes, this number is expected to climb dramatically. Nevertheless, long-distance rates are shrinking so rapidly that, eventually, phone call charges will be distance independent.[22] This highly efficient, low cost, telecommunications system has been made possible by the development of fiber optics. **Fiber optics** involves the use of thin glass fibers to transmit information, replacing the large copper wires that were the mainstay of our global telecommunications system for years.[23] With the use of fiber optics the calling capacity of the global telecommunications network jumps significantly. For example, in the 1950s only 12 simultaneous calls could be handled by the transatlantic cable.[24] By the 1980s the capacity was 4000 simultaneous calls.[25] With the use of fiber optics this number has now jumped to 80,000 simultaneous calls.[26] Moreover, the capacity for the telecommunications system between Japan and the United States installed by AT&T and Kokai Denshin Denwa (KDD) is 500,000 simultaneous calls.[27]

Economical long-distance voice transmission is only the beginning of the revolution in electronic communication. On the immediate horizon is the increased availability of videophones. At first sight, a videophone may appear to be nothing more than a frivolous, expensive plaything. Nothing could be further from the truth. With the widespread use of videophones, geography will no longer be the handicap that it is today. Videophones will allow business people to conduct complex meetings on a global basis as if the participants were in the same room. With the use of videophones and videoconferencing, individuals will be able to display graphs, charts, photos, video recordings, and blueprints to participants in a dozen different regions of the globe simultaneously, clearly improving the quality and the effectiveness of long-distance business meetings.[28] The use of fax machines also allows for long-distance communication of data that formerly required face-to-face interaction.[29] According to some experts, travel in the United States will decline by 15 percent or more once a total fiber optics telecommunications system is constructed. Plans are underway that predict the completion of this project within the next thirty years.[30]

Moreover, videoconferencing may soon become common place on the Internet. The **Internet** is a computerized collection of global networks. It was first established by the military in the 1960s but has been available to the public at large since the early 1990s. Videoconferencing on the Internet has been limited by the speed and capacity of current computer systems and by the need to develop sophisticated equipment that can transfer pictures to data and back to pictures once again. Using state-of-the-art advances in these areas, the Bailey Group, an Ohio corporation, has created software that allows individuals to access the Internet and to see and speak to one another in real time across long distances. The software package, known as Visual Care, has successfully linked four Ohio Country Club Retirement Centers for real-time videoconferencing. The Bailey Group has also successfully connected Ohio with centers in locations as far away as California.[31] Once all of these aspects of the telecommunications revolution are in place, the centralized urban commercial center that was characteristic of the industrial era will become a thing of the past.

The Uncertain Future of Telecommunications. On the other hand, some experts believe that the telecommunications revolution, as advanced and as promising as it appears, may never become a reality. Several factors may operate to preserve our present centralized economy. One primary factor involves the attitudes that many people bring to the workplace. For instance, some individuals may be reluctant to trade the professional, disciplined atmosphere of the workplace for the unstructured, frequently unpredictable conditions of a home-based office. Others may find that the professional activities that occur at the office are just as necessary for professional success as the electronic tools of the trade. Still others may see the commute from home to the office as a necessary separation between their professional and personal lives. Some people may discover that their jobs do not lend themselves to a home-based approach, while others may learn that their supervisors are unwilling or unable to deal with the adjustments that are required by a stay-at-home work force.[32]

These are just a few of the results uncovered by Patricia L. Mokhtarian, professor of civil and environmental engineering at the University of California. Mokhtarian has conducted a number of studies on the effects of telecommunications on commuting, land use, and the environment over the last 15 years. Among her discoveries are statistics that indicate that most people who choose to stay at home and take advantage of the telecommunications revolution, a group that Mokhtarian calls telecommuters, rarely do so on a full-time basis. Most telecommuters decide to stay at home only one or two days out of the entire work week. Consequently, a part-time shift to telecommuting will do little to alleviate urban congestion because part-time telecommuters must still remain close enough to the workplace to commute on those days that they do go to the office. Moreover, Mokhtarian's studies have indicated that telecommuters rarely assume that lifestyle on a permanent basis. After one year of adopting telecommuting as a lifestyle, over 50 percent of these new telecommuters abandon telecommuting and return to the workplace full-time. From these studies Mokhtarian has concluded that, under present conditions, on any single workday, only 2 percent of the labor force, and perhaps even less, is at home due to the telecommunications revolution.[33]

Mokhtarian also doubts that telecommuters will actually reduce urban congestion even when they do stay at home permanently. She predicts that most telecommuters will still live close enough to an urban center to shop, run errands, and fight cabin fever by traveling into the city, even on those days that they are scheduled to be at home. In addition, Mokhtarian speculates that in the future telecommuters may become less likely to car pool or use mass transportation on those days that they do head to the office, thus reducing the initial positive effect that their telecommuting has on traffic congestion and urban density. While Mokhtarian admits that this forecast does not necessarily preclude all future benefits from telecommuting, she does point out that grandiose predictions of a workplace dominated by telecommuting were made as long ago as the 1870s when the telephone made its first appearance in the office. Such prophecies did not come true at that time, and Mokhtarian sees little evidence to believe that they will come true in the future.[34] On the other hand, the telecommunications network that will develop in the future will be so unprecedented in scope that its actual effects defy accurate predictions. As such, it is best for us to reserve judgment until such time as future developments either vindicate or disprove Mokhtarian's speculations. In the meantime, we can hold on to the hope that the problems of urban congestion will be solved, at least partially, by the telecommunications revolution.

14-3   SCIENCE, TECHNOLOGY, AND THE THREAT OF WAR

At first glance the problem of nuclear war may seem beyond the saving power of science and technology. This is not the case, however. On the contrary, advances in science and technology might one day make nuclear war obsolete. As noted in Chapter 13, the peril of nuclear war is just as sharp today as it was during the height of the Cold War. Instead of ending the threat of a nuclear conflict, the end of the Cold War destabilized the delicate balance of power between the United States and the former Soviet Union. This destabilized condition led to a world situation in which many nations feel threatened, either from the activities of their neighbors or because of the unequal distribution of resources. If a certain level of stability can be restored, the threat of war could be significantly diminished. However, restoring stability means removing the perceived inequities in the distribution of resources among nations. This is how science and technology can come to the rescue.

Alternative Energy Sources

The practical ability of science and technology to achieve this redistribution of wealth is clear, as long as a properly controlled science and technology policy can be developed at a global level. For example, the Arab nations routinely threaten the existence of Israel. Much of the influence of the Arab states, however, is based on their control of a sizable percentage of the world's oil reserves. Those nations who are dependent on the good will of the Arab states for their oil are, therefore, susceptible to the undue influence of those states. This susceptibility may be the primary reason that both Germany and Japan refused to send troops into the Middle East during the Gulf War. If neither Germany nor Japan were vulnerable to such energy blackmail, real or imagined, they might have been more willing to join the allies against Iraq.

One way to discourage energy blackmail is to eliminate the leverage gained by the possession of oil. This leverage can be diminished by the development of alternative sources of energy, including solar power, wind power, geothermal energy, antimatter, and cold fusion. This is, in fact, the very scenario suggested by the noted science fiction writer Arthur C. Clarke, who, in a story entitled "The Hammer of God," predicts the demise of the Oil Age when a cold-fusion reactor is finally invented.[35] The intent here is not to cripple the Middle Eastern states, but to free those nations currently subject to the undue influence of the oil-producing nations. This freedom would allow those nations to develop a more unified front as they tackle the redistribution of the world's resources.

Biotechnology and the Green Revolution

Issues related to the redistribution of the world's resources are not limited to inequities in energy reserves. Similar problems can be seen in the area of agricultural production. At the present time, problems in agricultural production are most acute in Asia, where areas like South Korea, Taiwan, and Singapore have seen a decline of 20 percent in land devoted to grain production over the last decade.[36] China has also experienced a serious loss in productive acreage. Between 1987 and 1992, for instance, over 3.8 million hectares in China were lost to urbanization and industrialization. A fall like that means a loss of about 15 million tons of grain.[37] Certainly, these losses can be connected to China's deliberate policy of forcing the development of industry and housing. The loss of food, however, is not any less serious because it can be traced to industrialization and urbanization. Nor does the fact that the Chinese government is responsible for this drop in food production prevent that government from looking to greener pastures across its borders in an attempt to redistribute agricultural resources.

Could this need for increased agricultural output point a nation toward a war of redistribution? Perhaps. But again, science and technology, if properly managed, could handle this situation. To see how this might work, we must return temporarily to our study of genetic engineering. Recall that genetic engineering is the process by which modifications to living things are made by manipulating genes.[38] One of the tools of genetic engineering is the ability to produce recombinant DNA. This process is frequently called gene splicing.[39] Using gene splicing techniques, bioengineers can alter the genetic makeup of a strand of DNA by transferring a gene from one strand of DNA to a second one.[40] The ability to transfer genes in this way enables bioengineers to modify crops in a variety of ways that can be very effective in terms of food production.

In fact, it is possible that the use of genetic engineering in relation to crop production could lead to a green revolution that would allow us to reduce the effects of famine caused by the worldwide loss of agricultural acreage. For example, we have already seen that bioengineers have been successful in lengthening the shelf life of the average supermarket tomato by implanting a gene that inhibits the decomposition process. This allows more tomatoes to reach more consumers and to remain available to those consumers for a longer period of time.[41] Genetic engineering has also been used to make some plants more resistant to certain types of pests. Again, this has been done by inserting a gene from a plant with high resistance to the pest in question into a plant of a high-yield crop but that suffers from low resistance to that pest. Bioengineers are also working on ways to make some crops more nutritious. As noted earlier, one such project is designed to make corn a source of lysine, an amino acid that is needed by humans but that is not a part of zein, the protein produced by corn.[42]

Several other areas of bioengineering research promise to help us produce more abundant and stronger crops. One such project involves an activity known as **nitrogen fixation**, a process by which nitrogen in the atmosphere is changed into a more suitable form so that cells in plants can use the nitrogen to create amino acids and other beneficial molecules. Some plants, such as peas and beans, cooperate easily with nitrogen-fixing bacteria. This means that planting a crop of beans or peas will help maintain the soil's fertility level, lessening the need to rely on expensive and harmful chemical fertilizers. Using bioengineering techniques, scientists may be able to transfer the gene enabling peas and beans to cooperate with nitrogen-fixing bacteria into wheat or rice. Or, as an alternative, by splicing the bacterial gene onto the plants themselves, bioengineers might be able to create plants that can transform the nitrogen on their own.[43] Other promising bioengineering projects involve the attempt to make certain crops impervious to herbicides and to allow crops like wheat to grow in salt water.[44] Advances in biotechnology might also lead to "supercrops" that resist drought, temperature inversions, and disease. Such plants could also be genetically engineered to thrive in soil which, by today's standards, would be too nutrient-deficient or too saturated with salt to be fertile.[45] These advances would allow nations currently incapable of feeding themselves to produce more than enough food to support their growing populations.

Naturally, the genetic engineering of crops is not without its drawbacks. One argument often raised about the genetic engineering of plants is that many of the effects of altering the DNA of specific plants are unknown. Therefore, changing the genetic makeup of a crop to allow for nitrogen-fixing or to permit that crop to resist a certain type of pest might also trigger some toxic side effects within that plant of which scientists are unaware and which they cannot predict. Moreover, such effects may not be immediately obvious and thus may not be evident for a generation or two, at which point it may be too late for people who have ingested such crops all of their lives. An additional concern that flows from this initial problem is that the consumption of such crops by the public at large is primarily involuntary. Shoppers at the market may be completely unaware that an ear of corn or a tomato that they have purchased was produced by genetic alterations. Consumers have the right, the argument goes, to be informed of this fact as well as of the risks that go along with the consumption of such crops. Certainly, some of this concern can be alleviated by regulations demanding the proper labeling of such genetically enhanced crops. However, even the strictest labeling regulations will not eliminate the dangers of unknown and unpredictable side effects. On the other hand, such risks, precisely because they are unknown and unpredictable, cannot be allowed to halt research in this area entirely. It might be much more beneficial to use this concern as a reason for the implementation of tighter controls on such research now and in the future.[46]

Nanotechnology and the Redistribution of Wealth

Energy shortages and agricultural problems are only a small segment of a larger pattern of worldwide inequity. An unequal distribution of material wealth, including goods and services, among nations and between classes within a single nation, can also lead to global instability. For example, as we discovered in the preceding chapter, although the more developed nations of the world contain only 20 percent of the Earth's population, 80 percent of the world's goods and services are produced and consumed by them.[47] Because of this rate of production and consumption, of the 88 million people added to the world's population each year, the 4.5 million that join the more developed world have the same environmental impact as the 83.5 million added to the less developed world.[48]

However, even these figures mask the true extent of the disparity between the rich and the poor, because even within the nations of the less developed world, the wealth is also unevenly distributed. For instance, nations like Argentina, Brazil, Costa Rica, Uruguay, and Venezuela see the poorest 75 percent of the population economically exploited by the wealthiest 5 percent. Nor are these nations alone in such an unequal distribution of material wealth. Even more developed nations like the United States and Great Britain are experiencing a trend toward the same type of internal inequity seen among the classes in the nations of Latin America. Moreover, the division between the "haves" and "have nots" has not leveled off. On the contrary, the split grows wider with each passing year.[49]

How can science and technology deal with this unequal distribution of income, goods, services, medicine, and housing? The answer may lie in advances in nanotechnology. **Nanotechnology** will permit the construction of virtually any product be reconfiguring molecular structures atom by atom. This process permits scientists to reconfigure the basic molecular structure of an element to produce new materials and products.[50] If nanotechnology sounds similar to genetic engineering, that is because advances in bioengineering have pointed the way to nanotechnology. As noted above, bioengineers can now transfer a gene from one strand of DNA to a second one, thus altering the genetic makeup of that strand.[51] Such recombinant DNA techniques actually involve the manipulation of the DNA at the molecular level. Because of this, it is likely that the first generation of nanodevices will be protein machines. Eventually, protein machines will be used to make more advanced nanodevices that will be able to manufacture more than just protein.[52]

In this way, goods that are currently unavailable to the less developed world can be manufactured at the molecular level out of common elements like hydrogen, oxygen, nitrogen, and carbon.[53] Once nanotechnology makes it possible to manufacture those products that are needed by the less developed world, there will no longer be any need for those nations to look across their borders with envy or to take military action to seize what they believe is theirs by right. Nor will the impoverished lower class within a nation see the need to revolt against the elite within their own borders, as happened in Rwanda in 1994.[54] Instead, clothing, appliances, vehicles, medicine, housing, and many other products could be available to all. Levels of poverty, hunger, and unemployment can be reduced, and along with them, much of the motivation for war and revolution.

Universal Assemblers. Of course, there is the danger of being overly optimistic here. Nanotechnology is not without its shortcomings. First of all, like any technology, it is impossible to predict when nanotechnology will be available to the world at large. Although many firms, such as IBM, Genex, Hitachi, Sharp, and Toshiba, and government agencies such as the U.S. Naval Research Laboratory, are working on the development of nanotechnology, it could take years, perhaps decades, for the technology to come of age.[55] However, we can predict that a major breakthrough will occur when scientists and engineers develop **universal assemblers**, which are programmed nanodevices capable of using atoms and simple molecules to construct almost any complex molecular arrangement. Since universal assemblers will be capable of constructing complex molecular arrangements, they will be able to build almost anything imaginable.[56] Unfortunately, the operative words here are "almost anything imaginable." Universal assemblers will be faced with certain unavoidable limitations. They will not, for example, be able to alter the laws of physics. It is doubtful, for instance, that nanotechnology will allow us to construct devices that defy the laws of thermodynamics or that can escape the limitations imposed by the uncertainty principle.[57]

The Development of Nanocomputers and Disassemblers. Nevertheless, the universal assembler breakthrough will lead to developments of even more complex nanodevices. For example, with the advent of universal assemblers, **nanocomputers** could be designed that would be less than a micron wide and faster than the fastest of today's electronic microcomputers. Once nanocomputers are up and running, they will be able to improve the operation of the universal assemblers. Nanocomputers will program and direct the assemblers to manipulate atoms and molecules to manufacture products according to a complex set of instructions stored in the nanocomputer's memory. This could lead to the next generation of nanodevices, the disassemblers. **Disassemblers** will be able to analyze the existing molecular structure of an item so that an understanding of that structure can be stored in the memory of the nanocomputer, which could then direct the assemblers to make copies of the original.[58] Suppose, for example, that a pharmaceutical engineer wants to reproduce the flavor of a rare tropical fruit for use in creating a certain type of children's medicine. If the engineer has a sample of the crushed up remains of the fruit, that sample could be fed into the disassembler which could analyze its molecular structure and then transfer that data into the nanocomputer's memory so that the flavor could be duplicated by the assembler and added to the medicine.[59]

The Limits and Dangers of Nanotechnology. Although nanotechnology does have some very impressive supporters within the scientific community, notably, K. Eric Drexler, author of *Engines of Creation* and *Nanosystems*, Ralph C. Merkle of the Palo Alto Research Center, and Marvin L. Minsky, a leader in the field of artificial intelligence, there are other scientists of equal reputation who are not as optimistic about the practicality of nanotechnology in general and about the feasibility of universal assemblers, in particular.[60] One such expert, David E. H. Jones, of the University of Newcastle upon Tyne, has indicated that, as of yet, even the most dedicated disciples of nanotechnology have failed to answer some very basic questions about the viability of controlling matter at the atomic level.[61] Jones has pointed out that the supporters of nanotechnology have yet to explain how a universal assembler will be able to identify those atoms that are needed to construct a particular element or, once an assembler has located an atom, how the assembler will be able to transport that atom to the construction site. In addition, according to Jones the most crucial issue, that of the power supply for the assemblers, has yet to be addressed. Jones has noted that this question is extremely important because it calls into question the ability of the assemblers to break up existing material and make the complex computations needed for such an undertaking.[62]

Other critics have identified this lack of detail in the practical engineering aspects of nanotechnology as its major shortcoming. For instance, Philip W. Barth, a micromechanics expert at Hewlett Packard, points out that, thus far at least, some of the claims for nanotechnology are based primarily on computer simulations, which certainly lend plausibility to such ideas but fail to consider several fundamental engineering problems.[63] One such problem is the lack of stability in certain stages of the nanoconstruction process. Like Jones, Barth is also concerned with the lack of pragmatic detail in relation to the engineering practicalities of nanotechnology. The fear is that such problems may not be a simple gap in an otherwise sound technology, but instead, may reveal a fundamental defect in the very concept of nanotechnology, at least on the scale envisioned by Drexler, Merkle, and Minsky.[64]

If practical limitations were the only shortcoming of nanotechnology, many experts, even those who are skeptical, might maintain an attitude of cautious optimism about its future. After all, fifty years ago space flight also posed certain practical limitations that many people saw as insurmountable. Unfortunately, in addition to these pragmatic restrictions, the development of nanotechnology also carries certain risks, one of which is the accidental creation of a runaway replicator. Replicators represent a stage in the evolution of nanotechnology that is one step beyond the disassembler-nanocomputer-assembler interchange. Thus far, we have looked at nanodevices that manufacture products other than themselves. In contrast, a replicator would have the ability to reproduce duplicates of itself. More advanced replicators might even be able to evolve by reproducing duplicates of themselves complete with any variations and mutations in the original.[65]

This ability would make replicators completely self-contained and self-sustaining devices. However, it would also make them potentially dangerous. The danger lurks in their need to feed on raw materials in order to reproduce themselves. Such replicators would probably be stored in containers of chemicals that would provide that fuel and raw material. As long as this chemical bath is unavailable in the natural environment, the replicator will remain self-contained.[66] In effect, if the replicator were to escape from this carefully constructed and highly regulated environment, it would starve. At least, that is the plan. But what would happen if the replicator were able to reproduce itself using organic material available in the natural environment? If such a replicator escaped into the environment, it would find an abundant food supply. This could lead to an ecological catastrophe of monumental proportions as the replicators reproduce themselves by consuming any organic material that they encounter.[67] It is not difficult to imagine bacteria-size, renegade replicators "eating" their way through the Amazon rain forest or devouring the farmlands of the Great Plains states. Such a scenario could lead to so much destruction that it might very possibly lead to the breakdown of the entire biosphere.[68]

Of course, there are ways to lessen such risks. As noted, one way to lower the possibility of such a disaster is to ensure that the replicators cannot use any raw materials that are not artificially produced in the laboratory and are, therefore, unavailable in the natural environment. On the other hand, all the safeguards in the world are useless if an aggressor nation or a band of terrorists elects to use an advanced replicator as a weapon. Such a device could be employed to construct weapons of mass destruction in record time. Or the replicators themselves could be used as weapons if they were programmed to kill the population or destroy the environment of an enemy nation.[69] Many people would argue that such a risk is not balanced by the potential benefits of nanotechnology. On the other hand, if one of the benefits is a lessening of the threat of global warfare, then perhaps we should not dismiss nanotechnology until all the possible ramifications have been adequately explored.

The Dangers of a World without War

Of course, there are those who might argue that war can never be eliminated. Among the voices raised in dissent are those that insist that the complex nature of the international arena make technological advances, like nanotechnology, genetic engineering, and the development of alternate energy sources, impractical as means of avoiding war, at least by themselves. Although he does not deal with these three technological developments directly, Robert Borosage, the founder and director of the Campaign for America's Future, has pointed out that inaccurate predictions about the end of war have been made in the past—most recently, at the end of the nineteenth century, when the economic wealth produced by the industrial revolution seemed poised to eliminate those economic conditions that most frequently led to war. According to Borosage, technological achievements did not end the threat of war at that point in history, and they cannot, at least by themselves, dispose of war at the present time either. Borosage argues, rather convincingly, that war can be deterred only by the interaction of a number of diverse elements. To eradicate war he suggests that world leaders must pursue a policy of inspired diplomacy, promote peaceful solutions for all international disagreements, support the growth of international peacekeeping efforts, encourage the development of democratic regimes, preserve international principles of law, and eliminate the international tensions that most frequently lead to war.[70]

Although Borosage's position is disheartening, it is not completely without hope. His position is certainly not as discouraging as that held by experts who suggest that war *should not* be eliminated because of the economic, political, and sociological disruption that would result from dismantling the military-industrial complex. Following this line of argument, eliminating war would cause almost as much unemployment, poverty, and dislocation as war itself. This idea is, in fact, the premise of a book by writer/editor Leonard C. Lewin, entitled *Report from Iron Mountain on the Possibility and Desirability of Peace*. The book, which first appeared in 1967, was supposedly a report put together by a secret study group, convened for the purpose of examining the possibility of eliminating war as an instrument of foreign policy. Although Lewin eventually admitted that the entire report was fabricated as a way to focus attention on issues related to war and peace, it, nevertheless, remains a poignant and intriguing study.[71] One of the conclusions included in the report is that war provides certain economic, political, sociological, and cultural functions that cannot be adequately replaced by any other mechanism.[72]

This conclusion is a seductive one. However, there are many voices of dissent on this issue. At approximately the same time as Lewin's report, Edward Bernard Glick, an associate professor of political science at Temple University, conducted a study that suggests the opposite conclusion. It is Glick's contention that the military can be used for a variety of civic-action programs, from rebuilding cities and constructing power plants to providing medical care and training technicians. Glick's book, *Peaceful Conflict: The Non-Military Use of the Military*, is filled with historical examples of such civic action programs and humanitarian projects carried out by the military. Significantly, many of these examples include modern American military relief efforts in Latin America, the Middle East, Africa, and Southeast Asia.[73]

Nor is Glick alone in his evaluation of the many nonmilitary applications of military personnel and resources. In a recent report entitled *Our Global Neighborhood*, the Commission on Global Governance cites several examples of the nonmilitary use of the armed forces to provide humanitarian support primarily by supplying relief in areas like Sarajevo and Zaire.[74] The conclusion should be that the military need not be dismantled just because the planet has been demilitarized. The personnel, equipment, and resources formerly used only for war can be channeled into peaceful humanitarian activities. The potential is there. We need only the will and the organization to make it happen.

14-4   SCIENCE, TECHNOLOGY, AND THE ENVIRONMENT

The crisis caused by the deterioration of the environment is the most difficult of the three challenges facing humanity today. As noted by Heilbroner, although population growth can be diverted and wars can be avoided, the capacity of the planet to handle the pollutants being pumped into the environment has a finite upper limit. Moreover, as Heilbroner also points out, that limit may be just around the corner.[75] Can science and technology deal with this dangerous predicament? The temptation is to say, rather smugly, that, if science and technology caused the problems, then science and technology can solve them. Though there is much truth in this statement, science and technology did not themselves create the environmental dilemma. Rather, it is the mismanagement of science and technology that is to blame. So, if the question is whether corrective measures are possible, the answer is "yes"; but if the question is whether such measures will be implemented by the nations of the world, the answer becomes less definitive.

Scientific and Technological Practicality

As for the scientific practicality of correcting the damage already done to the environment, some of the technology needed to solve the problem is already in place, whereas other measures still in the developmental stage may one day become a reality. Recall that global warming caused by the emission of carbon dioxide and other harmful gases into the atmosphere is one of the most pressing environmental problems facing the Earth today. Because much of this carbon dioxide comes from the burning of fossil fuels, one way to attack the problem is to encourage the development of alternative sources of energy.[76] Fortunately, certain changes in energy consumption can be implemented right away using existing technology. For example, oil- and coal-burning furnaces now in operation can be converted to methane use. Moreover, any new factories, power plants, automobiles, and appliances that are planned for production in the future can be designed to use noncarbon-producing power sources, such as wind, batteries, and geothermal methods. Other techniques for generating power, such as cold fusion and the use of antimatter, may be available for homes and industry in the not-too-distant future.[77] Perhaps the most promising source of new energy, however, is solar power. Scientific and technological developments over the last four decades have vastly increased the efficiency of photovoltaic cells (PVs). In the 1950s most PVs could convert to useful form only 4 percent of the sunlight that fell on them. Now that percentage has tripled to 12 percent and in the next ten years the percentage will climb to 15 percent. Although this may not seem like much of an increase on the surface, it is very significant because the more efficient the PV, the lower the cost of the power that it produces. The outlook is so promising that Japan, Germany, and Italy currently offer incentives to households which adopt solar power. Moreover, many less developed nations, including India, Kenya, and Zimbabwe, have established long-range programs for the production of electricity based extensively on solar power.[78]

The development of alternative energy sources is not the only way that science and technology can help stop global warming. Another way is to halt the destruction of forests, which are threatened by disease, insect infestation, and fire and by the need to supply the industrial world with lumber, paper, and firewood. Once the trees are gone, they will no longer be able to absorb any excess carbon dioxide, thereby accelerating the global-warming pattern. If, however, we can create more durable trees and provide alternative sources of lumber, paper, and firewood, the trees would survive. This is where genetic engineering and, possibly nanotechnology come into play. Techniques in genetic engineering could create trees hearty enough to withstand the rapid temperature changes inherent in a global-warming pattern. These stronger trees would be able to resist disease and insect infestation, thus lessening their vulnerability to fire. Meanwhile, nanotechnology would permit the construction of almost any product by reconfiguring molecular structure atom by atom, which could eliminate the use of trees as a source of firewood, lumber, and paper.

The Earth Summit and Other Conferences

These are only a few of the ways that science and technology might be used to save the environment. As for the probability that any of these tactics will be implemented in time, corrective measures are already underway. Heilbroner's 1990 revision of *An Inquiry into the Human Prospect* makes this clear when he notes that, over the last two decades, the world community has begun to focus more intently on environmental issues.[79] Moreover, there are signs that world leaders understand what must be done, even if the current generation lacks the will to carry out effective programs. Evidence of this awareness can be seen in the U.N. Conference on Environment and Development held in Brazil in 1992. The delegates to this international conference, commonly referred to as the **Earth Summit**, produced **Agenda 21**, a global plan designed to combat environmental deterioration. This forty-chapter document outlines a strategy for rescuing the world's environment, while maintaining a sound global economy. As a part of this strategy, each nation must develop a national program to accomplish these two goals. Currently, over 100 nations have responded to this challenge by establishing national councils responsible for creating these programs. Unfortunately, much of the work done thus far by these groups has involved promoting already established internal programs that lack a global perspective. Still, it was a step in the right direction.[80]

Another step in the right direction was a second global summit, the Conference of the Parties to the Framework Convention on Climate Change, which was held in Berlin in 1995. This conference involved 120 nations working together to design an international plan to stabilize the rising levels of carbon dioxide and other gases in the atmosphere. The key to a successful plan will be to create a balance among three central factors: (1) a timetable that will not threaten existing ecosystems; (2) a scientifically sound approach that will allow for the adequate production of food; and (3) a fiscally responsible strategy that will permit economic development to continue in a productive manner.[81] This plan, known as the Berlin Mandate, directs national governments to set goals, objectives, and protocols for the reduction of carbon dioxide and similar greenhouse gases within specifically identified timetables.[82] Although the Berlin Mandate is a hopeful sign, it has its limitations. For one thing the mandate is not a legally binding document. This weakness may prevent many nations from implementing appropriate protocols despite the best intentions.[83]

A third global summit under the United Nations Framework Convention on Climate Change was held in Kyoto, Japan, in 1997. After eleven days of grueling negotiations which threatened to break down several times, 160 nations, including the United States, Japan, and the members of the European Union, reached an agreement on the lowering of greenhouse gas emissions during the early years of the twenty-first century. Under terms of the agreement, known as the Kyoto Protocol, the major industrial nations of the more developed world would be required to lower their emissions of carbon dioxide and other greenhouse gases below 1990 levels. The agreement calls for the European Union to lower emission rates by 8 percent, the United States by 7 percent, and Japan by 6 percent. These reductions must be made between 2008 and 2012.[84]

14-5 TOWARD A GLOBAL SCIENCE-AND-TECHNOLOGY POLICY

Thus far, except for our discussion of the Earth Summit, the Berlin Mandate, and the Kyoto Protocol, we have not looked at any of the practical ways for implementing the scientific and technological solutions offered for the challenges of overpopulation, nuclear war, and environmental deterioration. Instead, science and technology have been represented as autonomous beings capable of making progress on their own. Such is not the case. Science and technology as disciplines are created by people and can operate successfully only by the hard work and the dedication of individuals within those disciplines. Moreover, science and technology can only ameliorate global problems by becoming interconnected with governmental agencies and financial institutions that have the money to support their work and with educational institutions and research-and-development firms that have people to conduct the necessary scientific work. This, then, is the focus of the last part of this chapter. First, we look at the characteristics of an effective science-and-technology policy. We will then examine one practical suggestion for implementing a successful science-and-technology policy on a global basis.

Principles for the Creation of a Science-and-Technology Policy

The idea that the government of a nation or state should develop a coherent science-and-technology policy is not a new one. The Founding Fathers recognized the need for governmental involvement in science and technology, at least indirectly, when in the Constitution they gave the federal government the power to regulate weights and measures and the authority to grant patents to inventors for the creation of innovative devices. In the 1940s, Senator Harley M. Kilgore, a New Deal Democrat from West Virginia, proposed a social welfare style program for the development of a national science-and-technology policy. Kilgore's plan, which was far ahead of its time, promoted federal financial support for businesses involved in research and development. In addition, Kilgore believed that the driving force behind science and technology should be the requirements of society rather than the intellectual curiosity of scientists. It is probably for this reason that Kilgore advocated federal research projects to combat pollution and to supply affordable electrical service to rural areas. Perhaps the most interesting aspect of Kilgore's plan was his idea that federal spending on science and technology should be distributed on a geographic basis, so that all regions of the United States would benefit equally.[85]

Although Kilgore's social welfare plan was never adopted by the federal government, it, nevertheless, served as an effective counterpoint to another post–World War II formula for a national science-and-technology program. This formula was suggested in a report issued by a committee headed by Vannevar Bush, the director of the White House Office of Scientific Research and Development. The report was issued under the title *Science: The Endless Frontier*, but it is best known as the Bush Report. The Bush Report focused on pure research for human needs rather than applied research for social action programs.[86] Moreover, the report also opposed the geographical allocation of funding because the committee feared any geographical scheme might lower the overall quality of the scientific research.[87] In addition, whereas Kilgore's plan suggested that scientific funding decisions should be left up to nonscientists, the Bush Report promoted the establishment of a national research foundation, which would be run by a panel of scientists taken from universities and colleges.[88] Despite these differences, both Kilgore and Bush supported extensive governmental spending for scientific and technological research and development.[89]

In 1997, the New York Academy of Sciences published an extensive report that detailed over twenty case studies reviewing the development of government sponsored science-and-technology policies both in the United States and around the world.[90] Many of these policies met with success. Some were plagued with difficulties. Still, whether individual policies flourished or failed, an examination of these case studies can teach us many lessons about what science-and-technology policies need in order to succeed. In her article, "Lessons From Global Experience in Policy for Science-Based Development," Susan U. Raymond, Director, Policy Programs at the New York Academy of Science, identifies six issues that surfaced in the analysis of these case studies. We will focus on four of these issues: (1) the need for a long-range perspective, (2) the importance of partnerships involving the government, business, and the academic community, (3) the significance of education and human resource development, and (4) the requirement of imaginative capital arrangements.[91]

The Need for a Long-Range Perspective. The first lesson that must be taken from any examination of these science policies is that such policies show a higher rate of success when they are based on a long-range perspective.[92] One reason for this long-range perspective is that an extensive period of time often passes between an initial scientific discovery and the successful application of that discovery. This is especially true when the research being funded is the type of pure research advocated by the Bush Report. Pure research, or basic research as it is sometimes called, focuses on an understanding of nature rather than on technological objectives. In contrast, applied research concentrates on the direct application of a scientific discovery and is aimed at satisfying an identified human need.[93] The Bush Report supported pure research because the committee believed it would advance technology, promote economic independence, and preserve cultural and artistic values.[94]

Several studies conducted over the last thirty years have found that the Bush Report may have been overly generous in tracing technological innovations to pure research. One inquiry conducted for the National Science Foundation

(NSF) by Gellman Associates has revealed that basic research in colleges and universities accounts for only 10 percent of most technological advances.[95] Another assessment, this one conducted by Edwin Mansfield, targeted 76 separate manufacturing companies and discovered that only 11 percent of all new products developed by those companies came directly from pure research. Perhaps more to the point, as noted, studies have shown that even when there is a direct and identifiable connection between pure research and technological accomplishments, that connection takes quite a long time to develop. One study, conducted by IIT Research for the NSF, found that although pure research does contribute to major technological advances, the lead time needed for such a contribution is generally from two to three decades.[96] In fairness to the members of the Bush committee it is necessary to point out that they were well aware of this type of long-term time lag. In fact, the committee recommended that federal money should be earmarked explicitly for projects that would cut down on the time that it takes to move from pure research to technological progress.[97]

If nothing else, these studies emphasize the need to focus on long-term goals when developing a science-and-technology policy for the government. Clearly, if it takes two to three decades for pure research to lead to technological progress, then to be effective any science-and-technology policy must have a long-term perspective. Naturally, there are problems with any attempt to develop a science-and-technology policy with a long-term perspective. One difficulty is that such a point of view is based upon delayed gratification which many people find difficult, if not impossible, to accept. This problem is especially acute in democracies in which governmental officials depend upon the good will of their constituencies for reelection. Such constituencies are frequently self-centered and notoriously impatient and, therefore, demand immediate results that help them directly, generally from a financial perspective. In response, government officials adopt a similar shortsighted view. Frequently, the longest "long-term" view possible for such officials is the view that focuses on the next election.[98]

One way to offset this shortsightedness is to build short-range goals into the long-range plan. The theory is that, if the electorate can see some short-range advantage and can, therefore, experience some immediate gratification, they will be more likely to support programs that also have long-range goals built into their systems. Such was the case in Pennsylvania when, under its Ben Franklin Programs, the government built into its long-term planning projects that produced immediate results in training and entrepreneurship. These short-range goals then allow time for the long-term projects to reach completion.[99]

Many examples exist to support the notion that long-range planning is possible, given the right set of circumstances and sufficient motivation. For example, long-range planning was a key element in the establishment of Hong Kong's University of Science and Technology. The university was founded as a science magnet designed to attract the cooperation of other Chinese scientific centers.

The establishment of the university was specifically aimed at promoting such cooperation when China regained jurisdiction over Hong Kong in 1997. The plan, which was initiated in 1986, called for the university to see its full potential by the turn of the century.[100] Similar success stories involving long-range planning come from Japan in the creation of the Tsukuba Science Center and Singapore in its support for the National University of Singapore. Both of these projects resulted from long-range planning that extends back twenty to thirty years.[101]

Partnerships of Government, Business, and the Academic Community. A second lesson that must be absorbed from any examination of existing science-and-technology policies is that such programs must be based upon extensive partnerships that link three crucial elements of the social structure: the government, the business world, and the academic community.[102] One reason for establishing this type of partnership is to take advantage of the unique talents and resources peculiar to each of these elements. Researchers at universities are usually engaged in pure research, whereas those connected with industry are often involved in projects designed to answer the needs of the marketplace. Experience has shown that university researchers who are exposed to the market concerns of corporate scientists also tend to adopt a practical attitude toward their research. This results in a closer connection between research and the marketplace.[103]

Another, perhaps more fundamental, reason to establish a partnership approach to the development of a science-and-technology policy is that such partnerships create a feeling of "ownership" among the participants for any programs and projects that result from their cooperative efforts. This creates a sense of trust between institutions that traditionally see themselves at best as competitors and at worst as adversaries. Additionally, this feeling of ownership fosters a sense of accountability among the participants. Rather than becoming self-focused and narrow-minded, each group within a partnership strives to make certain that its role in the project is carried out properly according to the goals, objectives, and procedures set down as part of the project's operational plan.[104]

Such was the experience of participants in Ohio's Technology Transfer Organization (OTTO) that was established in that state in the 1980s. The idea behind the project was to link up the vast resources of Ohio's colleges and universities with the needs of the state's businesses and industries. Governmental representatives, called OTTO agents, were installed at key academic institutions within the state. Each agent was responsible for a specified geographic region within the state. According to the system, businesses that had ideas for projects but did not have the technical or scientific talent to complete those projects could contact their local OTTO agent who would link that business with an academic institution that had the resources and the personnel to help in that project. In that way, the government served as a liaison between businesspeople, on the one hand, and academic scientists and engineers, on the other. As a result, all the members of an individual project had a stake in the success of that project and, thus, developed the sense of ownership so crucial to the success of any science-and-technology policy.

Education and Human Resource Development. A third principle upon which to build a successful science-and-technology policy is the need to focus resources on education and human resource development. Experience has shown that the development of human potential will not succeed if it is limited to one or two narrowly focused, one-time programs. Instead, human resource development must be an ongoing process that is broadly based and able to adapt to advancements made in science and technology as well as to changes that occur in the marketplace.[105] Accordingly, programs in human resource development cannot be limited to a single educational level but must target all levels, including primary education, secondary education, and post–secondary education—especially continuing education. Moreover, such programs must also actively seek to shape curriculum to include updated topics in science, engineering, and technology at all levels.[106] Additionally, such changes must be made as part of an ongoing program. It is absolutely necessary to make certain that curriculum changes are reviewed frequently and with sufficient depth to ensure that each educational level keeps pace with changes in science and technology.[107]

Another concrete step that can be taken to ensure the continued effectiveness of an educational system is a conscious effort to improve the status of educators in relation to tenure, salary, prestige, and academic stature. Japan, for example, which has some of the highest paid science, engineering, and technology instructors in the world, continues to be a world leader in cutting-edge technology.[108] In contrast, the experience of Brazil demonstrates what can happen when a country relies on imported technology rather than on improvements produced by domestic human resource–development programs. Between 1965 and 1985, Brazil experienced a phenomenal rate of technological and industrial growth. However, most of this progress depended upon imported technology rather than on technology developed by domestic scientific and engineering talent. As a result, Brazil's internal growth in science and technology has been limited by its ability to attract and absorb outside talent.[109]

Creative Capital Arrangements. A fourth and final principle upon which to build a successful science-and-technology policy is the need to be creative in the formulation of financing plans. Scientific and technological projects cost money and that money must come from somewhere. Consequently, an effective science-and-technology policy must not only address the allocation of capital but must also find original techniques for obtaining such capital and for sustaining an appropriate level of support. Lessons from history have taught us that even the most innovative and beneficial scientific and technological projects may go unexploited if they are not sufficiently supported by an appropriate financial network.[110]

Such was the case with the American space program before the Russians launched *Sputnik* in 1957. Thanks to German scientists like Wernher von Braun, the United States had the expertise and the personnel to begin an extensive space program in the late 1940s. However, the program was not thought important enough for the type of funding necessary for its success. Consequently, the space program was delayed until the launch of Sputnik changed the attitude of those American politicians who controlled the national purse strings. Consequently, shortly after the advent of *Sputnik I* and *II*, a new White House post of Special Assistant to the President for Science and Technology was established, the National Aeronautics and Space Administration was set up, and spending on science and technology rose so rapidly that by 1967, it comprised 11 percent of the federal budget.[111]

A similar experience occurred during the Second World War involving the development of the atomic bomb. In 1939, the Hungarian-American physicists Leo Szilard and Eugene Wigner persuaded Albert Einstein to write a letter to President Roosevelt suggesting that an atomic bomb might be technologically feasible.[112] Legend has it that this letter led directly to one of the most ambitious scientific projects in the history of the planet.[113] Although it is true that Einstein's letter did spark Roosevelt's interest and did eventually lead to the Manhattan Project and the invention of the atomic bomb, the road to that goal was not as smooth as it is usually presented. In fact, the initial result of Einstein's letter was the formation of a committee with a budget of only $6000.[114] Moreover, the committee gave little direction and encouragement to the project because its members focused quite narrowly on the industrial application of the uranium project, rather than its military employment.[115]

Between 1939 and 1941, the project moved very slowly for several reasons, not the least of which was the personal prejudice of the director of the Office of Scientific Research and Development, Vannevar Bush, who was reluctant to commit needed resources to the program because he believed that the development of such a bomb was only a distant possibility. Bush was further persuaded in this direction by his deputy for nuclear projects, the chemist James P. Conant, who was interested only in projects that would produce immediate results. So, while the technological and scientific know-how necessary to build the bomb was present at the time, the financial support was lacking.[116]

This state of affairs continued until Conant and Bush were convinced by arguments made by the American physicists Arthur Holly Compton and Ernest O. Lawrence. Compton and Lawrence, prompted by information from the British scientist Marcus Oliphant, reported that the Germans were making great progress on their own atomic bomb project and predicted that, if the Germans were the first to develop such a bomb, they could easily conquer the world.[117] Only then did Bush persuade Roosevelt to commit more money to the project. Even so it was not until 1942, when the Italian-American physicist Enrico Fermi directed the first controlled nuclear chain reaction, that the full financial support of the American government was lent to the project. In that year, Roosevelt authorized the spending of over $400 million on the project, a far cry from the $6000 that had been allocated only three years before.[118]

There are those who might argue convincingly that it might have been better had the United States not entered the space race and not developed the atomic bomb. Certainly, such arguments are worthy of our attention, and nothing written here is intended to support the death and destruction caused by nuclear weapons, to say nothing of the threat that they continue to pose to the world today. That, however, is not the point of the examples. Rather, they are presented to demonstrate what can happen when accidental historical events and individual personal prejudices are allowed to dictate financial support of science-and-technology programs. We cannot possibly predict all of the events that would have occurred had a more systematic approach been used to finance the space program or the Manhattan Project. It is safe to say, however, that both projects would have started much sooner and might possibly have led to earlier spin-offs in a number of related areas, such as medicine, manufacturing, and health care.[119]

Fortunately, there are numerous examples that can serve as models for creative capital arrangements. What all of these models have in common is a belief that chance and personal bias should not be the driving force behind the funding of scientific and technological projects. Instead, these plans take a broad overview, looking not only at the projects that require financial support but also at a variety of ways to raise the money necessary to fund those projects. For instance, the approach taken in the establishment of the African Foundation for Research and Development (AFRAND) is to seek financial support from a number of diverse sources. Included on this list of potential financial backers are private foundations, individual investors, and corporations. AFRAND has also looked at a number of other creative financing propositions, including the possibility of forgiving a country's debts in exchange for the transfer of science and technology from that country to a creditor nation. Another example of variety in capital financing involves a creative strategic plan implemented by Cambridge University, which created two foundations, the Cambridge Capital Development Fund and Cambridge Research and Innovation. Institutions that have contributed funds to these two foundations include Cambridge, nine additional centers of higher education, the local government, and a number of financial institutions. Both foundations can provide financial support for university projects and to local businesses involved in research.[120]

Nor is it necessary for those who develop science-and-technology policies to look only within their own borders. It is conceivable that one nation, or several institutions within a single nation, might see that supporting the scientific development of another nation or nations would be to their own best interests, as well as the best interests of the entire world. This global perspective appears to be the motivation behind the recent creation of a consortium of Swiss entrepreneurs, who have joined forces to create a foundation named FUNDES. The objective of this foundation is to provide financial assistance to businesses in Latin America that are dedicated to new technological advances. In addition to providing direct assistance to these Latin American businesses, FUNDES also guarantees loans granted by other creditors. FUNDES has been so successful that over 750 small businesses in Latin America have received loans for the support of research-and-development programs.[121] Admittedly, the global perspective represented by the Swiss consortium is rare, as most science-and-technology policies encourage only internal national development. Nevertheless, the example of FUNDES does provide encouragement as we look for a model for the creation of a truly global science and technology policy.

Practical Steps toward a Global Science-and-Technology Policy

A properly established global science-and-technology policy will require a type of strategic planning which reaches across national borders and creates a program that allocates resources and finances projects which can improve the condition of the less developed world without destroying the environment, depleting the world's natural wealth, or disrupting the stability of the more developed world. This is a tall order, but it is not beyond our capability as a global community, given sufficient motivation and adequate will power. The seeds of such a program were planted in 1992 at the United Nations Conference on Environment and Development in Rio de Janeiro. Out of this conference came Agenda 21 and the Commission on Sustainable Development (CSD). The commission is the central coordinator of all efforts by the United Nations aimed at implementing Agenda 21. As such, it provides both administrative support and political leadership in the area of sustainable development.[122] Unfortunately, the CSD lacks the type of authority needed to enforce recommendations and suggestions that form the basis of Agenda 21. Moreover, the CSD is limited to specific environmental issues. Consequently, instead of relying on the CSD, it might be advisable for the United Nations to amend its charter to add a new element to its structure that would address issues beyond Agenda 21 and that would have the authority to enforce scientific and technological mandates.[123]

The Structure and Goals of the Economic Security Council. The Commission on Global Governance, made up of representatives from twenty-six different countries, including two such representatives from the United States, has recommended that just such an organization be created. This organization, called the **Economic Security Council** (ESC), would have a variety of responsibilities. For instance, it would assess the state of the international economy and implement long-term strategic planning that would encourage the goal of sustainable development throughout the globe. This strategic planning would necessarily include, as a major component, the development of a well-balanced science-and-technology policy. The principle job of the ESC would be to examine political, economic, environmental, and social trends throughout the international community and guide the global community to successful solutions to its problems. The idea behind the ESC is not to replace already established agencies but to provide guidance for agencies and to unravel overlapping and contradictory policies that might hurt the overall development of the planet.[124]

The Economic Security Council would exist as an autonomous body within the hierarchy of the United Nations, much as the Security Council exists today. However, unlike the Security Council, the ESC would not provide short-term damage control for existing crises. Instead, it would promote long-range policy implementation.[125] The membership of the ESC would include the largest economic blocs within the world, whether those blocs consist of a single nation, such as the United States or Japan, or unified regions, such as the European Union or the Association of South-East Asian Nations. Additionally, the U.N. charter amendment creating the ESC would insure that representation on the ESC is balanced in relation to the more developed and the less developed areas of the globe. Moreover, the ESC would be supported by an administrative division that would provide logistical support and research capabilities.[126]

The ESC would not necessarily replace present U.N. agencies and commissions, such as the Economic and Social Council (ECOSOC), the U.N. Conference on Trade and Development, and regional commissions, like the Economic Commission for Europe and the Economic Commission for Africa. Rather, the ESC would help consolidate the economic activity of the U.N., which is currently spread out among these diverse agencies, councils, and commissions.[127] Moreover, the Commission on Global Governance believes that the ESC should exercise more global authority than present institutions such as ECOSOC, which is currently charged with the task of conducting research, managing studies, and issuing recommendations to the General Assembly.[128] Similarly, the ESC would have responsibilities and authority beyond those that presently reside within the U.N. Commission on Science and Technology for Development, which, as a subdivision of ECOSOC, serves primarily as an advisory group for creating guidelines and making recommendations on science-and-technology development to U.N. agencies and national governments.[129]

The Role of the Economic Security Council. One task of the Economic Security Council would be to oversee the use of science and technology for the benefit of all the nations of the world. As we have seen, from a practical point of view, science and technology can provide many of the solutions necessary for dealing with the problems of population growth, war, and environmental deterioration. Projects in alternative-energy sources, telecommunications, transportation, and nanotechnology can be directed toward eliminating much of the inequity between the more developed and the less developed regions of the world.[130]

However, the solutions to these problems depend on the ability of the less developed world to acquire and utilize these new technological developments. Consequently, the less developed world must have access to the fruits of these technological advancements. A major stumbling block, however, is that the bulk of the research and development done in areas such as telecommunications, genetic engineering, and nanotechnology is produced in the more developed world. In fact, recent figures indicate that the more developed world is responsible for 97 percent of all the work currently being undertaken in biotechnology, telecommunications, nanotechnology, and other areas of advanced research.[131]

According to the Commission on Global Governance, the job of planning and directing the transfer of technology from the more developed to the less developed world would fall within the jurisdiction of the Economic Security Council.[132] Using a long-range outlook that takes into consideration the needs of the less developed world and the economic stability of the more developed world, the ESC could construct a policy that would support partnerships among the government, academia, and the business world to promote human resource development and seek out capital arrangements that would enhance technology transfer at all levels. There are a number of creative ways that such policy planning could be arranged. Following the Swiss model, the ESC could encourage foreign financing of programs in the less developed world.[133] Or it could develop a policy of technology transfer based on the Ohio Technology Transfer Organization that would allow governments, universities, and businesses of the less developed world to utilize the talents of scientists and engineers of the more developed world. The ESC might even provide a system of matching funds to supplement governments that provide their own initiatives to enhance technology transfer from the more developed to the less developed world.

This is not to suggest that the United Nations is today unaware of such technology transfer arrangements. In fact, in 1995, the United Nations Conference on Trade and Development commissioned an extensive study of technology partnerships that currently cross international boundaries. The results of that study reveal that most technology partnerships link the nations of the more developed world with one another, rather than with the nations of the less developed world. Consequently, one of the roles of the ESC could be to promote the transfer of science and technology from the more developed to the less developed world.[134]

Challenges Facing the Economic Security Council. Naturally, there are drawbacks to the establishment of an Economic and Security Council. One principal drawback would be convincing member nations that the ESC cannot be a substitute for the development of internal science-and-technology policies. One objective of the ESC is to provide guidance and financial support for the transfer of technology from the more developed to the less developed world. As we have already noted, the idea behind the ESC is not to replace already established policies or to set up policies for nations that do not have such plans in place. Rather, the goal is to motivate individual nations to adopt a global perspective as they attack the dual problems of sustainable development and technology transfer. The ESC also seeks to protect the development of the entire planet by eliminating national policies that either overlap or contradict one another.[135]

Equally as troublesome as those nations who seek to surrender their autonomy to the Economic Security Council would be those nations at the other end of the spectrum that might balk at any attempt by an international agency to influence their internal policies. This resistance is to be expected from some nations, at least initially, because the principle of sovereignty has been the basis of international relationships throughout the history of the modern world. The idea of sovereignty is both an international principle and a state of mind. All nations are entitled to see themselves as equal partners in the global community. Moreover, the political independence and the territorial integrity of all sovereign nations cannot be violated either by other sovereign states or by the United Nations itself.[136]

Since there is no intention on the part of the Commission on Global Governance to disrupt the internal affairs of any nation, the members of the commission stop short of suggesting that the pronouncements, policies, and recommendations of the ESC be made legally binding.[137] Nevertheless, it is interesting to suggest that, unlike similar agencies within the current hierarchy of the U.N., the ESC could be given the authority to compel nations to enter technology transfer partnerships as a way of redistributing the benefits of science and technology on a global basis. This authority would make the ESC much more effective than its present day counterparts. On the other hand, the suggestion that such an international agency be given any type of supernational power would almost certainly provoke strong opposition from many political leaders.

If past events can tell us anything about how nations will behave under these circumstances, the prospects for a powerful international ESC may be very dim. History indicates that international organizations that even appear to threaten national sovereignty frequently face rejection. This is precisely what happened in 1946 when the United States placed a proposal before the United Nations promoting the international control of nuclear weapons. The plan, which was based on the Acheson-Lilienthal Report, a strategy that had been created by a committee under the leadership of Dean Acheson, acting secretary of state, and David Lilienthal, the head of the Tennessee Valley Authority, proposed the creation of the Atomic Development Agency (ADA), a supernational body under the control of the U.N. that would have the authority to control all aspects of nuclear power.[138]

To ensure that the ADA maintained tight control over all aspects of nuclear research, development, and production, the proposal, which was presented to the United Nations Atomic Energy Commission in 1946, included provisions for the mandatory inspection of all laboratories, mines, and factories that might be involved in any aspect of nuclear power. Moreover, under the plan, the ADA would have been given the authority to order rapid and unavoidable punishment for any nation that violated the strict prohibitions against atomic research and development outside the control of the agency itself. Significantly, the prohibitions outlined in the proposal extended not only to the military applications of nuclear power, but also to peaceful uses.[139]

Predictably, the primary objection to the proposed ADA was that it violated national sovereignty because the plan mandated inspection within national borders and because it authorized the punishment of any nation that developed nuclear power on its own. Opposition to the ADA was led by the Soviet Union, which condemned the ADA's inspection authority as a violation of national integrity and which branded the entire plan as an attempt by the United States to maintain its monopoly on nuclear weapons.[140] Whether or not there was any truth to the second objection appears to be irrelevant because it was the first argument that won the day. Although the plan for the ADA was approved by the United Nations Atomic Energy Commission, it was never acted upon by the entire General Assembly.[141]

Nor is this driving need to preserve national sovereignty the only stumbling block standing in the way of a global effort to direct science and technology in ways that will help all nations. We need only recall the uproar in the United States when the federal government negotiated the North American Free Trade Agreement (NAFTA) with Canada and Mexico to see that self-centered protectionism still guides the thinking of many people in the international arena. NAFTA is a trading partnership set up among the governments of the United States, Canada, and Mexico, the purpose of which is to establish a North American trading market that would not be subject to internal tariff barriers. As such, NAFTA was designed to insure that the three neighbor nations could use this free market equally. Ideally, each nation would benefit by exchanging those goods that it produces for goods it does not produce.[142]

Even though NAFTA was negotiated by the U. S. government and was not imposed on the United States by the United Nations or any other outside agency, and even though it was designed to benefit the marketplace in all three countries, the treaty still caused an enormous amount of opposition within the United States. From this experience, it is not difficult to predict the type of opposition that might result should an outside international agency, like the ESC, attempt to enlist the cooperation of the United States, or any other developed nation with a similar disposition, with a global science-and-technology policy. Certainly, such nations can present convincing arguments against science-and-technology policies in general and in particular against a global policy imposed by an international organization. Aside from arguments of sovereignty, critics point out that the pursuit of a science-and-technology policy is beyond the scope of any government. Such critics would limit government action to the preservation of order, the pursuit of justice, and the protection of the health and welfare of the people. The direction of science and technology, the argument goes, should be left to the unregulated operation of the marketplace.[143] The same argument can be levied against the type of quasigovernmental agency that the ESC would represent.

Despite these arguments, it is not entirely clear that a government's role in establishing order and protecting the people would not include the development of a science-and-technology policy. Certainly, as we have seen here, many governments do engage in such activities on a regular basis. Indeed, there is evidence that such national policies can meet with at least modest success. Such was the experience of the Canadian government when it formed the Canadian NeuroScience Network, an organization that established an effective partnership among academic, industrial, and financial institutions for work in advanced technology.[144] Moreover, similar national programs designed to establish such partnerships have met with success in the Netherlands and in Portugal.[145]

The argument that the development of such policies should not be left to quasigovernmental agencies like the Economic Security Council is more problematic. Nevertheless, it should be clear that neither the scope nor the purpose of the ESC is meant to suggest that individual nations must give up their sovereignty. Nor is it necessary for individual nations to surrender the right to make their own policies in such matters. Indeed, as we have already noted, nothing developed by the proposed ESC would replace or supplant any existing or proposed national science-and-technology policy. In fact, any U.N. charter amendment that might implement an agency like the ESC should specify that nations cannot, indeed, must not abdicate their own responsibilities in this regard. However, an agency like the ESC could provide what individual nations cannot—a globally coordinated science-and-technology policy that unites governments, academic institutions, and businesses in developing scientific and technological solutions to the world's greatest problems.

CONCLUSION

It appears that the global challenges outlined in the preceding chapter are manageable given appropriate motivation and sufficient willpower. Moreover, we have seen numerous examples that can serve as models for meeting those challenges. For instance, the African Foundation for Research and Development was established to seek financial support for scientific research from private foundations, individual investors, and corporations. Another example we examined was the strategic plan implemented by Cambridge University, which created two foundations, the Cambridge Capital Development Fund and Cambridge Research and Innovation. Both foundations support research and development projects conducted by the university and local businesses. We also saw how a consortium of Swiss entrepreneurs created a foundation named FUNDES, the objective of which is to provide financial assistance to businesses in Latin America that are dedicated to new technological advances.

Certain recent technological advances can be used to solve specific global problems. For example, advances in genetic engineering should help us develop a way to feed the starving people of the less developed world. Developments in telecommunications and advanced transportation methods can help us alleviate many of the problems associated with the densely packed urban regions of the planet. The motivation for wars of redistribution can be lessened by the invention of alternate sources of energy, by developments in genetic engineering, and by the growth of nanotechnology. Alternative energy sources, biotechnology, and nanotechnology may also help us win the campaign against environmental decay. Whether the human race will survive the challenges presented by population growth, nuclear war, and environmental contamination will, therefore, depend less upon our development of such technological marvels and more on our ability to harness those marvels, not for our individual, selfish enjoyment, but for the betterment, indeed the survival, of the entire planet.

Review

14-1. The most significant intersection between the scientific and the nonscientific world is the link between science and the global challenges of population, nuclear war, and environmental deterioration. All three challenges are directly related to the advances made possible by science and technology. On the other hand, science and technology by themselves are not responsible for these problems. Rather, it is the uneven way that science and technology have been allowed to develop that lies at the basic core of the present crisis. Two questions are considered: (1) From a procedural perspective, can the global challenges of overpopulation, nuclear war, and environmental deterioration be solved by science and technology? (2) From a sociopolitical perspective, can the people of Earth redirect scientific and technological activities toward those solutions?

14-2. Recent developments in the area of reproductive technology can help alleviate uncontrolled population growth. One of these developments is the introduction of contragestive agents, which are effective because they prevent the implantation of a fertilized egg on the uterine wall. The morning-after pill, intrauterine devices, and low-dosage birth control pills are all examples of contragestive agents. Contragestive agents are more economical than other methods of reproductive regulation. Perhaps as important, contragestives offend fewer people than other reproductive control techniques because they are relatively safe and because they operate very early in a pregnancy. Another recent development in birth control technology is the Norplant system of contraception. Figures on the effectiveness of the Norplant system vary from a 1.5 to a 5 percent possibility of pregnancy over the five-year period of its effectiveness. Another problem associated with overpopulation involves the densely packed urban areas of the world. By the end of the twentieth century, millions of workers had abandoned agriculture and moved into the cities causing an urban crisis characterized by overcrowding, unsanitary conditions, inadequate waste disposal, and infectious diseases. However, if business and industry could be decentralized, people could move out of the cities. This type of decentralization has been made feasible by improvements in electronic communication.

14-3. The practical ability of science and technology to achieve a redistribution of wealth could help make war obsolete as long as a properly controlled science-and-technology policy can be developed at a global level. One way to redistribute the wealth is to develop alternative sources of energy, such as solar power, cold fusion, wind power, geothermal energy, and antimatter. Issues related to the redistribution of the world's resources are not limited to inequities in energy reserves. Similar problems can be seen in the area of agricultural production. The use of genetic engineering in relation to crop production could lead to a green revolution that would allow us to reduce the effects of famine caused by the worldwide loss of agricultural acreage. Energy shortages and agricultural problems are only a small segment of a larger pattern of worldwide inequity. An unequal distribution of material wealth, including goods and services, among nations and between the classes within a single nation can also lead to global instability. Science and technology may be able to deal with this unequal distribution of income, goods, services, and housing through advances in nanotechnology. Nanotechnology could conceivably permit the construction of virtually any product by reconfiguring molecular structures atom by atom. This process would permit scientists to reconfigure the basic molecular structure of an element to produce new materials and products that can be provided to the less developed world.

14-4. The crisis caused by the deterioration of the environment is the most difficult challenge facing humanity today. As for the scientific practicality of correcting the damage already done to the environment, some of the technology needed to solve the problem is already in place, whereas other technological measures still in the developmental stage may one day become a reality. As for the probability that any of these tactics will be implemented in time, corrective measures are already underway. Evidence of this can be seen in the U.N. Conference on Environment and Development held in Brazil in 1992. The delegates to this conference, referred to as the Earth Summit, produced Agenda 21, a global plan designed to combat environmental deterioration. Two other documents produced at global conferences include the Berlin Mandate and the Kyoto Protocol.

14-5. The characteristics of a successful global science-and-technology policy involve (1) the need for a long-range perspective, (2) the importance of partnerships involving the government, business, and the academic community, (3) the significance of education and human resource development, and (4) the creation of innovative capital arrangements. A properly established global science-and-technology policy will require a type of strategic planning that reaches across national borders and creates a program that allocates resources and finances projects that can improve the condition of the less developed world without destroying the environment, depleting the world's natural wealth, or disrupting the stability of the more developed world. The Commission on Global Governance has recommended the creation of the Economic Security Council (ESC). This organization would have a variety of responsibilities. For instance, it would assess the state of the international economy and implement long-term strategic planning that would encourage the goal of sustainable development throughout the globe. This strategic planning would necessarily include, as a major component, the development of a well-balanced science-and-technology policy. The job of the ESC would be to examine political, economic, environmental, and social trends throughout the international community and guide the global community to a successful solution of its problems.

Understanding Key Terms

Agenda 21
disassemblers
Earth Summit
Economic Security Council

fiber optics
hectare
Internet
nanocomputer
nanotechnology

nitrogen fixation
replicator
superconductivity
universal assembler

Review Questions

1. How will recent developments in reproductive technology help deal with overpopulation?
2. How may advancements in communication alleviate the population problem?
3. How may advancements in transportation solve a part of the population problem?
4. How may alternative sources of energy address the problem of war?
5. How will the threat of war be lessened by advancements in genetic engineering?

6. How may nanotechnology help eliminate inequities in resource allocation?
7. How may science and technology deal with environmental problems?
8. What are some recent international efforts to solve environmental problems?
9. What are the characteristics of an effective science-and-technology policy?
10. What are some practical steps for the establishment of a global science and technology policy?

Discussion Questions

1. The text suggests that some people oppose government regulation of birth control because they fear that the government may become abusive and dictatorial in such matters. Do you agree or disagree with this assessment? Explain your response.
2. Several strategies for eliminating the unequal distribution of wealth between the less developed and the more developed regions of the world are identified in this chapter. These include the development of alternative sources of energy, the genetic engineering of crops, and the development of nanotechnology. Which of these technologies do you think has the best chance of success in this regard? Explain. Which of these technologies would you oppose and why? Explain.
3. Some political experts have argued that war can never be eliminated. Part of this attitude comes from the idea that war should not be eliminated because of the economic, political, and sociological disruption that would result from any attempt to dismantle the military-industrial complex. Following this line of argument, eliminating war would cause almost as much unemployment, poverty, and dislocation as war itself. What do you think of this argument? Do you agree or disagree? Why?
4. Would you be in favor of or opposed to the establishment of a U.N. Economic Security Council of the type described in this chapter? Explain.
5. How successful do you think the people of Earth will be in redirecting science and technology to solve the challenges of population, nuclear war, and environmental deterioration? Explain.

ANALYZING *STAR TREK*

Background

The following episode from *Star Trek: The Next Generation* reflects some of the issues that are presented in this chapter. The episode has been carefully chosen to represent several of the most interesting aspects of the chapter. When answering the questions at the end of the episode, you should express your opinions as clearly and openly as possible. You may also want to discuss your answers with others and compare and contrast those answers. Above all, you should be less concerned with the "right" answer and more with explaining your position as thoroughly as possible.

Viewing Assignment—
Star Trek: The Next Generation**, "Evolution"**

In this episode Dr. Paul Stubbs has joined the crew of the Enterprise in order to perform a deep-space stellar experiment. The experiment must be timed perfectly because the star under study erupts every 197 years due to the build-up of stellar matter on its surface. As the crew prepares to launch a probe into the star, the computer system on board the Enterprise begins to collapse. As precious moments slip by, the computer mishaps increase, threatening Dr. Stubbs's experiment. Eventually, Wesley Crusher determines that the problem has been caused by one of his experiments. Wesley had been testing a new type of nanodevice of his own invention. He calls these nanodevices Nanites. The Nanites are supposed to operate as repair units that can work on the cells of organic beings. Wesley has improved their operation in two ways. First, he has taught the Nanites to work together. Second, he has allowed them to reproduce and to pass certain improvements on to their offspring. Unfortunately, in a moment of carelessness, Wesley fails to contain the Nanites properly. Consequently, they have escaped from the controlled environment of his laboratory. As a result they have entered the memory core of the ship's computer and have begun to feed on the computer's memory chips, causing massive systems failures on board the Enterprise. At first, Wesley tries to stop the Nanites on his own by setting up a series of traps around the ship. When this effort proves futile, he confesses his error and enlists the aid of LaForge and Data in his attempt to rid the computer of the Nanites. Dr. Stubbs takes it upon himself to stop the Nanites by bathing them in gamma radiation. The Nanites retaliate by flooding the bridge of the Enterprise with poisonous gas. This action convinces Captain Picard that the Nanites have developed intelligence. Consequently, he does not consent to having destroyed them. However, he does allow them to take over Data so that he can converse with them directly. Stubbs apologizes for attempting to annihilate the Nanites. The Nanites agree to help fix the damage to the computer in time for Stubbs's experiment, which is then performed as scheduled.

Thoughts, Notions, and Speculations

1. The Nanites are not only capable of reproduction, but can also pass on selected improvements or mutations to their offspring. In view of this ability, what type of nanodevice do they constitute? Explain your answer.
2. Wesley's Nanites escape from his laboratory and attack the memory core of the ship's computer system. Since the Nanites are designed to repair organic beings, explain why such an attack on the computer would be unlikely. Now identify a more likely victim or set of victims for the Nanites' dangerous feeding frenzy. Explain.
3. Based on what you learned in this chapter, speculate on what precautions Wesley should have taken to prevent the escape of his Nanites.
4. Imagine that you are a Starfleet officer and that you have the responsibility of investigating the escape of the Nanites. Would you choose

to prosecute Wesley for negligence? Why or why not? Now imagine that Wesley will be prosecuted, and that you are his defense counsel. How would you defend his actions? Explain.

5. Explain how this attack on the ship's computer could be used as a valid argument against the growth of nanotechnology? Could the same argument be used to oppose progress in genetic engineering? Could it be used against the invention of advanced methods of birth control? against the creation of alternative energy sources? against the development of an advanced fiber-optics telecommunications network? Explain each response.

NOTES

1. Delmore Schwartz, "The True-Blue American," 163.
2. Robert L. Heilbroner, *An Inquiry into the Human Prospect*, 57.
3. Paul Kennedy, *Preparing for the Twenty-First Century*, 24.
4. Ronald Dworkin, *Life's Dominion*, 177–78; John A. Robertson, *Children of Choice*, 63–64.
5. Carol Levine, "Should the Courts Be Permitted to Order Women to Use Long-Acting Contraceptives," 2–3; Robertson, 69.
6. United Nations Department for Economic and Social Information Analysis, *Concise Report on the World Population Situation in 1995*, 24–25.
7. Malcolm Potts, "Sex and Birth Rate," 14–15.
8. Ibid.
9. Levine, 3; See also Jim Persels, "The Norplant Condition," 4–13; For the opposing view see an article by the American Medical Association also in *Taking Sides: Clashing Views on Controversial Bioethical Issues*, ed. Carol Levine, 14–19. Another opposing view is offered in Rebecca Dresser, "Long-Term Contraceptives in the Criminal Justice System," 146–51.
10. Robertson, 84–89, 91–92.
11. Ibid., 25.
12. United Nations, *Concise Report*, 25.
13. K. Eric Drexler, Chris Peterson, and Gayle Pergamit, *Unbounding the Future*, 235; Richard L. Greaves, et al., *Civilizations of the World*, 796.
14. Drexler, Peterson, and Pergamit, 235.
15. Heilbroner, 37; Kennedy, 27.
16. Gary Gardner, "Preserving Global Cropland," 45–46.
17. Ibid., 46.
18. Ibid.
19. Drexler, Peterson, and Pergamit, 235.
20. Gregory Stock, *Metaman*, 200. For details on the Y2K situation see: Chris Clarke, "The Year 2000 Problem: An Environmental Impact Report," John Peterson, Margaret Whetley, and Myron Kellnez-Rogers, "The Y2K Problem: Social Chaos or Social Transformation?" William G. Phillips, "The Year 2000 Problem: Will the Bug Bite Back?" Richard Ravin, "Y2K and ADR: Get Ready for Midnight," and John L. Reed and Richard K. Herrmann, "An Alternative to the Not-Ready-for-the-Year 2000 Court System."
21. James B. Edwards, *The Great Technology Race*, 11.
22. Stock, 200.
23. Ibid., 130.
24. Ibid., 143–44, 308.
25. Ibid., 308.
26. Ibid., 143–44, 308.
27. Ibid., 308–309.
28. Ibid., 200.
29. Ibid., 198.
30. Ibid., 200–201.
31. Tom Tennant, "Technology Brings People Together," *Mount Vernon News*, 7 August 1997, sec. A, 4; Tom Tennant, "Resident Amazed by Internet Chatting," *Mount Vernon News*, 7 August 1997, sec. A, 4.
32. Patricia L. Mokhtarian, "Now That Travel Can Be Virtual, Will Congestion Virtually Disappear?" 93.
33. Ibid.
34. Ibid.
35. Arthur C. Clarke, "The Hammer of God," 84.
36. Gary Gardner, "Preserving Agricultural Resources," 80–81.
37. Gardner, "Preserving Global Cropland," 46.

38. Enzo Russo and David Cove, *Genetic Engineering*, 65.
39. Robert M. Hazen and James Trefil, *Science Matters*, 239.
40. Hazen and Trefil, 239; James Trefil and Robert Hazen, *The Sciences: An Integrated Approach*, 551–52.
41. Trefil and Hazen, 552; Russo and Cove, 74.
42. Russo and Cove, 82–83.
43. Alan Isaacs, John Daintith, and Elizabeth Martin, *Concise Science Dictionary*, 494; Russo and Cove, 82.
44. Russo and Cove, 82.
45. Edwards, 10–11.
46. Russo and Cove, 84–85.
47. Holmes Rolston, III, "People, Population, and Place," 36; Kennedy, 23.
48. Kennedy, 23; Rolston, 36.
49. Michael Renner, "Transforming Security," 121.
50. K. Eric Drexler, *Engines of Creation*, 14.
51. Hazen and Trefil, 239; Trefil and Hazen, 551–52.
52. Drexler, 8–14.
53. Edwards, 45–49.
54. Renner, 115–16.
55. Drexler, 11.
56. Ibid., 14.
57. Drexler, Peterson, and Pergamit, 228–29.
58. Drexler, 19–20.
59. Drexler, Peterson, and Pergamit, 120–30.
60. Gary Stix, "Trends in Nanotechnolgy," 94–97.
61. Ibid., 97–98.
62. Ibid., 98.
63. Ibid.
64. Ibid.
65. Drexler, 288.
66. Drexler, Peterson, and Pergamit, 253.
67. Drexler, 172–73.
68. Ibid.
69. Drexler, 173–74; Drexler, Peterson, Pergamit, 253–55.
70. Robert Borosage, "No Justice, No Peace," 10.
71. Leonard C. Lewin, *Report from Iron Mountain*, 150.
72. Ibid., 94–95.
73. Edward Bernard Glick, *Peaceful Conflict*, 45–66.
74. Commission on Global Governance, *Our Global Neighborhood*, 87.
75. Heilbroner, 47.
76. William K. Hartmann and Ron Miller, *The History of Earth*, 224.
77. Hartmann and Miller, 224; Edwards, 30.
78. Marvin Cetron and Owen Davies, *Probable Tomorrows*, 154–57.
79. Heilbroner, 76.
80. Christopher Flavin, "The Legacy of Rio," 3–5.
81. Christopher Flavin, "Facing Up to the Risks of Climate Change," 21, 36.
82. Ibid., 35.
83. Ibid., 36.
84. Thomas Richichi, "Environmental Law," sec. B, 4; William K. Stevens, "Tentative Accord Reached to Cut Greenhouse Gases," *New York Times*, 11 December 1997, sec A, 1; Traci Watson, "Global Warming Pact OK'd: 160 Nations at Conference Approve Emission Limits," *USA Today*, 11 December 1997, sec. A, 1.
85. Claude E. Barfield, "Introduction and Overview," 3–4; Daniel J. Kevles, "The Changed Partnership," 42.
86. Barfield, 3–4.
87. Ibid., 4.
88. Barfield, 3–4; David C. Mowery, "The Bush Report after Fifty Years—Blueprint or Relic?" 28–29.
89. Barfield, 3–4.
90. Susan U. Raymond, ed., *Science Based Economic Development*.
91. Susan U. Raymond, "Lessons from Global Experience in Policy for Science-Based Economic Development," 179–80.
92. Raymond, "Lessons from Global Experience," 181.
93. Richard Nelson, "Why the Bush Report Has Hindered an Effective Civilian Technology Policy," 43.
94. Barfield, 7; Mowery, 26.
95. William A. Niskanen, "R&D and Economic Growth—Cautionary Thoughts," 85.
96. Ibid.
97. Mowery, 34–36; See also note 19, page 98 in *Science for the 21st Century: The Bush Report Revisited*, ed. Claude E. Barfield.
98. Raymond, "Lessons from Global Experience," 182.
99. Walter H. Plosila and Susan U. Raymond, "Policies for Science-Based Development," 149.
100. Raymond, "Lessons from Global Experience," 181.
101. Ibid.
102. Plosila and Raymond, 149; Raymond, "Lessons from Global Experience," 182.
103. Plosila and Raymond, 149; Raymond, "Lessons from Global Experience," 182.
104. Plosila and Raymond, 149.
105. Raymond, "Lessons from Global Experience," 191.
106. Titus Adeboye, "Innovation without Science Policy," 201–2.

107. Raymond, "Lessons from Global Experience," 192.
108. Adeboye, 202.
109. Raymond, "Lessons from Global Experience," 192.
110. Ibid., 192–93.
111. Arthur C. Clarke, *The Exploration of Space*, 189–91; Kevles, "The Changed Partnership," 6. For a detailed look at the early years of the American space program see William E. Burrows, *This New Ocean: The Story of the First Space Age*; Kenneth W. Gatland, ed., *Project Satellite;* and Howard E. McCurdy, *Space and the American Imagination*. For a detailed look at the years of the Russian Space Program, see James Harford, *Korolev*.
112. Daniel J. Kevles, *The Physicists*, 324; Alexander Hellemans and Bryan Bunch, *The Timetables of Science*, 480. For a more detailed look at the history of the Manhattan Project see Los Alamos Scientific Laboratory, *Los Alamos*; Linda K. Woods, "The Men and Mission of the Manhattan Project," 39–45; Richard Rhodes, *The Making of the Atomic Bomb*; and Peter Wyden, *Day One*.
113. Hellemans and Bunch, 480.
114. Kevles, 324; Wyden 38–40.
115. Kevles, 324.
116. Ibid. 324–26; Wyden, 42–46.
117. Wyden, 42–46.
118. Kevles, 324–26.
119. For a creative look at what might have happened had the American space program begun in the 1940s see Allen Steele, *The Tranquility Alternative*. Similarly, for a look at what might have happened if the Nazis had the atomic bomb in 1942 and if America had delayed even further in its atomic bomb program, see James P. Hogan, *The Proteus Operation*.
120. Raymond, "Lessons from Global Experience," 193.
121. Ibid., 193–94.
122. Commission on Global Governance, 209–10.
123. Ibid., 160–61, 209–10, 215–16.
124. Ibid., 155–58.
125. Ibid., 158–59.
126. Ibid., 159–60.
127. Ibid., 162.
128. Commission on Global Governance, 162; New Zealand Ministry of Foreign Affairs and Trade, *United Nations Handbook 1997*, 77.
129. Commission on Science and Technology for Development, *Report on the Third Session*, 23.
130. Commission on Global Governance, 203–4.
131. Ibid.
132. Ibid., 204.
133. Raymond, "Lessons from Global Experience," 193–94.
134. United Nations Conference on Trade and Development, *Emerging Forms of Technological Cooperation: The Case for Technology Partnership*, xii-xiv.
135. Commission on Global Governance, 155–58.
136. Ibid.
137. Ibid., 155.
138. Richard Rhodes, *Dark Sun*, 229; John K. Jessup, "The Atomic Stakes," 77.
139. Margaret L. Coit, *Mr. Baruch*, 582–85; Herbert Feis, *From Trust to Terror*, 140–42; Rhodes, 229; Jessup, 77–78.
140. Coit, 569; Jessup, 78.
141. Feis, 153–54; Rhodes, 223. Eventually, an independent international agency under the sponsorship of the United Nations was established eleven years later, when the International Atomic Energy Agency (IAEA) was created in 1957. The IAEA, however, does not have the type of extensive regulatory power that the ADA would have had. Instead, the IAEA encourages the peaceful use of nuclear energy and recommends safety standards for the nuclear power industry. The IAEA is also responsible for promoting the transfer of scientific data among nations, supporting the education and exchange of nuclear scientists, assisting in nuclear research for peaceful objectives, and providing support for the exchange of material, equipment, and services. See New Zealand Ministry of Foreign Affairs and Trade, 299. See also "Then and Now: The IAEA Turns Forty," 672–73.
142. Gordon W. Brown and Paul A. Sukys, *Business Law with UCC Applications*, 672–73.
143. Susan U. Raymond, "Listening to Critics," 302–13.
144. Geraldine A. Kenney-Wallace and Warren C. Bull, "Partnerships for Science-Based Development," 302–13.
145. John J. Desmond, "Partnerships between Government, Industry, and Universities," 242–57.

Appendix I
Time Line of Scientific Thought

560 B.C. Pythagoras, the Greek mathematician and astronomer responsible for developing the science of acoustics and the theorem on right triangles, is born on the island of Samos.

500 B.C. Among the pre-Socratic philosophers are Parmenides of Eleas and Heraclitus of Ephesus, who represent opposing schools of thought on the underlying nature of the universe. Parmenides proposes the idea that a changeless unity dominates the universe. Heraclitus argues that the universe is in a constant state of change.

440 B.C. Democritus, a Greek philosopher, proposes the idea that matter is made of tiny indestructible particles. These particles are termed atoms, from the Greek word for "unbreakable" or "indivisible."

384 B.C. Aristotle, perhaps the most influential of all Greek philosophers, is born in Athens. Aristotle's views on physics, biology, and medicine are seen as absolute truth throughout the Middle Ages. His authority is broken only with the advent of the Scientific Revolution in the sixteenth century.

300 B.C.	Euclid, a Greek mathematician working in Egypt, writes *Elements of Geometry*, a textbook that will remain one of the primary authorities on mathematics for 2000 years.
287 B.C.	Archimedes, the Greek mathematician and inventor responsible for establishing the law of specific gravity and for developing an understanding of the applied mathematics of the lever, is born.
240 B.C.	Comet Halley is observed for the first time by astronomers in China.
60 B.C.	Like Democritus before him, Lucretius, a Roman philosopher and poet, speculates that atoms form the basic building blocks of matter.
A.D. 140	Alexandrian astronomer Ptolemy proposes a cosmology based on the premise that the Earth is the center of the universe.
160	Vitalism, which speculates that a force animates matter to create life, is proposed by Galen, a Greek physician.
500	Algebra and the decimal system are invented in India.
642	Alexandria is conquered by the Arabs. Greek scientific knowledge is preserved by the Arabs.
815	In Baghdad, a science library, known as Bayt al-Hikma, is established.
1075	Arzachel, an Arab astronomer, suggests that planetary orbits may be elliptical, rather than circular.
1543	Polish astronomer and priest Nicolaus Copernicus publishes *On the Revolution of the Heavenly Spheres*. In this work Copernicus suggests that the Sun is the center of the planetary system. This also signals the beginning of the Scientific Revolution, which runs until about 1700.
1574	Tycho Brahe's observatory is established.
1600	Johannes Kepler begins to work with Brahe.
1610	Galileo writes *The Starry Messenger* in which he offers certain observations made with a telescope as proof of the Copernican planetary system.
1616	The Copernican system is declared false by the Catholic church. *On the Revolution of the Heavenly Spheres* by Copernicus is placed on the index of forbidden books.
1628	William Harvey, a British physician, publishes *On the Motions of the Heart and Blood in Animals*, in which he explains the principles of circulation.
1632	Galileo's *Dialogue on the Two Great World Systems* is published. Although the work purports to be unbiased, it, nevertheless, portrays the anti-Copernican philosopher as a simpleton.

1633	Galileo is called before the Inquisition.
1637	Descartes publishes *Discourse on Method*, in which he supports the deductive method as a scientific procedure.
1660	The Royal Society of London is established. The Academie Royale des Sciences is established in Paris.
1665	Newton discovers calculus. In 1666, he writes a pamphlet now known as the *October 1666 Tract*, in which he discusses this discovery. Gottfried Wilhelm Leibniz also discovers calculus independently of Newton, but after 1665.
1687	*Philosophiae naturalis principia mathematica* is published by Sir Isaac Newton. His three laws of motion and the law of gravitation are included in this work.
1690	The speed of light is measured by Ole Romer.
1705	Edmund Halley suggests that a comet (now known as Comet Halley) which had appeared in 1682 will return 76 years later, in 1758.
1755	Immanuel Kant, a German philosopher, proposes the idea that distant "island universes," which are actually enormous groups of stars, may exist in deep space.
1758	The 1682 comet returns as predicted by Halley.
1798	Thomas Robert Malthus, a British economist, publishes his *Essay on the Principle of Population*. This essay later becomes the inspiration for Charles Darwin's theory of evolution based on natural selection.
1801	Karl Friedrich Gauss, a German mathematician, publishes *Researches into Arithmetic*, which signals the birth of modern number theory.
1808	*New System of Chemical Philosophy* is published by John Dalton.
1811	The complete skeleton of an Ichthyosaur is discovered by Mary Anning in Britain.
1812	Pierre Simon de Laplace publishes *Théorie analytique des probabilités* (*Analytic Probability Theory*.) In this work Laplace states that if a sufficiently great intelligence could know the mass, position, and velocity of every particle within the universe, that intelligence could predict all future events.
1831	The HMS *Beagle* sails on a five-year journey, the object of which is to chart the coast of South America. Charles Darwin is on board as the ship's naturalist. The voyage gives Darwin the chance to collect biological and botanical specimens from a variety of different locations.
1832	Michael Faraday, a British physicist, develops the basic laws of electrolysis.

1842	Christian Johann Doppler discovers a shift in frequency of sound waves as the source of the waves moves toward or away from a listener. The phenomenon has been called the Doppler effect.
1846	At the Royal Institution Faraday delivers a lecture on ray vibrations, which James Clerk Maxwell later points to as the inspiration for his electromagnetic theory.
1856	The first known skull of Neanderthal man is discovered in Dusseldorf by Johann C. Fuhrott.
1859	*The Origin of Species* is published by Charles Darwin. In this work Darwin proposes his theory of evolution by natural selection.
1865	*A Dynamical Theory of the Electrical Field* is published by James Clerk Maxwell. In this work Maxwell presents a set of equations that describes electromagnetism. Maxwell also proposes that, since the speed of light and electromagnetic radiation are virtually the same, light and electromagnetism are unified.
1866	The experiments of the Austrian botanist and monk Gregor Mendel establish the laws of heredity.
1869	A periodic table of elements is published by Dimitry Ivanovich Mendeleyev, a Russian chemist.
1871	*The Descent of Man* is published by Charles Darwin.
1873	At a site in Turkey, Heinrich Schliemann discovers the ruins of the ancient city of Troy.
1876	Cathode rays are discovered by Eugen Goldstein, a German physicist.
1879	Albert Einstein is born in Germany. James Clerk Maxwell dies in Cambridge, England.
1887	Albert Michelson and Edward Morley, at the Case School of Applied Science in Cleveland, perform an experiment that places the existence of ether in doubt. The photoelectric effect is discovered by Heinrich Hertz, a German physicist.
1897	The electron is discovered by J. J. Thomson. Marie Curie, a Polish physicist and chemist, identifies the uranium atom as the source of radiation, which was discovered by the French physicist, Antoine Becquerel in 1896. Massive, positively charged alpha rays and lighter, negatively charged beta rays are discovered by Ernest Rutherford, an English physicist.
1900	Max Planck, a German physicist, suggests that energy may be emitted only at certain levels which he calls "quanta." This gives birth to the quantum theory of physics. Also, Sigmund Freud publishes his seminal work, *The Interpretation of Dreams*.
1905	In *Annalen des Physik* Albert Einstein publishes a paper entitled "On the Electrodynamics of Moving Bodies," introducing the world to the special theory of relativity.

1907	Hermann Minkowski, a Russian-born German mathematician, proposes time as the fourth dimension of a four-dimensional universe. Minkowski's model is later used by Einstein in the development of the general theory of relativity.
1909	Burgess shale fossils are discovered in the Canadian Rockies by Charles Doolittle Walcott, an American paleontologist.
1910	*Principia Mathematica* is begun by English mathematicians and philosophers Bertrand Russell and Alfred North Whitehead.
1911	Ernest Rutherford proposes the idea that the atom is constructed like a miniature solar system. Also, the cloud chamber is invented by Charles Thomson Rees Williams, a Scottish physicist.
1913	Niels Bohr, a Danish physicist, suggests a solution to the weaknesses in Rutherford's solar system model of the atom. Bohr argues that electrons inhabit fixed orbitals and move from one orbital to another with the emission or the absorption of energy. Also, Sigmund Freud attempts to explain modern neuroses by analyzing the activities of primitive groups in his work *Totem and Taboo*.
1914	The proton is discovered by Ernest Rutherford.
1916	Einstein finalizes his general theory of relativity.
1919	Sir Arthur Eddington and Andrew Crommelin lead expeditions that reportedly confirm Einstein's theory of relativity.
1921	The Institute of Theoretical Physics, directed by Niels Bohr, opens in Copenhagen.
1922	The University of Göttingen installs Amalie Noether, a German mathematician, as the first woman member of the faculty.
1924	Louis de Broglie, a French physicist, proposes a wave theory of matter in his doctoral dissertation. Also, a fossil skull of *Australopithecus africanus* is discovered by Raymond Arthur Dart.
1925	Werner Heisenberg, a German physicist, creates the matrix mechanics technique for studying quantum physics.
1926	Erwin Schrödinger, a German physicist, applies de Broglie's theory that electrons can act as waves and develops wave mechanics. Also, the concept of the wave packet is proposed by German physicist Max Born. This concept is used to describe the probalistic aspect of electron orbits.
1927	The uncertainty principle of quantum physics is first articulated by Werner Heisenberg. The idea of complementarity in relation to quantum phenomena is developed by Niels Bohr. In October, the Solvay conference is held in Brussels. It is at this conference that an in-depth exploration of the Born-Heisenberg-Bohr interpretation of quantum theory takes place. This event also signals the beginning of the Einstein-Bohr debate.

1929	Edwin Hubble, an American astronomer, formulates what is now called Hubble's Law, which includes the Hubble constant that has been used to measure both the size and the age of the universe. Hubble's observations also verify the expanding-universe theory suggested earlier by Georges Lemaitre and Alexander Friedmann.
1930	The existence of the antielectron is predicted by Paul Dirac, a British physicist; its existence is demonstrated in 1932 by Carl Anderson, an American physicist.
1931	The incompleteness theorems are developed by Kurt Gödel, an American mathematician, born in Czechoslovakia. The incompleteness theorems demonstrate that any set of axioms used to establish a sufficiently complicated system of thought will contain statements that are true but unprovable within that system.
1936	*The Origin of Life on Earth* is published by Alexander Ivanovich Oparin, a Soviet biochemist. In this book, Oparin suggests that life on Earth originated in a "primeval soup" of organic compounds that resulted from the interaction of energy and the primitive atmosphere of methane, ammonia, and hydrogen.
1939	In response to a request from Leo Szilard, Edward Teller, and Eugene Paul Wigner, Albert Einstein signs a letter to Franklin Roosevelt in an attempt to persuade him to support the development of an atomic bomb.
1942	The first sustained nuclear chain reaction occurs on a converted squash court at the University of Chicago. The project is run under the direction of Enrico Fermi, an Italian-born, American physicist.
1945	At Alamogordo, New Mexico, the first atomic bomb is successfully tested on July 16.
1948	The steady-state theory of the universe is proposed by Hermann Bondi and Thomas Gold, two Austrian astronomers. The British astronomer Fred Hoyle becomes the most vocal supporter of this theory.
1949	The existence of background radiation as a remnant of the original big bang is predicted by the physicist George Gamow.
1950	The existence of a comet cloud beyond the orbit of Pluto is proposed by Jan Hendrik Oort, a Dutch astronomer. Also, the Turing test for determining the existence of artificial sentience is suggested by Alan Turing, a British mathematician.
1953	American chemists Stanley Miller and Harold Urey demonstrate that complex organic molecules such as amino acids can be formed from the interaction of electricity with ammonia, methane, hydrogen, and water vapor, thus lending support to Oparin's "primeval soup" theory. Also, the double-helix structure of DNA is discovered by Francis Crick and James Watson at Cambridge.

1957	On October 1, *Sputnik I*, the first artificial satellite, is launched into orbit around the Earth by the Soviet Union. The International Geophysical Year (IGY) begins on July 1, 1957. During the eighteen months that encompass the IGY, scientists from 64 nations coordinate their efforts in an attempt to learn as much as possible about the Earth, the Moon, and the Sun. Studies focus on climate, weather, the oceans, the atmosphere, sunspots, solar flares, cosmic rays, and the planned launch of artificial satellites.
1958	On January 31, *Explorer I*, the first American artificial satellite is launched into orbit by the United States. The satellite discovers the Van Allen Radiation Belt.
1959	Fossils now classified as either *Australopithecus robustus* or *Australopithecus boisei* are discovered in Tanzania by Mary and Louis Leakey, British anthropologists.
1964	Robert Wilson, an American astronomer, and Arno Penzias, a German-American physicist, discover the cosmic background radiation left over from the big bang. The existence of a spin-zero particle (the Higgs boson) with a mass of nonzero is predicted by Peter Higgs, a British physicist. Also, John S. Bell, a British physicist, formulates what is now called Bell's theorem, to determine a solution to the Bohr-Einstein debate on quantum physics.
1968	The electroweak theory is proposed by the American physicists, Steven Weinber and Sheldon Lee Glashow, along with Abdus Salam, a Pakistani physicist.
1969	The term "black hole" for a collapsed star five times as massive as the Sun is coined by John Archibald Wheeler. Also, on July 20 the American astronaut Neil Armstrong becomes the first human to set foot on the Moon.
1973	Proton decay is predicted as part of the grand unified theory proposed by Abdus Salam.
1975	At Asilomar, California, 150 molecular biologists meet to discuss critical guidelines for conducting genetic experimentation.
1977	The inflationary epoch of the early universe is proposed by Alan Guth, an American astrophysicist. Also, deep-sea ocean vents are discovered near the Galapagos Islands.
1979	Luis Walter Alvarez, an American physicist, proposes his theory that dinosaurs became extinct 65 million years ago because the Earth was struck by an asteroid or a comet.
1982	Alaine Aspect, a French physicist, conducts an experiment that demonstrates the existence of nonlocal influences within quantum systems.

1989	The largest structure yet discovered in the universe, the Great Wall, is found by astrophysicists. The Great Wall, which consists of an enormous sheet of thousands of galaxies, appears to be 15 million light years deep and 500 million light years in length.
1989	Stephen Jay Gould, a Harvard biologist, reinterprets Walcott's classification of the Burgess shale fossils to propose a new theory of evolution based upon a pattern of periodic extinctions.
1990	The existence of the Great Attractor, which consists of two dense superclusters of galaxies 300 million light years long, is confirmed.
1992	Small temperature variations in the cosmic background radiation are uncovered by the Cosmic Background Explorer. This discovery lends support to the big bang theory of the origin of the universe.
1994	Jupiter is impacted by the fragments of Comet Shoemaker-Levy 9. Also, the fossilized bones of an apelike hominid, *Australopithecus ramidus*, are discovered in Ethiopia.
1996	NASA research scientists uncover possible evidence of fossilized bacterialike organisms in a 4.5 billion-year-old meteorite discovered in Antarctica in 1985. The meteorite is believed to have originated on Mars.
1997	Ian Wilmut, a British embryologist at the Roslin Institute in Edinburgh, successfully clones a sheep using DNA from an adult animal. Also, the robot probe *Galileo* sends back images of Jupiter's moon, Europa, and NASA's *Pathfinder* lands on the surface of Mars.
1998	Scientists from twenty-three research centers working together in Japan discover that neutrinos have mass. Also in 1998 astronomers discover that the intense radiation produced by quasars may result from galactic collisions.
1999	The American Physical Society marks its 100th anniversary.
2000	The Human Genome Project is scheduled to have completed mapping 90 percent of the human genomes.
2004	Cassini, the largest deep-space craft sent aloft by NASA is scheduled to arrive at Saturn.
2005	The International Space Station is scheduled for completion sometime in 2005.

Appendix 2
Brief Biographies of Scientists Discussed

Hannes Alfven (1908–95) A Swedish physicist known principally for his work in plasma physics, Alfven proposed that the interaction of plasma and electromagnetism is responsible for the creation of the solar system, the galaxy, and the entire universe. The Nobel Prize was awarded to Alfven in 1970.

Carl Anderson (1905–91) An American physicist, he is known primarily for his discovery of the antielectron (positron) in 1932. While photographing cosmic ray tracks in a cloud chamber, Anderson noted that some of the tracks followed a pattern that confirmed Paul Dirac's 1928 prediction of the existence of the antielectron. In 1936, Anderson was awarded the Nobel Prize in physics for this discovery. He shared the prize that year with Victor Francis Hess, the discoverer of cosmic rays.

Aristotle (384–322 B.C.) A Greek philosopher who explored virtually every area of knowledge, his works on physics, metaphysics, biology, ethics, logic, and politics became the standard by which later scholarly works were judged well into the Middle Ages and beyond. In his work entitled *Physics*, Aristotle explored a wide variety of topics, including time, chance, change, infinity, and growth. He also explored the existence of the Prime Mover. His authority was broken only with the advent of the Scientific Revolution in the sixteenth century.

Niels Bohr (1885–1962) A Danish physicist who was one of the key figures in the development of quantum physics, he solved the instability problems of the Rutherford solar system model of the atom by suggesting the quantized nature of electron orbitals. He is also responsible for the complementarity principle of quantum physics. In 1921, the Institute of Theoretical Physics, which Bohr directed, opened in Copenhagen. In 1927, at the first Solvay conference in Brussels the famous Einstein-Bohr debate on the nature of quantum reality began. Einstein and Bohr continued this debate until Einstein's death in 1954. Bohr's position came to be known as the Copenhagen Interpretation because of his association with the Copenhagen Institute. He received the Nobel Prize for physics in 1922.

Hermann Bondi (1919–) An Austrian astronomer and physicist who is most well-known for his support of the steady-state theory of the universe, Bondi, along with Thomas Gold, another Austrian astronomer, and Fred Hoyle, a British astronomer, claim that the universe is eternal and that matter is continuously created to maintain a uniform density throughout the expanding universe. The rate of creation was determined to be 10^{-43} grams per cubic centimeter. This is an extremely small amount and accounts for the fact that the creation process has never been observed.

Max Born (1882–1970) A German physicist responsible for the establishment of a center for theoretical physics at Göttingen in 1926, Born proposed the concept of the wave packet, which is used to describe the probalistic aspect of electron orbits.

Tycho Brahe (1546–1601) A Danish nobleman and astronomer who performed some of the most accurate pretelescopic observations of the heavens, Brahe observed a supernova in the Cassiopeia constellation in 1572. Thanks to the patronage of Frederick II of Denmark, Brahe established an observatory on the island of Hveen, where he worked from 1574 to 1597. Later, he operated under the patronage of Rudolph II, Emperor of the Holy Roman Empire. During this later period, he was joined in his work by Johannes Kepler.

Louis de Broglie (1892–1987) A French physicist, de Broglie, in his doctoral thesis in 1924, introduced the idea that matter, as well as energy, has wavelike properties.

Nicolaus Copernicus (1473–1543) A Polish astronomer, Copernicus in *On the Revolution of the Heavenly Spheres*, proposed the idea that the Sun is the center of the planetary system. The idea was attacked by some very reputable scientists including Tycho Brahe. Nevertheless, the concept of the heliocentric planetary system was eventually proven to be correct. Its acceptance signaled the onset of the Scientific Revolution.

Francis Crick (1916–) A British biologist, Crick, along with James Watson, was responsible for discovering that DNA is structured as a double helix. The discovery explained the replication process by which genes carry information. The double-helix structure of DNA was discovered in 1953 while the two biologists were working at the Cavendish Laboratory at Cambridge.

Marie Curie (1867–1934) A Polish-French physicist and chemist, she was responsible for the discovery of radium. Along with her husband, Pierre, and Henri Becquerel, she was awarded the Nobel Prize for physics in 1903. In 1906, she became the first woman faculty member appointed to the Sorbonne. She received a second Nobel Prize in 1911.

John Dalton (1766–1844) A British chemist who linked quantitative chemistry and atomic theory, in 1808, he published *New System of Chemical Philosophy*. Ironically, the colorblind Dalton is also famous for a paper that he wrote in 1794 on the nature of color vision.

Charles Darwin (1809–82) A British naturalist who is responsible for the development of the theory of evolution based on natural selection, his landmark work, *On the Origin of the Species*, was published in 1859. Darwin's conclusions were reached after a five-year voyage on the HMS *Beagle*, during which he collected a vast variety of biological and botanical specimens from the Southern Hemisphere.

Democritus (470–400 B.C.) A Greek philosopher, Democritus introduced the idea that matter is composed of tiny indivisible particles, called atoms. The same idea was promoted in 60 B.C. by the Roman philosopher and poet Lucretius.

René Descartes (1596–1650) A French philosopher and mathematician, in 1637, he wrote *Discourse on Method*. In this work Descartes supports the deductive method as a scientific procedure. Descartes was also one of the foremost rationalists in the history of European philosophy. He is best known for his famous statement, "I think, therefore I am."

Paul Dirac (1902–84) A British physicist who, in his reformulation of Einstein's relativity equations, predicted the existence of the antielectron. His predictions were confirmed by Carl Anderson in 1932.

Arthur Eddington (1882–1944) A British astrophysicist who performed important work in the area of stellar structure, he also established the idea that the luminosity of a star must be related to its mass. In addition, he showed that the outward pressure of radiation in a star is balanced by the contracting force of gravity. Eddington also determined the upper limit to a star's mass. In 1919, Andrew Crommelin and Eddington led expeditions to confirm Einstein's theory of relativity. Crommelin's expedition went to Brazil, Eddington's to the island of Principe, off the coast of Spanish Guinea. The results of these expeditions confirming the theory of relativity were presented by Sir Frank Watson Dyson, Eddington, and Crommelin at a joint meeting of the Royal Society and the Royal Astronomical Society on November 6, 1919.

Albert Einstein (1879–1954) A German, Swiss, and American theoretical physicist, he was responsible for the development of the special and the general theories of relativity. In these theories Einstein showed the relationship between space and time, on one hand, and matter and energy on the other. Generally considered to be the most preeminent theoretician of modern times,

Einstein also contributed an enormous amount of work to the theory of quantum physics. Nevertheless, he and Niels Bohr kept up a running debate on this point throughout their careers. The debate began in earnest at the first Solvay conference in Brussels in 1927 and continued until Einstein's death. With regard to this debate, Einstein is reputed to have expressed his dissatisfaction with the probabilities of quantum theory by remarking that "God does not play dice." Also among his achievements are an understanding of the photoelectric effect and an explanation of Brownian motion. In 1939, in response to a request from Leo Szilard, Edward Teller, and Eugene Paul Wigner, Einstein signed a letter to President Franklin Delano Roosevelt to persuade him to support the development of the atomic bomb.

Enrico Fermi (1901–54) An Italian-American nuclear physicist, he was responsible for the creation of the first atomic reactor, which was built on a converted squash court at the University of Chicago in 1942 as a part of the Manhattan Project. Fermi continued his work, which led to the building of the atomic bomb, and eventually witnessed the first detonation of a nuclear device at Alamogordo, New Mexico, on July 16, 1945.

Alexander Friedmann (1888–1925) A Russian cosmologist, he demonstrated that Einstein's relativity theory could include the idea of an expanding universe. Friedmann's idea was a building block that led to the development of the big bang theory for the origin of the universe. Remarkably, he made his predictions in 1922, before the observations made by the American astronomer Edwin Hubble.

Kurt Gödel (1906–78) An American mathematician born in Czechoslovakia, Gödel developed the incompleteness theorems in the 1930s. These theorems demonstrate that within any system built upon a sufficiently complex set of axioms there will be statements that are true but unprovable.

Thomas Gold (1920–) An Austrian-American astronomer who, along with Hermann Bondi, an Austrian astronomer, and Fred Hoyle, a British astronomer, proposed the idea that the universe is eternal. As a part of this theory, Gold, Bondi, and Hoyle claimed that matter is continuously created to maintain a uniform density throughout the universe as it expands. This theory, known as the steady-state theory, was popular until the discovery of the cosmic background radiation in 1964. Gold has also done extensive work on pulsars.

Stephen Jay Gould (1941–) An American paleontologist who has suggested a theory of evolution based upon periodic extinctions, in 1989, Gould published *Wonderful Life*, which is based upon his reinterpretation of the Burgess shale fossils, first discovered by the American scientist Charles Doolittle Walcott in 1909.

Stephen Hawking (1942–) A British physicist who demonstrated that the universe must have begun as a space-time singularity, he has also done work on black holes and has demonstrated that under certain circumstances a black hole will radiate thermal energy. In 1988, Hawking published *A Brief History of Time*, which was written to introduce the lay person to the wonders of science.

Werner Heisenberg (1901–76) A German physicist best known for his formulation of the famous uncertainty principle of quantum physics, in 1925, Heisenberg created the matrix-mechanics technique for studying quantum physics, for which he received the Nobel Prize in 1932.

Heraclitus (sixth to fifth century B.C.) A pre-Socratic Greek philosopher from Ephesus, he proposed that the universe is in a constant state of change. He is reputed to have demonstrated the validity of this assertion by noting that a person cannot "step into the same river twice."

Fred Hoyle (1915–) A British astronomer famous for his support of the steady-state theory of the universe, ironically, Hoyle is also responsible for popularizing the label "big bang," a term used to describe the theory that he sees as the primary rival to his own. Hoyle later modified the steady-state theory by postulating the existence of a "creation field" that is responsible for the spontaneous creation of matter in extremely dense areas of the universe. This theory is known as quasi–steady state cosmology. Hoyle is also a prolific science fiction author, who has written such popular novels as *The Black Cloud*.

Edwin Hubble (1889–1953) An American astronomer who in the 1920s used the Mount Wilson Observatory to measure light coming from distant galaxies, Hubble discovered that the light coming from most of these galaxies is shifted to the red end of the spectrum, thus leading to speculation that the universe is expanding. Hubble's Law, which he formulated, includes the Hubble constant, which has been used to measure both the size and the age of the universe. Hubble's observations also verified the expanding universe theory suggested earlier by Georges Lemaitre and Alexander Friedmann. The orbiting Hubble space telescope is named for him.

Johannes Kepler (1571–1630) A German astronomer who, as a contemporary of Galileo, championed the heliocentric model of the planetary system, Kepler worked with Tycho Brahe for a time at Brahe's place of study in Prague. Kepler also solved one of the problems associated with the heliocentric model when he proposed that the orbits of the planets may be elliptical rather than circular.

Georges Lemaitre (1894–1966) A Belgian astronomer and Catholic priest, Lemaitre formulated a solution to Einstein's relativity equations that led Lemaitre to the conclusion that the universe is expanding. Lemaitre came up with this solution in 1927. In 1929, Edwin Hubble, working at the Mount Wilson Observatory, produced observational evidence to support Lemaitre's solution.

James Clerk Maxwell (1831–1879) A British physicist, in 1865, he published *A Dynamical Theory of the Electrical Field*, in which he presents a set of equations that describes electromagnetism. Maxwell also proposed that, since the speed of light and the speed of electromagnetic radiation are virtually the same, light and electromagnetism are unified.

Stanley Miller (1930–) An American chemist, Miller, along with Harold Urey, conducted an experiment at the University of Chicago in 1953 in an attempt to verify the "primeval soup" theory of the origin of life on the plant Earth. The experiment demonstrated that complex organic molecules, such as amino acids, could be formed from the interaction of electricity with ammonia, methane, hydrogen, and water vapor.

Isaac Newton (1642–1727) Newton, a British physicist and mathematician, was responsible for formulating a coherent theory of gravitation. He also developed a theory of optics and a theory of mechanics. In addition, he, and independently G. W. Leibniz, invented calculus. Newton's work, *Principia Mathematica*, is one of the most crucial works in the history of science. Newton's mathematical formulation of physical events gave a new direction to science. In 1672, he became a Fellow of the Royal Society. He was also president of the Royale Society from 1703 until 1727.

Robert Oppenheimer (1904–67) An American theoretical physicist, he was the civilian director of the Manhattan Project during World War II, which culminated in the detonation of the world's first atomic bomb at Alamogordo, New Mexico, on July 16, 1945. Oppenheimer was later an outspoken opponent of the hydrogen bomb. In 1947, he became the head of the Institute for Advanced Study at Princeton, New Jersey. Oppenheimer is also well known for his work on black holes.

Parmenides (fifth century B.C.) A pre-Socratic Greek philosopher, he proposed the idea that a changeless unity dominates the universe. Parmenides believed that reality exists as a single unified reality. As such, his ideas were directly opposed to those of Heraclitus, who claimed that change dominates the universe.

Max Planck (1858–1947) A German physicist, while attempting to solve the problem of black body radiation in 1900, he suggested that energy might be released only in packets, which he labeled quanta. This theoretical approach to the subject introduced the era of quantum physics.

Ernest Rutherford (1871–1937) A British physicist, he revamped the Thomson "plum pudding" model of the atom and in its place created the solar-system model of the atom. The revised model resulted from experiments that Rutherford had conducted with alpha particles. In 1908, he received the Nobel Prize for chemistry.

Erwin Schrödinger (1887–1961) An Austrian physicist, he was responsible for the establishment of the wave mechanics formulation of quantum theory. Later, Paul Dirac demonstrated that Schrödinger's formulation was the equivalent of matrix mechanics, which had been developed by Werner Heisenberg in 1925.

Pierre Teilhard de Chardin (1881–1955) A French geopaleontologist and priest, he advanced a theory of continuing human evolution leading to a future state of perfection, which he termed the Omega Point. He believed that humanity should be studied objectively as any natural phenomenon, yet he also attempted to harmonize the evolutionary process with religious truth. Teilhard's primary works include *The Divine Milieu*, *The Phenomenon of Man*, *Science and Christ*, and *Man's Place in Nature*.

Joseph John Thomson (1856–1940) A British physicist who discovered the existence of the electron, he used that knowledge to construct an early model of the atom. Thomson's model, which was called the "plum pudding" model, envisioned negatively charged electrons embedded within a cloud of positively charged energy.

Harold Urey (1893–1981) An American physical chemist, Urey, along with Stanley Miller, conducted an experiment at the University of Chicago in 1953 in an attempt to verify the "primeval soup" theory of the origin of life on the plant Earth. The experiment demonstrated that complex organic molecules, such as amino acids, could be formed from the interaction of electricity with ammonia, methane, hydrogen, and water vapor.

James Watson (1928–) An American molecular biologist, Watson, along with Francis Crick, was responsible for discovering that DNA is structured as a double helix. The discovery explained the replication process by which genes carry information The double-helix structure of DNA was discovered in 1953, while the two biologists were working at the Cavendish Laboratory at Cambridge. In 1988, Watson became the director of the Human Genome Project. He left the project in 1993 because of a dispute involving the patenting of genetic information.

Glossary

absolute magnitude. The magnitude of a star as it would appear to a hypothetical observer 32.6 light years away from it. *See also* apparent magnitude.

acceleration. The rate at which the velocity of an object changes.

accidents. According to Aristotle and Thomas Aquinas, the outward properties of an object such as the color, texture, aroma, and dimensions.

accretion disk. A spiral-shaped region surrounding a black hole, formed from dust and gas torn from disintegrating stars that have come under the gravitational influence of that black hole.

active galaxy. A galaxy that is a source of intense radiation, which is thought to result from the destruction of stars caused by the strong gravitational pull of centrally located massive black holes.

Administrative Procedure Act. A federal law that guarantees that all federal agencies will follow certain procedures that are set down by Congress. Many states have their own version of this law.

adversarial system. A method of settling legal disputes in which each side is championed by an attorney of record, who battles against the other attorney(s). The battle, either a civil lawsuit or a criminal prosecution, ensues, with victory going to the side whose attorney convinces the judge or jury of the veracity of his or her case. *See also* fact-finding system.

Agenda 21. A document produced at the United Nations Conference on Environment and Development, held in Brazil in 1992. The delegates to this international conference, commonly referred to as the Earth Summit, produced Agenda 21 as a global plan designed to combat environmental deterioration. Agenda 21 is a forty-chapter document outlining a strategy for rescuing the world's environment, while maintaining a sound global economy. As a part of this strategy each nation must develop a national program to accomplish these two goals.

allies. Within the context of the post–Cold War era, those nations, principally Germany and Japan, which have the technological ability to develop nuclear weapons, but which have refrained from doing so because they were under the protective mantle of the United States. *See also* orphans and rogues.

allometric relationship. Disparity between the overall size of an animal and the size of a particular organ, such as the brain, which indicates that the animal and its organ had developed at different rates. *See also* isometric relationship.

alpha particle. A helium-4 nucleus, consisting of two protons and two neutrons released during radioactive decay.

alternate time line. The passage of time in a parallel universe, whose history is similar to that found in the home universe, but which has been altered at a key branch point, often, but not necessarily, by a time traveler's intervention in the past. The belief in the existence of such hypothetical alternate time lines is based upon the premise that, instead of containing a single time stream, the universe contains an extradimensional time river with an infinite number of branching tributaries. According to this hypothetical concept, whenever a time traveler makes a trip into the past, he or she creates an alternate universe complete with its own alternate history. The term "alternate time line" is used to distinguish between the passage of time in the alternate universe from that of our own universe; it is generally associated with the many-worlds theory of quantum physics.

ambiplasma. A mixture of ordinary matter, also known as koinomatter, and antimatter. Ambiplasma is a key ingredient in the plasma cosmology theory pioneered by the Swedish theoretical physicist Hannes Alfven. *See also* koinomatter and antimatter.

amino acids. Complex molecules that serve as the basic building blocks of proteins and which are, therefore, essential to life as it exists on Earth. *See also* Miller-Urey experiment, protein, translation.

amplitude. The distance between the middle line and a crest or a trough of a wave.

anomaly. A natural event that seems to challenge the accepted scientific paradigm. The paradigm predicts that one thing should happen, and an experiment or an observation produces an unexpected, contradictory result, the anomaly.

anticodon. A three-nucleotide sequence along a strand of tRNA that complements the three-nucleotide sequence, or codon, on a strand of mRNA. *See also* codon.

antimatter. Matter whose atoms consist of particles with charges that are the opposite of those found in ordinary matter. The antielectron or positron, for example, which is the antimatter counterpart of the electron, has the same mass as an electron but has a positive rather than a negative charge. *See also* ambiplasma and koinomatter.

apeiron. According to the ancient Greek philosopher Anaximander, the most primeval substance that served as the basic material of the universe. This conclusion was based on his observation that each of the four most basic elements of the universe—fire, air, water, and earth—appear to be incompatible with one another. As a result of this observation Anaximander concluded that there must be a fifth substance, the apeiron, that served as the source of the other four.

ape-men. *See* Australopithecines.

apparent magnitude. A measure of the brightness of a star from our vantage point. *See also* absolute magnitude.

Ardipithecus ramidus. An early representative of the evolutionary line leading to homo sapiens. A. ramidus lived in Africa and may have been bipedal.

arithmetic progression. A numerical progression in which any pair of successive terms are separated by the same number. For example, in the arithmetic progression 2, 4, 6, 8 . . . 10, successive terms are separated by 2. *See also* geometric progression.

artificial selection. A process breeders use to develop a particular trait in a group of plants or animals. *See also* natural selection.

asteroid. An astronomical body consisting of rock and metal that circles the Sun generally between the orbits of Mars and Jupiter. Asteroids, sometimes called planetoids or minor planets, are believed to be the remnants of an unborn planet that failed to form during the early history of the solar system, probably due to the gravitational influence of the planet Jupiter.

asymmetric aging. A type of uneven aging that occurs among individuals because of the time-dilation effect.

atom. The smallest particle of an element that retains the characteristic properties of the element. The atom is itself composed of smaller, subatomic parts, including a central nucleus composed of protons and neutrons, and a set of electrons that inhabit regions of space around the nucleus. Since the electrons have a negative charge and the protons have a positive charge, the charge of an atom is neutral.

Australopithecines, ape-men, or **missing links.** The transitional apelike creatures that provide a bridge between humans and apes. This transitional group of ape-men appeared about four million years ago.

average life expectancy. The number of years a typical individual can be expected to live, measured from the time of birth.

axon. A long filament that extends from the main body of a neuron or brain cell and that carries nerve impulses from the main cell body to other organs or to other locations in the brain. The axon can be considered the sending wire of the neuron. Each axon starts as one long filament flowing from the central cell body of the neuron and ends by spreading out like the branches of a tree. *See also* dendrite.

balancing principle. A principle whose goal is to prevent any branch of the government from exercising absolute power within its sphere of influence. To prevent this, each branch shares some power with other branches. *See also* separation principle.

base. A chemical compound that can accept a hydrogen ion (proton) from another substance. Four nitrogenous bases are central components of DNA. These four bases are A (adenine), G (guanine), C (cytosine), and T (thymine).

base-paired sequence. The specific order in which pairs of bases along a DNA molecule will join. Thus, G (guanine) joins with C (cytosine), and vice versa, but not with A (adenine) or T (thymine); A joins with T, and vice versa, but not with G or C.

Bell's inequality. A limit to the interaction between duons in the measurement of a quantum system, given the existence of those local hidden variables. If the limit is not surpassed, then local hidden variables rule. However, if the interaction continues beyond Bell's limit, this would constitute a violation of Bell's inequality. Such a violation cannot be explained by local hidden variables but could be explained if we admit either to the indefinite nature of the quantum universe or to the existence of nonlocal effects.

Bell's interconnectedness theorem. A theorem that states that predictions within quantum theory cannot depend upon local hidden variables.

beta decay. The process by which a proton changes into a neutron.

big bang theory. The theory that the universe began some 15 to 20 billion years ago when an infinitely dense subatomic entity, known as the singularity, appeared, inflated suddenly and, as the inflation slowed and the universe cooled, coalesced into the cosmos as we know it today.

bilateral symmetry. An anatomical arrangement in which the body parts on one side of an organism are mirror images of those on the other side.

binding precedent. A precedent that must be followed by a court. *See also* case law, judicial decisions, persuasive precedent, precedent, stare decisis.

biotechnology. The use of living organisms in the manufacture of drugs for environmental management.

bipolar world. A term indicating that the international political arena is dominated by two major powers. During the Cold War, for instance, the international political arena was dominated by the United States and the Soviet Union.

black body radiation. The name given to the electromagnetic radiation released as an object heats up. The wavelength of the energy released depends upon the temperature of the object; the hotter the object the shorter the wavelength. At lower temperatures, bodies radiate infrared energy, which can be felt but not seen. As the temperature of the object increases, the object will begin to glow. This is because the wavelengths of most of the energy released at that point have passed into the visible spectrum. If the temperature of the object is permitted to increase, the object's glow will move to orange, yellow, white, and blue, and into the invisible range of ultraviolet light, X-rays, and gamma rays.

black hole. A hypothetical body in space that, under the influence of its own gravitational field has collapsed to such an extent that it has created a gravity well so powerful that not even light can escape. Some black holes are thought to result from the collapse of massive stars. Others, called primordial black holes, are believed to have formed under the extreme conditions that existed during the early moments of the universe. Black holes are also thought to exist at the heart of active galaxies. Such black holes might be so massive that they consist of billions of collapsed stars.

blastocyst. A specialized, multicellular structure that develops approximately two days after a human egg is fertilized. In the blastocyst cells are sufficiently differentiated to distinguish between those that will develop into the embryo and those that will attach to the uterine lining and develop into the placenta.

block universe. See space-time cube.

Bohr-Einstein debate. A disagreement between Einstein's classical approach and Bohr's Copenhagen Interpretation of quantum physics.

Boolean logic. A reasoning process that allows exactly two possibilities, and thus seeks an "either-or" solution to all problems.

brown dwarf. A nebula with less than 8 percent of solar mass (or less than 80 times the mass of Jupiter) that does not undergo nuclear fusion and cannot, therefore, be classified as a star.

Burgess Shale fossils. A group of strange fossils that display a widely diverse number of anatomical forms, most of which disappeared approximately 500 million years ago. The fossils, which were discovered in British Columbia in 1909, were recently recataloged in such a way as to support the decimation/contingency view of evolution.

case law. That body of law found in cases that are based on decisions made by judges as they apply common law, interpret statutes, determine issues of constitutionality, or any combination of the three. *See also* binding precedent, judicial decisions, persuasive precedent, precedent, stare decisis.

catalyst. A substance that expedites a chemical reaction without being altered permanently by the reaction. *See also* enzymes.

cathode rays. Electrons released at a negative electrode (or cathode) in a vacuum tube. The electrons flow to the positive electrode or anode.

Cepheid variable star. A star that varies in brightness at regular intervals, making comparison measurements easier.

cerebral cortex or **neocortex.** The outer layer of the brain, sometimes called the neocortex.

chemosynthesis. A synthesis of organic materials, using energy derived from inorganic molecules. This is the process by which life-forms on the ocean floor break down hydrogen sulfide into useful energy. The life-forms in the dark depths of the ocean use geothermal energy rather than sunlight as a source of food.

clarity. A characteristic of the scientific method which requires a successful theory to be concise enough to be tested according to objective criteria. Theories that demonstrate clarity will not require hidden knowledge, special talent, or mysterious magic to verify their validity.

chromosomes. Small, oblong structures, located within the nucleus of plant and animal cells, that house genes.

codon or **triplet.** A set of three nucleotides on a strand of mRNA. *See also* anticodon.

Coma cluster. A distant cluster of galaxies that has been used to measure the age of the universe.

comet. An extraterrestrial object found in the solar system, which consists of minerals, rock, dust, and frozen material, such as methane, water, and ammonia.

Comet Halley. A well-known comet that orbits the Sun every seventy-six years.

commerce clause. The clause of the U.S. Constitution which states that "Congress shall have Power . . . To regulate Commerce with foreign Nations and among the several States." The courts have gradually broadened Congressional power to regulate commerce under the commerce clause so that today, the national government can regulate just about any activity that somehow impacts upon commerce.

common law. English law based upon custom or court decisions, as opposed to statutes. Common law principles are also found at the heart of the American legal system.

compelling state interest. A principle applied by the courts to any situation in which a statute infringes upon a fundamental right guaranteed by the U.S. Constitution.

complementarity principle. The principle proposed by Danish physicist Niels Bohr, which states that either we describe quantum phenomena in the context of time and space, in which case, we must deal with the uncertainty principle, or we describe causal relationships mathematically, in which case, we lose the ability to visualize those relationships in terms of time and space.

complexification. The evolutionary process by which organic chemical processes developed complex molecules; this eventually led to biochemical processes.

conceptions. According to the philosophy of the German thinker Immanuel Kant, perceptions that have been shaped by the categories of quantity, quality, relation, and modality so that they follow certain predictable patterns. The categories operate in the following ways. Quantity embodies the concepts of unity, plurality, and totality. Quality embraces the notions of affirmation, negation, and limitation. Relation involves the ideas of substance-accidents, cause-effect, and causal reciprocity. Finally, modality includes the concepts of possibility, actuality, and necessity. Once perceptions have become conceptions they can be seen as universally valid and, therefore, objective. *See also* sensations and perceptions.

cone of increasing diversity. A model of the evolutionary process that depicts living things as simple at the beginning of life on Earth and more and more complex as the evolutionary process continued. The cone-of-diversity model is generally shown with the cone upside down, that is, with the point of the cone facing downward. At the inverted point of the cone a single, common ancestor exists. From this single, common ancestor all the complex organisms on the branches have evolved. This starting point subdivides, but only into a limited number of branches. Each branch diversifies over time, creating more groups that are subdivisions of earlier ones.

conservation paradox. A time-travel paradox which suggests that time travel may be impossible because it would violate the First Law of Thermodynamics. According to the First Law, energy can be converted from one form to another, but it can neither be created nor destroyed. The sudden and inexplicable appearance of a time traveler in the past would violate the law of the conservation of energy. Given this immutable law, the time traveler cannot disappear from the present and suddenly appear in the past because her disappearance from the present would represent the destruction of energy in the present, and her materialization in the past would represent the creation of energy in the past from nothing. If the total amount of energy cannot be changed within the closed system of the universe, there is no way to account for the energetic subtraction of the time traveler from the present and her energetic addition to the past.

constitution. The fundamental law according to which a nation or a state is governed. A constitution will establish the government of a nation, outline the rights and liberties of the people of that nation, and set down some basic principles that guide the operation of that nation.

constructivism. A sociological system which declares that all approximations of the truth are created within a particular social and cultural setting. According to constructivism, since the influence of a social and cultural setting is inescapable, all truth, even scientific truth, is subject to that social and cultural context. *See also* positivism.

contingency. The driving force behind the decimation model of evolution. Contingency is based on the assumption that any small, unpredictable event along the path of evolution might have deflected the selection process in a different direction. Thus, there is nothing inevitable about the particular evolutionary history that the Earth experienced.

continuity or **correspondence.** A characteristic of the scientific method which requires a successful theory to evolve logically from previously established scientific principles.

contradictory-event paradox. A time-travel paradox that occurs when incompatible incidents converge within the same time line. Such paradoxes are inherent in the inconsistency of the events themselves rather than in our understanding of those events. The three most famous contradictory time-travel paradoxes are the grandmother paradox, the conservation paradox, and the simultaneous temporal existence paradox. *See also* information paradox.

contragestive agent. A birth control technique that will prevent the implantation of a fertilized egg on the uterine wall. Contragestive agents include the morning-after pill, intrauterine devices, and low-dosage birth control pills.

Copenhagen Interpretation. An interpretation of quantum physics which recognizes that the deterministic picture of the universe which works so well for classical physics at the macroscopic level does not work at the quantum level. This conclusion is derived from two very critical aspects of quantum physics. First, the universe at the quantum level is predictable only in a statistical sense. It is impossible to predict the movement or the position of a single duon. Instead, we must always speak in terms of statistical distributions. Second, the physical state of quantum reality cannot be described without also specifying the observational technique that is used to conduct the observation. If the apparatus is constructed to observe the particle property of the duons, that is what will be observed. If the apparatus is constructed to observe the wave property, then that is what the observer will see. *See also* Bohr-Einstein debate.

Copernican system. A model that places the Sun, rather than the Earth, at the center of the planetary system.

correspondence. *See* continuity.

Cosmic Background Explorer (COBE). A satellite sent into a polar orbit by NASA in 1989, whose mission is to study the background radiation that provides scientists with evidence in support of the big bang. This radiation represents the leftover glow of the enormous heat that permeated the universe in the early moments of the big bang.

cosmic background radiation. Microwave radiation that permeates all space and which is believed to be a remnant of the big bang.

cosmic string. A slender, elongated strand of superenergy remaining as a by-product of the big bang. Cosmic strings are so massive (perhaps more than a thousand trillion tons for every inch) that they easily distort space-time within their vicinity.

cosmological constant. A repulsion factor or "negative pressure" that was added to the theory of relativity by Einstein to balance the attractive effects of gravity, thus creating the static or steady state universe that Einstein preferred.

creation center. An area of the universe characterized by a high density of matter, which, according to quasi–steady state cosmology, serves to trigger the creation field (C-field). Typical candidates for the location of a creation field would be in the heart of a single galaxy or near gigantic extragalactic structures.

creation field (C-field). A hypothetical energy field, the existence of which is predicted by certain alterations in Einstein's relativity equations dealing with space-time singularities and strong gravitational fields. The hypothetical C-field exists only in certain areas of the universe. These areas are characterized by a high density of matter, such as in the heart of a single galaxy or in regions of enormous extragalactic clusters. Such areas are labeled creation centers. A C-field is activated by the intense concentration of matter at a creation center. Creation centers are intrinsically unstable, and, therefore, produce an enormous burst of explosive energy, which act like a mini–big bang, creating matter and activating the expansion of the universe. This sudden, short-lived explosive event is followed by a long period of gradual expansion. As the universe expands, the strength of the C-field diminishes and the expansion gradually weakens. Eventually, conditions again favor the appearance of new creation centers until, once again, the C-field is activated. This new burst of energy creates matter and activates the expansion cycle of the universe once again. *See also* quasi–steady state cosmology, steady state cosmology.

creationism or **creation science.** An approach to the origin of the universe that requires a literal reading of the Bible. Thus, a literal reading of Genesis would indicate that the universe was created in six days, that the state of the universe has been unchanged since those original six days, and that the universe itself is a little over 6000 years old.

cyanobacteria. Unicellular organisms that have developed the ability to transform sunlight into energy via photosynthesis.

cytoplasm. That portion of a cell which exists outside the nucleus but inside the outer membrane of the cell wall. Cytoplasm is actually of two types: the outer ectoplasm, which is rather dense and which is involved in the movement of the cell, and endoplasm, which houses the majority of the cell's parts.

decimation. A model of evolution that is based upon the premise that the driving force behind evolution is not expansion through diversification, but diversification followed by elimination. According to the decimation model, when multicellular life forms first appeared on the planet, they burst into life in a widely diverse number of different anatomical forms so that at this early time life reached its maximum in terms of anatomical diversity. From that stage onward, many of these anatomical designs were eliminated, leaving the few remaining ones that exist today. The driving force behind the elimination process is contingency.

deductive logic. The system of logic that applies generalized principles to arrive at specific conclusions. *See also* inductive logic.

deep sea vents. Openings in the sea floor that emit the volcanic gas hydrogen sulfide.

dendrite. A filament that extends from the main cell body of a neuron or brain cell. A dendrite will receive impulses from the axons and will send them to the main cell body of the neuron. For this reason the dendrite can be considered the receiving antenna of the neuron. *See also* axon.

density. Mass per unit volume.

deoxyribonucleic acid (DNA). The complex molecule that provides the basic material of chromosomes.

determinism. A philosophical system based on the belief that human decision-making is subject to forces that predetermine those decisions and compel humans to act according to a preordained pattern.

device. An invention or contrivance, often mechanical or electrical, which qualifies for a patent. The device stipulation does not necessarily have to be an artifact. The courts, for example, have approved patents for processes that involve computer programs. However, an inventor cannot patent a law of nature, a mathematical formula, or an abstract theory.

differential reproduction. *See* natural selection.

disassemblers. Hypothetical nanodevices that will be able to analyze the existing molecular structure of an item so that an understanding of that structure can be stored in the memory of a nanocomputer, which could then direct assemblers to make copies of the original.

DNA. *See* deoxyribonucleic acid.

DNA mapping. The process of determining the sites of all genes within a genome.

DNA sequencing. The process by which the sequence of all base pairs for a genome are determined.

doctrine of identity. The Vedantic belief that the universe exists as an undivided whole.

Doppler effect. An explanation of the nature of sound waves as an object approaches, passes, and then moves away from a fixed position. As the object moves toward a fixed position, the frequency with which the sound waves reach that position increases. This causes the pitch of the sound emitted from the object to rise. As the object passes and moves away from the fixed position, the frequency with which the sound waves reach the position falls. Consequently, anyone located at the fixed position will experience a corresponding fall in the pitch of any sound emitted from the object as it moves away. A similar effect is noted with light waves. If a light source, such as a galaxy, is moving toward an observer, the frequency with which the light waves reach the observer increases. This will result in light shifted toward the blue end of the spectrum. If the light source is moving away from the observer, then the frequency with which the waves reach the observer decreases. The result is light shifted towards the red end of the spectrum. This latter shift convinced the American astronomer Edwin Hubble that the galaxies he had observed were moving away from the Earth's position in the universe.

double helix. The basic structure of DNA, which looks something like a spiral staircase, but which is often simplified so that the spiral staircase is flattened out to look like a ladder. The double helix consists of sets of precisely matched bases. These matching bases make up the rungs of the ladder. There are four such bases. They are A (adenine), G (guanine), C (cytosine), and T (thymine). Each base is supported by a sugar-phosphate group. The sugar-phosphate groups would correspond to the supporting sides or railings of the ladder. A base plus its supporting sugar-phosphate group is a nucleotide. The bases will join only in specifically ordered pairs. This pattern is a base sequence. G joins with C, and vice versa, but not with A or T; A joins with T, and vice versa, but not with G or C. This is critical because as the double helix splits (imagine the ladder being divided down the rungs), each side becomes a mold for the production of a matching strand. Moreover, an exact duplicate of the strand of DNA is assured because of the restricted pattern in which the bases can join. The double helix must split to perform replication and the synthesis of protein. *See also* deoxyribonucleic acid.

due process of law. The properly regulated operation of the law according to accepted practices that safeguard the rights of individuals and insure the administration of justice.

duon. A generic name used for any subatomic particle.

Earth Summit. The name generally given to the United Nations Conference on Environment and Development held in Brazil in 1992. The delegates to this international conference produced Agenda 21, a global plan designed to combat environmental deterioration.

Economic Security Council (ESC). A proposed subdivision of the United Nations somewhat similar to the present Security Council in function. The ESC would have a variety of responsibilities. For instance, it would assess the state of the international economy and implement long-term strategic planning that would encourage the goal of sustainable development throughout the globe. This strategic planning would include the development of a well-balanced science-and-technology policy. The principle job of the ESC would be to examine political, economic, environmental, and social trends throughout the international community and guide the global community to successful solutions to its problems. The idea behind the ESC is not to replace already established agencies, but to provide guidance for agencies and to unravel overlapping and contradictory policies that may hurt the overall development of the planet.

Einstein separability. A distance established in an experiment involving subatomic phenomena to ensure that the devices used for measuring the duons are far enough apart from each other that all chance of a causal connection is removed. The establishment of an Einstein separability in an experiment would mean that any measured correlation between the duons could not be caused by any type of local event.

electromagnetic field. A condition in space that results when a changing magnetic field creates an electric field and/or a changing electric field creates a magnetic field.

electromagnetic force. One of the four fundamental forces of nature. It is manifested as disturbances in the electric and magnetic fields. *See also* gravity, strong nuclear force, weak nuclear force.

electrons. Freely moving, negatively charged subatomic matter particles belonging to the lepton family. If we maintain the solar system model of the atom, we might say that the electrons "orbit" the nucleus. This description, of course, is inaccurate because the electrons do not orbit the nucleus like the planets orbit the Sun. According to Heisenberg's uncertainty principle, it is impossible to determine the exact position of an electron. Focusing on the energy of the electron rather than on the position means that we cannot know the exact location of the electron. We can know only the probability of finding an electron within a region of space within the atom. This is known as the probability distribution of the electron. The region of space where the probability of finding the electron is high is called its orbital. Electrons can exist only at certain predetermined energy levels within the space of the atom. Moreover, the maximum number of electrons that can exist at each level is severely limited. For instance, the limit at the closest level to the nucleus is two, whereas at the second closest level, the limit is eight. To move from one energy level to the next, the electron must emit or absorb energy in the form of a photon.

elsewhen or **elsewhere.** The region of a Minkowski space-time diagram that represents everything outside the past and future light cones.

embryo. An animal in its early developmental stages, generally from the instant that a fertilized ovum or egg cell begins to subdivide until the time of birth or hatching. In humans, however, some prefer to label the fertilized ovum a pre-embryo up until the fifteenth day after conception. This is because during the first fifteen days the function of the fertilized ovum is to create the placenta; after that time, according to this view, the formation of the true embryo commences. After the eighth week the embryo is called a fetus.

emergentism. A theory that is based on the premise that life should be viewed as an emergent quality of the complex organization of matter, from the molecular stage on up. The emergent theory of life finds its roots in the holistic approach to scientific investigation.

encephalization quotient (EQ). The relationship between the actual size of an organism's brain and its expected size. The higher the EQ, the larger the brain in relation to what is expected, given the overall size of the animal.

entity theory. The theory of human evolution which states that Homo sapiens alone possessed the predisposition that would lead to the level of intelligence found in modern humans. *See also* tendency theory.

enumerated powers. Those governmental powers that are actually granted to the federal government in the Constitution.

enzymes. Proteins that act as catalysts to facilitate biochemical reactions within living things.

EPR. The name given to a thought experiment described in an article written by Albert Einstein, Boris Podolsky, and Nathan Rosen, while the three of them were at the Institute for Advanced Research at Princeton. The article, "Can Quantum-Mechanical Description of Physical Reality Be Considered Complete?," which was published in the May 15, 1935 issue of *Physical Review*, introduced a thought experiment that purported to demonstrate the incomplete nature of quantum theory. The experiment, which is now known as EPR, after its authors, challenged the very heart of the Copenhagen Interpretation.

equity. The fair and equal administration of the law.

ergosphere. The region located between the static limit and the event horizon of a rotating black hole.

escape velocity. The lowest speed necessary for an object to move out of the gravitational influence of a massive astronomical body. For example, if an astronaut wants to leave the Earth, her ship must reach a speed of 6.8 miles per second; thus, the escape velocity of Earth is 6.8 miles per second. Escape velocity is related to the size and the mass of an astronomical object.

ether. *See* lumniferous ether.

ethical relativism or **subjective ethics**. A set of ethical concepts based upon the belief that right and wrong can change from person to person or from situation to situation.

eukaryotes. Cells that contain an independent, self-contained nucleus that houses the essential mechanism of the cell's operation. *See also* prokaryotes, protocells.

event horizon. A boundary in space surrounding a black hole within which the gravitational attraction of the black hole is so powerful that even light cannot escape.

evolving universe theory. A theory based on the belief that the origin of the universe was triggered by the collapse of a massive star in a previous parent universe.

existentialism. A philosophy which emphasizes existence over essence and, as such, focuses on the responsibility that each individual has in shaping his or her own identity.

extreme Kerr object. An object that would result if the spin of a rotating black hole increases so that its angular momentum is equal to its mass. *See also* rotating black hole, inner event horizon, outer event horizon, superextreme Kerr object.

fact-finding system. A method of settling legal disputes in which the major objective of a trial is to uncover the truth. Many people find this approach preferable to the adversarial system, which has victory as its primary objective.

fetus. An embryo after some point near the eighth week following conception. *See* embryo, preembryo.

fiber optics. The branch of optics that involves the use of thin glass fibers to transmit information, instead of the large copper wires that were the mainstay of the global telecommunications system for years.

field. A region encircling an object in which a force operates.

First Law of Thermodynamics. The principle that energy can be neither created nor destroyed but can only be changed from one form to another.

foreknowledge. Knowledge of an event before it happens that does not interfere with that event.

Fornax cluster. A distant galactic cluster, located south of the galactic plane, that has been used to measure the age of the universe.

free will. *See* volition.

frequency. The number of crests of a wave that move by a specified point within a unit of time.

fundamental rights. Rights of Americans that are guaranteed by the Constitution.

fusion. A nuclear reaction during which small atomic nuclei fuse into one large atomic nucleus. If the mass of the final nucleus is less than that of the original constituent nuclei, energy is released. Stars, including the Sun, are powered by the fusion of hydrogen into helium.

future light cone. That portion of a Minkowski space-time diagram that represents the future of an individual as measured from a central point designated "here and now." *See also* past light cone, elsewhen.

galactic cluster. A collection of galaxies that are comparatively near one another and that generally relate to one another gravitationally.

galaxy. An enormous grouping of stars. Galaxies frequently contain billions of stars. The Milky Way Galaxy, for example, consists of approximately 200 billion stars.

Galilean relativity. The assertion that everything in the universe is moving relative to everything else.

gedanken experiment or **thought experiment.** An experiment that is visualized in the mind of a scientist rather than one that is actually performed. Generally, thought experiments cannot be carried out because of certain practical limitations. Gedanken experiments are frequently used in quantum physics because such experiments help relate the mathematical abstractions inherent in quantum theory to the physical universe.

gene. The hereditary unit found within the cells of living organisms.

general welfare clause. That clause of the U.S. Constitution which states that Congress has the "Power To lay and collect Taxes, Duties, Imposts and Excises, to pay the Debts and provide for the common Defence and general welfare of the United States."

gene therapy. Therapy which involves curing and perhaps preventing those diseases that may have a basis within the genetic code.

genetic code. The code in DNA which governs the production of protein within the cell.

genetic engineering. The process by which modifications to living things are made by manipulating genes.

genetic variant. A random change in the genetic code.

genome. The sum total of all genetic information located within a particular group of chromosomes.

geometric progression. A numerical progression in which each term after the first is a constant multiple of the preceding term. For instance, in the geometric progression, 1, 5, 25, 125, 625, 3125, the constant multiple is five. *See also* arithmetic progression.

glial cells. A network of cells that provide a support system for the neurons of the nervous system.

global warming. The gradual rise in the overall temperature of Earth caused primarily by the increased use of fossil fuels. This usage results in an escalation in the level of carbon dioxide and other gases in the atmosphere. *See also* greenhouse effect.

gluons. The force-carrying particles that convey the strong nuclear force. Gluons bind quarks together making protons and neutrons.

golden rule or **principle of reciprocity.** The primary guiding principle for sentient beings to treat others as they would like to be treated themselves.

grandmother (grandfather, grandparent) paradox. A time-travel paradox that results when a retrograde time traveler visits the past and meets her grandmother. Through a series of misadventures, the time traveler prevents her grandmother from meeting her grandfather before the birth of the time traveler's mother. As a result, the time traveler's mother is never born. Herein lies the convergence of incompatible events. Since the time traveler's mother does not exist, it is impossible for the time traveler to exist. Therefore, the time traveler could not, and, therefore, did not, travel back in time to prevent her mother's birth. Therefore, her mother was born and the time traveler was also born, and she did exist to travel into the past to prevent her mother's birth and thereby eliminate her own existence. *See also* contradictory event paradox.

grand unified theory. A theory which seeks to understand the unification of three of the fundamental forces of the universe: electromagnetism, the weak nuclear force, and the strong nuclear force.

gravitational field equations. Einstein's equations that purportedly determine the relationship between the power of a gravitational field and the matter and energy that produced that field.

gravitational radius. *See* Schwarzschild radius.

gravitons. The force-carrying particles responsible for gravity. Since the gravitational attraction exerted by the graviton is extremely weak, gravitons are difficult to detect. For this reason, up until the present at least, scientists have been unable to detect gravitons. Nevertheless, so much is known about how gravity operates that it is possible for scientists to predict what properties gravitons would possess. According to these predictions gravitons should have neither mass nor charge and they should move at light speed.

gravity. One of the four fundamental forces of nature. Gravity is distinguished from the other three forces because it always attracts and because, although it manages to operate over great distances, it is relatively weak. It may seem strange to describe gravity as a relatively weak force, because, from our perspective at least, gravity appears to be quite strong. At the subatomic level, however, gravity is very weak when compared with the power exerted by the electromagnetic force.

Great Attractor. The purported unification of two enormous superclusters of galaxies that have a combined mass exceeding that of the Milky Way by a factor of 200,000. The Great Attractor occupies a portion of space beyond Hydra-Centaurus that covers 300 million light years. At least three galactic clusters, the Local Group, Virgo, and Hydra-Centaurus, are moving at 600 kilometers per second toward this Great Attractor.

Great Wall. A colossal extragalactic configuration consisting of thousands of galactic structures and stretching some 500 million light years in length and 15 million light years in thickness. The discovery of the Great Wall led astronomers to a new understanding of the structure of the intergalactic universe. Instead of existing in a random but relatively even distribution throughout the entire universe, as was previously imagined, the galaxies now appear to be collected into unimaginably huge extragalactic sheets or bubbles consolidating thousands of galaxies together. In between these monumental extragalactic sheets, such as the Great Attractor and the Great Wall, lie vast deserts of barren space.

greenhouse effect. An overall environmental trend toward global warming because of an escalation in the level of carbon dioxide and other gases in the atmosphere caused by the increased use of fossil fuels. The net effect is a gradual warming of the upper atmosphere, which, in turn, results in an overall warming of the entire planet.

ground state. The lowest energy state possible for an atom.

handmade cell. An experimental attempt to create a self-replicating strand of RNA that can act as both enzyme and RNA.

hectare. A land measurement of 10,000 square meters.

holism. The theory that whole entities, as fundamental components of reality, have an existence other than as the sum of their parts. Holism seeks to comprehend the complex interaction of the interconnected parts of the universe. In a sense what the reductionist takes apart, the holist puts back together.

holomovement. A term used by physicist David Bohm to refer to the continuous unfolding of the enfolded universe.

hominids. A subdivision of the hominoids which includes ape-men and humans. The family is referred to as the Hominidae.

hominisation. The process by which protohumans reached the point of acquired humanness. The term was used by the noted paleontologist and Catholic priest Pierre Teilhard de Chardin in his seminal work *The Phenomenon of Man*. Teilhard calls the acquisition of humanness by several different names, including the "leap of intelligence" and the "threshold of thought."

hominoids. A term used by zoologists and anthropologists as a broad classification which includes apes (gibbons, orangutans, gorillas and chimpanzees), Australopithecines, and humans.

Homo. The evolutionary line leading to modern humans.

Homo sapiens. The designation given to modern humans.

Hubble constant. The ratio of velocity to distance, as stated in Hubble's law.

Hubble's law. The assertion that the velocity of a galaxy as it moves away from the Milky Way increases in proportion to its distance from our galaxy.

Human Genome Project (HGP). A global project, the immediate objective of which is to obtain a complete set of human genetic information, that is, to create a comprehensive, sequential blueprint of human DNA.

Hydra-Centaurus supercluster. An immense group of galactic clusters.

hypothesis. A systematic attempt to explain a hitherto unobserved or unexplained phenomenon. A hypothesis is generally stated in the form of an unproven empirical principle. The goal of a hypothesis is to impose order on the observed phenomenon so that it no longer appears as a random event but instead becomes part of a system. A genuine hypothesis will also make testable predictions about the observed phenomenon.

idealism. The philosophical belief that all reality is simply the product of the mind. The origin of idealism can be traced to the eighteenth-century British philosopher, Bishop George Berkeley. According to Berkeley, it is not possible for the human mind to perceive physical objects. All the mind can perceive are the qualities of those objects. Thus, a thinking subject may know the color, size, shape, and texture of an object, but not its substance. With this premise in mind, Berkeley concludes that the substance of physical objects and, therefore, the physical objects themselves do not exist. Berkeley further believed that unexperienced objects cannot exist because they cannot be perceived.

implicate order. A term used by physicist David Bohm to refer to the enfolded nature of the universe.

inductive logic. The system of logic that involves observing specific instances and individual pieces of evidence to arrive at a general conclusion. In the construction of a mathematical model according to inductive logic, a theoretician observes examples of real-world phenomena and, from those observations, constructs a mathematical model that reflects those observations. *See also* deductive logic.

inertial system. A system moving at an undisturbed, constant speed.

infant mortality rate. A measurement of the death rate of children who do not reach their first birthday.

information or **knowledge paradox.** A paradox that occurs when events in the past are caused by information brought back from the future. A true information paradox occurs only if the knowledge brought from the future is used to cause itself to occur. *See also* contradictory event paradox.

inner event horizon. The second of two event horizons produced by the rotation of a Kerr black hole. Beyond the inner event horizon space and time resume their normal roles. *See also* outer event horizon.

Internet. A computerized collection of global networks. It was first established by the military in the 1960s but has been available to the public at large since the early 1990s.

isometric relationship. A relationship between the overall size of an animal and the size of a particular organ, such as the brain, which indicates that both the animal and the organ developed at the same rate. See also *allometric relationship*.

judicial decisions. Decisions made by judges as they interpret statutory law, create and interpret common principles, and determine the constitutionality of statutes, regulations, and executive actions. *See also* case law, common law, constitution, stare decisis, statutory law.

juriscience. The treatment of issues that stand at the intersection of science and the law.

jurisdiction. The authority of the court.

justice. The result of applying the law with impartiality and equity.

Kerr black hole. *See* rotating black hole.

Kerr-Newman black hole. A black hole that is electrically charged, rather than electrically neutral. Some physicists believe that Kerr-Newman black holes are simply mathematical constructs and, therefore, may not exist within the physical universe.

knowledge paradox. *See* information paradox.

koinomatter. According to the plasma cosmology theory of Hannes Alfven, a term meaning ordinary matter. *See also* ambiplasma and antimatter.

laser interferometer gravity observatory (LIGO). An instrument which may be able to detect gravity waves.

law. A set of rules and regulations established by the government of a nation or one of its subdivisions in order to maintain order and justice.

law of nature or **natural law.** A hypothesis that has been successfully verified repeatedly by experimental observation.

leptons. Elementary matter particles that are distinguished from other matter particles, such as quarks, because the leptons are usually not associated with the nucleus of an atom. Prominent members of the lepton family include electrons and neutrinos.

less developed world. Those economically and industrially challenged regions of the world designated by the United Nations as including Africa, Latin America, and that part of Asia which is not within the former Soviet Union. *See also* more developed world.

lifeline, lifeworm, or **worldline.** A person's pattern of perceptions nestled within the space-time cube. The concept is based on the notion that human beings inhabit a universe that consists of the totality of their perceptions laid out like a long line of successive events within a four-dimensional space-time cube. *See also* space-time cube.

light speed. The speed of light in a vacuum measured at 186,000 miles per second or 300,000 kilometers per second.

light year. The distance that light travels in a year.

linear. The quality of a straight line in a process, graph, or image.

lines of force. Invented by the nineteenth-century physicist Michael Faraday, they help visualize how a field moves between and among bodies in space. These lines of force can depict the direction of a force in a field as well as the strength of that force.

Local Group. The collection of galaxies which consists of two large spiral galaxies, the Milky Way and Andromeda, along with twenty-four other galactic structures. Most of these other galaxies are grouped around either the Milky Way or Andromeda.

Lorentz contraction. A term used to identify the fact that the length of an object shrinks in the direction that the object is moving. Another way to say this is to note that the space that the object occupies will contract in the direction of the moving object.

Lorentz factor. A mathematical formula that can determine exactly how slowly a moving clock will run relative to an Earth-bound clock.

Lorentz light clock. An imaginary device used in a thought experiment designed to demonstrate the nature of time dilation.

Lorentz transformation. An explanation of how the relationships among space, time, and mass are altered by relative motion. The Lorentz transformation is named for the Dutch physicist Hendrick Lorentz who, before Einstein, devised the transformation to explain the problems of the absolute speed of light in relation to the ether. *See also* lumniferous ether.

luminosity. The brightness of a star compared to the Sun.

luminiferous ether or **ether.** A hypothetical substance once believed to fill all space, and thought to be a substance necessary for the movement of electromagnetic radiation through space.

Malthusian check. Any event or set of circumstances that impacts upon the ability of a population group to compete for the limited resources of a particular area.

Manhattan Project. The code name for the project dedicated to the development of the atomic bomb by the United States during World War II.

many-worlds, many-universes, or **parallel universes theory.** A theory that postulates the creation of a new universe every time a quantum measurement is made.

mass. A measure of an object's inertia. Inertia is an object's capacity to obstruct any alteration in its movement.

mass extinction. An event that destroys many species in their entirety in a relatively short period of time. Although the development of life on Earth does involve a progression of ever-increasing complexity, that complexification process has at times been interrupted by the occurrence of such mass extinctions. Some evidence indicates that these mass extinctions occur at regular intervals of every 26 to 30 million years.

mathematical model. A simulation of nature made from mathematical symbols.

matrix mechanics. An understanding of quantum mechanics formulated by the German physicist Werner Heisenberg in 1925 and later shown to be equivalent to Schrödinger's equation.

messenger RNA (mRNA). A type of RNA charged with the job of moving genetic data from the DNA in the nucleus of a cell to the ribosomes. The ribosomes are located within the cytoplasm of the cell. The objective of this activity is to produce protein. *See also* ribosomal RNA.

metabolism. The complete system of chemical reactions taking place within a living system.

metagalaxy. A term used by the Swedish theoretical physicist Hannes Alfven in his theory of plasma cosmology to refer to our local region of space.

meteorite. An extraterrestrial object made of rock or metal that has passed through the atmosphere and hit Earth.

methodological reductionism. A time-honored approach to scientific investigation that insists on explaining any phenomenon by examining its parts. Many highly respected scientists, including British astronomer Fred Hoyle and British biologist Francis Crick, champion the reductionist approach to science.

Miller-Urey experiment. A project conducted at the University of Chicago in 1953 by the chemists Stanley Miller and Harold Urey, who subjected a mixture of ammonia and methane to a series of electrical discharges designed to duplicate the intense energy discharges of the early Earth and to stimulate the photochemical processes that may have been a part of the environment at that time. In this pool of chemical soup after several days, Miller and Urey observed the appearance of a number of organic compounds. Included in these compounds were at least twenty-five amino acids and urea.

mini–black hole. *See* primordial black hole.

Minkowski diagram. A two-dimensional graphic representation of the geometry of four-dimensional space-time. A typical Minkowski diagram is laid out like a simple graph. The vertical axis is used to represent time, while the horizontal axis represents space. Thus, vertical movement up the graph without any corresponding horizontal movement represents a traveler remaining still in space but moving in time. In contrast, on a horizontal line, movement in space is generally represented from left to right without a corresponding movement in time. The movement of light is generally represented by lines at 45° angles from the horizontal. The light leaving an event on the lifeline of a traveler angles upward to the left and to the right of a central point, which represents the "here and now" for that particular traveler. This creates the future light cone of that traveler, whereas the light that converges on that point of "here and now" from below represents the past light cone. Everything outside the two cones exists in a space-time region known as elsewhen or elsewhere. Elsewhen represents those areas of the spacetime cube that are inaccessible. However, elsewhen is inaccessible only as long as we are limited by the speed of light. If the barrier represented by the speed of light could be broken, all points within elsewhen might be accessible. Traveling at the speed of light would allow us to reach the outer limit of the future light cone.

missing link. *See* Australopithecines.

modality. One of four categories within Immanuel Kant's philosophy of mind; the other three are possibility, actuality, and necessity.

model. A representation that helps promote the understanding of a concept because it simplifies the complex aspects of that concept and in this way makes it easier for us to comprehend its intricacies.

momentum. A measurement of the motion of an object determined by the product of the object's mass and its velocity.

more developed world. Those economically and industrially advanced regions of the world designated by the United Nations as including northern America, Europe, Oceania, and the former Soviet Union. *See also* less developed world.

multipolar world. A term that describes a global situation in which the international political arena is dominated by several major powers. A multipolar world is inherently unstable because in such a world the activities within the global arena are beyond the control of any single nation.

muons. Subatomic particles of the lepton family that are created when cosmic rays hit molecules located high in the Earth's atmosphere. Muons are especially useful for time-dilation experiments because they undergo radioactive decay at an exceedingly rapid pace.

mutation. An alteration in the quantity or the pattern of DNA in the chromosomes of an organism. This alteration is generally caused by an outside environmental factor, including but not limited to such things as radiation, chemicals, or heat.

naked singularity. A singularity that results in a super–extreme Kerr object when the event horizon has disappeared, leaving the singularity exposed to the rest of the universe. Such a situation is untenable to most physicists, who believe that naked singularities have no real physical existence. *See also* rotating black hole, inner event horizon, outer event horizon, extreme Kerr object.

nanocomputer. A nanodevice that will program and direct assemblers to manipulate atoms and molecules to manufacture products according to a complex set of instructions stored in memory.

nanotechnology. A process that will permit the construction of virtually any product by reconfiguring molecular structures, atom by atom.

natural law. *See* law of nature.

natural selection or **differential reproduction.** The mechanism behind evolution suggested by the nineteenth-century British naturalist Charles Darwin, according to which, in the struggle for existence those individuals with advantages in the local environment will survive long enough to pass their characteristics on to their offspring. In contrast, those which do not have those advantages will be unable to compete and, their descendants will eventually disappear. This process has also been labeled transmutation and "survival of the fittest." *See also* artificial selection.

Neanderthals. A subspecies group of people who lived between 35,000 and 150,000 years ago. They had a remarkable culture that involved toolmaking, cloth making, herbal medicine, and elaborate burial ceremonies that indicated some concern for the spiritual world. There is also some evidence that the Neanderthals also had a brain that was some 10 percent larger than that found in modern humans. The Neanderthals have been plagued with an undesirable image as stooped-over hulking brutes, with massive skulls, beetle brows, and malevolent dispositions. This inaccurate image almost certainly resulted from the fact that some of the first Neanderthal remains that had been unearthed belonged to individuals who suffered from severe arthritis. The resulting reconstructions of these arthritic individuals contributed to this incorrect image. Although recent comparisons of human and Neanderthal DNA have provided evidence that the Neanderthals were actually part of a different species, the question has yet to be answered conclusively.

nebula. A large interstellar cloud of dust and gas.

Nemesis. The name given to the Sun's hypothetical companion star. Nemesis is thought to orbit the Sun every 26 million years. Moreover, as the companion star passes through the Oort comet cloud, the gravitational shock caused by its passage disturbs thousands, probably millions, of comets, kicking them into the inner solar system. The passage of Nemesis through the comet cloud might last as long as one million years. During this time, the inner solar system would be subjected to a virtual storm of comets, drastically increasing the chance that the planets in the solar system, including Earth, might collide with one or more of these comets. The devastation caused by such an impact would account for the regular occurrence of mass extinctions on Earth every 26 million years.

neocortex. *See* cerebral cortex.

neurons. Cells that function as the basic building blocks of the nervous system. Each neuron consists of a main cell body from which extend two types of filaments. One filament, the axon, can be considered the sending wire of the neuron. The other filament, the dendrite, is roughly analogous to a receiving antenna. Each axon starts as one long filament flowing from the central cell body of the neuron. Although the axon begins as a single filament, it ends by spreading out like the branches of a tree. At the end of each branch is a synaptic terminal, which houses tiny bubbles filled with neurotransmitter molecules. When stimulated by an electric current flowing down the axon, the bubbles, or synaptic vesicles, as they are called, travel to the edge of the synaptic terminal, where they attach themselves and release the neurotransmitter molecules into the synaptic cleft, a space between the axon and the dendrite of another neuron. The axon and the dendrite do not actually touch each other. Instead, the connections are made by the release of the neurotransmitter molecules into the synaptic cleft.

neurotransmitter molecules. Molecular structures contained within the synaptic vesicles that are released into the synaptic cleft in response to impulses from the axon of a neuron.

neutrinos. Nearly massless subatomic matter particles of the lepton family particles that have no electric charge but which swarm about the universe, rarely interacting with any other particles. A neutrino can fly though a million miles of solid iron and never touch another particle.

neutron. Neutrally charged subatomic particles that make up part of the nucleus of an atom. The neutron was discovered by the British physicist James Chadwick in 1932.

neutron star. A collapsed stellar body made primarily of neutrons. Neutron stars are similar to but not identical with black holes. Both of these astronomical bodies are actually stars that have used up their nuclear fuel and collapsed in upon themselves. The difference between these two types of stars is that the neutron star is much less massive.

nitrogen fixation. A process by which nitrogen in the atmosphere is changed onto a more suitable form so that cells in plants can use the nitrogen to create amino acids and other beneficial molecules. Some plants, such as peas and beans, cooperate easily with nitrogen-fixing bacteria. This means that planting a crop of beans or peas helps maintain the soil's fertility level, lessening the need to rely on chemical fertilizers.

node. The immovable points at each end of and within the length of a vibrating string.

nonlocality. A reference to the fact that the measurement of a duon will affect the state of a second duon despite the fact that the measurement of the first is a nonlocal, that is a distant event in relation to the second.

normality. A condition characterized by the absence of disease-causing genes.

normal science. The type of scientific activity that is carried on during a period dominated by a paradigm that is generally accepted as correct by a majority of the scientific community. The term was introduced by the American philosopher Thomas Kuhn of the Massachusetts Institute of Technology in his seminal work *The Structure of Scientific Revolutions*. *See also* paradigm and revolutionary science.

Novikov consistency conjecture. Named after its originator Igor Novikov of the University of Copenhagen, the conjecture states, in its simplest form, that the past cannot be altered no matter what a time traveler knows or how hard he or she tries to make such things happen. It is even possible, perhaps likely, that the time traveler's trip into the past is a catalyst that causes the very event that the time traveler seeks to alter.

nucleotide. A base plus its supporting sugar-phosphate group. *See also* base, deoxyribonucleic acid.

nucleus. In biology, the central body of a cell, housing the cell's control system. In physics, the nucleus is the central core of an atom, consisting of positive protons and neutral neutrons.

objective standards. Impartial operating norms endorsed by the scientific method in order to ensure that observations and experiments conducted by scientists produce unbiased results.

omniscience. The divine ability to know everything.

one solar mass. A measurement equal to the mass of the Sun.

ontological reductionism. *See* reductionism.

Oort comet cloud. An enormous storehouse of comets located about ½ to 1 light year from the Sun. The cloud contains at least 100 billion comets. *See also* Nemesis.

orbit. The course which an object in space follows around a second object.

orbital. The area within an atom where an electron may be located.

order. A legal principle that exists when there is harmony and stability within a nation or state.

orphans. Within the context of the post–Cold War era, those nations which during the Cold War lacked the type of firm military commitment given to the allies, but which, nevertheless, looked to the United States for protection. Such nations have the added disadvantage of being threatened by neighboring states that do not recognize their legitimate status as independent sovereign nations. The two primary candidates for orphan status are Israel and Pakistan. These nations are likely to feel more threatened by the end of the Cold War precisely because they never had the luxury of a defense treaty with the United States nor the comforting presence of American troops on their home soil. *See also* allies, rogues.

outer event horizon. The more exterior of two event horizons produced by the rotation of a Kerr black hole. Within the outer event horizon of a Kerr black hole there exists a space-time inversion zone where the roles of space and time are reversed. *See also* inner event horizon.

ozone layer. A region in the upper atmosphere, approximately 60 to 120 kilometers above the surface of the Earth, which contains ozone (O_3). The layer blocks harmful ultraviolet radiation.

panspermia. A theory which proposes that life exists throughout the universe and that the first organic molecules on Earth had an extraterrestrial origin.

paradigm. In science, a scientific world-view that is generally accepted by a scientific community as the unspoken set of assumptions under which that community operates. The term paradigm was introduced by the American philosopher Thomas Kuhn of the Massachusetts Institute of Technology in his seminal work *The Structure of Scientific Revolutions*. According to Kuhn, a successful paradigm demonstrates two characteristics. First, it must be so innovative that it causes a large number of researchers to abandon any previous, rival achievements and adhere firmly to the new one. Second, it must be incomplete enough to give those researchers the chance to work on a variety of problems aimed at developing a more complete understanding of the new achievement. A good example of a successfully introduced paradigm occurred when Copernicus proposed the idea that the Sun, rather than the Earth, might be at the center of the planetary system. *See also* normal science and revolutionary science.

parallel universe theory. *See* many-worlds theory.

particle accelerator. A device designed to accelerate subatomic particles, such as electrons or protons, and smash those particles either into a stationary barrier or into one another. The resulting flash of energy creates new particles, some of which no longer exist in nature but exist only in the high-energy arena of the particle accelerators. The objective of the accelerators is to explore as closely as possible the early moments of the creation of the universe.

particle physics. A branch of physics which seeks to understand the nature of subatomic particles, the most fundamental building blocks of the universe.

past light cone. That portion of a Minkowski space-time diagram that represents the place where light converges at the central point known as "here and now." *See also* future light cone, elsewhen.

patent. A right granted by the federal government giving an inventor the exclusive rights to an invention, which must be novel and nonobvious, that is, it must be unique and must be not so self-evident that a person of ordinary skill could have come up with the same idea. The invention must also be useful.

peer review. The process of letting experts in a field judge a project. The rationale behind the peer-review process for the funding of scientific projects is the conviction that such decisions generally require scientific knowledge that is beyond the background of most members of Congress. Delegating the decision-making process to experts helps prevent Congress from attempting to control the projects, the universities, and the individuals who receive such funding; it places the decisions on science funding in the hands of those who should know best where the money should be spent.

perceptions. According to the belief system of the German philosopher Immanuel Kant, unstructured sensations become perceptions after they have been organized by the mind. *See also* conceptions, sensations.

perfect cosmological principle. A principle which states that from a large-scale point of view, the universe should look the same regardless of the observer's position in space and time. The perfect cosmological principle also supports the assumption that the laws of physics operate in the same way at all positions in space and at all points in time throughout the universe.

periodic table of elements. A standard chart which arranges the elements according to proton number.

persuasive precedent. A legal precedent that can either be followed (or ignored) by a court. *See also* case law, binding precedent, judicial decision, precedent, stare decisis.

photoelectric effect. The effect that occurs when light hits metal and deflects electrons from the surface of that metal.

photons. The force-carrying particles responsible for the transmission of electromagnetic radiation or light.

photosynthesis. The process which allows plants to use sunlight in the creation of carbohydrates from water and carbon dioxide.

placenta. The organ that connects the embryo to the pregnant woman so that the embryo can receive nourishment from and eliminate waste into the pregnant woman's blood stream.

planetary or **solar system model.** A model, first proposed by the British physicist Ernest Rutherford in the nineteenth century, that depicts the atom as a miniature solar system. At the center of the atom is the nucleus, which is many times more massive than the electrons. The electrons, in turn, exist outside the nucleus in a spheroid space surrounding the massive nucleus.

plasma. A gaslike state of matter consisting of charged particles.

plasma cosmology. A theoretical explanation for the existence of an eternal universe that was proposed by the Swedish theoretical physicist Hannes Alfven, according to which colossal blankets of electromagnetic radiation converge and intersect throughout the universe like a vast layer of energy. These blankets of electromagnetic radiation interact with clouds of matter and antimatter, creating enormous bursts of energy that cause the expansion of certain regions of the universe. *See also* ambiplasma, antimatter, koinomatter, megagalaxy.

plasmid. A form of DNA which carries a particular gene.

plum pudding or **raisin bread model.** A model that depicts the atom as a positive cloud in which negative electrons are located like the plums in a plum pudding. This model was proposed by the British physicist J. J. Thomson in the nineteenth century after his discovery of the electron.

pongids. The subdivision of the hominoids that includes the apes.

population. In ecology, a group of interbreeding individuals of the same species.

positivism. A sociological belief system which declares that science can be distinguished from law, literature, and most other areas of knowledge because scientific truth is derived from an objective evaluation of carefully controlled observation and experimentation. According to positivism, objective truth exists and is discoverable by properly executed scientific techniques. Scientific theories are considered accurate reflections of the real world precisely because they have been tested according to objective techniques that can be duplicated and verified. *See also* constructivism.

precedent. In law, a previous case that is used as authority in a present one. *See also* case law, binding precedent, judicial decisions, persuasive precedent, stare decisis.

precellular organisms. *See* protocells.

predestination. A religious doctrine promoted by John Calvin and his successor, Theodore Beza, in the sixteenth century. According to Calvin and Beza, God, in His justice, has already determined who will be saved and who will be damned, and it does little good for anyone to try to substitute his or her will for that of God. In its most basic form, the doctrine of predestination eliminates the idea that individuals have the ability to exercise free will. Predestination, then, is the theological equivalent of determinism.

predictability. A principle that requires that a theory make verifiable forecasts about phenomena associated with that theory.

preembryo. A fertilized ovum or egg cell during the time that the placenta is formed up until about day fifteen. *See also* embryo, fetus.

prelife or **primeval soup theory.** A theory that explains the move from prelife to life as a series of progressively more complex leaps in organization. During the first stage in the development of life from prelife material, beginning between 3500 and 4200 million years ago, the primeval soup came under the energetic influence of ultraviolet radiation, volcanic heat, and/or lightning. The interaction of these forces created more and more complicated molecules, leading inevitably to such molecules as RNA, DNA, amino acids, and proteins. These complex molecules then developed the capacity to replicate themselves and to pass their characteristics to their offsprings. This pattern of replication provided the underlying mechanism for heredity, mutation, and natural selection, all of which furnish the driving force of the evolution from prelife to life.

premise. One of the basic assumptions from which a logical argument can be built.

pressure. A force that acts uniformly on the surface of an object. Also, a force that opposes an opposite force.

preventative war. A war that results when one nation attacks another to prevent the second nation from taking an action considered detrimental to the best interests of the first.

primeval atom. Conceived of by the Belgian scientist Georges Lemaitre in the early twentieth century as the condensed initial condition of the expanding universe.

primordial black hole or **mini-black hole.** A black hole that may have been created in the high-density, high-temperature environment of the first moments of the expanding universe.

principle of the conservation of energy. The principle which holds that energy can be neither created nor destroyed. Changes that occur are only changes of form.

principle of equivalence. A principle which states that acceleration and gravity are equivalent.

principle of the fixity of the past. A philosophical position which holds that the past exists as a tapestry of events that are fixed and immutable.

principle of the fixity of time. A philosophical position which holds that all of time—past, present, and future—exists simultaneously and therefore, time is fixed and immutable.

probability wave. An idea proposed by the German physicist Max Born that suggests that the solutions to Schrödinger's equation did not demonstrate the location of the wave. Instead, these solutions, now referred to as wave functions, predicted only the probability of pinpointing an electron (or any subatomic entity for that matter) at a specific time and in a particular place.

Proconsul. A possible ancestor of both modern apes and humans. Proconsul, whose fossilized remains were found in Africa, was relatively small by today's standards, perhaps as tall as a modern baboon; but it had a skull and teeth that resembled those found in apes. Proconsul also displayed an interesting combination of ancient and advanced traits, including a hand that was proportioned like a human hand and that displayed an opposable thumb. In contrast, the ankles and feet remained apelike, and Proconsul moved in a four-legged style.

procreative freedom. Various rights of all people, including the right to have children, to have an abortion, to use contraception, to adopt a child, to give up a child for adoption, to seek treatment for infertility, and to seek assisted reproduction.

prokaryotes. Protocells which appeared between 2500 and 3000 million years ago and which developed a membrane that allowed them to thrive in the prevalent environment. They lacked a self-contained nucleus, however. *See also* eukaryotes and protocells.

proofreading. In genetics, a process that occurs during DNA replication to prevent mistakes that may have been made in the matching of pairs of bases. *See also* base sequence.

propagating waves. Waves that travel through a medium such as water or air. The particles constituting the water or air move up and down, or right and left, crossing a middle line. The highest point of this line is called the crest, the lowest point, the trough. The distance between the midline and a crest is called the amplitude, and the number of crests to move by one point within a time unit is designated as the wave's frequency.

protein. Complicated molecules consisting of smaller groups of molecules called amino acids. Proteins, which are necessary for life, perform a variety of functions for the body. They act as the catalysts responsible for chemical reactions in the cells as messengers and as building blocks for cells and tissues within the body.

protocells. Precellular organisms that appeared about 3.5 billion years ago. They provided a bubblelike environment for the protection and replication of the complex molecular structures that were present in the primeval environment; they also developed a transport system allowing the molecular structures within the "bubble" to take advantage of the protective nature of the membrane while interacting with and feeding upon the environment. *See also* eukaryotes and prokaryotes.

protons. Positively charged subatomic particles that make up part of the nucleus of an atom. *See also* neutrons.

psychological relativity. A phenomenon that describes how the same experience can be seen from different perspectives.

psychological or **subjective time.** A perception of time relative to an individual's experience. If the individual is enjoying a series of events, then time passes very quickly. In contrast, if the experience is boring, painful, or unpleasant, time will drag interminably for that individual.

quality. One of four categories within Immanuel Kant's philosophy of mind, the other three being affirmation, negation, and limitation.

quantity. One of four categories within Immanuel Kant's philosophy of mind, the other three being unity, plurality, and totality.

quantum. The smallest amount by which the energy of a system can be altered. The existence of such quantum events was first postulated by the German physicist Max Planck, who solved the problem of black-body radiation by suggesting that energy was released not as a continuous wave but in discrete packets of energy that he labeled quanta.

quantum physics. The area of physics that attempts to explain the laws of nature as they affect matter and energy at the atomic and subatomic levels. *See also* quantum.

quantum weirdness. A reference to those quantum phenomena that appear to defy common experience when explained in terms of everyday life.

quarks. The basic building blocks of protons and neutrons. Protons consist of two up and one down quark, whereas neutrons consist of two down and one up quark. Quarks are thought to be indivisible. However, some physicists believe that quarks may someday be found to consist of even smaller subatomic entities referred to as preons. Quarks were named by Murray Gell Mann of the California Institute of Technology, who took the term from "Three quarks for Muster Mark," a line in James Joyce's *Finnegans Wake*.

quasars. Highly luminous, deep-space objects that probably lie at the centers of distant galaxies.

quasi–steady state cosmology. A theory which proposes that a cyclical series of mini–big bangs, rather than a single big bang event, is responsible for the continual creation of matter within the universe. *See also* creation field, steady-state cosmology.

radiation. Energy transmitted as photons or electromagnetic waves.

radioactive decay. The process by which certain atomic nuclei disintegrate because of the unstable nature of their atomic structure.

raisin bread model. *See* plum pudding model.

rational basis. A legal principle used by the courts to judge the constitutionality of statutes that involve the equal protection clause of the Constitution. Under this standard the government has to show that the statute in question bears a rational basis to a legitimate governmental interest for the statute to be constitutional.

reciprocity. *See* golden rule.

recombinant DNA. A stretch of DNA that has been altered by genetic-engineering procedures.

red shift. A movement of the spectral lines of light coming from distant galaxies in the direction of the red end of the spectrum of visible light.

reductionism or **ontological reductionism.** An approach to a complex situation by electing to divide it into smaller and smaller parts, the object being to understand the whole by seeing how the parts operate by themselves.

regulations. Rules created by the administrative and regulatory agencies of the government.

relation. One of four categories within Immanuel Kant's philosophy of mind, the other three being the ideas of substance-accidents, cause-effect, and causal reciprocity.

relativity, theory of. A theory proposed by Albert Einstein to explain certain discrepancies between the motion of material objects as described by classical Newtonian physics and the motion of electromagnetic waves.

relativity of simultaneity. A term used to indicate the relative nature of a moment for a moving object compared with that of a stationary observer.

replication. A procedure by which copies of DNA molecules are produced by a parent DNA molecule.

replicator. A nanodevice that has the ability to produce duplicates of itself.

reproducing-universe theory. The theory which holds that the present universe began with a quantum fluctuation in a previously existing universe. Additionally, the present universe may give birth to its own offspring universes through similar quantum fluctuations.

retrovirus. A virus that contains RNA and which, by using an enzyme known as reverse transcriptase, can change RNA into DNA, thus invading the DNA of the host organism and producing copies of the retrovirus.

revolutionary science. The type of scientific activity that is carried on during a period which is dominated by a crisis that threatens the existing paradigm. The term was introduced by the American philosopher Thomas Kuhn of Massachusetts Institute of Technology in his seminal work *The Structure of Scientific Revolutions*. According to Kuhn, revolutionary science occurs during a creative period in which the conventional procedures and the customary rules of normal science are relaxed. New and different approaches to the crisis are encouraged. Radical solutions are entertained as the most creative minds of science tackle the difficulties in an attempt either to alter the old paradigm radically or to introduce a new one. *See also* normal science, paradigm.

restriction enzyme. A biochemical substance with the capacity to make a very precise incision in a strand of DNA.

ribonucleic acid (RNA). A complex molecule involved in translating the genetic information carried in DNA into proteins.

ribosomal RNA (rRNA). One of the key ingredients of the ribosomes.

ribosome. An enormous molecule consisting of a large number of proteins and a type of RNA called ribosomal RNA (rRNA). Protein synthesis takes place within the ribosomes.

ring envelope. A region in space that surrounds a ring singularity. *See also* rotating black hole, ring singularity.

ring singularity. A singularity at the heart of a rotating black hole. The rotation of the black hole alters the singularity so that it becomes shaped like a ring. *See also* rotating black hole, ring envelope.

RNA. *See* ribonucleic acid.

RNA polymerase molecule. A molecule that splits the appropriate segment of the double helix during transcription, revealing the sequence of bases (the As, Ts, Cs, and Gs). The RNA-polymerase molecule uses free-floating RNA nucleotides to match the nucleotides on the exposed DNA strand. The only exception here is that instead of using thymine (T), the base uracil (U) is used. When the process is complete, the result is a working copy of that DNA strand, that is, of the gene. *See also* deoxyribonucleic acid (DNA).

rogues. Within the context of the post–Cold War era, an aggressive group of nations with expansive goals that run counter to the interests of most other nations and that threaten the stability of the global community. The two primary rogue nations presently are Iraq and North Korea; in the past, Iran, Syria, Libya, and Algeria have also acted as rogues.

rotating, spinning, or **Kerr black hole.** A black hole that is spinning on its axis. The singularity at the heart of a rotating black hole is shaped like a ring.

RU486. A contragestive agent.

Schrödinger's cat. A thought experiment, proposed by the German physicist Erwin Schrödinger, to demonstrate the limitations of Niels Bohr's position with regard to quantum theory.

Schrödinger's equation. A mathematical formulation, devised by the German physicist Erwin Schrödinger, describing the wave function of a particle. One crucial feature of Schrödinger's equation is that it clearly supports the wave interpretation of matter. It can be seen as the mathematical support for de Broglie's wave model of the electron within the structure of the atom.

Schwarzschild black hole. *See* stationary black hole.

Schwarzschild radius, gravitational radius. The distance between the center of a body and a point at which the escape velocity reaches light speed.

science. A body of knowledge aimed at understanding how nature works, through research based on an objective method which includes observation, experimentation, and verification.

science court. A proposed legal body that would be charged with, among other things, the responsibility of reviewing the scientific merits of grant proposals. The science court was first advocated in 1976 in a report entitled *The Science Court Experiment: An Interim Report*, issued by the Task Force of the Presidential Advisory Group on Anticipated Advances in Science and Technology. The report suggests the establishment of a science court that would employ the traditional adversarial procedure to settle disagreements over issues of science. The science court would provide a forum for advocates on all sides of a scientific controversy, who would present their arguments before a panel of neutral judges with scientific backgrounds.

scientific method. An objective process in which scientists use observation and experimentation to verify theoretical explanations about the operation of nature.

Second Law of Thermodynamics. The principle which demonstrates that over time the entropy of a closed system will increase.

sensations. According to the philosophy of the German thinker Immanuel Kant, unorganized sense perceptions as they enter the mind. *See also* conceptions and perception.

separation principle. A legal device designed to prevent governmental power from falling into the hands of a single person or a small group of people. The legislative branch makes the law; the executive branch enforces it; the judicial branch interprets it. *See also* balancing principle.

simultaneous-temporal-existence paradox. The paradox in which a time traveler journeys into the past (or future) and meets a younger (or older) version of himself in that era.

singularity. According to the big bang theory, an infinitely small, infinitely dense, symmetrical, subatomic "entity," thought to be the initial state of the expanding universe. The singularity is not really a "thing" or an "entity." Instead, it should be viewed as that point of existence at which the entire universe was crystallized into a uniform, symmetrical dividing line between existence and nonexistence. It is at this boundary that cosmology and physics merge.

solar system model. *See* planetary system model.

solipsism. A philosophical system based on the belief that only the self exists. According to solipsism, all other existents exist only in the mind of the self.

space-time continuum. A geometry, implied by the theory of relativity, which consists of four dimensions, three of space and one of time.

space-time cube or **block universe.** A space-time concept based on the theory of relativity, which says that space and time are connected as a single, unified reality. The space-time cube can be thought of as one enormous four-dimensional object. Three of the four dimensions are reserved for space, while one is for time.

spacialization. A concept first promoted by the Russian-born, German mathematician, Hermann Minkowski, who, in 1908, proposed the idea that time should be seen in much the same way that space is visualized. Time, Minkowski said, must be "spacialized."

speciation. The formation of a new species.

species. A classification of living things which includes a group of interbreeding individuals.

spectrum. The arrangement of electromagnetic radiation according to wavelength or frequency.

speed. The rate at which an object moves without regard to direction.

spinning black hole. *See* rotating black hole.

standing waves. Stationary waves. The vibrating strings of a violin or a guitar are examples of standing waves. The opposite tips of the string are fastened, while the central part of the string vibrates up and down.

stare decisis. The process of relying on past legal authority as a guideline in present cases. Literally translated, stare decisis means "let the decision stand." *See also* case law, binding precedent, judicial decisions, persuasive precedent, precedent.

static or **stationary limit.** An outer boundary created by a rotating black hole. The static limit has its greatest separation at the equator. However, it will touch the event horizon at each pole. The term is used to describe this boundary because once anything, even light, passes that boundary, it will not be able to stop. Between the static limit and the event horizon of a spinning black hole there is an area called the ergosphere.

stationary or **Schwarzschild black hole.** A black hole that is spherical and that is not rotating.

stationary limit. *See* static limit.

statutory law. The law made by a legislature, such as Congress or a state general assembly. Generally, statutory law will command, prohibit, regulate, or declare something.

steady-state cosmology. A theory that promotes the idea that the universe is eternal, having no beginning and no end. To account for the fact that the universe in a steady state would look the same at all times to all observers, the steady-state theory holds that matter is continuously created spontaneously throughout all regions of space. *See* quasi–steady state cosmology.

streaming motion. Motion that occurs when galaxies exhibit a movement which appears to run counter to the gradual expansive movement of the universe.

strong nuclear force. Quantum "glue" that holds the quarks together within the protons and neutrons. The strong nuclear force is one of the four fundamental forces of nature.

strong tendency theory. The theory of evolution which states that the tendency toward intelligence was present in all forms of primitive life. Accordingly, the evolutionary rise of intelligent life was inevitable once a certain level of development was reached.

subjective ethics. *See* ethical relativism.

subjective time. *See* psychological time.

subquantal reality. A hypothetical dimension that exists beneath the quantum level and is responsible for all quantum phenomena.

substance. According to Aristotle and Thomas Aquinas, the metaphysical essence of an object.

supercluster. A collection of galactic clusters.

superconductivity. A condition of some materials that are characterized by a lack of measurable electrical resistance at 0 degrees Kelvin.

super-extreme Kerr object (SEKO). An object that would result if the rotation of an extreme Kerr object increases so that its angular momentum exceeds its mass. Under these conditions, the merged event horizons would disappear completely. Once they have vanished, the internal ring singularity would no longer be hidden. Some physicists believe that SEKOs are simply mathematical constructs and, therefore, may not exist within the physical universe. *See also* rotating black hole, inner event horizon, outer event horizon.

super-quantum theory. A theory that would represent a step beyond the present theoretical understanding of quantum phenomena. Such an advanced theory, once understood, would reveal any hidden variables that would be needed to forecast the results of any quantum measurement. Once these local hidden variables were in hand, physicists would be able to figure out by direct measurement the simultaneous position and momentum of any duon.

superstring theory. A theory which replaces the idea that subatomic particles are one-dimensional points with the notion that they are multidimensional strings. The vibration of these multidimensional strings can account for the existence of matter and energy in the universe.

synaptic cleft. The space between an axon of one neuron and the dendrite of another.

synaptic terminal. A terminal at the end of each branch axon, which houses tiny vesicles, or bubbles, filled with neurotransmitter molecules.

synaptic vesicles. Tiny bubbles that contain neurotransmitter molecules. They are located within the synaptic terminal until stimulated by an impulse that causes the vesicles to travel to the edge of the terminal and release the neurotransmitter molecules into the synaptic cleft.

synthetic protolife. An experimental attempt to create animate matter in a laboratory by permitting a set of complex chemical reactions to reinforce one another until a self-sustaining chemical network arises from those reactions.

tendency theory. The theory of evolution which states that the tendency toward intelligence was present in all forms of primitive humanity. Accordingly, the evolutionary rise of Homo sapiens or some similar cousin with the same intellectual predisposition was inevitable once a certain level of development was reached. *See also* entity theory.

terraforming. The process of transforming a planet with an environment that is hostile to life into one with an Earthlike and, therefore, life-supporting environment.

theorem. A generalized principle upon which a mathematical model is based.

theory. A coherent group of natural laws that work together to predict intricate patterns within a large portion of the natural world. A theory will also make predictions that can be tested by observation and experiment.

Theory of Everything. An explanation of how the four fundamental forces of nature—the electromagnetic force, the strong nuclear force, the weak nuclear force, and gravity, are all manifestations of one single force.

thought experiment. *See* gedanken experiment.

time dilation or **time-dilation effect.** A term used to identify the fact that the faster an object travels, the slower time will move for that object relative to a stationary observer.

time loop. A cycle in time which repeats itself endlessly.

transcription. In genetics, the process by which a working copy of a gene is made.

transfer RNA (tRNA). A type of RNA charged with the responsibility of moving amino acids to the ribosomes for protein synthesis.

translation. In biochemistry, the joining of amino acids to make proteins.

transubstantiation. The doctrine which asserts that, during the performance of the sacrament of the Eucharist, the priest actually changes the bread and wine into the body and the blood of Christ. Transubstantiation is based upon the belief that the substance, that is, the fundamental nature or metaphysical essence, of the bread and wine are altered, although the outer properties such as its color, texture, aroma and so on, are not.

triplet. *See* codon.

ultraviolet (UV) radiation. Energy transmitted as electromagnetic waves at wavelengths between violet light and X-rays.

uncertainty principle. The principle which states that an experimenter can never simultaneously pinpoint both the momentum and the position of a subatomic particle.

unidirectional flow. The notion that time flows in one direction only, from the present to the future.

unipolar world. A global situation in which the international political arena is dominated by a single major power. After the Cold War, for instance, the international political arena was dominated by the United States.

universal assembler. Programmed nanodevices which, presumably, will be capable of using atoms and simple molecules to construct almost any complex molecular arrangement. They would therefore be able to build almost anything imaginable. *See* nanotechnology.

utilitarianism. An ethical theory which states that right and wrong can be determined by measuring an action in terms of the good or evil that it produces. If the action produces more good than evil in relation to the people affected, then the action is right. On the other hand, if the action produces more evil than good, it is wrong.

velocity. Speed in a specific direction.

verifiability. A principle in science that requires theories to be verified by experimental methods.

viability. That point at which a fetus can survive outside the womb, albeit with artificial assistance.

Virgo cluster. A group of galaxies, located north of the galactic plane, that has been used to measure the age of the universe.

vitalism. A theory which states that a life force or a life-giving energy is responsible for animating matter, thus transforming it from nonliving to living matter. Vitalism is closely associated with the theory of dualism, which asserts that human beings consist of two parts, the physical body and the metaphysical soul.

volition or **free will.** The innate ability of human beings to make decisions for themselves without being controlled by outside influences.

W^+, W^-, and Z^0 bosons. The force-carrying particles responsible for the weak nuclear force.

war of redistribution. A war in which an attacking nation expressly intends to reallocate the wealth held by the nation under attack for its own benefit.

wave function. A mathematical formulation, using Schrödinger's equation, that predicts the probability of pinpointing an electron, or any subatomic entity for that matter, at a specific time and in a particular place.

wave-particle duality. The idea in quantum theory that subatomic phenomena appear with either wavelike or particlelike properties.

weak nuclear force. The force that transforms protons into neutrons during the process of fusion within stars. This transformation is called beta decay. The weak nuclear force is one of the four fundamental forces of nature.

white dwarf. The collapsed state of a star which had a mass roughly equivalent to the mass of the Sun.

white hole. A theoretical region in space that may exist at the other end of a black hole. Since a black hole pulls in matter and radiation because of its enormous gravitational field, a white hole would serve as that region at the opposite end of the black hole where the matter and radiation are ejected.

worldline. *See* lifeline.

wormhole. A tunnel in space that could theoretically connect distant regions of the universe, even regions that are billions of miles apart.

Zeno's paradoxes. A series of arguments set forth by the ancient Greek philosopher Zeno, the object of which was to defend the position proposed by Parmenides, another Greek philosopher, that the universe is unchanging. With these paradoxes, Zeno hoped to demonstrate that the common-sense view of motion and space could be shown to be just as absurd as the position proposed by Parmenides. One of the most famous of Zeno's paradoxes suggests that it is not possible for a runner to cover the distance between two points, A and B, because before the runner can cross the entire distance to B, he must cross half that distance. However, to reach that half, he must cross half the distance to the midpoint, and so on. Since the runner must cross an infinite number of midpoints, he can never reach the second point. Zeno's goal was not to attack the world as it exists, but to show that any counterargument to Parmenides could also be reduced to absurdity.

zygote. The single cell that exists after an egg has been fertilized.

Bibliography

Adeboye, Titus. "Innovation without Science Policy." In *Science Based Economic Development: Case Studies around the World*, edited by Susan U. Raymond. New York: New York Academy of Sciences, 1996.

Aiken, Henry. "The Fate of Philosophy in the Twentieth Century." *Kenyon Review* 24 (Spring 1962): 233–52.

Alfven, Hannes. *Cosmic Plasma*. Boston: D. Reidel, 1981.

———. "Cosmology and Recent Developments in Plasma Physics." *Australian Physicist* 17 (1980): 162.

———. "Plasma Physics Applied to Cosmology." *Physics Today*, February 1971, 28.

———. *Structure and Evolutionary History of the Solar System*. Boston: D. Reidel, 1975.

Alfven, Hannes, and Per Carlqvist. "Interstellar Clouds and the Formation of Stars." *Astrophysics and Space Science* 55 (1978): 487–509.

Allen, E. L. *From Plato to Nietzsche: An Introduction to the Great Thoughts and Ideas of the Western Mind*. New York: Fawcett Premier, 1993.

Allen, Thomas B. *War Games: The Secret World of the Creators, Players, and Policy Makers Rehearsing World War III Today*. New York: Berkley, 1989.

American Medical Association. "Board of Trustees Report: Requirements or Incentives by Government for Use of Long-Acting Contraceptives." In *Taking Sides: Clashing Views on Controversial Bioethical Issues*, edited by Carol Levine. Guilford, CT: Dushkin, 1993. First published in *Journal of the American Medical Association*, 1 April 1992.

Anderson, W. French. "Human Gene Therapy," *Science*, 8 May 1992, 808–13.
Antia, Meher. "Lost Horizon: A Trapdoor to Other Unverses Slams Shut," *The Sciences*, November/December 1998, 6.
Appleyard, Bryan. *Understanding the Present: Science and the Soul of Modern Man*. New York: Doubleday, 1992.
Apter v. Richardson, 510 F. 2d 351 (7th Cir. 1975).
Aquinas, Thomas. *Summa Theologiae: A Concise Translation*. Edited by Timothy McDermott. London: Eyre and Spottiswoode, 1989.
Areen, Judith, ed., et al. *Law, Science and Medicine*. Mineola, NY: Foundation, 1984.
Aristotle. *Metaphysics*. Translated by Richard Hope. Ann Arbor: Univ. of Michigan Press, 1966.
———. *Physics*. Translated by Richard Hope. Lincoln: Univ. of Nebraska Press, 1961.
Armstrong, Karen. *A History of God*. New York: Alfred A. Knopf, 1993.
Asimov, Isaac. *Atom: Journey across the Subatomic Cosmos*. New York: Truman Talley, 1992.
———. *The Collapsing Universe: The Story of Black Holes*. New York: Simon and Schuster, 1977.
Aspect, Alaine. "Alaine Aspect." Interview in *The Ghost in the Atom*, edited by P. C. W. Davies and J. R. Brown. Cambridge: Cambridge Univ. Press, 1991.
Bahm, Archie J. *The World's Living Religions*. Berkeley: Asian Humanities, 1992.
Baldwin, Simeon E. "Liability for Accidents in Aerial Navigation." *Michigan Law Review* 9 (1910): 20–28.
Barfield, Claude E. "Introduction and Overview." In *Science for the 21st Century: The Bush Report Revisited*, edited by Claude E. Barfield. Washington D.C.: AEI, 1997.
Barnes, Jonathan. *The Presocratic Philosophers*. 2 vols. London: Routledge and Kegan Paul, 1979.
Barrow, John D. *The World within the World*. Oxford: Oxford Univ. Press, 1988.
Barrow, John, and Joseph Silk. *The Left Hand of Creation: The Origin and Evolution of the Expanding Universe*. Oxford: Oxford Univ. Press, 1983.
Bartusiak, Marcia. *Through a Universe Darkly: A Cosmic Tale of Ancient Ethers, Dark Matter, and the Fate of the Universe*. New York: HarperCollins, 1993.
Basinger, David. "Simple Foreknowledge and Providential Control: A Response to Hunt." *Faith and Philosophy*, July 1993, 421–27.
Bazelon, David. "Coping with Technology through the Legal Process." In *Law, Science and Medicine*, edited by Judith Areen, et al. Mineola, NY: Foundation, 1984. First published in *Cornell Law Review* 62 (1977): 817.
Belkin, Lisa. "Splice Einstein and Sammy Glick." *New York Times Magazine*, 23 August 1998, 28–31, 56–61.
Bell, J. S. "On the Einstein-Podolsky-Rosen Paradox." *Speakable and Unspeakable in Quantum Mechanics*. Cambridge: Cambridge Univ. Press, 1987.
Berg, Paul, David Baltimore, Sydney Brenner, Richard O. Roblin III, and Maxine F. Singer. "Summary Statement of the Asilomar Conference on Recombinant DNA Molecules." *Science*, 6 June 1975.
Bergson, Henri. *Time and Free Will: An Essay on the Immediate Data of Consciousness*. Translated by F. L. Pogson. New York: Harper and Row, 1910.
Berkman, Harvey. "High Court Defers to Judge on Scientific Evidence." *The National Law Journal*, 29 December 1997–5 January 1998, sec. A, p. 10.
Bernstein, Jeremy. *Cranks, Quarks, and the Cosmos*. New York: HarperCollins, 1993.
Berra, Tim M. *Evolution and the Myth of Creationism: A Basic Guide to the Facts in the Evolution Debate*. Stanford: Stanford Univ. Press, 1990.

Birx, H. James. *Interpreting Evolution: Darwin and Teilhard de Chardin*. Buffalo: Prometheus, 1991.
Bishop, J. Michael. "Enemies of Promise." *Wilson Quarterly* 19 (Summer 1995): 61–65.
Black, Henry Campbell. *Black's Law Dictionary*. St. Paul, MN: West, 1968.
Bohm, David. *Unfolding Meaning*. London: Ark, 1987.
Bohm, David, and B. J. Hiley. *The Undivided Universe: An Ontological Interpretation of Quantum Theory*. New York: Routledge, 1993.
Bohr, Niels. *Atomic Physics and Human Knowledge*. New York: John Wiley and Sons, 1958.
———. "Can Quantum-Mechanical Description of Physical Reality Be Considered Complete?" Partially reprinted in *Niels Bohr: A Centenary Volume*, edited by A. P. French and P. J. Kennedy. Cambridge: Cambridge Univ. Press, 1985. First published in *Physical Review* 48 (1935): 696.
Bondi, Hermann. *Cosmology*. Cambridge: Cambridge Univ. Press, 1952.
Born, Max. *Atomic Physics*. Translated by John Dougall. New York: Dover, 1969.
Borosage, Robert. "No Justice, No Peace." *Boston Review*, October/November 1997, 10.
Boslough, John. *Masters of Time: Cosmology at the End of Innocence*. Reading, MA: Addison-Wesley, 1992.
Bova, Ben. *Brothers*. New York: Bantam, 1996.
Brody, Baruch. "On the Humanity of the Fetus." In *Contemporary Issues in Bioethics*, edited by Tom L. Beauchamp and LeRoy Walters. Belmont, CA: Wadsworth, 1978. First published in Robert L. Perkins, ed., *Abortion: Pro and Con* (Cambridge, MA: Schenkman, 1974).
Bromley, D. Allan. *By the Year 2,000: First in the World*. Report of the Federal Coordinating Council for Science, Engineering and Technology, Committee on Education and Human Resources. Washington D.C.: FCCSET, 1991.
Brown, Betty, and John Clow. *Introduction to Business: Our Business and Economic World*. New York: Glencoe/McGraw-Hill, 1992.
Brown, Gordon W., and Paul A. Sukys. *Business Law with UCC Applications*. 9th ed. New York: Glencoe/McGraw-Hill, 1997.
———. *Understanding Business and Personal Law*. 10th ed. New York: Glencoe/McGraw-Hill, 1998.
Brown, Lester R. "The Acceleration of History." In *State of the World 1996*, edited by Lester R. Brown. New York: W. W. Norton, 1996.
Bucher, Martin A. and David N. Spergel. "Inflation in a Low-Density Universe." *Scientific American*, January 1999, 62–69.
Burrows, William E. *This New Ocean: The Story of the First Space Age*. New York: Random House, 1998.
Byrne, John. "The Man Who Made Tomorrow." *OMAC: One Man Army Corps*. New York: Warner, 1991.
Cahn, Steven M., ed. *Classics of Western Philosophy*. Indianapolis: Hackett, 1977.
Cairns-Smith, A. G. *Seven Clues to the Origin of Life*. Cambridge: Cambridge Univ. Press, 1986.
Calder, Nigel. *Einstein's Universe: A Guide to the Theory of Relativity*. London: Penguin, 1979.
Cameron, George D. *The Legal and Regulatory Environment of Business*. Cincinnati: South-Western, 1994.
Cappadora v. Celebreze, 356 F. 2d 1 (2d Cir. 1966).
Capra, Fritjof. *The Tao of Physics: An Exploration of the Parallels between Modern Physics and Eastern Mysticism*. 3rd. ed. Boston: Shambhala, 1991.

Carey, Diane, and James I. Kirkland. *First Frontier*. New York: Pocket Books, 1995.

Carey v. Population Services International, 431 U.S. 678, 97 Sup. Ct. 2010 (1977).

Carr, Orson Scott. *Pastwatch: The Redemption of Christopher Columbus*. New York: Tom Doherty Associates, 1996.

Casper, Barry M. "Technology Policy and Democracy: Is the Proposed Science Court What We Need?" *Science*, 1 October 1976, 29–35.

Casper, Barry, and Paul Wellstone. *Powerline: The First Battle of America's Energy War*. Amherst: Univ. of Massachusetts Press, 1981.

———. "Science Court on Trial in Minnesota." In *Science in Context: Readings in the Sociology of Science*, edited by Barry Barnes and David Edge. Cambridge, MA: MIT Press, 1982.

Cassidy, David. *Einstein and Our World*. Atlantic Highlands, NJ: Humanities, 1995.

Casti, John L. *Searching for Certainty: What Scientists Know about the Future*. New York: William Morrow, 1990.

Cetron, Marvin, and Owen Davies. *Probable Tomorrows: How Science and Technology Will Transform Our Lives in the Next Twenty Years*. New York: St. Martin's, 1997.

Chaisson, Eric. *Relatively Speaking: Relativity, Black Holes, and the Fate of the Universe*. New York: W. W. Norton, 1988.

Chapman, Barry. *Reverse Time Travel: The Exciting Revelation that Traveling Backwards through Time Is Possible*. London: Cassell, 1996.

Charo, R. Alta. "A Political History of RU-486." In *Biomedical Politics*, edited by Kathi E. Hanna. Washington D.C.: National Academy, 1991.

Clark, Bruce, "NATO Survey: Knights in Shining Armour?" *The Economist*, 24–30 April 1999, 1–18.

Clarke, Arthur C. *The Exploration of Space*. New York: Harper, 1951.

———. "The Hammer of God." *Time*, Fall 1992, 83–87.

Clarke, Chris. "The Year 2000 Problem: An Environmental Impact Report." *Earth Island Journal*, Fall 1998, 33–35.

Coit, Margaret L. *Mr. Baruch*. Boston: Houghton Mifflin, 1957.

Cole, Marcia. "25 Years of 'Roe v. Wade.'" *The National Law Journal*, 26 January 1998, sec. A, p. 24.

Cole, Stephen. *Making Science: Between Nature and Society*. Cambridge: Harvard Univ. Press, 1992.

Collinson, Diane. *Fifty Major Philosophers*. London: Routledge, 1987.

Commission on Global Governance. *Our Global Neighborhood: The Report of the Commission on Global Governance*. Oxford: Oxford Univ. Press, 1995.

Commission on Science and Technology for Development. *Report on the Third Session*. New York: United Nations, 1997.

Constable, George, ed., *The Cosmos*. Alexandria, VA: Time-Life, 1988.

Cooper, Necia Grant, ed. *The Human Genome Project: Deciphering the Blueprint of Heredity*. Mill Valley, CA: University Science, 1994.

Coveney, Peter, and Roger Highfield. *The Arrow of Time: A Voyage through Science to Solve Time's Greatest Mystery*. New York: Fawcett Columbine, 1990.

Cowen, Ron. "New Challenge to the Big Bang?" *Science News*, 9 October 1993, 236–37.

Crick, Francis. *The Astonishing Hypothesis: The Scientific Search for the Soul*. New York: Scribner's, 1994.

Croswell, Ken. "A Milestone in Fornax." *Astronomy*, October 1995, 42–47.

———. "Epsilon Eridani: The Once and Future Sun." *Astronomy*, December 1995, 46–49.

Darlton, John. *Neanderthal*. New York: Random House, 1996.

Darwin, Charles. Autobiography in *The Life and Letters of Charles Darwin*. Edited by F. Darwin. London: John Murray, 1887.

———. *The Origin of Species*. Oxford: Oxford Univ. Press, 1996.

Daubert v. Merrell Dow Pharmaceuticals, 113 Sup. Ct. 2786 (1993).

Daudel, Raymond. *The Realm of Molecules*. New York: McGraw-Hill, 1993.

Davies, Owen. "Volatile Vacuums," *Omni*, February 1991, 50–56, 72.

Davies, Paul. *About Time: Einstein's Unfinished Revolution*. New York: Simon and Schuster, 1995.

———. *The Cosmic Blueprint: New Discoveries in Nature's Creative Ability to Order the Universe*. New York: Simon and Schuster, 1988.

———. *God and the New Physics*. New York: Simon and Schuster, 1983.

———. *The Mind of God: The Scientific Basis for a Rational World*. New York: Simon and Schuster, 1992.

Davies, P. C. W., and J. Brown, eds. *Superstrings: A Theory of Everything*? Cambridge: Cambridge Univ. Press, 1988.

———. "The Strange World of the Quantum." *The Ghost in the Atom*. Cambridge: Cambridge Univ. Press, 1991.

Davis, Joel. *Alternate Realities: How Science Shapes Our Vision of the World*. New York: Plenum, 1997.

de Duve, Christian. *Vital Dust: Life as a Cosmic Imperative*. New York: HarperCollins, 1995.

Deleuze, Gilles. *Bergsonism*. Translated by Hugh Tomlinson and Barbara Habberjam. New York: Zone, 1991.

Desmond, John J. "Partnerships between Government, Industry, and Universities." In *Science Based Economic Development: Case Studies around the World*, edited by Susan U. Raymond. New York: New York Academy of Sciences, 1996.

Deutsch, David, and Michael Lockwood. "The Quantum Physics of Time Travel." *Scientific American*, March 1994, 68–74.

Diamond v. Chakrabarty, 447 U.S. 303, 100 Sup. Ct. 2204 (1980).

Dickinson, Terence. *From the Big Bang to Planet X*. Buffalo: Camden House, 1993.

Dingle, Herbert. "Scientific and Philosophical Implications of the Special Theory of Relativity." In *Albert Einstein: Philosopher and Scientist*. Library of Living Philosophers, vol. 17, ed. Paul Arthur Schlipp. LaSalle, IL: Open Court, 1970.

Disney, Michael. "A New Look at Quasars." *Scientific American*, June 1998, 52–57.

Downey, James Patrick. "On Omniscience." *Faith and Philosophy*, April 1993, 230–34.

Dozier, Rush W., Jr. *Codes of Evolution: The Synaptic Language Revealing the Secrets of Matter, Life, and Thought*. New York: Crown, 1992.

Drees, Willem B. *Beyond the Big Bang: Quantum Cosmologies and God*. LaSalle, IL: Open Court, 1990.

Dresser, Rebecca. "Long-Term Contraceptives in the Criminal Justice System." In *Taking Sides: Clashing Views on Controversial Bioethical Issues*, edited by Carol Levine. Guilford, CT: Dushkin/McGraw-Hill, 1997. First published in *Hastings Center Report* 20 (January–February 1995).

Drexler, K. Eric. *Engines of Creation: The Coming Era of Nanotechnology*. New York: Bantam Doubleday Dell, 1986.

Drexler, K. Eric, Chris Peterson, and Gayle Pergamit. *Unbounding the Future: The Nanotechnology Revolution*. New York: William Morrow, 1991.

Duff, Michael J. "The Theory Formerly Known as Strings." *Scientific American*, February 1998, 64–69.

Dunn, James, "Can the Greens Destroy Nature?" *21st Century Science and Technology*, Winter 1998–1999, 8–11, 13.

Durant, Will. *The Story of Philosophy: The Lives and Opinions of the Greater Philosophers*. New York: Simon and Schuster, 1926.

Dworkin, Ronald. *Life's Dominion: An Argument about Abortion, Euthanasia, and Individual Freedom*. New York: Random House, 1994.

Earman, John. "Recent Work on Time Travel." In *Time's Arrows Today: Recent Physical and Philosophical Work on the Direction of Time*, edited by Steven F. Savitt. Cambridge: Cambridge Univ. Press, 1995.

Edwards, James B. *The Great Technology Race*. Norfolk, VA: Hampton Roads, 1993.

Einstein, Albert. "Autobiographical Notes." In *Albert Einstein: Philosopher-Scientist*. Library of Living Philosophers, vol. 7, ed. Paul Arthur Schlipp. LaSalle, IL: Open Court, 1970.

———. *Ideas and Opinions*. New York: Crown, 1954, 1982.

———. *The Meaning of Relativity: Including the Relativistic Theory of the Non-Symmetric Field*. Princeton: Princeton Univ. Press, 1956.

———. *Out of My Later Years*. New York: Carol, 1956, 1995.

———. *Relativity: The Special and General Theory*. New York: Crown, 1961.

———. *Sidelights on Relativity*. New York: Dover, 1983.

Einstein, Albert, Boris Podolsky, and Nathan Rosen. "Can Quantum-Mechanical Description of Physical Reality Be Considered Complete?" Partially reprinted in *Niels Bohr: A Centenary Volume*, edited by A. P. French and P. J. Kennedy. Cambridge: Cambridge Univ. Press, 1985. First published in *Physical Review* 47 (1935): 777.

Eisenstadt v. Baird, 405. U.S. 438, 92 Sup. Ct. 1029 (1972).

English, Jane. "Abortion and the Concept of a Person." In *Contemporary Issues in Bioethics*, edited by Tom L. Beauchamp and LeRoy Walters. Belmont, CA: Wadsworth, 1978. First published in *Canadian Journal of Philosophy* 5 (October 1975).

Enslin, Morton Scott. *Christian Beginnings*. New York: Harper Torchbooks, 1956.

Epstein, Lewis Carroll. *Relativity Visualized*. San Francisco: Insight, 1991.

Erni, John Nguyet. "AIDS Science: Killing More than Time." *Science as Culture* 5 (1996): 400–430.

Eubank, John Augustine. "Land Damage Liability in Aircraft Cases." *Dickinson Law Review* 57 (1953): 188–97.

Ewing, A. C. "Idealism." In *The Concise Encyclopedia of Western Philosophy and Philosophers*, edited by J. O. Urmson and Jonathan Ree. London: Unwin Hyman, 1991.

Falthammar, Carl-Gunne. "Hannes Alfven." *Physics Today*, September 1995, 118–19.

Feis, Herbert. *From Trust to Terror: The Onset of the Cold War 1945–1950*. New York: W. W. Norton, 1970.

Ferre, Frederick. *Philosophy of Technology*. Athens: Univ. of Georgia Press, 1995.

Ferris, Timothy. *Coming of Age in the Milky Way*. New York: Doubleday, 1988.

———. *The Whole Shebang: A State of the Universe(s) Report*. New York: Simon and Schuster, 1997.

Feynman, Richard P. *Six Easy Pieces: Essentials of Physics Explained by Its Most Brilliant Teacher*. New York: Addison-Wesley, 1994.

———. *Six Not-So-Easy Pieces: Einstein's Relativity, Symmetry, and Space-Time*. Reading, MA: Addison-Wesley, 1997.

Finkelstein, Sidney. *Existentialism and Alienation in American Literature*. New York: International Publishers, 1965.

Fishbein, Morris, ed. *Medical and Health Encyclopedia*. Westport, CT: H. S. Stuttman, 1978.

Flamsteed, Sam. "Crisis in the Cosmos." *Discover*, March 1995, 66–77.

Flavin, Christopher. "Facing Up to the Risks of Climate Change." In *State of the World 1996*, edited by Lester R. Brown. New York: W. W. Norton, 1996.

———. "The Legacy of Rio." In *State of the World 1997*, edited by Lester R. Brown, Christopher Flavin, and Hilary French. New York: W. W. Norton, 1997.

Florman, Samuel C. *The Introspective Engineer*. New York: St. Martin's Press, 1996.

Foley, Robert. *Humans before Humans*. Oxford: Blackwell, 1995.

Foster, Kenneth R., and Peter W. Huber. *Judging Science: Scientific Knowledge and the Federal Courts*. Cambridge: MIT Press, 1997.

Fowler, H. Ramsey, and Jane E. Aaron. *The Little, Brown Handbook*. New York: HarperCollins, 1995.

Fredrickson, Donald S. "Asilomar and Recombinant DNA: The End of the Beginning." In *Biomedical Politics*, edited by Kathi E. Hanna. Washington, D.C.: National Academy, 1991.

Freedman, David M. "The Handmade Cell." *Discover*, August 1992, 46–52.

———. "Molding the Metabolism." *Discover*, August 1992, 36–45.

Friedman, Alan J., and Carol C. Donley. *Einstein as Myth and Muse*. Cambridge: Cambridge Univ. Press, 1985.

Frye v. United States, 54 U.S. App. D.C. 46, 293 Fed. 1013 (1923).

Fuller, Graham E. *The Democracy Trap: Perils of the Post–Cold War World*. New York: Penguin, 1991.

Gallagher, Paul. "Implosion of Population Growth Rate Continues Through 1998." *21st Century Science and Technology*, Winter 1998–1999, 19–23.

Gardner, Gary. "Preserving Agricultural Resources." In *State of the World 1996*, edited by Lester R. Brown. New York: W. W. Norton, 1996.

———. "Preserving Global Cropland." In *State of the World 1997*, edited by Lester R. Brown, Christopher Flavin, and Hilary French. New York: W. W. Norton, 1997.

Gatland, Kenneth W., ed. *Project Satellite*. New York: British Book Centre, 1958.

Gell-Mann, Murray. *The Quark and the Jaguar: Adventures in the Simple and the Complex*. New York: W. H. Freeman, 1994.

Gibbins, Peter. *Particles and Paradoxes: The Limits of Quantum Logic*. Cambridge: Cambridge Univ. Press, 1987.

Gilligan, Carol. *In a Different Voice: Psychological Theory and Women's Development*. Cambridge: Harvard Univ. Press, 1982.

Ginsparg, Paul, and Sheldon Glashow. "Desperately Seeking Superstrings?" *Physics Today*, May 1986, 7–9.

Gjertsen, Derek. *Science and Philosophy: Past and Present*. New York: Penguin, 1989.

Glanz, James, "Cosmic Motion Revealed." *Science*, 18 December 1998, 2156–57.

Glick, Edward Bernard. *Peaceful Conflict: The Non-Military Use of the Military*. Harrisburg, PA: Stackpole, 1967.

Gödel, Kurt. "A Remark about the Relationship between Relativity Theory and Idealistic Philosophy." In *Albert Einstein: Philosopher-Scientist*. Library of Living Philosophers, vol. 7, ed. Paul Arthur Schlipp. LaSalle, IL: Open Court, 1949, 1970, 1995.

Goldberg, Steven. *Culture Clash: Law and Science in America*. New York: New York Univ. Press, 1994.

Goodstein, David L. "After the Big Crunch." *Wilson Quarterly* 19 (Summer 1995): 53–60.

Gorbachev, Mikhail. "U.S.S.R. Arms Reduction." *Vital Speeches of the Day*, 1 February 1989, 230.

Goswami, Amit. *The Self-Aware Universe: How Consciousness Creates the Material World*. New York: Putnam, 1993.

Gott, J. Richard. "Closed Timelike Curves Produced by Pairs of Moving Cosmic Strings: Exact Solutions." *Physical Review Letters* 66 (1991): 1126–29.

Gould, Stephen Jay. *Ever Since Darwin: Reflections in Natural History*. New York: W. W. Norton, 1977.

———. *Wonderful Life: The Burgess Shale and the Nature of History*. New York: W. W. Norton, 1989.

Grassetti v. Weinberger, 408 F. Supp. 142 (1976).

Greaves, Richard L., et al. *Civilizations of the World: The Human Adventure*. New York: Addison-Wesley Longman, 1997.

Greeley v. Miami Valley Maintenance, 49 Ohio St. 3d 229 (1990).

Gribbin, John. *In the Beginning: The Birth of the Living Universe*. New York: Little, Brown, 1993.

———. *The Omega Point: The Search for the Missing Mass and the Ultimate Fate of the Universe*. New York: Bantam Books, 1988.

———. *Schrödinger's Kittens and the Search for Reality: Solving the Quantum Mysteries*. New York: Little, Brown, 1995.

———. *Time-Warps*. New York: Delacorte, 1979.

———. *Unveiling the Edge of Time: Blackholes, White Holes, Wormholes*. New York: Crown, 1992.

Grimes, Lee. *Dinosaur Nexus*. New York: Avon, 1994.

Griswold v. Connecticut, 381 U.S. 479, 85 Sup. Ct. 1678 (1965).

Guthrie, W. K. C. *A History of Greek Philosophy*. 2 vols. Cambridge: Cambridge Univ. Press, 1971.

Haisch, Bernard, Alfonso Rueda, and H. E. Puthoff. "Beyond $E=mc^2$." *The Sciences*, November–December 1994, 26–31.

Halberstam, David. *The Fifties*. New York: Villard, 1993.

Halpern, Paul. *Cosmic Wormholes: The Search for Interstellar Shortcuts*. New York: Dutton, 1992.

Han, M. Y. *The Probable Universe*. Blue Ridge Summit, PA: TAB, 1993.

Harford, James. *Korolev*. New York: John Wiley, 1997.

Hartmann, William K. *The Cosmic Voyage through Time and Space*. Belmont, CA: Wadsworth, 1992.

Hartmann, William K., and Ron Miller. *The History of Earth: An Illustrated Chronicle of an Evolving Planet*. New York: Workman, 1991.

Hawking, Stephen. *A Brief History of Time: From the Big Bang to Black Holes*. New York: Bantam, 1988.

———. *The Illustrated A Brief History of Time*. New York: Bantam, 1996.

Hazen, Robert M., and James Trefil. *Science Matters: Achieving Scientific Literacy*. New York: Doubleday, 1990.

Heilbroner, Robert L. *An Inquiry into the Human Prospect: Looked at Again for the 1990s*. New York: W. W. Norton, 1991.

Heisenberg, Werner. *The Physical Principles of Quantum Theory*. New York: Dover, 1949.

Hellegers, André E. "Fetal Development." In *Contemporary Issues in Bioethics*, edited by Tom L. Beauchamp and LeRoy Walters. Belmont, CA: Wadsworth, 1978. First published in *Theological Studies* 31 (March 1970): 3–9.

Hellemans, Alexander, and Bryan Bunch. *The Timetables of Science: A Chronology of the Most Important People and Events in the History of Science*. New York: Simon and Schuster, 1991.

Hennessy, James J., and Clarence H. A. Romig. "A Review of the Experiments Involving Voiceprint Identification." *Journal of Forensic Science* (April 1971) 183–98.

Herbert, Nick. *Faster than Light: Superluminal Loopholes in Physics.* New York: New American Library, 1988.

———. *Quantum Reality: Beyond the New Physics.* New York: Anchor, 1985.

Hey, Tony, and Patrick Walters, *The Quantum Universe.* Cambridge: Cambridge Univ. Press, 1987.

Higher Catechetical Institute at Nijmergen. *A New Catechism.* New York: Herder and Herder, 1967.

Hinson, Ed. *The New World Order.* Wheaton, IL: Victor, 1991.

Hod, Shahar, and Tsui Piran. "Mass Inflation in Dynamical Gravitational Collapse of a Charged Scalar Field." *Physical Review Letters* (24 August 1998): 1554–57.

Hofstadter, Douglas R. *Gödel, Escher, Bach: An Eternal Golden Braid.* New York: Vintage, 1979.

Hogan, Craig J., Robert P. Kirshner, and Nicholas B. Suntzett. "Surveying Space-Time with Supernovae." *Scientific American*, January 1999, 47–51.

Hogan, James P. *The Genesis Machine* New York: Ballantine, 1978.

———. *The Proteus Operation.* New York: Bantam, 1985.

Horgan, John. *The End of Science: Facing the Limits of Knowledge in the Twilight of the Scientific Age.* Reading, MA: Helix, 1996.

Hoyle, F., G. Burbidge, and J. V. Narlikar. "A Quasi-Steady State Cosmological Model with Creation of Matter." *Astrophysical Journal*, 20 June 1993, 437–57.

Huber, Peter W. *Galileo's Revenge: Junk Science in the Courtroom.* New York: Basic Books, 1993.

Hunt, David P. "Divine Providence and Simple Foreknowledge." *Faith and Philosophy*, July 1993, 394–414.

———. "Prescience and Providence: A Reply to My Critics." *Faith and Philosophy* (July 1993) 428–38.

Hussein, Saddam. *On Current Affairs in Iraq.* Baghdad: Translation and Foreign Languages Publications, 1981.

Huxley, Aldous. *Brave New World.* New York: Harper and Row, 1932.

———. *Brave New World Revisited.* New York: Harper and Row, 1958.

Huxley, Julian. Introduction to *The Phenomenon of Man*, by Pierre Teilhard de Chardin. New York: Harper and Row, 1959.

Icke, Vincent. *The Force of Symmetry.* Cambridge: Cambridge Univ. Press, 1995.

Isaacs, Alan, ed. *A Dictionary of Physics.* Oxford: Oxford Univ. Press, 1996.

Isaacs, Alan, John Daintith, and Elizabeth Martin. *Concise Science Dictionary.* Oxford: Oxford Univ. Press, 1996.

Jaki, Stanley. *Cosmos and Creator.* Edinburgh: Scottish Academic Press, 1980.

———. *Is There a Universe?* Liverpool: Liverpool Univ. Press, 1993.

Jammer, Max. *The Philosophy of Quantum Mechanics: The Interpretations of Quantum Mechanics in Historical Perspective.* New York: John Wiley and Sons, 1974.

Jastrow, Robert. *God and the Astronomers.* New York: W. W. Norton, 1978.

Jessup, John K. "The Atomic Stakes." *Life*, 11 November 1946, 77.

Jones, Roger S. *Physics for the Rest of Us.* Chicago: Contemporary Books, 1992.

Kahn, P., and A. Gibson, "DNA from Extinct Human." *Science*, 11 July 1997, 5323.

Kaku, Michio. *Hyperspace: A Scientific Odyssey through Parallel Universes, Time Warps, and the 10th Dimension.* New York: Oxford Univ. Press, 1994.

Kane, Gordon. *The Particle Garden: Our Universe as Understood by Particle Physicists.* Reading, MA: Addison-Wesley, 1995.

Kanipe, Jeff. "Beyond the Big Bang." In *The New Cosmos*, edited by Robert Burnham. Waukesha, WI: Kalmbach, 1992.

Kant, Immanuel. "Prolegomena to Any Future Metaphysics." In *Ten Great Works of Philosophy*, edited by Robert Paul Wolff. New York: Penguin, 1969.

Kapitan, Tomis. "Providence, Foreknowledge, and Decision Procedure." *Faith and Philosophy* (July 1993) 415–20.

Karl, Frederick R., and Leo Hamalian, eds. *The Existential Imagination*. Greenwich, CT: Fawcett, 1963.

Kaufmann, Stuart. *At Home in the Universe: The Search for Self-Organization and Complexity*. Oxford: Oxford Univ. Press, 1995.

———. " 'What is Life?': Was Schrödinger Right?" In *What Is Life? The Next Fifty Years: Speculations on the Future of Biology*, edited by Michael P. Murphy and Luke A. J. O'Neill. Cambridge: Cambridge Univ. Press, 1995.

Kaufmann, Walter, ed. *Existentialism from Dostoevsky to Sartre*. Cleveland: World, 1964.

Kaufmann, William J., III. *Relativity and Cosmology*. New York: Harper and Row, 1973.

Keller, Evelyn Fox. "Nature, Nurture, and the Human Genome Project." In *Taking Sides: Clashing Views on Controversial Bioethical Issues*, edited by Carol Levine. Guilford, CT: Dushkin Publishing, 1993. First published in Daniel J. Kevles and Leroy Hood, eds., *The Code of Codes: Scientific and Social Issues in the Human Genome Project* (Cambridge: Harvard Univ. Press, 1992).

Kennedy, Paul. *Preparing for the Twenty-First Century*. New York: Random House, 1993.

Kennedy, Robert F. *To Seek a Newer World*. New York: Doubleday, 1968.

Kenney-Wallace, Geraldine A., and Warren C. Bull. "Partnerships for Science-Based Development: NeuroScience Network as a Practical Model." In *Science Based Economic Development: Case Studies around the World*, edited by Susan U. Raymond. New York: New York Academy of Sciences, 1996.

Kevles, Daniel J. "The Changed Partnership." *Wilson Quarterly* 19 (Summer 1995): 41–52.

———. *The Physicists: The History of a Scientific Community in Modern America*. Cambridge: Harvard Univ. Press, 1995.

Kevles, Daniel, and Leroy Hood. "The Code of Codes." In *Taking Sides: Clashing Views on Controversial Bioethical Issues*, edited by Carol Levine. Guilford, CT: Dushkin, 1993. First published in Daniel J. Kevles and Leroy Hood, eds., *The Code of Codes: Scientific and Social Issues in the Human Genome Project* (Cambridge: Harvard Univ. Press, 1992).

Kirk, G. S., J. E. Raven, and M. Schofield. *The Presocratic Philosophers: A Critical History with a Selection of Texts*. Cambridge: Cambridge Univ. Press, 1983.

Kletschka v. Driver, 411 F. 2d 436 (2d Cir. 1969).

Klyshko, D. N. "Quantum Optics: Quantum, Classical, and Metaphysical Aspects." In *Fundamental Problems in Quantum Theory: A Conference Held in Honor of Professor John A. Wheeler*, edited by Daniel M. Greenberger and Anton Zeilinger. New York: New York Academy of Sciences, 1995.

Korner, Stephen. "Immanuel Kant." In *The Concise Encyclopedia of Western Philosophy and Philosophers*, edited by J. O. Urmson and Jonathan Ree. London: Unwin Hyman, 1991.

Kragh, Helge. *An Introduction to the Historiography of Science*. Cambridge: Cambridge Univ. Press, 1991.

Krauss, Lawrence M. "Cosmological Antigravity." *Scientific American*, January, 1999, 53–59.

———. *The Physics of Star Trek*. New York: HarperCollins, 1995.

Kuhn, Karl F. *Basic Physics*. New York: John Wiley and Sons, 1996.

Kuhn, Thomas S. *The Structure of Scientific Revolutions*. 2nd ed. Chicago: Univ. of Chicago Press, 1970.

Lacourt, Dominique. Introduction to *The Chemistry of Life*, by Martin Olomucki. New York: McGraw-Hill, 1993.

Laplace, Pierre Simon, Marquis de. *A Philosophical Essay on Probabilities*. Translated by Frederick Wilson Truscott and Frederick Lincoln Emory. New York: Dover, 1951.

Lasota, Jean-Pierre. "Unmasking Black Holes." *Scientific American*, May 1999, 41–47.

Lavine, T. Z. *From Socrates to Sartre: The Philosophic Quest*. New York: Bantam, 1984.

LaViolette, Paul A. *Beyond the Big Bang*. Rochester, VT: Park Street, 1995.

Leakey, Richard. *The Origins of Humankind*. New York: Basic Books, 1994.

Leakey, Richard E., and Roger Lewin. *Origins: What New Discoveries Reveal about the Emergence of Our Species and Its Possible Future*. New York: E. P. Dutton, 1979.

Lederman, Leon M., and David N. Schramm. *From Quarks to the Cosmos: Tools of Discovery*. New York: Scientific American Library, 1995.

Leftow, Brian. "Eternity and Simultaneity." *Faith and Philosophy*, April 1991, 148–79.

Leibowitz, Wendy R. "Tracking Those Year-2000 Legal Issues on the Net: Attorneys Turn to Web to Stay Current on Latest Y2K Developments." *National Law Journal*, 8 February 1999, Sec. A, p1; Sec. A. p14.

Leinster, Murray. "Dear Charles." In *Twists in Time*. New York: Avon, 1964. First published in *Fantastic Story Monthly* (May 1953).

Lemonick, Michael D. "How to Go Back in Time." *Time*, 13 May 1991, 74.

———. *The Light at the Edge of the Universe*. New York: Villard, 1993.

Leonard, Peter J. T., and Jerry T. Bonnell. "Gamma-Ray Bursts of Doom." *Sky and Telescope*, January/February 1998, 28–34.

Lerner, Eric J. *The Big Bang Never Happened*. New York: Random House, 1991.

Lestienne, Remy. *The Children of Time: Causality, Entropy, Becoming*. Chicago: Univ. of Chicago Press, 1990.

Levine, Carol. "Should the Courts Be Permitted to Order Women to Use Long-Acting Contraceptives." In *Taking Sides: Clashing Views on Controversial Bioethical Issues*, edited by Carol Levine. Guilford, CT: Dushkin, 1993.

Lewin, Leonard C. *Report from Iron Mountain: On the Possibility and Desirability of Peace*. New York: Free Press, 1996.

Lewis, C. S. *The Dark Tower and Other Stories*. New York: Harcourt Brace, 1977.

Lightman, Alan. *Time for the Stars: Astronomy in the 1990's*. New York: Warner, 1992.

Linde, Andrei. "The Self-Reproducing Inflationary Universe." *Scientific American*, November 1994, 48–55.

Lindley, David. *The End of Physics: The Myth of a Unified Theory*. New York: HarperCollins, 1993.

———. *Where Does the Weirdness Go? Why Quantum Mechanics Is Strange, but Not as Strange as You Think*. New York: HarperCollins, 1996.

Los Alamos Scientific Laboratory. *Los Alamos: Beginning of an Era*. Los Alamos, NM: Los Alamos Public Relations Office, n.d.

Lusso, Paolo, et al. "Expanded HIV-1 Cellular Tropism by Phenotyopic Mixing with Murine Endogenums Retroviruses." *Science*, 16 February 1990, 848–52.

Mabbott, J. D. "Thomas Hill Green." In *The Concise Encyclopedia of Western Philosophy and Philosophers*, edited by J. O. Urmson and Jonathan Ree. London: Unwin Hyman, 1989, 1991.

McCune, Joseph M., et al. "Phenotypes in HIV-Infected Mice." *Science*, 23 November 1990, 1152–53.

Mach, Ernst. *The Science of Mechanics: A Critical and Historical Account of Its Development*. LaSalle, IL: Open Court, 1960.

Macklin, Ruth. "Personhood and the Abortion Debate." In *Abortion: Moral and Legal Perspectives*, edited by Jay L. Garfield and Patricia Hennessey. Amherst: Univ. of Massachusetts Press, 1984.

Macvey, John W. *Time Travel*. Chelsea, MI: Scarborough House, 1990.

Madsen, Axel. *Unisave*. New York: Grosset and Dunlap, 1980.

Mandelbaum, Michael. "Lessons of the Next Nuclear War." *Foreign Affairs* 74 (March/April 1995): 22–37.

March, Robert H. *Physics for Poets*. New York: McGraw-Hill, 1992.

Maria, Julian. *History of Philosophy*. New York: Dover, 1967.

Marshall, Eliot, and Elizabeth Pennisi. "Hubris and the Human Genome." *Science*, 15 May 1998, 994–95.

Martin, James A. "The Proposed Science Court." *Michigan Law Review* 75 (1977): 1058–91.

Matthews, John. *The Arthurian Tradition*. Rockport, MA: Element, 1994.

Maxwell, James Clerk. *A Treatise on Electricity and Magnetism*. 2 vols. New York: Dover, 1891, 1954.

McClintock, Jefferey E. "A Black Hole Caught in the Act." *Scientific American*, May 1999, 45.

McCurdy, Howard E. *Space and the American Imagination*. Washington: Smithsonian Institution, 1997.

Meredith, Richard C. *At the Narrow Passage*. New York: Berkley, 1973.

Millar, David, et al., eds. *The Cambridge Dictionary of Scientists*. Cambridge: Cambridge Univ. Press, 1996.

Miller, Jon D. *The Public Understanding of Science and Technology in the United States, 1990: A Report to the National Science Foundation*. DeKalb: Northern Illinois Univ. Press, 1991.

Milman, Gregory. "HIV Research in SCID Mouse: Biosafety Considerations." *Science*, 23 November 1990, 1152.

Mokhtarian, Patricia L. "Now That Travel Can Be Virtual, Will Congestion Virtually Disappear?" *Scientific American*, October 1997, 93.

Morowitz, Harold J., and James S. Trefil. *The Facts of Life: Science and the Abortion Controversy*. Oxford: Oxford Univ. Press, 1992.

Morris, Richard. "The Perils of Time Travel." *The Futurist*, September–October, 1994, 60.

Morris, Simon Conway. "Showdown on the Burgess Shale: The Challenge." *Natural History*, December 1998/January 1999, 48–51.

Mowery, David C. "The Bush Report after Fifty Years—Blueprint or Relic?" In *Science for the 21st Century: The Bush Report Revisited*, edited by Claude E. Barfield. Washington, D.C.: AEI, 1997.

Mueller, John, and Karl Mueller. "Sanctions of Mass Destruction." *Foreign Affairs* 78 (May/June 1999): 43–53.

Muller, Richard. *Nemesis: The Death Star*. New York: Weidenfeld and Nicolson, 1988.

Musser, George. "Here Come the Suns: Stars With Planets Seem to Harbor 'Heavy' Metals." *Scientific American*, May 1999, 20.

Naeye, Robert. "Was There Life on Mars?" *Astronomy*, November 1996, 46–53.

National Commission on Excellence in Education. *A Nation at Risk: The Imperative for Educational Reform*. Washington, D.C.: National Commission on Excellence in Education, 1983.

Nelson, Richard. "Why the Bush Report Has Hindered an Effective Civilian Technology Policy." In *Science for the 21st Century: The Bush Report Revisited*, edited by Claude E. Barfield. Washington, D.C.: AEI, 1997.

News and Editorial Staffs. "Breakthrough of the Year: A Glimpse of Neanderthal DNA." *Science*, 19 December 1997, 2041.
Newton, Roger. *What Makes Nature Tick?* Cambridge, MA: Harvard Univ. Press, 1993.
Newton, Sir Isaac. *Mathematical Principles*. Berkeley: Univ. of California Press, 1960.
———. *Opticks or a Treatise of the Reflections, Inflections and Colours of Light*. New York: Dover, 1952.
New Zealand Ministry of Foreign Affairs and Trade. *United Nations Handbook 1997*. Wellington, New Zealand: New Zealand Ministry of Foreign Affairs and Trade, 1997.
Niskanen, William A. "R&D and Economic Growth—Cautionary Thoughts." In *Science for the 21st Century: The Bush Report Revisited*, edited by Claude E. Barfield. Washington, D.C.: AEI, 1997.
Nusseibeh, Sari. "Can Wars Be Just?" In *But Was It Just? Reflections on the Morality of the Persian Gulf War*, edited by David E. Decosse. New York: Bantam Doubleday Dell, 1992.
Olomucki, Martin. *The Chemistry of Life*. New York: McGraw-Hill, 1993.
Pagels, Heinz R. *The Cosmic Code: Quantum Physics as the Language of Nature*. New York: Bantam, 1982.
Pais, Abraham. *Subtle Is the Lord: The Science and the Life of Albert Einstein*. Oxford: Oxford Univ. Press, 1982.
Paque, Julie. "What Makes a Planet a Friend for Life?" *Astronomy*, June 1995, 47–51.
Parker, Barry. *Cosmic Time Travel: A Scientific Odyssey*. New York: Plenum, 1991.
Parker, Sybil P., ed. *McGraw-Hill Concise Encyclopedia of Science and Technology*. New York: McGraw-Hill, 1994.
Parsons, Paul. "A Warped View of Time Travel." *Science*, 11 October 1996, 202–3.
Pendleton, Yvonne, and Dale P. Cruikshank. "Life from the Stars." *Sky and Telescope*, March 1994, 36–42.
Pennisi, Elizabeth. "Academic Sequences Challenge Celera in a Sprint to the Finish." *Science*, 19 March 1999, 1822–23.
Penrose, Roger. *The Emperor's New Mind: Concerning Computers, Minds, and the Laws of Physics*. New York: Penguin, 1989.
Persels, Jim. "The Norplant Condition: Protecting the Unborn or Violating Fundamental Rights?" In *Taking Sides: Clashing Views on Controversial Bioethical Issues*, edited by Carol Levine. Guilford, CT: Dushkin Publishing, 1993. First published in *Journal of Legal Medicine* 13 (1992).
Peterson, Aage. "The Philosophy of Niels Bohr." In *Niels Bohr: A Centenary Volume*, edited by A. P. French and P. J. Kennedy. Cambridge: Harvard Univ. Press, 1985.
Peterson, Ivars. "Timely Questions: Visiting the Past by Whipping around Mobile Cosmic Strings." *Science News* 141 (1992): 202–3.
Peterson, John L., Margaret Whetley, and Myron Kellner-Rogers. "The Y2K Problem: Social Chaos or Social Transformations?" *The Futurist*, October 1998, 21–27.
Phillips, William G. "The Year 2000 Problem: Will the Bug Bite Back?" *Popular Science*, October 1998, 88–93.
Pickover, Clifford A. *Time: A Travelers Guide*. New York: Oxford Univ. Press, 1998.
Planned Parenthood of Pennsylvania v. Casey, 112 Sup. Ct. 2791 (1992).
Platt, Anne E. "Confronting Infectious Diseases." In *State of the World 1996*, edited by Lester R. Brown. New York: W. W. Norton, 1996.
Plosila, Walter H., and Susan U. Raymond. "Policies for Science-Based Development: The Experiences of Pennsylvania and Ohio." In *Science Based Economic Development: Case Studies around the World*, edited by Susan U. Raymond. New York: New York Academy of Sciences, 1997.

Polkinghorne, J. C. *The Quantum World*. Princeton: Princeton Univ. Press, 1984.
Pollack, Rachel, and Chris Weston. *Time Breakers: Mind Out of Time*. New York: Time Warner, 1997.
Pollack, Robert. *Signs of Life: The Language and Meanings of DNA*. New York: Houghton Mifflin, 1994.
Pool, Robert. "Score One (More) for the Spooks." *Discover*, January 1998, 53.
Posner, Richard A. *The Problems of Jurisprudence*. Cambridge: Harvard Univ. Press, 1990.
Postel, Sandra. "Forging a Sustainable Water Strategy." In *State of the World 1996*, edited by Lester R. Brown. New York: W. W. Norton, 1996.
Potts, Malcolm. "Sex and Birth Rate: Human Biology, Demographic Change, and Access to Fertility-Regulation Methods." *Population and Development Review* 23 (1997): 14–15.
Powell, Corey S. "The Golden Age of Cosmology." *Scientific American*, July 1992, 17–22.
Rae, Alastair. *Quantum Physics Illusion or Reality*. Cambridge: Cambridge Univ. Press, 1986.
Ramet, Sabrina P. "The Breakup of Yugoslavia." *Global Affairs* 6 (Spring 1991): 93–110.
Rampino, Michael R. "The Shiva Hypothesis: Impacts, Mass Extinctions, and the Galaxy." *The Planetary Report* 18 (January/February 1998): 6–11.
Ratzsch, Del. *Battle of Beginnings: Why Neither Side Is Winning the Creation-Evolution Debate*. Downers Grove, IL: InterVarsity, 1996.
Raup, David M. *The Nemesis Affair: A Story of the Death of Dinosaurs and the Ways of Science*. New York: W. W. Norton, 1986.
Ravin, Richard L. "Y2K and ADR: Get Ready for Midnight," *Dispute Resolution Journal*. (November 1998): 8–15.
Raymond, Susan U. "Lessons from Global Experience in Policy for Science-Based Economic Development." In *Science Based Economic Development: Case Studies around the World*, edited by Susan U. Raymond. New York: New York Academy of Sciences, 1996.
———. "Listening to Critics: Enlarging the Discussion of Policy for Science-Based Development." In *Science Based Economic Development: Case Studies around the World*, edited by Susan U. Raymond. New York: New York Academy of Sciences, 1996.
Raymond, Susan U., ed. *Science Based Economic Development: Case Studies around the World*. New York: New York Academy of Sciences, 1996.
Redhead, Michael. *From Physics to Metaphysics*. Cambridge: Cambridge Univ. Press, 1995.
Reed, John L. and Richard K. Herrmann. "An Alternate to the Not-Ready-for-the-Year-2000 Court System," *Dispute Resolution Journal*. (November 1998): 16–18.
Reichenbach, Hans. "The Philosophical Significance of the Theory of Relativity." In *Albert Einstein: Philosopher-Scientist*. Library of Living Philosophers, vol. 7, ed. Paul Arthur Schlipp. LaSalle, IL: Open Court, 1970.
Renner, Michael. "Transforming Security." In *State of the World 1997*, edited by Lester R. Brown, Christopher Flavin, and Hilary French. New York: W. W. Norton, 1997.
Rensberger, Boyce. *Instant Biology: From Single Cells to Human Beings and Beyond*. New York: Fawcett Columbine, 1996.
"Reproduction." *Collier's Encyclopedia*, 1990.
Rhodes, Richard. *Dark Sun: The Making of the Hydrogen Bomb*. New York: Simon and Schuster, 1995.
———. *The Making of the Atomic Bomb*. New York: Simon and Schuster, 1986.
Richichi, Thomas. "Environmental Law: Although Storm Clouds Threaten throughout Global Warming Conference in Kyoto, the Conferees Reached an Agreement on Greenhouse Gas Emissions." *National Law Journal*, 29 December 1997–5 January 1998, sec. B, p. 4.
Riordan, Michael, and David Schramm. *The Shadows of Creation*. New York: W. H. Freeman, 1991.

Robertson, John A. *Children of Choice: Freedom and the New Reproductive Technologies.* Princeton: Princeton Univ. Press, 1994.

Rockwood, Roy. *By Air Express to Venus.* Racine, WI: Whitman, 1929.

———. *Through Space to Mars.* New York: Cupples and Leon, 1910.

Roe v. Wade, 410 U.S. 113, 93 Sup. Ct. 705 (1973).

Rollin, Bernard E. *The Frankenstein Syndrome: Ethical and Social Issues in the Genetic Engineering of Animals.* Cambridge: Cambridge Univ. Press, 1995.

Rolston, Holmes, III. "People, Population, and Place." In *Ethics and Agenda 21*, edited by Noel J. Brown and Pierre Quiblier. New York: United Nations, 1994.

Ronan, Colin A. *The Natural History of the Universe from the Big Bang to the End of Time.* New York: Macmillan, 1991.

Rothman, Tony. *Instant Physics: From Aristotle to Einstein, and Beyond.* New York: Fawcett Columbine, 1995.

Rowan-Robinson, Michael. *Ripples in the Cosmos: A View behind the Scenes of the New Cosmology.* Oxford: W. H. Freeman, 1993.

Rucker, Rudy. *The Fourth Dimension: A Guided Tour of the Higher Universes.* Boston: Houghton Mifflin, 1984.

Ruse, Michael. *Philosophy of Biology Today.* Albany: State Univ. of New York Press, 1988.

Russell, Bertrand. *The ABC of Relativity.* New York: New American Library, 1985.

Russo, Enzo, and David Cove. *Genetic Engineering: Dreams and Nightmares.* Oxford: W. H. Freeman, 1995.

Sachs, Mendel. *Einstein versus Bohr: The Continuing Controversy in Physics.* LaSalle, IL: Open Court, 1988.

———. *Relativity in Our Time: From Physics to Human Relations.* London: Taylor and Francis, 1993.

Saferstein, Harvey. "Nonreviewability: A Functional Analysis of 'Committed to Agency Discretion,'" *Harvard Law Review* 82 (1968): 367–98.

Sagan, Carl. *Cosmos.* New York: Random House, 1980.

———. *The Dragons of Eden: Speculations on the Evolution of Human Intelligence.* New York: Ballantine, 1977.

Schatzman, Evry. *Our Expanding Universe.* New York: McGraw-Hill, 1989.

Schlipp, Paul Arthur, ed. *Albert Einstein: Philosopher-Scientist.* Library of Living Philosophers, vol. 7. LaSalle, IL: Open Court, 1970.

Schrödinger, Erwin. *My View of the World.* Woodbridge, CT: Ox Bow, 1983.

Schwartz, Bernard. *Constitutional Law.* New York: Macmillan, 1972.

Schwartz, Delmore. "The True-Blue American." In *Selected Poems: Summer Knowledge.* New York: New Directions, 1959.

Scott-Kakures, Dion, et al. *History of Philosophy.* New York: HarperCollins, 1993.

Scully, Marlan O., Ulrich W. Rathe, and Susanne F. Yelin. "Second-Order Photon-Photon Correlations and Atomic Spectroscopy." In *Fundamental Problems in Quantum Theory: A Conference Held in Honor of Professor John A. Wheeler*, edited by Daniel M. Greenberger and Anton Zeilinger. New York: New York Academy of Sciences, 1995.

Shapiro, Robert. *The Human Blueprint: The Race to Unlock the Secrets of Our Genetic Code.* New York: Bantam, 1992.

Shapiro v. Thompson, 394 U.S. 618, 89 Sup. Ct. 1322, 22 L.Ed. 600 (1969).

Sheldrake, Rupert. *A New Science of Life: The Hypothesis of Morphic Resonance.* Rochester, VT: Park Street, 1995.

Shepherd, Gordon M. *Neurobiology.* Oxford: Oxford Univ. Press, 1994.

Shreeve, James. "Sunset on the Savanna." *Discover*, July 1996, 116–25.

Silk, Joseph. *The Big Bang.* New York: W. H. Freeman, 1980.

Smith, Len Young, and G. Gale Roberson. *Smith and Roberson's Business Law*. St. Paul, MN: West, 1985.

Smith, Len Young, Richard Mann, and Barry Roberts. *Essentials of Business Law*. St. Paul, MN: West, 1986.

Smith, L. Neil. *The Gallatin Divergence*. New York: Ballantine, 1985.

Smoot, George, and Keay Davidson. *Wrinkles in Time*. New York: William Morrow, 1993.

Sobel, Dava. "Man Stops Universe, Maybe." *Discover*, April 1993, 20–21.

Spielberg, Nathan, and Byron D. Anderson. *Seven Ideas that Shook the Universe*. New York: John Wiley and Sons, 1987.

Stebbing, L. Susan. *Philosophy and the Physicists*. New York: Penguin, 1944.

Steele, Allen. *The Tranquility Alternative*. New York: Ace, 1996.

Sternbach, Rick, and Michael Okuda. *Star Trek: The Next Generation Technical Manual*. New York: Pocket, 1991.

Stevens, Leonard A. *The Case of Roe v. Wade*. New York: Putnam, 1996.

Stevenson, Leslie, and Henry Byerly. *The Many Faces of Science*. Boulder, CO: Westview, 1995.

Stix, Gary. "Trends in Nanotechnology: Waiting for Breakthroughs." *Scientific American*, April 1996, 94–97.

Stock, Gregory. *Metaman: The Merging of Humans and Machines into a Global Superorganism*. New York: Simon and Schuster, 1993.

Strong, John William. "Questions Affecting the Admissibility of Scientific Evidence." *University of Illinois Law Forum* (1970): 1–22.

Sullivan, J. W. N. *The Limitations of Science*. New York: Mentor, 1949.

Summers, Harry G. *A Critical Analysis of the Gulf War*. New York: Bantam Doubleday Dell, 1992.

Talbot, Michael. *Beyond the Quantum: How the Secrets of the New Physics Are Bridging the Chasm between Faith and Science*. New York: Bantam, 1986.

———. *Mysticism and the New Physics*. London: Penguin, 1992.

Talcott, Richard. "COBE's Big Bang!" In *The New Cosmos*, edited by Robert Burnham. Waukesha, WI: Kalmbach, 1992.

Tarnas, Richard. *The Passion of the Western Mind*. New York: Ballantine, 1991.

Teilhard de Chardin, Pierre. *The Phenomenon of Man*. New York: Harper and Row, 1959.

———. *Science and Christ*. New York: Harper and Row, 1965.

Teitelman, Robert. *Profits of Science: The American Marriage of Business and Technology*. New York: HarperCollins, 1994.

"Then and Now: The IAEA Turns Forty." *IAEA Bulletin: Quarterly Journal of the International Atomic Energy Agency*, September 1997.

Thomson, J. J. "Cathode Rays." *Philosophical Magazine and Journal of Science* 44 (1897): 293–318.

Thorne, Kip S. *Black Holes and Time Warps: Einstein's Outrageous Legacy*. New York: W. W. Norton, 1994.

Thuan, Trinh Xuan. *The Birth of the Universe: The Big Bang and After*. New York: Abrams, 1993.

———. *The Secret Melody: And Man Created the Universe*. Translated by Storm Dunlop. Oxford: Oxford Univ. Press, 1995.

Tiller, William. "The Positive and Negative Space/Time Frames as Conjugate Systems." In *Future Science*, edited by John White and Stanley Krippner. Garden City, NY: Anchor, 1977.

Tipler, Frank J. *The Physics of Immortality: Modern Cosmology, God and the Resurrection of the Dead*. New York: Doubleday, 1994.

Tooley, Michael. "A Defense of Abortion and Infanticide." In *The Problem of Abortion*, edited by Joel Feinberg. Belmont, CA: Wadsworth, 1973. First published in *Philosophy and Public Affairs* 2 (1972).

Traweek, Sharon. *Beamtimes and Lifetimes: The World of High Energy Physics*. Cambridge: Harvard Univ. Press, 1988.

Trefil, James. *The Dark Side of the Universe: A Scientist Explores the Mysteries of the Cosmos*. New York: Doubleday, 1988.

———. *From Atoms to Quarks: An Introduction to the Strange World of Particle Physics*. New York: Doubleday, 1994.

———. *Reading the Mind of God: In Search of the Principle of Universality*. New York: Doubleday, 1989.

Trefil, James, and Robert Hazen. *The Sciences: An Integrated Approach*. New York: John Wiley and Sons, 1995.

United Nations Conference on Trade and Development. *Emerging Forms of Technological Cooperation: The Case for Technology Partnership*. New York: United Nations, 1996.

United Nations Department for Economic and Social Information Analysis. *Concise Report on the World Population Situation in 1995*. New York: United Nations, 1995.

———. *Long Range World Population Projections*. New York: United Nations, 1992.

United Nations Fund for Population Affairs. *The State of World Population, 1998*. New York: United Nations, 1998.

United States v. Butler, 297 U.S. 1, 56 Sup. Ct. 312 (1936).

Urmson, J. O. "John Ellis McTaggart." In *The Concise Encyclopedia of Western Philosophy and Philosophers*, edited by J. O. Urmson and Jonathan Ree. London: Unwin Hyman, 1989, 1991.

Urmson, J. O., and Jonathan Ree, eds. *The Concise Encyclopedia of Western Philosophy and Philosophers*. London: Unwin Hyman, 1989, 1991.

Ushenko, Andrew Paul. "Einstein's Influence upon Philosophy." In *Albert Einstein: Philosopher-Scientist*. Library of Living Philosophers, vol. 7, edited by Paul Arthur Schlipp. LaSalle, IL: Open Court, 1949, 1970, 1995.

von Baeyer, Hans Christian. *Taming the Atom: The Emergence of the Visible Microworld*. New York: Random House, 1992.

Waldrop, M. Mitchell. *Complexity: The Emerging Science at the Edge of Order and Chaos*. New York: Simon and Schuster, 1992.

———. "The Quantum Wave Function of the Universe." *Science*, December 1988, 1248–50.

Walker, E. H., and Nick Herbert. "Hidden Variables: Where Physics and the Paranormal Meet." In *Future Science*, edited by John White and Stanley Krippner. Garden City, NY: Anchor, 1977.

Warren, Mary Anne. "On the Moral and Legal Status of Abortion." In *Contemporary Issues in Bioethics*, edited by Tom L. Beauchamp and LeRoy Walters. Belmont, CA: Wadsworth, 1978.

Watson, James D. *The Double Helix: A Personal Account of the Discovery of the Structure of DNA*. New York: Penguin, 1969.

Weinberg, Steven. *Dreams of a Final Theory*. New York: Pantheon, 1992.

West, Robin. "Forward: Taking Freedom Seriously." *Harvard Law Review* 104 (1990): 43–106.

Wheeler, J. Craig. "Of Wormholes, Time Machines, and Paradoxes." *Astronomy*, February 1996, 52–57.

Wheeler, John A. *At Home in the Universe*. Woodbury, NY: American Institute of Physics, 1994.

Whipple, Chris. "Can Nuclear Waste Be Stored Safely in Yucca Mountain." *Scientific American*, June 1996, 72–79.

Whiteman, J. H. M. "The Convergence of Physics and Psychology." In *Future Science*, edited by John White and Stanley Krippner. Garden City, NY: Anchor, 1977.

Widerker, David. "Providence, Eternity, and Human Freedom." *Faith and Philosophy* (April 1994) 242.

Will, Clifford M. *Was Einstein Right? Putting General Relativity to the Test*. New York: Basic Books, 1986.

Wills, Christopher. *The Runaway Brain: The Evolution of Human Uniqueness*. New York: Basic Books, 1993.

Witten, Edward. "Reflections on the Fate of Spacetime." *Physics Today*, April 1996, 24–30.

Wolf, Fred Alan. *Parallel Universes: The Search for Other Worlds*. New York: Simon and Schuster, 1988.

———. *Taking the Quantum Leap: The New Physics for Non-Scientists*. New York: Harper and Row, 1981, 1989.

Wolf, Jonathan S. *Physics*. Hauppauge, NY: Barron's Educational Series, 1996.

Wolff, Robert Paul. "Immanuel Kant: Prolegomena to Any Future Metaphysics." In *Ten Great Works of Philosophy*, edited by Robert Paul Wolff. New York: Penguin, 1969.

———. ed. *Ten Great Works of Philosophy*. New York: Penguin, 1969.

Woods, Linda K. "The Men and Mission of the Manhattan Project." *World War II*, July 1995, 39–45.

Wright, John, ed. *New York Times 1998 Almanac*. New York: Penguin, 1997.

Wyden, Peter. *Day One: Before Hiroshima and After*. New York: Warner, 1985.

Ziman, John. *Teaching and Learning about Science and Society*. Cambridge: Cambridge Univ. Press, 1980.

Zimmerman, Barry, and David Zimmerman. *Why Nothing Can Travel Faster than Light*. Chicago: Contemporary Books, 1993.

Zohar, Danah. *The Quantum Self: Human Consciousness Defined by the New Physics*. New York: Quill/William Morrow, 1990.

Index

abortion:
  acquired humanness and, 312–13
  acquisition-of-humanness standard and, 357–59
  compulsory, 453
  contragestives versus, 367–68
  law and, 347–50, 354–55
  unsafe, 452
absolute magnitude, 75;
  definition of, 504
academic community, and science-and-technology policy, 471–72
acceleration, 210;
  definition of, 504
accidents, 110;
  definition of, 504
accretion disk, 217;
  definition of, 504
acquired immunodeficiency syndrome (AIDS), 425
acquisition of humanness, 312–13;
  emergentism and, 313–14
  progression of, 323–24
  reductionism and, 313
  vitalism and, 312–13
acquisition-of-humanness standard, 357–59;
  advantages of, 361–62
  problems with, 362–65
active galaxy(ies), 217–19;
  definition of, 504
ADA. *See* Americans with Disabilities Act
adenosine triphosphate, 298
Administrative Procedure Act, 387;
  definition of, 504

adversarial system, 8;
    definition of, 504–5
AFRAND. *See* African Foundation for Research and Development
African Foundation for Research and Development (AFRAND), 474
Agenda 21, 467;
    definition of, 505
AIDS. *See* acquired immunodeficiency syndrome
Aiken, Henry, 275–76
Alfven, Hannes, 86–87, 89, 497
allies, 430–31;
    definition of, 505
allometric relationship, definition of, 505
alpha particle(s), 115;
    definition of, 505
Alpher, Ralph, 59
ALS. *See* Assembly of Life Sciences
alternate time line(s), 237–40;
    definition of, 505
Alvarez, Luis, 332–33
ambiplasma, 87;
    definition of, 505
American law, contemporary sources of, 346–48
Americans with Disabilities Act (ADA), 407
amino acids, 296;
    definition of, 505
amplitude, 122;
    definition of, 505
Anaxagoras, 109
Anaximander, 107, 176–77
Anaximenes, 107
ancient Greeks:
    on reality, 106–8
    on stability, 175–77
Anderson, Carl, 497
Anderson, W. French, 401–2
Andromeda galaxy, 42*f*, 43
anomaly, 15–17;
    and big bang theory, 72
    definition of, 505
anticodon, 398;
    definition of, 505

antielectron, 87
antimatter, 87;
    definition of, 506
    in *Star Trek: The Next Generation,* 28
apeiron, 107, 176–77;
    definition of, 506
ape-men, 321;
    definition of, 506
Apollo program, 4*f*
apparent magnitude, 75;
    definition of, 506
Ardipithecus ramidus, 322;
    definition of, 506
Aristotle, 177, 497
arithmetic growth, 425
arithmetic progression, definition of, 506
artificial selection, 317;
    definition of, 506
asexual reproduction, 318
Asilomar conference, 405
Aspect, Alaine, 20–21, 153–54
Aspect verification experiment, 154–55
Assembly of Life Sciences (ALS), 405
asteroid(s), 301;
    definition of, 506
asymmetric aging, 199;
    definition of, 506
atom, 109;
    definition of, 506
    plum pudding model of, 114–15, 116*f*
    size of, 51
    solar system model of, 48, 49*f*, 115–16, 117*f*
atomic structure:
    Bohr on, 120–21, 121*f*
    debate on, 113–18
    de Broglie on, 121–23, 123*f*
    Schrödinger on, 123–24
atomic theory:
    development of, 110–13
    Greeks on, 109
At the Narrow Passage (Meredith), 239
Australopithecines, 321–23;
    definition of, 506
Australopithecus afarensis, 322
Australopithecus africanus, 322

Australopithecus anamensis, 322
Australopithecus boisei, 322
average life expectancy, 422;
    definition of, 506
axon, 358;
    definition of, 506–7

Back to the Future, 206, 232
Baker, James, 426
balancing principle, 382;
    definition of, 507
Baltimore conference, 20–21
Barrow, John, 71–72, 142
Barth, Philip W., 462
baseball analogy for scientific method, 11–12;
    shortcomings of, 22–23
base-paired sequence, 396;
    definition of, 507
bases, 395;
    definition of, 507
Beagle voyage, 315–16
Becquerel, 499
Bell, John, 152–53
Bell's inequality, 153;
    definition of, 507
Bell's interconnectedness theorem, 153–54;
    definition of, 507
Berg, Paul, 402, 405
Bergson, Henri, 198–99, 259–60, 289
Berkeley, George, 258
Berlin Mandate, 467
Bernstein, Jeremy, 13
Berra, Tim, 326
Berzelius, Jöns Jakob, 289
beta decay, 52;
    definition of, 507
Beza, Theodore, 272
big bang, 33–67;
    causes of, 53–57
    chronology of, 56f, 57–58
    description of, 52–58
    nature of, 41–52
    reasons for study of, 34–36
big bang theory, 33;
    alternatives to, 82–89
    challenges to, 68–101
    definition of, 507
    evidence for, 58–60
    misconceptions about, 36–40
    novelty of, 70–74
    problems with, 74–82
bilateral symmetry, 330;
    definition of, 507
binding precedent, 347;
    definition of, 507
biotechnology, 405;
    definition of, 507
    government regulation of, 406–7
    and green revolution, 458–60
    private regulation of, 405–6
bipolar world, 427;
    definition of, 507
birth control technologies, 451–53;
    difficulties with, 452–53
Birx, H. James, 293–94
black body radiation, 119;
    definition of, 507–8
black hole(s), 55;
    definition of, 508
    and future time travel, 213–16
    types of, 216–20
Black Skull, 322
blastocyst, 348;
    definition of, 508
block universe. *See* space-time cube
blue shift, 79
Bode's rule, 66n38
Boethius, 272
Bohm, David, 155–56, 273
Bohr-Einstein debate, 147–52;
    definition of, 508
    experimental verification and, 154
Bohr, Niels, 118, 139–40, 154, 273, 498;
    and atomic theory, 120–21, 121f
    and Copenhagen Interpretation, 140–44, 159
    on EPR experiment, 150–52
Bondi, Hermann, 39, 82–84, 498
Boolean logic, 231;
    definition of, 508

Born, Max, 136–37, 498
Borosage, Robert, 464
Boscovich, Ruder, 181
bosons, 52
Brahe, Tycho, 16, 498
Brave New World (Huxley), 366
Briggs, Derek, 329
Brody, Baruch, 360–61
brown dwarf, 302;
　definition of, 508
Bryan, William Jennings, 310
Burbidge, Geoffrey, 85–86
Burgess Shale fossils, 328–31;
　definition of, 508
Bush, George, 426
Bush, Vannevar, 468–69, 473–74
business, and science-and-technology policy, 471–72
By Air Express to Venus (Rockwood), 68–69
Byrne, John, 173–74

Cairns-Smith, A. G., 289
Calvin, John, 272
Cameron, Donald, 133
Canadian NeuroScience Network, 480
Capra, Fritjof, 157
Card, Orson Scott, 239
Carey, Diane, 239
Carey v. Population Services International, 366
carousel theory, 334
case law, 347;
　definition of, 508
Casti, John, 23
catalyst, definition of, 508
cathode rays, 114;
　definition of, 508
causality, in quantum physics, 138–39
Celera Genomics, 407–8
cellular organisms, 295–96
Cepheid variable star(s), 75;
　definition of, 508
cerebral cortex, 324;
　definition of, 508
　and humanness, 358, 363–64

origin of, 358
size of, 325
CERN, European Organization for Nuclear Research, 152, 195
C-field, 85–86;
　definition of, 511–12
Chadwick, James, 116
CHD. *See* Commission on Hereditary Disorders
chemistry, and atomic theory, 112–13
chemosynthesis, 298;
　definition of, 508–9
Christianity, and atomism, 109–10
chromosomes, 395;
　definition of, 509
clarity, 13;
　definition of, 509
　in law, 350–51
Clarke, Arthur C., 458
classical school, 155, 157–58
Clauser, John, 153
climate of extremes, 438
COBE. *See* Cosmic Background Explorer
codon, 398;
　definition of, 509
Cole, Stephen, 19–20, 386
Coma cluster, 74;
　definition of, 509
comet(s), 334;
　definition of, 509
Comet Halley, 300*f*, 300–301;
　definition of, 509
commerce clause, 383;
　definition of, 509
Commission on Global Governance, 477–78
Commission on Hereditary Disorders (CHD), 407
Commission on Sustainable Development (CSD), 475
common law, 346–47;
　definition of, 509
communication, and decentralization, 453–57
companion-star hypothesis, 333–34
compelling interest, 349

compelling state interest, definition of, 509
complementarity principle, 139–40;
   definition of, 509
complexification, 292–94;
   definition of, 509
complexity, 286
Compton, Arthur Holly, 473
Compton, Karl, 429
Conant, James P., 473
conceptions, 254;
   definition of, 509
cone of increasing diversity, 329;
   definition of, 510
Conference of the Parties to the Framework Convention on Climate Change, 467
consequentialism, 346
conservation paradox, 233–35;
   definition of, 510
Constitution, 347–48;
   basic plan of, 382–84
   definition of, 510
constructivism:
   definition of, 510
   versus positivism, 18–22, 116–18
contingency, 330;
   definition of, 510
continuity, 12–13;
   definition of, 510
   in law, 350–51
contraception, compulsory, 452–53
contradictory-event paradox(es), 231–36;
   definition of, 510
contragestive agent(s), 367–71, 451;
   definition of, 511
Copenhagen Interpretation, 140–48;
   Bohr-Einstein debate on, 147–48
   children of, 159–63
   definition of, 511
   essentials of, 141–44
Copernican system, definition of, 511
Copernicus, Nicolaus, 15, 498
correspondence, 12–13;
   definition of, 510
cortex. *See* cerebral cortex

Cosmic Background Explorer (COBE), 33, 35, 81–82;
   definition of, 511
cosmic background radiation, 59–60;
   definition of, 511
cosmic string (CS), 225–26;
   definition of, 511
cosmological constant, 46;
   definition of, 511
   rebirth of, 94–95
Coveney, Peter, 159
craft, 391
creation center(s), 86;
   definition of, 511
creation field, 85–86;
   definition of, 511–12
creationism (creation science), 1940 presidential election;
   and big bang theory, 34–35, 38–40
   definition of, 512
   and evolution, 310–11
creative capital arrangements, in science-and-technology program, 472–75
creativity, in science, 14–18
Crick, Francis, 290, 301, 395, 498
crisis, 17;
   and emergence of quantum physics, 17–18
Crommelin, Andrew, 499
CS. *See* cosmic string
CSD. *See* Commission on Sustainable Development
culture, and science, 18–22
Curie, Marie, 499
cyanobacteria, 295;
   definition of, 512
cytoplasm, 398

Dalibard, Jean, 154
Dalton, John, 112, 499
Daniels, David, 238
Dar, Arnon, 335
The Dark Tower (Lewis), 233
Darrow, Clarence, 310
Darwin, Charles, 315–18, 319f–320f, 328, 499

Daudel, Raymond, 288
Davies, Paul, 37, 162, 286–88
"Dear Charles" (Leinster), 232–33
de Broglie, Louis Victor, 498;
   and atomic theory, 121–23, 123f
decentralization, 453–57
decimation, 329;
   definition of, 512
decimation/contingency view of evolution, 328–31
deductive logic, 23;
   definition of, 512
deep sea vents, 298;
   definition of, 512
Democritus, 109, 499
dendrite, 358;
   definition of, 512
density, 46;
   definition of, 512
deoxyribonucleic acid (DNA), 298;
   definition of, 512
   double helix structure of, 395–97, 396f
   and humanness, 363
   and life, 293
   recombinant, 399
Descartes, René, 249, 289, 499
descent with modification, rationale for, 316–18
determinism, 270–73;
   definition of, 512
device, 392;
   definition of, 512–13
DeWitt, Bryce, 162, 237–38
Diamond v. Chakrabarty, 380–81, 392–93, 400
differential reproduction, 318;
   definition of, 524
Dingle, Herbert, 195
Dinosaur Nexus (Grimes), 239
dinosaurs, extinction of, 332
Dirac, Paul, 136, 499
disassemblers, 461–62;
   definition of, 513
disease, population growth and, 424–26
DNA. *See* deoxyribonucleic acid

DNA mapping, 401;
   definition of, 513
DNA sequencing, 401;
   definition of, 513
doctrine of identity, 144–45;
   definition of, 513
Doppler effect, 41;
   definition of, 513
Doppler, J. Christian, 41
double helix, 395–97, 396f;
   definition of, 513
Dressler, Alan, 80
Drexler, K. Eric, 462
drug approval process, 369
due process of law, definition of, 514
duon, 137;
   definition of, 514
duration, as product of human consciousness, 273–74
Dyson, Frank Watson, 499

Earman, John, 234–35
Earth, in prelife era, 294
Earth Summit, 466–67;
   definition of, 514
Economic and Social Council (ECOSOC), 476
Economic Security Council (ESC):
   challenges facing, 478–80
   definition of, 514
   role of, 476–77
   structure and goals of, 475–76
economic warfare, 434
ECOSOC. *See* Economic and Social Council
Eddington, Arthur, 47, 212–13, 247, 256, 499
education, in science-and-technology program, 472
Einstein, Albert, 1–2, 18, 46, 70, 186f, 251f, 499–500;
   and atomic bomb, 473
   on cosmological constant, 94
   general theory of relativity, 210–11
   on gravity, 212
   and philosophy, 247–53

Einstein, Albert *(continued)*
  and photoelectric effect, 120
  problem facing, 210–11
  on psychological time, 175
  on quantum weirdness, 145–47
  special theory of relativity, 185–88
  on uniform motion, 180
Einstein-Kant debate, 255–56
Einstein-Podolsky-Rosen thought experiment (EPR), 148–52;
  definition of, 515
Einstein separability, 154;
  definition of, 514
Eisenhower, Dwight D., 435
Eisenstadt v. Baird, 365–66
electricity, 181–82
electromagnetic field, 113;
  definition of, 514
electromagnetic force, 51;
  definition of, 514
electromagnetism, 113–14
electrons, 48;
  definition of, 514–15
  discovery of, 114–15
  size of, 51
elliptical galaxy, 44f
elsewhen (elsewhere), 267f, 268;
  definition of, 515
embryo, 348;
  definition of, 515
emergentism, 291–94;
  and acquisition of humanness, 313–14
  definition of, 515
  definition of life in, 292–94
  versus reductionism, 291–92
encephalization quotient (EQ), 324–25;
  definition of, 515
encephalographic activity, and humanness, 360–61
energy sources, alternative, 457–58, 466
entity theory, 331;
  definition of, 515
entropy, 83
enumerated powers, 382;
  definition of, 515

environment, 436–39;
  conferences on, 466–67
  science and technology and, 465–67
Environmental Protection Agency (EPA), 353
enzymes, 299–300;
  definition of, 515
EPA. *See* Environmental Protection Agency
EPR. *See* Einstein-Podolsky-Rosen thought experiment
Epsilon Eridani, 302–3
Epstein, Lewis Carroll, 187
EQ. *See* encephalization quotient
equal protection clause, 349
equity, definition of, 515
equivalence, principle of, 211
ergosphere, 222;
  definition of, 515
ESC. *See* Economic Security Council
escape velocity, 213;
  definition of, 515
Escher, M. C., 290
ether. *See* luminiferous ether
ethical relativism, 346;
  definition of, 516
ethics:
  and law, 346
  and science, 344–45
eukaryotes, 296;
  definition of, 516
event horizon, 215;
  definition of, 516
  inner, 222
  outer, 222
Everett, Hugh, 162, 237–38
Evin, Claude, 379n110
evolution, 310–42;
  concept of, 315–18
  decimation/contingency view of, 328–31
  evidence for, 315–16
  mass extinction view of, 331–35
  theory of, 315–28
evolving universe theory, 55–57;
  definition of, 516
existentialism, definition of, 516

expanding universe theory, 37, 70–71. *See also* big bang theory
experiments, 10
extraterrestrial life, 283–84
extraterrestrial theory, 332
extreme Kerr object, 224;
   definition of, 516

Faber, Sandra, 80
fact-finding system, definition of, 516
falsification, 12–13
Faraday, Michael, 181
faster than light (FTL) travel, 208–9
FDA. *See* Food and Drug Administration
Fermi, Enrico, 474, 500
fetus, 348;
   definition of, 516
Feynman, Richard, 11
fiber optics, 454–55;
   definition of, 516
field(s), 53, 181;
   definition of, 516
Finnegan's Wake (Joyce), 48
First Frontier (Carey & Kirkland), 239
First Law of Thermodynamics, 233;
   definition of, 516
Foley, Robert, 325–26
Food and Drug Administration (FDA), 353;
   and biotechnology, 369
foreknowledge, definition of, 516
Fornax cluster, 76–77;
   definition of, 516
Fox, Sydney, 295
Franck, James, 136
Freedman, Wendy, 74–76
free will. *See* volition
frequency, 122;
   definition of, 516
Friedmann, Alexander, 44–47, 500
Frye v. United States, 353
FTL. *See* faster than light
Fuller, Graham E., 427–28
fundamental rights, definition of, 516
FUNDES, 475
fusion, definition of, 516

future, 448–88;
   science and, 417–47
future light cone, 267*f*, 268;
   definition of, 517
future time travel, 207–20;
   shortcomings of, 208–10
   twins paradox and, 192–95

galactic cluster(s), 47, 74–75, 79*f*;
   definition of, 517
galactic groups, 99n19
galaxy(ies), 41;
   active, 217–19
   definition of, 517
   elliptical, 44*f*
   irregular, 45*f*
   movement of, 77–81
   spiral, 42*f*, 43
Galilean relativity, 178;
   definition of, 517
Galileo, on relativity, 178–79
The Gallatin Divergence (Smith), 238–39
gamma-ray burst theory, 335
gedanken experiment, 149;
   definition of, 517
Geller, Margaret, 80
Gell-Mann, Murray, 158
gene, 395;
   definition of, 517
general-acceptance standard, 353
general theory of relativity, 210–13;
   predictions made by, 212–13
general welfare clause, 383;
   definition of, 517
The Genesis Machine (Hogan), 90
gene splicing, 399
gene therapy, 401–2;
   definition of, 517
genetic code, 397;
   definition of, 517
genetic engineering, 351, 399–404;
   benefits of, 399–401
   concerns about, 402–4
   definition of, 517
   patents and, 392–94
   regulation of, 380–408

genetics, introduction to, 395–99
genetic variant(s), 318;
    definition of, 517
genome, 401;
    definition of, 517
geometric growth, 425
geometric progression, definition of, 517
Gerhardt, Charles, 289
Germany, 430–31
Gilligan, Carol, 367
Glashow, Sheldon, 91–92
glial cells, 358;
    definition of, 517
Glick, Edward Bernard, 464–65
global warming, 436;
    consequences of, 438–39
    definition of, 517
    evidence for, 436–38
gluon(s), 52;
    definition of, 517
Gödel, Kurt, 253, 257–61, 270, 273, 289, 500
golden rule, 346;
    definition of, 518
gold foil experiment, 115–16
Gold, Thomas, 39, 82, 84, 500
Gonzales, Guillermo, 303
Goodstein, David L., 5
Gorbachev, Mikhail, 426
Goswami, Amit, 162–63
GOTT/Closed Time Curve/Cosmic String/Temporal Distortion Twist, 225–27, 226f
Gott, J. Richard, 225–27
Gould, Stephen Jay, 326, 328–31, 500
government, and science-and-technology policy, 471–72
Grail, 103
grandmother (father, parent) paradox, 231–33;
    definition of, 518
grand unified force, 89
grand unified theory, definition of, 518
Grassetti v. Weinberger, 387
gravitational field equations, 214;
    definition of, 518

gravitational radius, 213–14;
    definition of, 534
graviton(s), 51;
    definition of, 518
gravity, 46, 51–52, 93;
    definition of, 518
    Einstein's theory of, 212
    and time, 210
Great Attractor, 80;
    definition of, 518
Great Wall, 80;
    definition of, 518–19
greenhouse effect, 436;
    definition of, 519
green revolution, 458–60
Green, Thomas Hill, 258
Grimes, Lee, 239
Griswold v. Connecticut, 365
ground state, 50;
    definition of, 519
Groves, Leslie, 435
Gulf War, 426

Halpern, Paul, 238
"The Hammer of God" (Clarke), 458
handmade cell, 299–300;
    definition of, 519
Hartmann, William K., 292, 326
Hawking, Stephen, 105, 219, 500
Hazen, Robert, 4, 51, 93
hectare, 454;
    definition of, 519
Heilbroner, Robert L., 343–44, 419–20, 423, 425–26, 432, 436, 450, 465–67
Heisenberg, Werner, 135–38, 501. *See also* uncertainty principle
helium, and big bang theory, 59
Heraclitus, 107–8, 177, 501
Herbert, Nick, 156–57, 160–61
Herman, Robert, 59
Hess, Victor Francis, 497
Higgs field, 53, 91
Highfield, Roger, 159
HIV. *See* human immunodeficiency virus
Hod, Shahar, 246n91
Hogan, James P., 90, 238

holism, 291;
  definition of, 519
holomovement, 156;
  definition of, 519
hominids, 322;
  definition of, 519
hominisation, 314;
  definition of, 519
hominoids, 321;
  definition of, 519
Homo, 322;
  definition of, 519
Homo erectus, 324;
  encephalization quotient of, 325
  Gould on, 331
Homo habilis, 323
Homo sapiens, 323–24;
  definition of, 519
Homo sapiens neanderthalensis, 326
Hoyle, Fred, 37, 39–40, 82, 85–86, 301, 501
Hubble constant, 43
Hubble, Edwin, 38, 41, 43, 501
Hubble's law, 43;
  definition of, 519
Huchra, John, 80
human consciousness, duration as product of, 273–74
Human Genome Project (HGP), 401–2;
  definition of, 519
  private industry and, 407–8
human immunodeficiency virus (HIV), 425;
humanity
  appearance of, 321–28
  emergence of, 310–42
human life, nature of, 312–15;
humanness:
  acquired. *See* acquisition of humanness
  other standards of, 359–60
human resource development, in science- and-technology program, 472
Hussein, Saddam, 426, 434
Huxley, Aldous, 366
Hydra-Centaurus supercluster, 80;
  definition of, 519

hydrogen, and big bang theory, 59
hypothesis, 10;
  definition of, 519–20

IAEA. *See* International Atomic Energy Agency
idealism, 258;
  definition of, 520
  and relativity, 258–60
  and simultaneous existence of time, 257–69
implicate order, 156;
  definition of, 520
Incompleteness Theorem, 289–90
inductive logic, 23;
  definition of, 520
inertial system(s), 180;
  definition of, 520
infant mortality rate(s), 422;
  definition of, 520
inflationary epoch, 57
information paradox, 236–37;
  definition of, 520
Inherit the Wind, 310–11
inner event horizon, 222;
  definition of, 520
International Atomic Energy Agency (IAEA), 488n141
Internet, 455;
  definition of, 520
Iraq, 432–33
irregular galaxy, 45*f*
isometric relationship, definition of, 520
Israel, 431–32

Jaki, Stanley, 39
Jammer, Max, 136, 138, 140–41
Japan, 430–31
Jastrow, Robert, 39
Jeans, James, 260
John Paul II, pope, 99n14
Jones, David E. H., 462
Joyce, James, 48
judicial decisions, 347;
  definition of, 520

juriscience, 343–79;
  definition of, 520
  frontiers of, 394–408
  and procreative freedom, 354–65
  and procreative rights, 365–71
  use of term, 346
jurisdiction, 347;
  definition of, 521
justice, 346;
  definition of, 521

Kaku, Michio, 93–94
Kane, Gordon, 124
Kant, Immanuel, 254–56, 258
Kauffman, Stuart, 292
Kelvin, Lord, 92–93
Kennedy, John F., 3, 19, 427
Kennedy, Paul, 422–26
Kepler, Johannes, 15–16, 501
Kerr black hole. *See* rotating black hole
Kerr-Newman black hole, 224;
  definition of, 521
Kerr, Roy, 220
Kevles, Daniel, 18, 148
Kilgore, Harley M., 468
Killian, James, 2
Kirkland, James, 239
Kirshner, Robert, 76
Klyshko, David, 21, 157
knowledge paradox, 236–37;
  definition of, 520
koinomatter, 87;
  definition of, 521
Kragh, Helge, 106
Kuhn, Thomas, 14–18, 20, 52, 70;
  and big bang theory, 71–74
Kyoto Protocol, 467

Lacourt, Dominique, 288–89
Laplace, Pierre Simon, Marquis de, 112, 170n14
laser interferometer gravity observatory (LIGO), 86;
  definition of, 521
law:
  definition of, 521
  ethical roots of, 346
  goals and process of, 7–8
  interaction with science, 352–54
  and manufactured life, 380–81
  nature of, 345–48
  and peer review, 386–88
  on procreative freedom, science and, 355–57
  and regulation of science, 381–90
  and regulation of technology, 390–94
  and science, 344–45, 348–54
  similarities with science, 350–52
law of nature, 10;
  definition of, 521
Lawrence, Ernest O., 473
laws of physics, and uniform motion, 179–80
Leakey, Richard, 323, 326–27
Leftow, Brian, 282n91
Leibniz, G. W., 502
Leinster, Murray, 232–33
Lemaitre, Georges, 38, 44–46, 501
LeMay, Curtis, 429, 435
Lenzen, V. F., 249
leptons, 50–51;
  definition of, 521
less developed world, 420;
  definition of, 521
Leucippus, 109
Lewin, Leonard C., 464
Lewis, C. S., 233
life:
  as accidental versus inevitable, 301–3
  characteristics of, 286–87
  extraterrestrial, 283–84
  manufactured, law and, 380–81
  nature of, theories on, 288–94
  from nonlife, 299–301
  versus nonlife, 285–88
  origin of. *See* origin of life
  at subatomic level, 287–88
lifeline (lifeworm, worldline), 261, 262f;
  definition of, 521
  example of, 261–63, 264f, 266f

light:
  conflict caused by, 184–85
  nature of, 181–82
  and relative motion, 180–85
light cones, 265–68, 267f
light speed, 52, 185, 244n4;
  definition of, 521
  and space-time cube, 265
light year, 75;
  definition of, 521
LIGO. *See* laser interferometer gravity observatory
Linde, Andrei, 53–55, 57
Lindley, David, 93
linear, 209;
  definition of, 521
lines of force, 181;
  definition of, 521
literary criticism, goals and process of, 8–9
Local Group, 79;
  definition of, 521
locality, classical school on, 157–58
Lorentz contraction, 189–90;
  definition of, 521–22
Lorentz factor, 190;
  definition of, 522
Lorentz, Hendrik Antoon, 136, 189–90
Lorentz light clock, 191;
  definition of, 522
Lorentz transformation, 189–90;
  definition of, 522
Lowell, Percival, 1
Lucretius, 109, 499
Lucy, 322
luminiferous ether, 113, 182;
  definition of, 522
  mystery of, 182–84
luminosity, 75;
  definition of, 522

Mach, Ernst, 210–11
Madsen, Axel, 343, 366
magnetism, 181–82
Malthusian check, 425–26;
  definition of, 522

Malthus, Thomas Robert, 316, 425
Mandelbaum, Michael, 430–32
Manhattan Project, 474;
  definition of, 522
Mann, Murray Gell, 48
Mansfield, Edwin, 470
"The Man Who Made Tomorrow" (Byrne), 173–74
many-worlds theory (parallel universe theory), 162–63;
  and alternate time lines, 237–38
  definition of, 522
March, Fredric, 310
March, Robert, 160
Mars, 285f;
  Kepler on, 15–16
  life on, 283–84
  Lowell on, 1
mass, 52;
  definition of, 522
mass extinctions, 331–32;
  definition of, 522
  and evolution, 331–35
  interval of, 333
massive black holes, 217–19;
mathematical model(s):
  definition of, 522
  role of, 22–24
matrix mechanics, 136;
  definition of, 522
matter, continuous creation of, 83–84
Matthews, Clifford, 301
Maxwell, James Clerk, 112, 139, 180–82, 501;
  on electromagnetism, 113–14
  on ether, 182–83
McTaggart, John Ellis, 258
Mearsheimer, John, 446n70
mechanism, 290
membrane theory, 101n103
Mendeleev, Dimitri, 112
Mendel, Gregor, 395
Meredith, Richard C., 239
Merkle, Ralph C., 462
messenger RNA (mRNA), 398;
  definition of, 522

Messier, Charles, 43
metabolism, 299;
   definition of, 522
metagalaxy, 87;
   definition of, 522
meteorite, 283;
   definition of, 523
methodological reductionism, 290;
   definition of, 523
Michelson, Albert, 184
Michelson-Morley experiment, 139, 183–84
Miller, Ron, 292, 326
Miller, Stanley, 296, 501–2
Miller-Urey experiment, 296–98, 297*f*;
   definition of, 523
Milosevic, Slobodan, 428
mini-black hole, 219;
   definition of, 530
Minkowski diagram(s), 265–68, 267*f*;
   definition of, 523
Minkowski, Hermann, 261
Minsky, Marvin L., 462
missing links, 321;
   definition of, 506
modality, 254;
   definition of, 523
model, 22;
   definition of, 523
   mathematical, role of, 22–24
Mohktarian, Patricia L., 456–57
momentum, 137;
   definition of, 523
Moon, 4*f*
more developed world, 420;
   definition of, 523
Morley, Edward, 184
Morowitz, Harold J., 312, 314, 325, 357
Morris, Michael, 227
Morris, Simon Conway, 329, 331
Morris/Thorne Traversable Wormhole, 227–30, 228*f*–229*f*
mRNA. *See* messenger RNA
M-theory, 101n103
Mueller, John, 432–34
Mueller, Karl, 432–34

Muller, Richard, 333
multicellular organisms, 295–96
multipolar world, 427;
   alignments in, 430–34
   dangers of, 427–28
   definition of, 524
   strategic options in, 434–36
muons, 195;
   definition of, 524
mutation, 318;
   definition of, 524

NAFTA. *See* North American Free Trade Agreement
naked singularity, 225;
   definition of, 524
nanocomputer(s), 461–62;
   definition of, 524
nanotechnology, 460;
   definition of, 524
   limits and dangers of, 462–63
   and redistribution of wealth, 460–63
   in *Star Trek: The Next Generation*, 28
Narlikar, Jayant V., 85–86
National Academy of Sciences (NAS), 405
National Aeronautics and Space Administration, 3–4, 473
National Institutes of Health (NIH), 384;
   Recombinant DNA Advisory Committee (RAC), 406
National Science Foundation (NSF), 3, 104–5, 469–70
nations, global realignment of, 430–34
natural law, 10;
   definition of, 521
natural selection, 317;
   definition of, 524
Neanderthals, 325–28;
   definition of, 524–25
   Gould on, 331
nebula:
   definition of, 525
   explosion of, 302
Nemesis, 334;
   definition of, 525

neocortex, 324;
   definition of, 508
neurons, 358;
   creation of, 358–59
   definition of, 525
neurotransmitter molecules, 358;
   definition of, 525
neutrino(s), 51, 288;
   definition of, 525
neutron(s), 48;
   definition of, 525
   in solar system model, 116
neutron star(s), 55;
   definition of, 525–26
   and future time travel, 219–20
Newman, Ezra, 224
Newton, Isaac, 502;
   and atomic theory, 110–12
   on gravity, 212
   laws of motion, 132n95
   on relativity, 177–80
new world order, 426–30
New York Academy of Sciences, 469
NIH. See National Institutes of Health
nitrogen fixation, 459;
   definition of, 526
node(s), 122;
   definition of, 526
nonlocality, 151;
   Baltimore conference on, 20–21
   classical school on, 157–58
   definition of, 526
   experimental verification of, 154–55
   implications of, 155–58
   quantum school on, 155–57
normality, 404;
   definition of, 526
normal science, 15;
   definition of, 526
Norplant, 370–71, 451–53
North American Free Trade Agreement
   (NAFTA), 479–80
North Korea, 432–33
Novikov consistency conjecture, 271;
   definition of, 526
Novikov, Igor, 271

NRC. See Nuclear Regulatory Commission
NSF. See National Science Foundation
Nuclear Regulatory Commission (NRC),
   353
nuclear weapons, dangers of, 429–30
nucleotide, 396;
   definition of, 526
nucleus, 48;
   definition of, 526
   size of, 51
   in solar system model, 116

objective standards, 9;
   definition of, 526
observation, 10
Ohio's Technology Transfer Organization
   (OTTO), 471–72
Olumucki, Martin, 289
omniscience, 272;
   definition of, 526
one solar mass, 302;
   definition of, 526
ontological reductionism, 290;
   definition of, 532
Oort comet cloud, 334;
   definition of, 526
open-mindedness, 13, 24
Oppenheimer, Robert, 502
orbit, 48;
   definition of, 526
orbital, 49;
   definition of, 526
order, 346;
   definition of, 526
organization, 286
Orgel, Leslie, 301
origin of life, 283–309;
   as accidental versus inevitable, 302–3
   alternative theories on, 298–99
   experimental verification of, 296–98
orphans, 431–32;
   definition of, 527
OTTO. See Ohio's Technology Transfer
   Organization
outer event horizon, 222;
   definition of, 527

ozone layer, 296;
  definition of, 527

Pagels, Heinz, 124, 141, 158
Pakistan, 431–32
panspermia, 300–301;
  definition of, 527
paradigm, 15–17;
  definition of, 527
parallel universe theory (many-worlds/universes theory), 162–63;
  and alternate time lines, 237–38
  definition of, 522
paranormal, quantum physics and, 161–62
Parker, Barry, 227–30
Parmenides, 108, 502
particle accelerator(s), 91;
  definition of, 527
particle physics, 91;
  definition of, 527
past light cone, 267f, 268;
  definition of, 527
past time travel, 220–30
Pastwatch: The Redemption of Christopher Columbus (Card), 239
Pasynskii, A. G., 296–98
patent(s), 353;
  definition of, 528
  and genetic engineering, 392–94
Pauli, Wolfgang, 18
Pavlovskais, T. E., 296–98
peer review, 384–85;
  advantages and disadvantages of, 385–86
  definition of, 528
  law and, 386–88
Penzias, Arno, 38, 59, 85
perceptions, 254;
  definition of, 528
perfect cosmological principle, 83;
  definition of, 528
periodic table of elements, 112;
  definition of, 528
persuasive precedent, 347;
  definition of, 528

PFP. *See* principle of the fixity of the past
PFT. *See* principle of the fixity of time
The Philadelphia Experiment, 206
The Philadelphia Experiment: Part II, 232
philosophy:
  Bohr and, 142
  Einstein and, 248–53
  physics and, 252–53
  and quantum physics, 105–5
  and relativity, 247–82
  and relativity of simultaneity, 198–99
  and space-time cube, 270–74
photoelectric effect, 120;
  definition of, 528
photon(s), 49–50, 120;
  definition of, 528
  in *Star Trek: The Next Generation*, 28
photosynthesis, 295;
  definition of, 528
photovoltaic cells (PVs), 466
physics, and philosophy, 252–53
Pierce, Michael, 76
Piran, Tsui, 246n91
Pius XII, pope, 72
placenta, 348;
  definition of, 528
Planck, Max, 50, 118–20, 502
planetary model, 48, 49f, 115–16, 117f;
  definition of, 528
Planned Parenthood of Pennsylvania v. Casey, 354–57, 367–68
plasma, 86;
  definition of, 528
plasma cosmology, 86–89;
  definition of, 528–29
plasmid, 380;
  definition of, 529
Plato, 289
plum pudding model, 114–15, 116f;
  definition of, 529
Podolsky, Boris, 149;
  and EPR experiment, 148–52
Pollack, Rachel, 237
Pollack, Robert, 293

pongids, 322;
    definition of, 529
Ponnamperuma, Cyril, 298
population, 320;
    causes of explosion of, 422–23
    definition of, 529
    growth of, 420–22
    issues regarding, 420–26
    science and technology and, 450–57
    unchecked, consequences of, 423–26
Population Council, 369
positivism, 13–14;
    constructivism versus, 18–22, 116–18
    definition of, 529
    Einstein and, 250–52
    and relativity, 254–57
positron, 87
Posner, Richard, 351
precedent, 347;
    definition of, 529
precellular organisms, 295
predestination, 272;
    definition of, 529
predictability, 13;
    definition of, 529
    in law, 350–51
preembryo, 348;
    definition of, 529
prelife, 294–99;
    to life, 294–96
prelife era, conditions in, 294
prelife soup theory, 294;
    definition of, 529
    and evolution, 318–19
premise, 289;
    definition of, 530
pressure, 46;
    definition of, 530
preventative war, 434–35;
    definition of, 530
primeval atom, definition of, 530
primeval soup theory, 294;
    definition of, 529
    and evolution, 318–19
primordial black hole, 219;
    definition of, 530

Princeton Interpretation, 160
principle of equivalence, 211;
    definition of, 530
principle of infinite divisibility, 109
principle of reciprocity, definition of, 518
principle of the conservation of energy, definition of, 530
principle of the fixity of the past (PFP), 270–71;
    definition of, 530
principle of the fixity of time (PFT), 271;
    definition of, 530
probabilities, 135–40
probability wave(s), 136–37;
    definition of, 530
Proconsul, 321;
    definition of, 530
procreative freedom, 343–79;
    definition of, 530
    juriscience and, 354–65
    and responsibilities, 366–67
    and right to privacy, 365–66
    types of, 345
procreative rights, juriscience and, 365–71
prokaryotes, 295;
    definition of, 530
proofreading, 397;
    definition of, 531
propagating waves, 122;
    definition of, 531
protein(s), 298;
    definition of, 531
    synthesis of, 397–99
proteinoid microspheres, 295
The Proteus Operation (Hogan), 238
protocells, 295;
    definition of, 531
proton(s), 48;
    definition of, 531
    in solar system model, 116
psychological relativity, 175;
    definition of, 531
psychological time, 175;
    definition of, 531
PVs. *See* photovoltaic cells

QSSC. *See* quasi-steady state cosmology
quality, 254;
   definition of, 531
quantity, 254;
   definition of, 531
quantum (quanta), 50, 119–20;
   definition of, 531
quantum foam, 215
Quantum Leap, 236
quantum physics, 50, 105;
   definition of, 531
   emergence of, crisis and, 17–18
   UMBC conference on, 20–21, 24
quantum school, 155–57
quantum theory:
   advent of, 118–24
   development of, 106–10
   reasons for study of, 104–6
   versus relativity theory, 272–74
quantum universe, 103–32
quantum weirdness, 133–72;
   complementarity and, 138–40
   definition of, 531
   Einstein on, 145–47
   experimental verification of, 152–55
   puzzle of, 134–35
quarks, 48;
   definition of, 531–32
quasars, 88*f*, 89, 216–17;
   definition of, 532
quasirigid, 183
quasi-steady state cosmology (QSSC), 85–86;
   definition of, 532

RAC. *See* Recombinant DNA Advisory Committee
radiation, 145;
   definition of, 532
radioactive decay, 145;
   definition of, 532
raisin bread model, 114–15;
   definition of, 529
Rampino, Michael R., 334
Rasmussen, Steen, 299

rational basis, 349;
   definition of, 532
Ratzsch, Del, 316
Raup, David, 333
Raymond, Susan U., 469
reality:
   ancient Greeks on, 106–8
   classical school on, 157–58
   Copenhagen Interpretation and, 159
   Princeton Interpretation and, 160
   quantum physics and, 142–43
   subquantal theory of, 160–62, 536
reciprocity, 346
recombinant DNA, 399;
   definition of, 532
Recombinant DNA Advisory Committee (RAC), 406
redistribution of wealth:
   nanotechnology and, 460–63
   war and, 432–34, 528
red shift, 41–43, 46*f*;
   definition of, 532
   and galactic movement, 78
   implications of, 43–48
reductionism, 290–91;
   and acquisition of humanness, 313–14
   definition of, 532
   emergentism versus, 291–92
regulation(s), 383;
   of biotechnology, 405–7
   definition of, 532
   of genetic engineering, 380–416
   of research funding, 384–88
   of science, 381–90
   of technology, 390–94
regulatory agencies, and technology, 394
Reichenbach, Hans, 250–53, 256–57, 275
relation, 254;
   definition of, 532
relativity:
   Galileo and Newton on, 177–80
   idealism and, 258–60
relativity of simultaneity, 189–90, 195–98;
   definition of, 532

relativity theory, 46–46;
　assumptions of, 70
　definition of, 532
　general, 210–13
　philosophy and, 247–82
　positivism and, 254–57
　versus quantum theory, 272–74
　relative nature of, 174–80
　special, 173–205
religion:
　and atomic theory, 109–10
　and big bang theory, 34–35, 38–40, 72
　and evolution, 310–11
　Newton and, 111–12
replication, 397;
　definition of, 532
replicator(s), 463;
　definition of, 532
　in *Star Trek: The Next Generation,* 28
reproducing-universe theory, 53–54;
　definition of, 532–33
reproductive technologies, 451–53;
　difficulties with, 452–53
research funding, regulation of, 384–88
restriction enzyme, 399;
　definition of, 533
retrovirus, 402;
　definition of, 533
revolutionary science, 17;
　and big bang theory, 71–74
　definition of, 533
rhenium, 334
ribonucleic acid (RNA), 298, 398;
　definition of, 533
ribosomal RNA (rRNA), 398;
　definition of, 533
ribosome, 398;
　definition of, 533
Ridgway, Matthew, 435
right to privacy, procreative freedom and, 365–66
ring envelope, 224;
　definition of, 533
ring singularity, definition of, 533
Riordan, Michael, 47–48, 51
RNA. *See* ribonucleic acid

RNA polymerase model, definition of, 533
RNA-polymerase molecule, 398
Rockwood, Roy, 68–69, 114
Roe v. Wade, 347–50, 355, 357
Roger, Gerard, 154
rogues, 432–33;
　definition of, 533
Roosevelt, Franklin D., 473–74
Rosen, Nathan, 149;
　and EPR experiment, 148–52
rotating black hole, 221;
　dangers within, 222–24
　definition of, 533
　and past time travel, 218*f*, 220–25
rRNA. *See* ribosomal RNA
RU-486, 368–69, 451;
　definition of, 533
Rucker, Rudy, 260–61, 263–65, 270–71, 273
Running Against Time, 206–7, 235
Russell, Bertrand, 281n11
Rutherford, Ernest, 115–18, 502

Sachs, Mendel, 199
Sakiz, Edouard, 379n110
scalar field, 53
Schatzman, Evry, 41–43
Schmitt, Harrison "Jack," 4*f*
Schramm, David, 47–48, 51
Schrödinger, Erwin, 502;
　and atomic theory, 123–24
Schrödinger's cat, 144–45, 146*f*;
　definition of, 534
Schrödinger's equation, 123–24;
　definition of, 534
　Heisenberg and, 135–36
Schwartz, Delmore, 448–49
Schwarzschild black hole:
　definition of, 535
　and time travel, 214–16
Schwarzschild, Karl, 213–14
Schwarzschild radius, 213–14;
　definition of, 534
SCID. *See* severe combined immunodeficiency

science, 5, 7–14, 391;
  creativity in, 14–18
  cultural and social influence on, 18–22
  definition of, 534
  end of, 91–94
  and environment, 465–67
  ethics and, 344–45
  and future, 417–47, 449–50
  global policy on, 468–80
  goals and process of, 9
  interaction with law, 352–54
  law and, 344–45, 348–54
  and law on procreative freedom, 355–57
  and population problem, 450–57
  reasons for study of, 2–7
  regulation of, 381–90
  similarities with law, 350–52
  versus technology, 391–92
  and war, 457–65
science appreciation:
  goal of, 6–7
  need for, 1–31
science court, 388–90;
  definition of, 534
science education, 5;
  paradox of, 3–5
science fiction, and alternate universes, 238–40
science illiteracy, 3;
  causes of, 5
  consequences of, 5–6
scientific evidence, admissibility of, 353
scientific method:
  baseball analogy for, 11–12, 22–23
  cycle of, 11f
  definition of, 534
  traditional, 9–14
Scientific Revolution, 110
scientific thought, timeline of, 489–96
scientific truth, nature of, 7–9
Scully, Marlan, 21
Second Law of Thermodynamics, definition of, 534
SEKO. *See* super-extreme Kerr object
self-awareness, and humanness, 361

sensations, 254;
  definition of, 534
separation principle, 382;
  definition of, 534
Sepkoski, J. John, 333
severe combined immunodeficiency (SCID), 401–2
sexual reproduction, 318
Shapiro v. Thompson, 349
Shaviv, Nir, 335
Sheldrake, Rupert, 290
Silk, Joseph, 71–72
simultaneous-temporal-existence paradox, 235–36;
  definition of, 534
singularity, 37, 47;
  definition of, 534
  naked. *See* naked singularity
skepticism, 13, 24
Slipher, Vesto, 43
Smith, L. Neil, 238–39
Smith, William, 316
Smolin, Lee, 55, 57
Smoot, George, 33, 35, 82
society, and science, 18–22
solar mass, 302
solar power, 466
solar system model, 48, 49f, 115–16, 117f;
  definition of, 528
solipsism, 249;
  definition of, 535
Solvay conference, 148
Soylent Green, 417–18
space:
  contents of, 53
  expansion of, 37
  Kant on, 254–55
  in quantum physics, 139
space-time continuum, 54, 190;
  definition of, 535
space-time cube, 260–65;
  definition of, 535
  lifeline in, 261, 262f
  philosophy and, 270–74
  volition and, 271–72

spacialization, 261;
  definition of, 535
special theory of relativity, 173–205;
  account of, 185–88
  and previous assumptions, 189–99
speciation, 320;
  definition of, 535
species:
  definition of, 535
  new, appearance of, 320–21
spectrum, 41;
  definition of, 535
  full, 47*f*
speed, 178;
  definition of, 535
spinning black hole. *See* rotating black hole
spiral galaxy, 42*f*, 43
Sputnik, 2, 473
stability, 55;
  ancient Greeks on, 175–77
standing waves, 122;
  definition of, 535
star, birth of, 302
stare decisis, 347, 351;
  definition of, 535
Star Trek, 239
Star Trek: The Next Generation:
  background on, 27–29
  "Cause and Effect," 279–81
  "Contagion," 29–30
  "Datalore," 98–99
  "Encounter at Farpoint," 443–44
  "Evolution," 485–86
  FTL travel in, 209–10
  "Home Soil," 287, 307–8
  "The Measure of a Man," 339–41
  "The Offspring," 375–76
  "Sarek," 168–69
  "Unnatural Selection," 413–14
  "We'll Always Have Paris," 243–44
  "Where No One Has Gone Before," 64–65
  "Who Watches the Watchers?," 129–30
  "Yesterday's Enterprise," 202–4

static limit, 222;
  definition of, 535
stationary black hole, definition of, 535
stationary limit. *See* static limit
statutory law, 347;
  definition of, 535
steady-state cosmology, 37, 39–40, 82–85;
  definition of, 535–36
sterilization, compulsory, 453
Stoney, George Johnstone, 115
Stothers, Richard, 334
streaming motion, 78–81;
  definition of, 536
strong nuclear force, 51–52;
  definition of, 536
strong tendency theory, 331;
  definition of, 536
subatomic universe, 48–52, 49*f*
subjective ethics, 346;
  definition of, 516
subjective time, 175;
  definition of, 531
subquantal reality, 160–62;
  definition of, 536
substance, 110;
  definition of, 536
Sullivan, J. W. N., 290–91
Summers, Harry G., 426–27
supercluster, 80;
  definition of, 536
superconductivity, definition of, 536
super-extreme Kerr object (SEKO), 224–25;
  definition of, 536
supermassive black holes, 216–17
super-quantum theory, 151;
  definition of, 536
superstring theory, 92;
  definition of, 536
synaptic cleft, 358;
  definition of, 536
synaptic terminal, 358;
  definition of, 536
synaptic vesicles, 358;
  definition of, 537

synthetic protolife, 299;
  definition of, 537
Szilard, Leo, 473
Szostak, Jack, 299–300

Talbot, Michael, 156
technology, 391;
  and environment, 465–67
  and future, 449–50
  global policy on, 468–80
  and population problem, 450–57
  regulation of, 390–94
  regulatory agencies and, 394
  versus science, 391–92
  and war, 457–65
Teilhard de Chardin, Pierre, 39, 292–93, 314, 502
telecommunications:
  future of, 455–57
  revolution in, 454–55
temporal distortion twist theory, 225–27, 226f
tendency theory, 331;
  definition of, 537
terraforming, 307;
  definition of, 537
Thales of Miletus, 107
theorem(s), 23;
  definition of, 537
theory, 10;
  definition of, 537
theory of everything, 89–90, 104;
  definition of, 537
  reality of, 91–94
theory of relativity. See relativity theory
Thomas Aquinas, 110, 177, 272, 289
Thomson, J. J., 114–15, 118
Thomson, Joseph John, 503
Thorne, Kip, 215, 227
thought experiment, 149;
  definition of, 517
  Einstein-Podolsky-Rosen, 149–52
Through Space to Mars (Rockwood), 114
Tifft, William, 78
time:
  versus causation, 239–40
  gravity and, 213
  Kant on, 254–55
  in quantum physics, 139
  Rucker on, 261–63, 264f, 266f
  simultaneous existence of, 257–69
  unidirectionality of, 257
Time Bandits, 206
Time Breakers: Mind Out of Time (Pollack & Weston), 237
Timecop, 235–36
time dilation (effect), 189–95;
  definition of, 537
  and time travel, 207–10
  and twins paradox, 192, 193f
  verification of, 191f
time loop, 232–33;
  definition of, 537
The Time Machine, 206
time tornado. *See* temporal distortion twist theory
time travel, 206–46;
  future, twins paradox and, 192–95
  paradoxes in, 230–37
Tipler, Frank, 162
Tonry, John, 76
Tooley, Michael, 361
Tracy, Spencer, 310
transcription, 398;
  definition of, 537
transfer RNA (tRNA), 398;
  definition of, 537
translation, 398;
  definition of, 537
transmutation, 316
transportation, and decentralization, 453–57
transubstantiation, 109–10;
  definition of, 537
Trefil, James, 4, 13, 51, 93, 252, 312, 314, 325, 357
triplet, 398;
  definition of, 509
tRNA. *See* transfer RNA
"The True-Blue American" (Schwartz), 448–49

twins paradox, 192, 193*f*;
   and future time travel, 192–95
   Minkowski diagram and, 268, 269*f*
   verification of, 194t

ultraviolet (UV) radiation, 294–95;
   definition of, 537
uncertainty, 135–40
uncertainty principle, 49, 137–38;
   definition of, 528
undue burden standard, 356–57
unemployment, population growth and, 424–26
unidirectional flow, 257;
   definition of, 528
uniform motion, 179–80
unipolar world, 427;
   definition of, 528
Unisave (Madsen), 343, 366
United States v. Butler, 383
unity, quantum school on, 155–57
universal assembler(s), 461;
   definition of, 528
universe:
   age of, 74–77
   Einstein's restructuring of, 187–88
   origin of, nonscientists and, 89–94
University of Maryland Baltimore County, conference on quantum physics, 20–21, 24
urban population growth, 423–24
Urey, Harold, 296, 503
Ushenko, Andrew Paul, 249
Ussher, James, 310
U.S. Supreme Court, 347–49, 351, 353–57, 356*f*
utilitarianism, 346;
   definition of, 528
UV. *See* ultraviolet radiation

Van Vleck, John, 17
velocity, 74, 179;
   definition of, 528
Venter, J. Craig, 407–8
verifiability, definition of, 528
verification, 12–13
viability, 348–49;
   definition of, 528
viability standard, 357
videophones, 455
Viking lander, 284
Virgo cluster, 75;
   definition of, 528
vitalism, 288–90;
   and acquisition of humanness, 313–14
   definition of, 528
volition, 270, 273;
   definition of, 528
   and space-time cube, 271–72
von Liebig, Justus, 289
von Neumann, John, 160

Walcott, Charles Doolittle, 329
Walker, E. H., 160–61
Waltz, Kenneth N., 446n70
war, 426–36;
   dangers of elimination of, 464–65
   science and technology and, 457–65
Warren, Mary Anne, 361
wars of redistribution, 432–34;
   definition of, 528
Watson, James, 293, 395, 503
wave function(s), 137;
   definition of, 528
wave-particle duality, 17–18, 123;
   definition of, 528
wave theory of atomic structure, 121–23, 123*f*
W+ bosons, 52;
   definition of, 538
W− bosons, 52;
   definition of, 538
weak nuclear force, 51–52;
   definition of, 539
weapons of mass destruction (WMDs), 432–33
Weinberg, Steven, 92
Weston, Chris, 237
West, Robin, 366–67
Wheeler, J. Craig, 271
Wheeler, John A., 160, 227

white dwarf, 55;
  definition of, 539
white hole, 223;
  definition of, 539
Whittington, Harry, 329
Wickramasinghe, Chandra, 301
Wigner, Eugene, 160, 473
Wills, Christopher, 326
Wilson, Robert, 38, 59, 85
Witten, Edward, 92
WMDs. *See* weapons of mass destruction
worldline. *See* lifeline
wormhole(s), 227;
  definition of, 539
  traversable, 227–30, 228f–229f

Z0 bosons, 52;
  definition of, 538
Zeno, 108
Zeno's paradoxes, 108;
  definition of, 539
zero-point field (ZPF) theory, 233–34
Ziman, John, 6
ZPF. *See* zero-point field theory
zygote, 348;
  definition of, 539
  and humanness, 363

About the Author

Paul A. Sukys is a professor of humanities, literature, law, and legal studies at North Central State College in Mansfield Ohio. He is coauthor of *Understanding Business and Personal Law*, *Business Law with UCC Applications* and *Civil Litigation* and sole author of *Ohio Supplement to Civil Litigation*. At North Central State he teaches a variety of humanities courses including "The Philosophy of Technology," "History of the Future," "Great Ideas in Western Civilization," "Contemporary Ethical Issues," and "Legal and Ethical Issues in Health Science." In 1998 he was named North Central's Outstanding Faculty Member of the Year. In the past, he has also taught at John Carroll University, the Fairmount Center for Creative and Performing Arts, and Cuyahoga Community College. Professor Sukys has written numerous articles and has made a wide variety of presentations including "Quantum Mechanics and the Humanities," which was presented at the General Educators of Ohio Conference in the spring of 1990. He has been involved in several projects for the National Business Education Association and was a member of the DANTES *Ethics in America* National Test Writing Committee sponsored by the Chauncey Group International in 1996. Professor Sukys received his bachelor's and master's degrees from John Carroll University in Cleveland and his law degree from Cleveland State University and has attended additional classes at New York University and Kenyon College. He is currently pursuing his doctorate at The Union Institute in Cincinnati and is working on a new book entitled *Quantum Relativity: The Philosophical, Artistic, and Literary Dimensions of the Theory of Everything*. Sukys, his wife Susan, and their youngest daughters, Megan and Ashley, reside in Gambier, Ohio. Their oldest daughter, Jennifer, and her family live in Pennsylvania.